Society of Earth Scientists Series

Series Editor

Satish C. Tripathi, Lucknow, India

The Society of Earth Scientists Series aims to publish selected conference proceedings, monographs, edited topical books/text books by leading scientists and experts in the field of geophysics, geology, atmospheric and environmental science, meteorology and oceanography as Special Publications of The Society of Earth Scientists. The objective is to highlight recent multidisciplinary scientific research and to strengthen the scientific literature related to Earth Sciences. Quality scientific contributions from all across the Globe are invited for publication under this series. Series Editor: Dr. Satish C. Tripathi

More information about this series at http://www.springer.com/series/8785

Jyotiranjan S. Ray · M. Radhakrishna
Editors

The Andaman Islands and Adjoining Offshore: Geology, Tectonics and Palaeoclimate

The Society of Earth Scientists
2008

Springer

Editors
Jyotiranjan S. Ray
Geosciences Division
Physical Research Laboratory
Ahmedabad, Gujarat, India

M. Radhakrishna
Department of Earth Sciences
Indian Institute of Technology Bombay
Mumbai, Maharashtra, India

ISSN 2194-9204 ISSN 2194-9212 (electronic)
Society of Earth Scientists Series
ISBN 978-3-030-39842-2 ISBN 978-3-030-39843-9 (eBook)
https://doi.org/10.1007/978-3-030-39843-9

This Springer imprint is published by the registered company Springer Nature Switzerland AG
The registered company address is: Gewerbestrasse 11, 6330 Cham, Switzerland

Series Editor Foreword

The picturesque Andaman and Nicobar Islands of India is a chain of N-S trending islands at the juncture of the Bay of Bengal in the west and the Andaman Sea in the east. This archipelago attracted the geoscientists because the Andaman Sea is an active backarc basin lying above and behind the Sunda subduction zone where convergence between the overriding Southeast Asian plate and the subducting Australian plate is highly oblique. The geological and tectonic history of the Andaman Sea cannot be separated from Myanmar (Burma) on the north and Sumatra on the south, as such representing an important sector of India–Asia collision and its timing and evolution when compared with the Himalayan belt. The book on geology, tectonics and Palaeoclimate of the Andaman Islands and adjoining offshore will certainly open new vistas for future researches.

The mega-event of 36th International Geological Congress 2020 in India opened a new chapter on the Geology of India. On such an occasion Society of Earth Scientist Series by Springer decided to bring out 36th IGC Commemorative Volumes on various recent geological and geophysical studies of India. As such veteran geoscientists were requested to prepare comprehensive accounts as monographs or edited volumes. I am personally thankful to all the editors and authors for the timely submission of high-quality manuscripts for inviting interest of the global community of geoscientists.

Lucknow, India

Satish C. Tripathi
Series Editor

Preface

The Andaman and Nicobar Islands and the adjoining ocean basins in the northeast Indian Ocean contain valuable records of some of the most important geological events of South Asia since the Late Cretaceous. These N-S elongated islands form a part of the outer arc accretionary prism at the convergent boundary between the India and Burma plates, called the Andaman Subduction Zone, which acts as a transitional tectonic link between the Indo-Myanmar orogenic belt in north and the Sumatra–Java subduction system towards south. Some of the fascinating tectonic features of the Andaman arc include the convergence–transpressional transition, a thick accretionary prism, an insignificant volcanic arc and a thickly sedimented backarc. The accretionary prism, mainly characterized by the subaerial islands, is home to obducted or thrusted ophiolites, trench–forearc sediments, volcaniclastic deposits, mud volcanoes, tectonic landforms and tsunami deposits. The volcanic arc is characterized by two subaerial volcanoes and possible many submarine eruptive centres. The Andaman Sea, believed to have formed subsequent to the backarc opening about 4 million years ago, contains several bathymetric highs whose origins remain elusive. The islands and the adjoining ocean basins contain thick post-Eocene fan sediments and abundant coral reefs which serve as important proxies for continental weathering and erosion linked to the climate and tectonics of South Asia, which have implications for global climatic fluctuations and changes to ocean circulations.

Although these islands and adjacent oceans basins have remained focus of geoscientific research for a long time, the last two decades have seen a surge in global efforts to understand the geological records of this region, resulting in many interesting studies. However, the evolutionary aspects of the region with particular emphasis on the tectonics, sedimentation and Palaeoclimate have not been fully explored. Therefore, we decided to bring out this edited volume in order to provide space for active researchers working in the Andaman–Nicobar region to publish some of their new results, provide comprehensive reviews and present new ideas in these disciplines that would likely give an impetus to the future research.

The first five chapters of this book deal with the geological evolution of the Andaman accretionary prism with a special emphasis on the ophiolite suite and the sedimentary deposits. The first chapter "Andaman Ophiolite: An Overview" (Bandyopadhyay et al.) reviews the current understanding of the origin of the Andaman ophiolite, and the second "Timing of Formation and Obduction of the Andaman Ophiolite" (Bhattacharya et al.) provides new geochemical and geochronological data for these rocks and determines the tectonic nature of the ophiolite, age of its formation and timings of the ophiolite obduction events. The third chapter "Facies Stacking of a Deep Sea Depositional Lobe: Case Study from Palaeogene Andaman Flysch, South Andaman Islands, India" (Roy and Banerjee) presents sedimentological aspects of the siliciclastic turbidite deposits of the islands and the fourth "Spatial Variation in the Composition of Andaman Flysch Across the Andaman Islands in Relation to Source of Sediments and Tectonics" (Roy et al.) and the fifth "Tracing the Sources and Depositional Pathways for the Oligocene Sediments in the Andaman Forearc" (Awasthi et al.) make an attempt to determine the provenances of the sediments and their transport pathways into the Palaeogene Andaman Basin, using geochemical and isotopic tracers. Chapters 6 to 12 deal with the geophysical data interpretation and provide several new insights on the structure, deformation and crustal evolution that are critical to our understanding of the ongoing tectonics and seismogenesis of the region. The sixth chapter "Seismicity, Lithospheric Structure and Mantle Deformation in the Andaman Nicobar Subduction Zone" (Srijayanthi and Ravi Kumar) brings out slab configuration below the islands and highlights the significance of slab geometry, and identify tear fault and subslab anisotropy that control the rupture characteristics and deformation patterns along the arc. The seventh chapter "Lithospheric Framework of Sumatra–Andaman Subduction Zone—A Review" (Yadav and Tiwari) gives an overview of the lithosphere structure below the Andaman–Sumatra region and highlighted the significance of gravity–geoid modelling for better imaging of structure in this region. The eighth chapter "Crustal Model for the Andaman Outer Arc: Constraints from Earthquake, Gravity and Receiver Function Data" (Haripriya et al.) deals with seismologically constrained gravity interpretation for delineating crustal structure below the Andaman outer arc region which suggested a thicker oceanic crust approaching the subduction zone; the ninth chapter "Geomagnetic Deep Sounding in Andaman Islands" (Subba Rao et al.) presents the modelling of geomagnetic depth sounding data acquired study over the islands which gave rise to anomalous conductivity zones associated with the Jarawa thrust and the eastern margin fault areas. The tenth chapter "On the Trail of the Great 2004 Andaman-Sumatra Earthquake: Seismotectonics and Regional Tsunami History from the Andaman-Nicobar Segment" (Rajendran and Rajendran) gives a detailed overview of seismotectonics of the Andaman arc with special emphasis on the past tsunamigenic earthquakes in the region, identifying seismicity pattern and associated land-level changes in the Andaman and Nicobar segments of the arc before and after the 2004 megathrust earthquake. In the eleventh chapter "A Glimpse of Crustal Deformation Through Earthquake Supercycle in the Andaman Region Using GPS Measurements" (Dadhich et al.), the authors present the crustal deformation pattern in the Andaman arc region that spans the complete supercycle corresponding to 2004 megathrust earthquake using the long-term geodetic measurements. The study indicated significant post-seismic deformation which is still continuing in the region. The twelfth chapter “Tectonics of the Andaman Backarc Basin—Present Understanding and Some Outstanding Questions” (Kamesh Raju et al.) provides a detailed synthesis of geophysical data in the Andaman backarc basin and highlighted various issues related to the present understanding of the tectonic history of the region. The last three chapters are devoted to the Neogene–Quaternary Palaeoceanography and Palaeoclimate of the Andaman region based on sediment core studies. The thirteenth chapter “Miocene to Pleistocene Palaeoceanography of the Andaman Region: Evolution of the Indian Monsoon on a Warmer-Than-Present Earth” (Kuhnt et al.) reviews the information available from two International Ocean Discovery Program (IODP) cores from the Andaman Sea and predicts the future evolution of the Indian monsoon in a global warming scenario. The fourteenth chapter “Late Quaternary Chronostratigraphy, Carbonate Mass Accumulation Rates and Palaeoceanography of the Andaman Sea” (Sijinkumar et al.) sheds light on the Quaternary carbonate accumulation rates and Palaeoceanography of the Andaman Sea by reviewing existing data from a large number of cores. The last chapter (Nagasundaram et al.), based on geochemical proxies from a core, collected from the Bay of Bengal near the northernmost island of the Andamans, attempts to reconstruct the late Holocene Indian monsoon.

This book would not have been possible without the invitation of Satish C. Tripathi, the Series Editor for the Society of Earth Scientists Series/Springer, to come up with a special volume for the International Geological Congress 2020. We express our sincere appreciation to all the authors of this volume for contributing their valuable research here. We offer our special gratitude to the reviewers who provided their precious time to critically review the manuscripts. People who served as reviewers for this volume are B. R. Arora, Neeraj Awasthi, P. C. Bandopadhyay, Ravi Bhushan, Rajneesh Bhutani, Abhisek Chatterjee, Anil Earnest, V. K. Gahalaut, Biswajit Ghosh, Oinam Kingson, G. Mohan, Basab Mukhopadhyay, Manoj Mukhopadhyay, B. P. K. Patro, Debajyoti Paul, Kamesh Raju, Rajeev Saraswat, Subir Sarkar, Hetu C. Sheth, A. V. Sijinkumar, A. D. Singh, Bijendra Singh, P. B. V. Subba Rao, P. S. Sunil and Manish Tiwari.

Ahmedabad, India — Jyotiranjan S. Ray
Mumbai, India — M. Radhakrishna

Contents

About the Editors

Dr. Jyotiranjan S. Ray is a Professor at the Physical Research Laboratory, Ahmedabad, India. He is a geologist by training and a geochemist by profession. He uses principles of elemental and isotope geochemistry to address some of the important questions related to the evolution of the silicate Earth and its various reservoirs. His current research interests include the evolution of the Earth's mantle through time, subduction zone processes and arc magmatism, Proterozoic isotope stratigraphy and evolution of Quaternary landscapes. He has been working in the Andaman region for more than a decade. His focus has been on establishing chronology of volcanism in this region, understanding slab–mantle wedge interaction and its implication for the chemistry of lavas in the Barren–Narcondam arc, deciphering the nature of tectonic and climatic controls on the sedimentation in the Andaman Basin since Palaeogene and determining timings of major tectonic events on the accretionary prism. He is an alumnus of the Indian Institute of Technology Roorkee, Roorkee, India, and the Physical Research Laboratory, Ahmedabad, India.

Dr. M. Radhakrishna is presently Institute Chair Professor in the Department of Earth Sciences, IIT Bombay, Mumbai, India. He has been teaching geophysics for the last 27 years and has guided 13 Ph.D. students. His research interest mainly includes the gravity and magnetic data acquisition, analysis and interpretation with special reference to their application for delineating the crustal structure, rheology, basin formation mechanisms and tectonic evolution of the Indian continental margins. He also made significant investigations towards understanding the structure and rheological characteristics of the northern Indian Ocean lithosphere using potential field modelling, gravity–topography transfer function analysis and flexural modelling studies. His integration of seismotectonics and constrained gravity interpretation along the Andaman-Sumatra-Java subduction zone gave rise to several valuable insights on the along and across arc variation in mass heterogeneities and flexure that have implications on the seismogenesis of the region. He is an alumnus of Andhra University, Visakhapatnam, and the Indian School of Mines (IIT (ISM)), Dhanbad, India.

Andaman Ophiolite: An Overview

Debaditya Bandyopadhyay, Douwe J. J. van Hinsbergen, Alexis Plunder, Pinaki C. Bandopadhyay, Eldert Advokaat, Soumi Chattopadhaya, Tomoaki Morishita and Biswajit Ghosh

Abstract The Andaman Ophiolite remained little explored for long but recent studies yielded important first-order findings that hold promise for further research. Here we summarise these first-order constraints on the structure, geochemistry, and evolution of the Andaman Ophiolite and identify key frontiers for future research. The uniqueness of Andaman Ophiolite is their petrological and geochemical diversity in a

D. Bandyopadhyay · P. C. Bandopadhyay · S. Chattopadhaya · B. Ghosh (✉)
Department of Geology, University of Calcutta, 35 Ballygunge Circular Road, Kolkata 700019, India
e-mail: bghoshgeol@caluniv.ac.in

D. Bandyopadhyay
e-mail: dbgeol_rs@caluniv.ac.in

S. Chattopadhaya
e-mail: scgeol_rs@caluniv.ac.in

D. Bandyopadhyay
Department of Geology, University of North Bengal, Darjeeling 734013, India

D. J. J. van Hinsbergen · E. Advokaat
Department of Earth Sciences, Utrecht University, Princetonlaan 8A, 3584 CB Utrecht, The Netherlands
e-mail: d.j.j.vanhinsbergen@uu.nl

E. Advokaat
e-mail: e.l.advokaat@uu.nl

A. Plunder
BRGM, 3 avenue Claude Guillemin, BP 36009, 45060 Orléans Cedex 2, France
e-mail: a.plunder@brgm.fr

E. Advokaat
Department of Physical Geography, Utrecht University, Princetonlaan 8A, 3584 CD Utrecht, The Netherlands

T. Morishita
Faculty of Geosciences and Civil Engineering, Institute of Science and Engineering, Kanazawa University, Kanazawa 920-1192, Japan
e-mail: moripta@se.kanazawa-u.ac.jp

J. S. Ray and M. Radhakrishna (eds.), *The Andaman Islands and Adjoining Offshore: Geology, Tectonics and Palaeoclimate*, Society of Earth Scientists Series,
https://doi.org/10.1007/978-3-030-39843-9_1

close spatial association. Among many unresolved issues the age, stratigraphic coherence, and emplacement mechanisms of Andaman Ophiolite remain enigmatic. In particular, although many of the elements of the typical ophiolitic pseudostratigraphy are present, they are not found in structural coherence, and available geochronological constraints suggest a much longer than normal time window for the amalgamation of the ophiolite's crust. We raise questions on the plate tectonic setting and evolution of the ophiolite, suggesting that a multi-phase history of magmatism and ophiolite evolution is one of the key questions that deserves international geoscientific attention.

Keywords Andaman Ophiolite · Subduction initiation · SSZ setting · Arc maturity

1 Introduction

Ophiolite investigations have always played a pivotal role in testing and establishing hypotheses and theories in plate tectonics, both on the evolution and dynamics of spreading ridges as well as on the juvenile stages of subduction. The suboceanic origin of ophiolites in various tectonic settings and their emplacement onto continental margins or accretionary prisms have been the topics of a dynamic and continually evolving concept since long (Coleman 1971; Anonymous 1972; Dewey 1976; Shervais 2001; Dilek and Furnes 2011). Earlier debates were centred on establishing the relationship between ophiolites and modern oceanic lithosphere analogues, specific to tectonic settings. The apparent conflict between geochemical-petrologic evidence for derivation of magmas from highly depleted mantle similar to that found at modern subduction zones, and structural-stratigraphic evidence for seafloor spreading in a non-island-arc environment led to the so-called *ophiolite conundrum* (Dilek and Moores 1990; Moores et al. 2000).

However, global investigations in last two decades have demonstrated that fossil oceanic lithosphere preserved as ophiolites in most orogenic belts, is petrologically and geochemically diverse. Oceanic crust may form in any tectonic setting, at mid-ocean ridges, and in the upper plate close to, or distal from subduction zones and based on their geochemical composition they are classified into *subduction-related* and *subduction-unrelated* ophiolites (Dilek and Furnes 2011). Suprasubduction zone (SSZ) ophiolites, those formed close to subduction zones have the highest likelihood of becoming thrusted upon continental margins or accretionary prisms and are thus by far the dominant ophiolite-type, constituting ~75% of the Phanerozoic ophiolites. Geochemical fingerprinting clearly substantiates the role of subduction zone derived components in development of SSZ ophiolitic magmas (Beccaluva et al. 1994; Bedard et al. 1998; Dilek et al. 1999; Shervais 2001; Dilek and Flower 2003; Dilek and Furnes 2009, 2014). On the other hand, mid-ocean ridge (MOR) ophiolites are relatively lower in abundance, but not rare, making up ~20% of the population

(Furnes et al. 2014). Exclusive MOR ophiolitic crust may be found where intra-oceanic subduction was not associated with upper plate extension to form SSZ ophiolites. It may also be found together with SSZ ophiolitic crust where it represents the pre-existing lithosphere within which SSZ spreading occurred (e.g., Maffione et al. 2015). Nevertheless, forearc, embryonic arc, and backarc settings in suprasubduction zones are the most widely documented tectonic environments for the origin of ophiolites. Magmatism during each stage of subduction, from infancy to maturity produces spatially and temporally associated, mafic-ultramafic to highly evolved rock assemblages. These rock units, which have varying internal structures, geochemical affinities, and age ranges, and originally formed in different geodynamic settings, constitute discrete ophiolite complexes and can become tectonically juxtaposed in collision zones (Dilek 2003).

In the present geodynamic configuration of SE Asia, the modern northern collisional plate boundary of the Indian plate along the Main Frontal Thrust of the Himalaya extends south from the eastern Himalayan syntaxis through Nagaland, Manipur and the western Burma (Myanmar) region and merges with an oceanic subduction system along the Andaman–Sumatra–Java (ASZ) trench where the Indian plate is downgoing below the intensely deformed SE Asian tectonic collage in the upper plate (e.g., Hall 2002). The Andaman and Nicobar Islands form the outer arc high, and are composed of remnant ophiolites and underlying mélange, a Palaeogene–Neogene clastic sedimentary cover, overlying a Neogene accretionary prism (Bandopadhyay and Carter 2017a). To the west, this outer arc is bordered by the ASZ trench and to the east there occur two volcanic islands (Barren and Narcondam) that belong to a volcanic arc extending from central Myanmar Basin in northern Myanmar and continues eastwards into Java (Acharyya 2007) (Fig. 1).

Discontinuous occurrences of ophiolitic rocks in the Andaman and Nicobar Islands have been known for over fifty years (Karunakaran et al. 1964, 1967) and were described as dismembered bodies in reviews of "Ophiolites in SE Asia" (Hutchison 1975). Although the geographic positions of these ophiolites are strategic from the petrological point of view as they provide an important transitional link between the Himalayan collision zone and the Indonesian arc system, they have remained unexplored and elusive for long because of the natural isolation and inaccessibility to these islands. However, the active seismicity all along this convergent boundary as manifested by the 2004 Sumatra earthquake, and hazards posed by volcanisms at Barren Island have received attention of many workers, particularly in the last decade (Pal and Bhattacharya 2011; Sheth 2014; Sheth et al. 2009). Recent and active studies have focused on the ophiolites to establish their geodynamic setting of formation and evolution, and to fit them in Tethyan plate tectonic context. The first order findings have been documented well, yet the critical questions are still unresolved. This chapter examines the current state of knowledge on Andaman Ophiolite and discusses the scope of future research.

2 What Are the First Order Findings?

Ophiolites crop out along the east coast of North, Middle, South Andaman and Rutland Islands. An idealized coherent ophiolite stratigraphy is nowhere exposed in the islands, different elements of the Penrose sequence (Anonymous 1972) are arguably present as dismembered units, scattered in isolated occurrences throughout the islands. The exposed lithological units in South Andaman (and Rutland), Middle, and North Andaman comprise a tectonized, restitic mantle sequences (>700 m thick in north- Andaman) hosting chromitite pods, an intrusive crustal cumulate sequence (up to 150 m) and volcanic rocks (up to 400 m thick in south-Andaman) (Fig. 2).

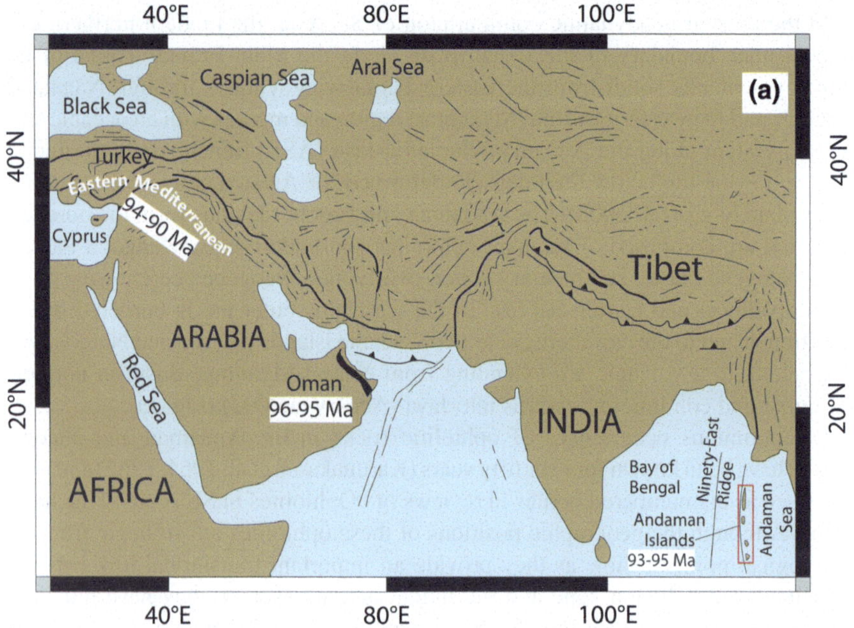

Fig. 1 **a** Distribution of eastern Neotethyan ophiolitic suture zones (thick black lines) (modified after Pedersen et al. 2010). Ophiolite locations of Eastern Mediterranean, Oman and Andaman along with U–Pb zircon dates [in white box] are marked; **b** Colour-shaded relief image showing eastern Himalayan syntaxis, the western Burma (Myanmar) region and the Andaman–Sumatra–Java (ASZ) trench in the regional geodynamic framework. Major tectonic features are redrawn from literatures (Awasthi 2017; Liu et al. 2016; Imsong, et al. 2016; Sheth 2014). Relief data are from ETOPO1 Global Relief Model (http://www.ngdc.noaa.gov/mgg/global/). Pink triangles represent Holocene volcanoes (after Sheth 2014). LA: Little Andaman; CN: Car Nicobar; NI: Nicobar islands; GN: Great Nicobar; EMF: Eastern Margin Fault; DF: Diligent Fault; WAF: West Andaman Fault; BI: Barren island; NI: Narcondam island; AR: Alcock Rise; SR: Sewell Rise; STDS: South Tibetan Detachment System, MCT: Main Central Thrust, MBT: Main Boundary Thrust, MFT: Main Frontal Thrust. Geochronological results available in the regions (yellow stars with red outline), are also shown for reference (see text for details) **c** Geologically mapped ophiolitic regions (marked with thick black lines) of Andaman Islands over processed-hillshade map (generated using SRTM digital elevation data with a resolution of 1 arc-second from USGS Earth Explorer https://earthexplorer.usgs.gov/). North-South extending Jarwa Thrust is also shown

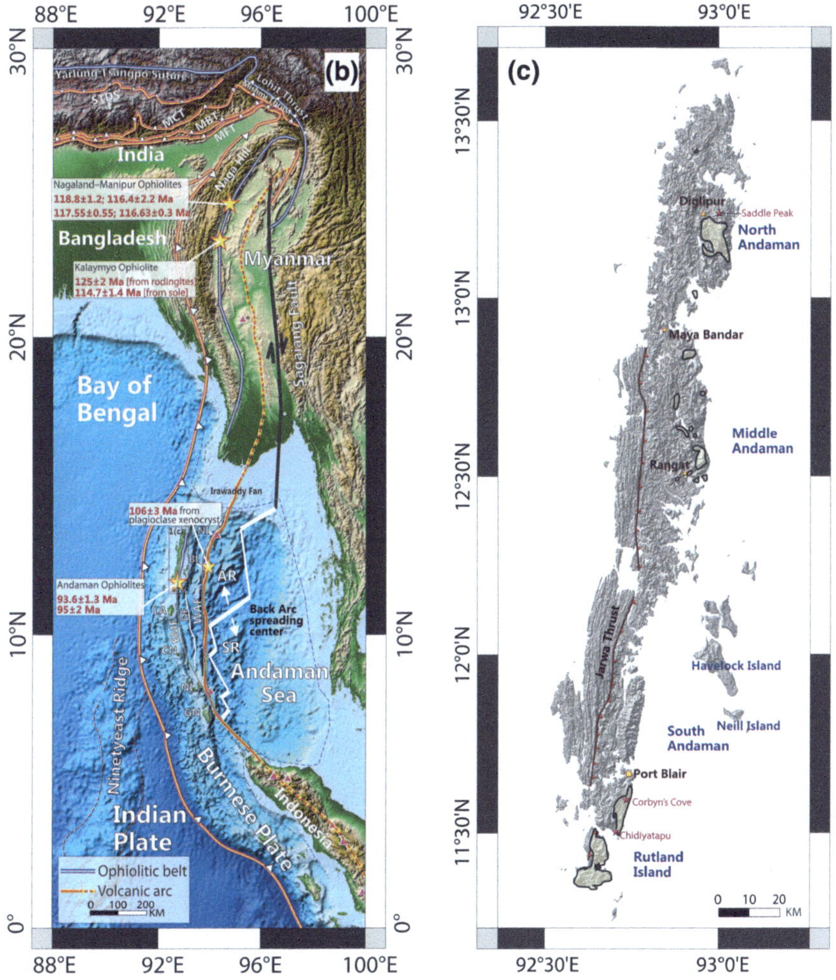

Fig. 1 (continued)

The Andaman Ophiolite and overlying Palaeogene stratigraphy have been thrusted and shortened, and are collectively thrust westwards upon an accretionary prism of deep-marine post-Eocene turbidites along the Jarawa (or Jarwa) Thrust that is traced over all islands (Bandopadhyay and Carter 2017a) (Fig. 1c).

Detailed field, petrographic and petrologic description of the various lithological units are beyond the scope of this chapter but may be found in the cited literature. Here we present a summary of the work done based on geochemical fingerprinting of the Andaman ophiolitic rock units. Since most of these works are pivoted around either the mantle rocks or the volcanics, we excluded other rock types in this section.

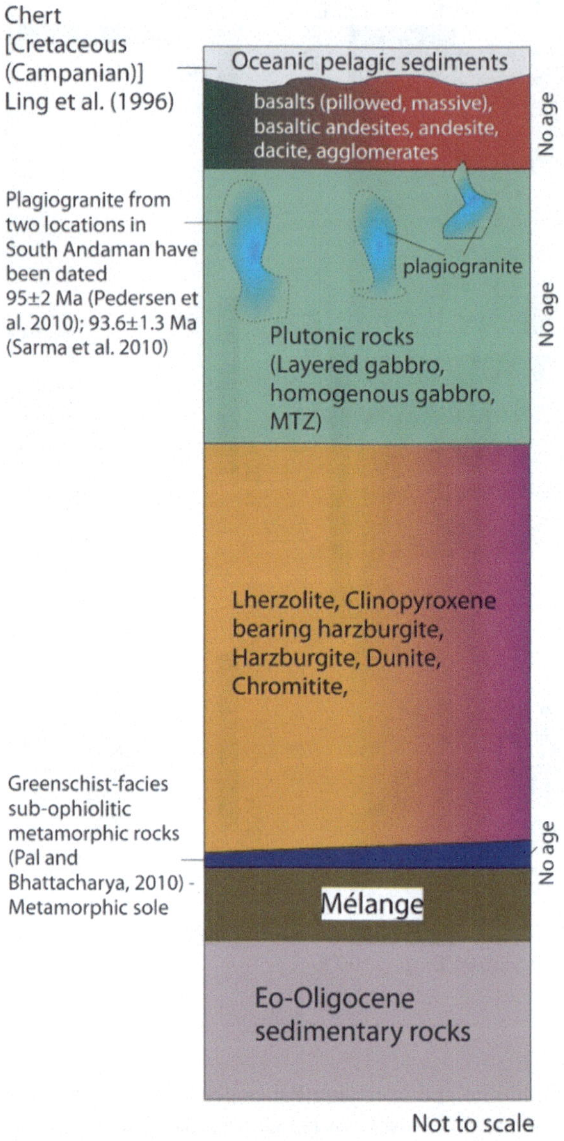

Fig. 2 Schematic (pseudo)stratigraphy of Andaman Ophiolite (not to scale). Various lithomembers belonging to volcanic rocks and mantle peridotites are shown in their respective positions, however their true genetic and stratigraphic interrelationship are ambiguous. Geochronogical information of the corresponding rock types, where available, have been summarized, thus showing potential gap areas in the age constraints

2.1 Contrasting Mantle Rocks

In terms of petrography and mineral chemistry of the ophiolitic rocks, especially the mantle sequence, a pronounced variation exists between Rutland Island in the south and Middle and North Andaman towards the north (Ghosh et al. 2013). The composition of chromian spinels and clinopyroxenes from the harzburgite-dominated mantle rocks of the Andaman ophiolite on Rutland Island was inferred to be residue of 14–18% melting of a fertile mantle (e.g., DMM; Workmann and Hart 2005) in a supra-subduction zone environment (Ghosh et al. 2009, 2018).On the other hand, the mantle peridotites in Middle and North Andaman Island are distinctive, corresponding to <10% mantle melting with minor flux infiltration (Morishita et al. 2018), and mostly belong to relatively fertile lherzolite that occasionally grades to clinopyroxene-bearing harzburgite with a geochemistry that instead suggests formation in a backarc setting (Ghosh et al. 2018) (Fig. 3).

Finally, the peridotites on Middle and North Andaman are underlain by a serpentinite-hosted melange that contains metavolcanic rocks and metacherts with greenschist facies to amphibolite-facies metamorphic grade attributed to a dismembered metamorphic sole (Pal and Bhattacharya 2010).

Spatial distribution of these mantle rocks with contrasting compositions opens up a debate on whether the distinct geochemical signatures are a reflection of switchover of tectonic setting. Are we looking at two different mantle domains? If yes, does this spatial distribution reflect variations linked to the melting history where the same suboceanic mantle underwent different styles of melting in different sub-arc domain? And were they tectonically juxtaposed, or is one older than, and the protolith of the other?

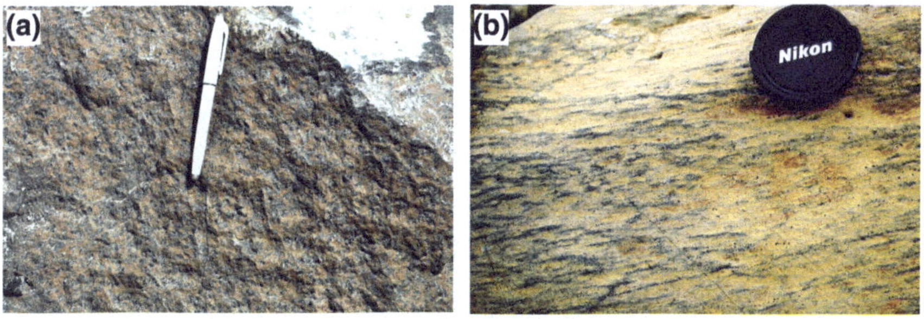

Fig. 3 Collage of field photographs showing variation of mantle peridotites. **a** Lherzolite from Panchawati, Middle Andaman; **b** Foliated harzburgite from Rutland Island

2.2 Volcanic Rocks

These are the most extensively studied rock types of the Andaman Ophiolite suite (Ray et al. 1988; Jafri et al. 1990, 2010; Srivastava et al. 2004; Pal 2011; Bhattacharya et al. 2013; Jafri and Sheikh 2013; Ghosh et al. 2014). The volcanic rocks of the Andaman Ophiolite are divided into (a) pillow lava (basaltic) and (b) East Coast volcanics (basalts, basaltic andesites and acid differentiates) (Ray et al. 1988), each having their own field and geochemical characteristics (Fig. 4). The East Coast volcanics are commonly brecciated and accompanied by a pyroclastic (agglomerates) facies (Pal et al. 2003; Sarma et al. 2010). Agglomerates are mostly of andesitic composition, either include abundant fragments of diorite-plagiogranite suite of rocks (at the southeast coast of South Andaman) and or occur in proximity (in Panchawati Coast, Middle Andaman). Vohra et al. (1989) described these two volcanic groups as upper lava and lower lava respectively. The two volcanic rock groups, upper and lower lava found in different thrust slices, often as large blocks along the east coast, overthrust by ophiolitic melange and the ophiolites proper, and are compositionally distinct. The pillow basalts are MORB like (Jafri et al. 2010; Srivastava et al. 2004) while the East Coast volcanics have an island arc affinity (Ray 1985). Further, they have been intruded by plagiogranites which also have an island arc affinity (Jafri et al. 1995). Pal (2011) described the upper lava as island arc tholeiitic (IAT) basalt and the lower lava as low-Ca boninite and correlated them with two distinct types of subduction related magmatism. The author interpreted stratigraphically older boninitic rocks as derived from a source depleted in Nb and Y in the first phase, and stratigraphically younger IAT magmas were generated from a less depleted source, contaminated by slab-derived components in a suprasubduction zone setting in the second phase. Jafri and Sheikh (2013) correlated the pillow basalts from Bompoka Island, Nicobar with the pillows of South Andaman and described them as geochemically akin to backarc basin basalts.

Fig. 4 Collage of field photographs showing variation of volcanic rocks. **a** Pillow basalts from Corbyn's Cove, South Andaman; **b** Agglomerate facies of the East Coast volcanics (containing fragments of felsic rock, volcanic rock, embedded in Andesitic matrix) from Chidiyatapu, South Andaman

3 Discussion

Tethyan ophiolites are generally regarded as subduction-related ophiolites (Wakabayashi and Dilek 2003; Dilek and Furnes 2009) and ophiolites of Andaman Islands are no exception (e.g., Ghosh et al. 2009, 2018; Pal 2011). The internal structure and geochemistry of the Phanerozoic ophiolites in the Indonesian orogenic belt show a complex pattern of igneous accretion that involved multiple stages and sources of melt evolution and life cycles in suprasubduction zone environments (Shervais 2001; Dilek and Flower 2003). A stocktaking of the recent researches on Andaman Ophiolite highlights the gap areas where the future research should be directed into.

3.1 Age of the Ophiolite?

The diversity of Andaman Ophiolite in terms of petrological and geochemical variability of mantle peridotites and volcanics is quite distinctive. Plagiogranites from two nearby locations on South Andaman, with their island arc geochemical signatures and intimate association with the East Coast volcanics, have been dated at two locations using U/Pb on zircon: A 93.6 $\pm$ 1.3 Ma age was derived from a plagiogranite dyke intruded the East Coast volcanics (Sarma et al. 2010) and a 95 $\pm$ 2 Ma age was derived from a plagiogranite sample (Pedersen et al. 2010) which the authors described as 'in situ' whereas at the sampling location, there is only plagiogranite in volcanic agglomerate present, so presumably it was an agglomerate block. These ages were interpreted as the spreading age of Andaman ophiolitic crust (Pedersen et al. 2010), and correlated with other classical Neotethyan ophiolites in Oman (96–95 Ma; Rioux et al. 2012, 2013) and the ophiolites of the Eastern Mediterranean region (~ 94–90 Ma; Dilek and Furnes 2011; van Hinsbergen et al. 2016, and references therein). The compositional similarity between the plagiogranites and the East Coast volcanics in terms of major, minor, trace elements, REE and the Rb/Sr ratio coupled with their close spatial association, however, establishes their genetic linkage (Jafri et al. 1995). The lithology and geochemistry of the East Coast volcanics with their abundant pyroclastic components and the plagiogranites of Andaman Ophiolite, with their common island arc geochemical affinity may instead represent a matured arc magmatism that intruded and overlies the Andaman Ophiolite. So, what does this ~95 Ma age really indicate? Does it represent the true age of the ophiolite and if it does, then how does it fit to the regional geodynamic setting where the zircons of plagiogranites from Nagaland-Manipur Ophiolite and Kalaymyo Ophiolite (Myanmar), located farther north along the same ophiolite belt gave ages of 116.4 $\pm$ 2.2 to 118.8 $\pm$ 1.2 Ma (Singh et al. 2017); 117.55 $\pm$ 0.55 Ma and 116.63 $\pm$ 0.30 Ma (Aitchison et al. 2019) and 127 Ma (Liu et al. 2016) respectively?

Alternatively, if the ~95 Ma event in the evolutionary history of Andaman Ophiolite represents arc maturity, then when did the subduction begin? If we consider Izu-Bonin-Mariana (IBM) as a modern analogue of SSZ ophiolite as pointed out by

Stern et al. (2012), there is an age gap of nearly 8 Myr between the forearc basalts and contemporary bottom cumulates (gabbros) and normal arc magmatism (Ishizuka et al. 2011). This evolution from subduction initiation to arc normalcy occurred all along the entire length of the IBM subduction system. In Andaman, we have not dated the layered gabbros or any volcanic unit yet other than that they are overlain by upper Cretaceous (Campanian) cherts (Ling et al. 1996). In absence of such age constraints, we do not know the true age of subduction initiation. One way forward is studying the sub-ophiolitic metamorphic rocks identified by Pal and Bhattacharya (2010). Do these have mid-pressure and high-temperature metamorphic paragenesis typical of metamorphic soles (e.g., Agard et al. 2016)? If so, their cooling ages may help identify forearc extension and magmatism—which is typically synchronous and likely caused by suprasubduction zone spreading (van Hinsbergen et al. 2015). If present, garnet-bearing sole rocks may allow dating the inception of metamorphic sole growth that may predate ophiolite spreading and sole exhumation by >10 Myr (Guilmette et al. 2018). Zircon from the metamorphic sole of the Kalaymyo Ophiolite, Myanmar yielded a ca. 115 Ma age (Liu et al. 2016) which shows that subduction and upper plate extension were long underway along-strike to the north well before 95 Ma. We can also note the existence of the $^{40}Ar/^{39}Ar$ plateau age of 106 ± 3 Ma (Ray et al. 2015) for the plagioclase xenocryst, hosted in Barren lava flow and interpreted as pieces of the Barren island Basement. Does this age reflect the age of the ocean floor below Barren Island? If so, Andaman ophiolite must have been attesting for some ophiolite spreading as early as 106 Ma.

3.2 Emplacement Mechanism?

Two contrasting models have been proposed for uplift and emergence of the Andaman Ophiolite in absence of obduction onto a major continent. One model suggested that the subduction has been continuing along the western margin of the island arc since the late Cretaceous, and that the ophiolites represent thrust wedges offscraped from a downgoing plate, forming accreted nappes during this prolonged period of subduction (Curray et al. 1979; Mukhopadhyay and Dasgupta 1988; Pal et al. 2003; Curray 2005). By analogy, this model postulates an accretionary prism setting for the uplift and emergence of Andaman Ophiolite. In this context what does the geochemical fingerprinting of the magmatic rocks and presence of metamorphic sole indicate? Do they support the above model? On the contrary, the flat-lying mode of occurrence of the Andaman Ophiolite and their close spatial relationship to a zone of negative gravity anomalies led to an alternative model. This model regarded the Andaman Ophiolite as rootless subhorizontal bodies without much extension at depth. They are interpreted to occur as westward propagated nappes from the eastern ophiolite belt where they were accreted during Middle Eocene to Late Oligocene, much before the currently active subduction (Acharyya et al. 1990; Sengupta et al. 1990; Acharyya 2007). However, field observations certainly challenge the flat-lying mode of occurrence of Andaman Ophiolite in many places, specially in South Andaman. While the recent literatures discuss the SSZ origin of the Andaman Ophiolite (Ghosh et al. 2009, 2013; Pal 2011), none of the above models considered them derived from the upper plate.

Emplacement of SSZ ophiolites could be either *Cordilleran-type,* involving underplating of material in the accretionary prism (accretionary uplift), or *Tethyan-type,* where the upper plate is thrusted over passive continental margin (Wakabayashi and Dilek 2003; Dilek and Furnes 2009) by obduction (Coleman 1971; Moores 1982; Shervais 2001; Dilek 2003). In the present context the geophysical evidences like a prominent gravity anomaly (Kumar et al. 2013) and a seismic profile (Singh et al. 2013) across the Andaman-Nicobar accretionary ridge suggest the presence of (micro)continental body derived from the Indian plate below the ophiolites, which may have contributed to the uplift, although its underthrusting left no accretionary record (Fig. 5). What is at stake in resolving this emplacement issue?

The heart of the problem lies in how we interpret the ophiolite record, in particular, with regard to plate tectonic reconstructions. Recently, Ghosh et al. (2017) proposed a model where the morphology of the present Andaman-Nicobar outer arc high is explained as the resultant of coalescence of two accretionary prisms, each belonging to a temporally and spatially different subduction system. The initial subduction formed the varieties of petrologically and geochemically distinct oceanic lithosphere in different sub-arc domains (Fig. 5a, b). The emplacement of this oceanic lithosphere was unlike typical Tethyan-type ophiolites because before its uplift due to underthrusting of the geophysically imaged microcontinental block (Fig. 5c) the upper plate was shortened by thrusting and the subduction margin was charged with sediments that accreted at the leading age of the overriding plate, resembling to some extent a *cordilleran-type* ophiolites. This is also supported by the E-W swath profile across Andaman-Nicobar accretionary ridge (Bandopadhyay and Carter 2017a).

Following this, a second subduction was initiated by induced nucleation to the west of the microcontinent in response to strong compressional force. This mechanism might have emplaced the segment of backarc lithosphere formed by ea0rlier subduction at the top and also produced thrust imbrication within the upper plate (Fig. 5d). This is in accord with the findings of various thrust contacts within the ophiolitic lithounits and other accreted sediment packages (Ray et al. 1988; Pal et al. 2003). The new induced subduction with its gradual maturity started arc volcanism at Barren and Narcondam, and finally resulted in opening of the East Andaman Basin, the backarc basin in early Miocene. Major sediment inputs from the river delta system(s) to the north together with the offscraped sediments of the Indian ocean floor constitutes the younger accretionary prism belonging to the Miocene subduction system. Thus Andaman Ophiolite formed by spreading and subsequent arc magmatism above a subduction zone in the Cretaceous, were overlain by forearc sediments in the Palaeocene-Eocene, and subsequently thrusted in the Eo-Oligocene by upper plate shortening, after which they were underthrusted and uplifted by the Bengal-Nicobar fan that accreted as a lower-plate derived accretionary prism exposed west of the Jarawa thrust.

A little different view may be developed from a recent postulation of Advokaat et al. (2018) who linked the Cretaceous initiation of subduction and the subsequent

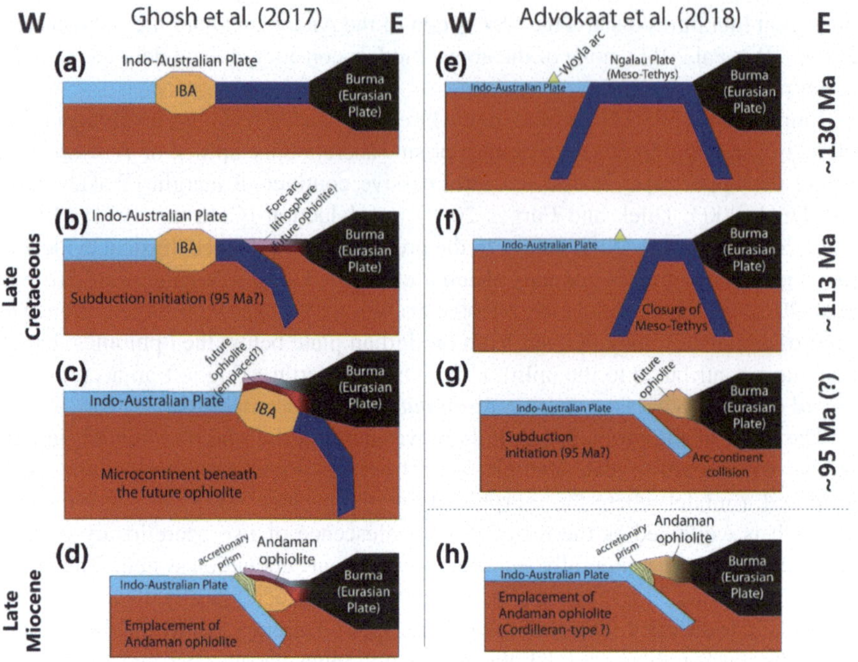

Fig. 5 Schematic diagram illustrating views on possible geodynamic evolution including the origin, evolution and emplacement of Andaman ophiolite at stages. Left panel (**a**–**d**) is a simplified version of Fig. 7.7 of Ghosh et al. (2017). The right panel (**e**–**g**) is drawn based on reconstruction and kinematic scenario proposed by Advokaat et al. (2018); Fig. 7.C). (**a**–**b**) Subduction initiation and formation of forearc lithosphere (future Andaman ophiolite). **c** Strong slab pull dragged the Indo-Burma-Andaman (IBA) microcontinent beneath the upper plate (future ophiolite), similar to *Tethyan-type* ophiolite emplacement mechanism. **d** Final emplacement more akin to *Cordilleran-type*, might be associated with the formation of accretionary prism due to the second phase of subduction beneath the microcontinent. (**e**–**f**) Double sided subduction of Ngalau plate leading to closure of Meso-Tethys and formation of Woyla Arc. **g** Collision of Woyla Arc and Eurasian plate leading to east verging subduction (subduction polarity reversal). (h) Hypothetical *Cordilleran-type* emplacement of Andaman ophiolite. Please note, Ghosh et al. (2017) did not consider the Woyla history in their model. Similarly, geophysically imaged Indo-Burma-Andaman (IBA) microcontinent was not addressed by Advokaat et al. (2018) and does not exclude the microcontinent being later subducted. So, in our point of view, these two models are not mutually exclusive and hence should not be considered as alternatives

formation of the Andaman ophiolitic crust to a subduction polarity reversal following collision of the Woyla arc of west Sumatra with Sundaland (Eurasia) (Fig. 5e–g). This would place the Andaman lithosphere in an original backarc position prior to inception of subduction at the western edge of the Woyla arc, opening opportunities to explain the contrasting geochemical signatures. Uplift and subsidence of the ophiolites recorded in shallow marine Palaeogene and deep-marine Oligo-Miocene sediments (Bandopadhyay and Carter 2017b, c) and alluded to above may result from the interplay of sediment or microcontinental crustal accretion, and subduction erosion and upper plate shortening. The detailed analysis of deformation, magmatism, metamorphism, and sedimentation of the Andaman Ophiolite in context of Myanmar and Sumatra geology and India-Australia-Asia plate motion will allow the further unlocking of the forearc archive of subduction in the Andaman-Nicobar archipelago.

4 Conclusion

Studies reveal that Andaman Ophiolite plays a crucial role in unravelling the India-Eurasia collisional history and global geodynamics as well. Petrochemical studies on the various litho-member confirm its suprasubduction zone origin (Ghosh et al. 2013; Pal 2011), however, some key questions have not yet been critically addressed. This contribution summarizes the major findings and highlights the gap areas.

(1) Volcanic rocks of Andaman Ophiolite are geochemically the most diverse rock types and they have been divided into—(i) Pillow basalts, akin to MORB characters and (ii) East coast volcanics with island arc affinity. The origin of these two volcanic groups of rocks has been interpreted variously by earlier workers who attempted to fit them in various tectonic settings. However, a comprehensive geochemical study of the volcanic rocks with reference to their field dispositions are still lacking.

(2) The mantle sections of the Andaman Ophiolite also demonstrate their variation over the islands. This has been explained in terms of variation in degree of melting and mode of melting specific to tectonic settings. Mantle peridotites of Rutland Island dominantly represent arc peridotites (Ghosh et al. 2009, 2013) whereas that in Middle and North Andaman are akin to backarc basin peridotites (Ghosh et al. 2018). Spatial distribution of these two contrasting mantle rocks and their association with other rock types point towards a complex evolution of the Andaman Ophiolite.

(3) U-Pb zircon ages from plagiogranites of Andaman Ophiolite yielded 93–95 Ma age (Pedersen et al. 2010; Sarma et al. 2010) which has been correlated with global Neotethyan subduction initiation (ca. 95 Ma). In case this age is assigned to the age of Andaman Ophiolite it is difficult to fit in the regional geodynamic setting of this region because two neighbouring ophiolites from Nagaland-Manipur and Myanmar, both lying farther north along the same ophiolite belt date ~117 Ma (Singh et al. 2017, Aitchison et al. 2019) and ~127 Ma (Liu et al. 2016) respectively. Field and geochemical studies of Andaman plagiogranites indicate that the ~95 Ma age should be assigned to arc maturity. If so, this age may not approximate the age of subduction initiation, because in Oman subduction initiation dated from prograde garnet of metamorphic sole predate the forearc spreading by at least 8 Ma (Guilmette et al. 2018). Sub-ophiolitic metamorphic sole from Andaman Ophiolite is reported, but not dated yet. Detailed thermobarometric, geochemical, and geochronological studies in particular of the

sole rocks and gabbroic cumulates might unravel the real story of the evolution of Andaman Ophiolite.

Acknowledgements This work is part of the doctoral research (PhD) program of the first author, supported with DST INSPIRE Fellowship (IF 130148). BG acknowledges financial support received from Science and Engineering Research Board, DST, India (EMR/2017/000929). DJJvH acknowledges a Netherlands Organization for Scientific Research (NWO) VICI grant 865.17.001. Authors acknowledge the help received from Debojit Talukdar and Ashish Raul in preparing the DEM map. J.S. Ray and M. Radhakrishna, Editors of this monograph are thankfully acknowledged for their invitation to submit this article. Hetu Sheth and one anonymous reviewer are thanked for their insightful comments and suggestions that significantly improved the quality of the manuscript.

References

Acharyya SK (2007) Collisional emplacement history of the Naga-Andaman Ophiolite and the position of the eastern Indian suture. J Asian Earth Sci 29(2):229–242

Acharyya S, Ray K, Sengupta S (1990) Tectonics of the ophiolite belt from Naga Hills and Andaman Islands, India. Proc Indian Acad Sci-Earth Planet Sci 99(2):99–187

Advokaat EL, Bongers MLM, Rudyawan A, BouDagher-Fadel MK, Langereis CG, van Hinsbergen DJJ (2018) Early Cretaceous origin of the Woyla Arc (Sumatra, Indonesia) on the Australian plate. Earth Planet Sci Lett 498:348–361

Agard P, Yamato P, Soret M, Prigent C, Guillot S, Plunder A, Dubacq B, Chauvet A, Monié P (2016) Plate interface rheological switches during subduction infancy: control on slab penetration and metamorphic sole formation. Earth Planet Sci Lett 451:208–220. https://doi.org/10.1016/j.epsl.2016.06.054

Aitchison JC, Ao A, Bhowmik S, Clarke GL, Ireland TR, Kachovich S, Lokho K, Stojanovic D, Roeder T, Truscott N, Zhen Y, Zhou R (2019) Tectonic evolution of the Western Margin of the Burma microplate based on new fossil and radiometric age constraints. Tectonics 38(5):1718–1741. https://doi.org/10.1029/2018tc005049

Anonymous (1972) Penrose field conference on ophiolites. Geotimes 17:24–25

Awasthi N (2017) Provenance and paleo–weathering of Tertiary accretionary prism–forearc sedimentary deposits of the Andaman Archipelago, India. J Asian Earth Sci 150:45–62

Bandopadhyay PC, Carter A (2017a) Chapter 2 Introduction to the geography and geomorphology of the Andaman–Nicobar Islands. In: Bandopadhyay PC, Carter A (eds) The Andaman–Nicobar accretionary ridge: geology, tectonics and hazards, vol 47. Geological Society, London, Memoirs, pp 9–18

Bandopadhyay PC, Carter A (2017b) Chapter 8 Mithakhari deposits. In: Bandopadhyay PC, Carter A (eds) The Andaman–Nicobar accretionary ridge: geology, tectonics and hazards, vol 47. Geological Society, London, Memoirs, pp 111–132

Bandopadhyay PC, Carter A (2017c) Chapter 9 Submarine fan deposits: petrography and geochemistry of the Andaman Flysch. In: Bandopadhyay PC, Carter A (eds) The Andaman–Nicobar accretionary ridge: geology, tectonics and hazards, vol 47. Geological Society, London, Memoirs, pp 133–140

Beccaluva L, Coltorti M, Galassi B, Macciotta G, Siena F (1994) The Cainozoic calcalkaline magmatism of the western Mediterranean and its geodynamic significance. Bollettino di Geofisica Teorica ed Applicata 36:293–308

Bédard JH, Lauzière K, Tremblay A, Sangster A (1998) Evidence for forearc seafloor-spreading from the Betts Cove ophiolite, Newfoundland: oceanic crust of boninitic affinity. Tectonophysics 284(3):233–245

Bhattacharya A, Pal T, Ghosh B (2013) Characterization of the accreted ophiolite slices of Rutland Island, Andaman Sea: evolution in a suprasubduction zone setting. Ofioliti 38(2):121–142

Coleman RG (1971) Plate tectonic emplacement of upper mantle peridotites along continental edges. J Geophys Res 76(5):1212–1222

Curray JR (2005) Tectonics and history of the Andaman Sea region. J Asian Earth Sci 25(1):187–232

Curray JR, Moore DG, Lawver LA, Emmel FJ, Raitt RW, Henry M, Kieckhefer R (1979) Tectonics of the Andaman Sea and Burma. In: Watkins JS, Montadert L, Dickerson PW (eds) Geological and geophysical investigations of continental margins, AAPG Memoir, vol 29. American Association of Petroleum Geologists, pp 189–198

Dewey JF (1976) Ophiolite obduction. Tectonophysics 31(1):93–120

Dilek Y (2003) Ophiolite concept and its evolution. In: Dilek Y, Newcomb S (eds) Ophiolite concept and the evolution of geological thought, GSA Special Papers, vol 373. Geological Society of America, pp 1–16

Dilek Y, Flower MFJ (2003) Arc-trench rollback and forearc accretion: 2. A model template for ophiolites in Albania, Cyprus, and Oman. In: Dilek Y, Robinson PT (eds) Ophiolites in earth history, vol 218. Geological Society, London, Special Publications, pp 43–68

Dilek Y, Furnes H (2009) Structure and geochemistry of Tethyan ophiolites and their petrogenesis in subduction rollback systems. Lithos 113(1):1–20

Dilek Y, Furnes H (2011) Ophiolite genesis and global tectonics: geochemical and tectonic fingerprinting of ancient oceanic lithosphere. GSA Bull 123(3–4):387–411

Dilek Y, Furnes H (2014) Ophiolites and their origins. Elements 10(2):93–100

Dilek Y, Moores E (1990) Regional tectonics of the eastern Mediterranean ophiolites. In: Malpas J, Moores E, Panayiotou A, Xenophontos C (eds) Ophiolites, oceanic crustal analogues, proceedings of the symposium "Troodos 1987", pp 295–309

Dilek Y, Thy P, Hacker B, Grundvig S (1999) Structure and petrology of Tauride ophiolites and mafic dike intrusions (Turkey): implications for the Neotethyan ocean. GSA Bull 111(8):1192–1216

Furnes H, de Wit M, Dilek Y (2014) Precambrian greenstone belts host different ophiolite types. In: Dilek Y, Furnes H (eds) Evolution of Archean crust and early life. Springer Netherlands, Dordrecht, pp 1–22

Ghosh B, Bandyopadhyay D, Morishita T (2017) Andaman-Nicobar ophiolites, India: origin, evolution and emplacement. In: Bandopadhyay PC, Carter A (eds) The Andaman-Nicobar accretionary ridge: geology, tectonics and hazards, vol 47. Geological Society, London, Memoirs, pp 95–110

Ghosh B, Morishita T, Bhatta K (2013) Significance of chromian spinels from the mantle sequence of the Andaman ophiolite, India: paleogeodynamic implications. Lithos 164–167:86–96

Ghosh B, Morishita T, Gupta BS, Tamura A, Arai S, Bandyopadhyay D (2014) Moho transition zone in the Cretaceous Andaman ophiolite, India: a passage from the mantle to the crust. Lithos 198–199:117–128

Ghosh B, Mukhopadhyay S, Morishita T, Tamura A, Arai S, Bandyopadhyay D, Chattopadhaya S, Ovung TN (2018) Diversity and evolution of suboceanic mantle: constraints from Neotethyan ophiolites at the eastern margin of the Indian plate. J Asian Earth Sci 160:67–77

Ghosh B, Pal T, Bhattacharya A, Das D (2009) Petrogenetic implications of ophiolitic chromite from Rutland Island, Andaman—a boninitic parentage in supra-subduction setting. Mineral Petrol 96(1):59

Guilmette C, Smit MA, van Hinsbergen DJJ, Gürer D, Corfu F, Charette B, Maffione M, Rabeau O, Savard D (2018) Forced subduction initiation recorded in the sole and crust of the Semail ophiolite of Oman. Nat Geosci 11(9):688–695

Hall R (2002) Cenozoic geological and plate tectonic evolution of SE Asia and the SW Pacific: computer-based reconstructions, model and animations. J Asian Earth Sci 20(4):353–431

Hutchison CS (1975) Ophiolite in Southeast Asia. GSA Bull 86(6):797–806

Imsong W, Choudhury S, Phukan S (2016) Ascertaining neotectonic activities in the southern part of Shillong plateau through geomorphic parameters and remote sensing data. Curr Sci 110(1):91–98

Ishizuka O, Tani K, Reagan MK, Kanayama K, Umino S, Harigane Y, Sakamoto I, Miyajima Y, Yuasa M, Dunkley DJ (2011) The timescales of subduction initiation and subsequent evolution of an Oceanic Island arc. Earth Planet Sci Lett 306(3):229–240
Jafri SH, Balaram V, Ramesh SL (1990) Geochemistry of Andaman-Nicobar Island basalts: a case for a possible plume origin. J Volcanol Geoth Res 44(3):339–347
Jafri SH, Charan SN, Govil PK (1995) Plagiogranite from the Andaman ophiolite belt, Bay of Bengal, India. J Geol Soc 152(4):681–687
Jafri SH, Sarma DS, Sheikh JM (2010) Hyaloclastites in pillow basalts, South Andaman Island, Bay of Bengal, India. Curr Sci 99(12):1825–1829
Jafri SH, Sheikh JM (2013) Geochemistry of pillow basalts from Bompoka, Andaman-Nicobar Islands, Bay of Bengal, India. J Asian Earth Sci 64:27–37
Karunakaran C, Pawde MB, Raina VK, Ray KK, Saha SS (1964) Geology of South Andaman Island, India. Reports of 22nd international geological congress, New Delhi, vol 11, pp 79–100
Karunakaran C, Ray KK, Saha SS (1967) A revision of the stratigraphy of Andaman and Nicobar Islands, India. Bull Natl Inst Sci India 38:436–441
Ling HY, Chandra R, Karkare SG (1996) Tectonic significance of Eocene and Cretaceous radiolaria from South Andaman Island, northeast Indian Ocean. Island Arc 5(2):166–179
Liu C-Z, Chung S-L, Wu F-Y, Zhang C, Xu Y, Wang J-G, Chen Y, Guo S (2016) Tethyan suturing in Southeast Asia: Zircon U-Pb and Hf-O isotopic constraints from Myanmar ophiolites. Geology 44(4):311–314
Maffione M, Thieulot C, van Hinsbergen DJJ, Morris A, Plümper O, Spakman W (2015) Dynamics of intraoceanic subduction initiation: 1. Oceanic detachment fault inversion and the formation of supra-subduction zone ophiolites. Geochem Geophys Geosyst 16:1753–1770
Moores EM (1982) Origin and emplacement of ophiolites. Rev Geophys 20(4):735–760
Moores EM, Kellogg LH, Dilek Y (2000) Tethyan ophiolites, mantle convection, and tectonic "historical contingency": A resolution of the "ophiolite conundrum". In: Dilek Y, Moores EM, Elthon D, Nicolas A (eds) Ophiolites and oceanic crust: new insights from field studies and the Ocean Drilling Program, GSA Special Papers, vol 349. Geological Society of America, pp 3–12
Morishita T, Yoshikawa M, Tamura A, Guotana JM, Ghosh B (2018) Petrology of peridotites and Nd-Sr isotopic composition of their clinopyroxenes from the middle Andaman ophiolite, India. Minerals 8(9):410. https://doi.org/10.3390/min8090410
Mukhopadhyay M, Dasgupta S (1988) Deep structure and tectonics of the Burmese arc: constraints from earthquake and gravity data. Tectonophysics 149(3):299–322
Pal T (2011) Petrology and geochemistry of the Andaman ophiolite: melt–rock interaction in a suprasubduction-zone setting. J Geol Soc 168(4):1031–1045
Pal T, Bhattacharya A (2010) Greenschist-facies sub-ophiolitic metamorphic rocks of Andaman Islands, Burma-Java subduction complex. J Asian Earth Sci 39(6):804–814
Pal T, Bhattacharya A (2011) Block-and-ash flow deposit of the Narcondam Volcano: product of dacite–andesite dome collapse in the Burma-Java subduction complex. Island Arc 20(4):520–534. https://doi.org/10.1111/j.1440-1738.2011.00782.x
Pal T, Chakraborty PP, Gupta TD, Singh CD (2003) Geodynamic evolution of the outer-arc–forearc belt in the Andaman Islands, the central part of the Burma-Java subduction complex. Geol Mag 140(3):289–307
Pedersen RB, Searle MP, Carter A, Bandopadhyay PC (2010) U-Pb zircon age of the Andaman ophiolite: implications for the beginning of subduction beneath the Andaman-Sumatra arc. J Geol Soc 167(6):1105–1112
Ratheesh Kumar RT, Windley BF, Rajesh VJ, Santosh M (2013) Elastic thickness structure of the Andaman subduction zone: implications for convergence of the Ninetyeast Ridge. J Asian Earth Sci 78:291–300
Ray JS, Pande K, Bhutani R (2015) $^{40}Ar/^{39}Ar$ geochronology of subaerial lava flows of Barren Island volcano and the deep crust beneath the Andaman Island Arc, Burma Microplate. Bull Volcanol 77:57. https://doi.org/10.1007/s00445-015-0944-9

Ray KK (1985) East Coast Volcanics: a new suite in the ophiolite of Andaman Islands. Rec Geol Surv India 116(2):83–87

Ray KK, Sengupta S, van Den Hul HJ (1988) Chemical characters of volcanic rocks from Andaman ophiolite, India. J Geol Soc 145(3):393–400

Rioux M, Bowring S, Kelemen P, Gordon S, Dudás F, Miller R (2012) Rapid crustal accretion and magma assimilation in the Oman-U.A.E. ophiolite: high precision U-Pb zircon geochronology of the gabbroic crust. J Geophys Res Solid Earth 117(B7):B07201. https://doi.org/10.1029/2012jb009273

Rioux M, Bowring S, Kelemen P, Gordon S, Miller R, Dudás F (2013) Tectonic development of the Samail ophiolite: high-precision U-Pb zircon geochronology and Sm-Nd isotopic constraints on crustal growth and emplacement. J Geophys Res Solid Earth 118(5):2085–2101

Sarma DS, Jafri SH, Fletcher IR, McNaughton NJ (2010) Constraints on the tectonic setting of the Andaman ophiolite, Bay of Bengal, India, from SHRIMP U-Pb Zircon geochronology of plagiogranite. J Geol 118(6):691–697

Sengupta S, Ray KK, Acharyya SK, de Smeth JB (1990) Nature of ophiolite occurrences along the eastern margin of the Indian plate and their tectonic significance. Geology 18(5):439–442

Shervais JW (2001) Birth, death, and resurrection: the life cycle of suprasubduction zone ophiolites. Geochem Geophys Geosyst 2(1):2000GC000080. https://doi.org/10.1029/2000gc000080

Sheth H (2014) What drives centuries-long polygenetic scoria cone activity at Barren Island volcano? J Volcanol Geoth Res 289:64–80. https://doi.org/10.1016/j.jvolgeores.2014.10.019

Sheth HC, Ray JS, Bhutani R, Kumar A, Smitha RS (2009) Volcanology and eruptive styles of Barren Island: an active mafic stratovolcano in the Andaman Sea, NE Indian Ocean. Bull Volcanol 71(9):1021. https://doi.org/10.1007/s00445-009-0280-z

Singh AK, Chung S-L, Bikramaditya RK, Lee HY (2017) New U-Pb zircon ages of plagiogranites from the Nagaland-Manipur Ophiolites, Indo-Myanmar Orogenic Belt, NE India. J Geol Soc 174(1):170–179

Singh SC, Moeremans R, McArdle J, Johansen K (2013) Seismic images of the sliver strike-slip fault and back thrust in the Andaman-Nicobar region. J Geophys Res Solid Earth 118(10):5208–5224

Srivastava RK, Chandra R, Shastry A (2004) High-Ti type N-MORB parentage of basalts from the south Andaman ophiolite suite, India. Proc Indian Acad Sci—Earth Planet Sci 113(4):605–618

Stern RJ, Reagan M, Ishizuka O, Ohara Y, Whattam S (2012) To understand subduction initiation, study forearc crust: to understand forearc crust, study ophiolites. Lithosphere 4(6):469–483. https://doi.org/10.1130/L183.1

van Hinsbergen DJJ, Maffione M, Plunder A, Kaymakcı N, Ganerød M, Hendriks BWH, Corfu F, Gürer D, de Gelder GINO, Peters K, McPhee PJ, Brouwer FM, Advokaat EL, Vissers RLM (2016) Tectonic evolution and paleogeography of the Kırşehir Block and the Central Anatolian Ophiolites, Turkey. Tectonics 35(4):983–1014

van Hinsbergen DJJ, Peters K, Maffione M, Spakman W, Guilmette C, Thieulot C, Plümper O, Gürer D, Brouwer FM, Aldanmaz E, Kaymakcı N (2015) Dynamics of intraoceanic subduction initiation: 2. Suprasubduction zone ophiolite formation and metamorphic sole exhumation in context of absolute plate motions. Geochem Geophys Geosyst 16(6):1771–1785. https://doi.org/10.1002/2015gc005745

Vohra CP, Haldar D, Roy AKG (1989) The Andaman-Nicobar ophiolite complex and associated mineral resources—current appraisal. In: Ghose NC (ed) Phanerozoic ophiolites of India. Sumna Publishers and Distributers, India, pp 281–315

Wakabayashi J, Dilek Y (2003) What constitutes 'emplacement' of an ophiolite?: Mechanisms and relationship to subduction initiation and formation of metamorphic soles. In: Dilek Y, Robinson PT (eds) Ophiolites in Earth History, vol 218. Geological Society, London, Special Publications, pp 427–447

Workman RK, Hart SR (2005) Major and trace element composition of the depleted MORB mantle (DMM). Earth Planet Sci Lett 231(1–2):53–72. https://doi.org/10.1016/j.epsl.2004.12.005

Timing of Formation and Obduction of the Andaman Ophiolite

Shrema Bhattacharya, Kanchan Pande, Alok Kumar, Oinam Kingson and Jyotiranjan S. Ray

Abstract The Andaman ophiolite forms part of the ophiolite chain that encircles the northern and eastern boundaries of the Indian plate. Unlike its northern counterparts, it occurs on an accretionary prism, located at the convergent margin of Indian and Burma plates. It not only forms part of the basement of the accretionary complex but also occurs as thrust sheets on outer-arc sediments. The timings of formation and emplacement/obduction of this ophiolite had remained tentative. Here, we present results of our effort to date these events by ^{147}Sm–^{143}Nd and ^{40}Ar–^{39}Ar methods and understand their implications for the evolution of the Andaman subduction zone. Our geochemical data suggest that the ophiolite had formed in a supra subduction zone setting. The whole-rock ^{147}Sm–^{143}Nd isochron age of the ophiolite suggests a formation age of 98 ± 8 (2σ) Ma, which is consistent with the existing age data and with the age of the oceanic crust located beneath the Andaman volcanic arc. The ^{40}Ar–^{39}Ar apparent age spectra for rocks that occur along the faulted margin of the east coast of the south and middle Andaman Islands, though disturbed, yield plateaus for low temperature steps (<900 °C) having concordant ages with an average of 0.9 ± 0.3 (2σ) Ma, the timing of argon resetting. This timing points to thermal metamorphism during a thrusting event, which most likely was responsible for obduction of the entire eastern belt ophiolite. Based on our results and stratigraphy of the Andaman Islands, we propose a tectonic evolutionary model for the Andaman Accretionary Prism that involves twofold ophiolite obduction, one prior to 55 Ma and the other at ~0.9 Ma.

Keywords Ophiolite · Obduction · Andaman Islands · ^{147}Sm–^{143}Nd dating · ^{40}Ar–^{39}Ar dating

---

S. Bhattacharya · A. Kumar · O. Kingson · J. S. Ray (✉)
Physical Research Laboratory, Navrangpura, Ahmedabad 380009, India
e-mail: jsray@prl.res.in

K. Pande
Department of Earth Sciences, Indian Institute of Technology Bombay, Powai, Mumbai 400076, India

A. Kumar
Department of Geology, Banaras Hindu University, Varanasi 221005, India

J. S. Ray and M. Radhakrishna (eds.), *The Andaman Islands and Adjoining Offshore: Geology, Tectonics and Palaeoclimate*, Society of Earth Scientists Series,
https://doi.org/10.1007/978-3-030-39843-9_2

1 Introduction

Ophiolites are remnants of ancient oceanic crust and parts of upper mantle exposed on land. Although they are widely distributed world over, most of them occur in orogenic belts or at continental margins. They are the only remnants of the seafloor older than ~170 Ma; therefore, present important records of tectonic/magmatic/hydrothermal processes in ancient oceanic lithospheres. They serve as the best archives of the evolution of ocean basins beginning with oceanic spreading through subduction and ultimate closure (Dilek and Furnes 2014). They have also been utilized to understand processes at mid-oceanic ridges, evolution of mantle chemistry, initiation of subduction zones, tectonic evolution of orogenic belts and paleogeography of ocean basins etc. (Dilek and Furnes 2014). Ophiolites can vary a great deal in terms of their structure, chemistry and emplacement mechanisms. They are usually exposed along thrusted oceanic-oceanic or oceanic-continental plate margins. The manner by which ophiolites get emplaced or obducted is unresolved and the determination of timings of their emplacements remains tricky. The latter is so because in many cases there is no suitable material for dating the obduction. In addition, radioisotope dating of ophiolitic rocks often yields either the age of original magmatic crystallization or subsequent metamorphism on sea-floor or during orogenesis (Vaughan and Scarrow 2003). However, at times it may be possible to determine the timing of an obduction event through the age of thermal metamorphism or magmatism during the obduction (Chew et al. 2010), if the records of the event are well preserved and the dating technique is appropriate. In this work, we have made an attempt to date the timings of both the formation and obduction of the Andaman ophiolite belt that discontinuously crop out along the convergent margin of the Indian and Burma plates (Fig. 1).

One of the extensive subduction systems of the Mesozoic existed during the closure of the Tethys Ocean and its evidence is found across the west and south-east Asia as ophiolite remnants along the convergent margin of Indian and Eurasian plates. The history of the Tethyan subduction can be traced back in time by studying these ophiolite sections. This subduction system continued from Cyprus in the west through Oman, northern margin of the Indian plate, and India-Myanmar boundary in the east into the Andaman-Sumatra sector, where it is currently active (Pedersen et al. 2010). The archipelago of Andaman and Nicobar Islands of India is located on the forearc of the Andaman Subduction Zone (ASZ), where the Indian plate is subducting obliquely beneath the Burma microplate, which is a sliver of the Eurasian Plate (Fig. 1a). The Andaman-Nicobar Ridge represents the accretionary wedge/prism and is made up of imbricate stack of thrust slices (Figs. 1 and 2) comprising segments of oceanic lithosphere (ophiolite), pelagic sediments (chert and carbonates) and trench-forearc sediments (turbidites). The knowledge about the timing of formation and mode of emplacement of the Andaman Ophiolite remains equivocal. Some believe that the ophiolite is part of the subducting late Cretaceous oceanic slab that got obducted onto the accretionary wedge sometime during the Paleocene (e.g., Pal et al. 2003; Curray 2005; Pal 2011), whereas others propose a supra subduction zone (SSZ) origin and

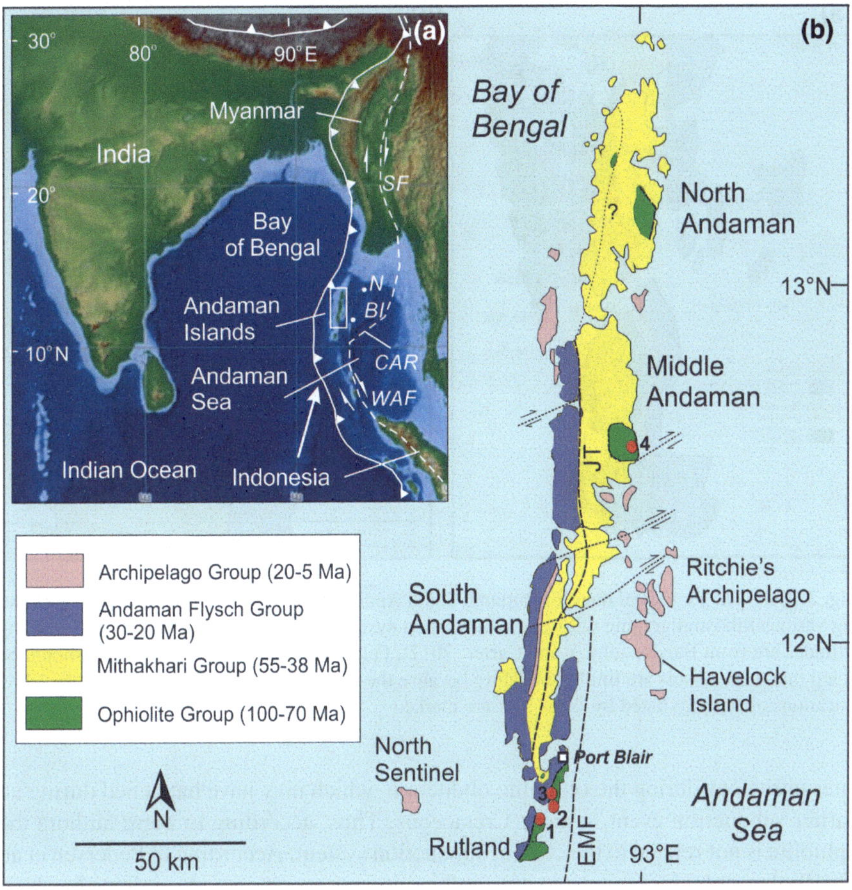

Fig. 1 Geological map of the Andaman Islands showing various lithologies. Inset shows the geographic position of these islands and the regional geo-tectonic framework. N = Narcondam and BI = Barren Island are two subaerial volcanoes of the active Andaman volcanic arc. CAR = Central Andaman Ridge; WAF = West Andaman Fault; EMF = Eastern Margin Fault; JT = Jarwa Thrust; SF = Sagaing Fault. The figure uses our field data in conjunction with the lithological, stratigraphic and structural and details given in Allen et al. (2008), Bandopadhyay and Carter (2017), Curray (2005), Pal et al. (2003) and Pal (2011). The locations of the samples dated by ^{40}Ar–^{39}Ar method are marked: 1–4 (sample IDs are given in Fig. 2)

subsequent thrusting onto the eastern margin of the prism at some unspecified time (e.g., Ghosh et al. 2017; Saha et al. 2018 and references therein).

To understand the sequence of events during the formation of the Andaman Ophiolite, Sarma et al. (2010) and Pedersen et al. (2010) dated zircons from plagiogranites exposed along the east coast of South Andaman. These studies, though suggest identical formation ages of 93.6 ± 1.3 Ma and 95 ± 2 Ma, respectively, propose contradicting models for ophiolite origin and emplacement. Sarma et al. (2010) suggest that the plagiogranites are not cogenetic with the ophiolite sequence, rather represent

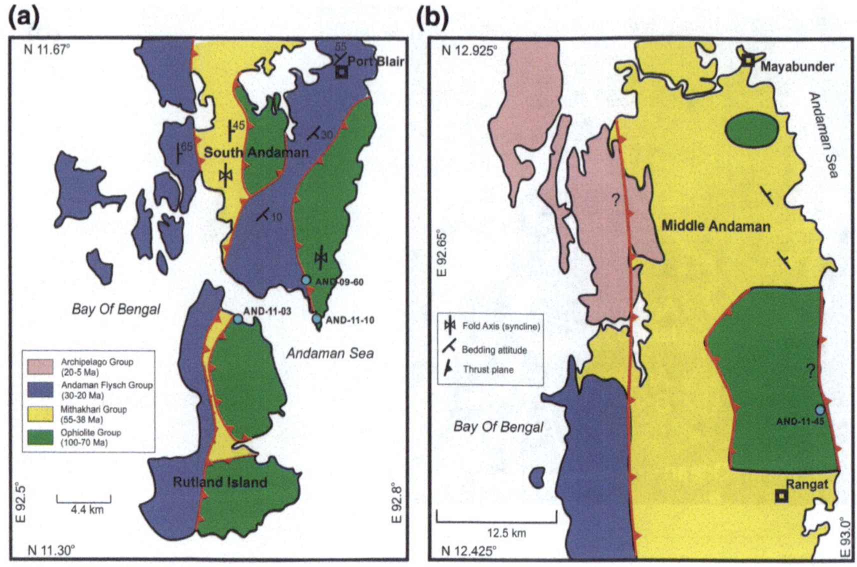

Fig. 2 Enlarged geological maps of Rutland, South Andaman and Middle Andaman Islands showing various lithostratigraphic units and major thrust systems. Structural details including bedding attitudes are from Bandopadhyay and Carter (2017), Pal et al. (2003) and Pal (2011). It should be noted that these details are limited in nature because the islands for the most part are inaccessible. Locations of samples dated by ^{40}Ar-^{39}Ar are marked

later intrusions during the ophiolite obduction, which may have happened during an earlier subduction event in early Cretaceous. Thus, according to these authors the ophiolite is not related to the current subduction system. According to Pedersen et al. (2010) these plagiogranites are of trondhjemitic composition and originated as late-stage differentiates of gabbroic melts, notwithstanding the fact that field relationship shows the latter to intrude the former. They propose that the Andaman ophiolite, on which the volcanic arc was built, had formed during the initiation of subduction during Cenomanian and is contemporaneous with most Neo-Tethyan ophiolites. Since the above studies are limited only to one type of lithology from one location and do not suggest much about the timing and mechanism of ophiolite obduction, it is imperative that a detailed investigation on the origin and emplacement of the Andaman ophiolite is done by studying both the crustal and mantle sections from multiple outcrops. This work is an effort in this direction. We studied geochemistry of these rocks to establish the nature of the ophiolite and dated rocks from both the mantle and crustal sections by ^{147}Sm–^{143}Nd and ^{40}Ar–^{39}Ar methods to determine the age of origin and timing of obduction, respectively.

2 Geology and Samples

In the ASZ, the main chain of islands of the Andaman and Nicobar archipelago forms the outerarc of the Andaman forearc, the volcanic islands of Narcondam and Barren represent a poorly developed volcanic arc, and the Central Andaman Ridge is a backarc spreading centre that is believed to have begun operating at ~4 Ma (Fig. 1a; Curray 2005). The outer arc ridge is made up of thrust slices of ophiolite and a thick sequence of trench-forearc sediments of Paleogene-Neogene antiquity. The lithostratigraphy of the Andaman Accretionary Prism (AAP) is divided into five groups which in ascending chronological order are: (1) the Ophiolite Group (100–70 Ma), (2) the Mithakhari Group (55–35 Ma), (3) the Andaman Flysch Group (35–20 Ma), (4) the Archipelago Group (20–5 Ma) and (5) the Nicobar Group (5–0 Ma) (Curray 2005; Allen et al. 2008; Awasthi and Ray 2019). The ophiolitic rocks are believed to form the basement for the later sedimentary deposits (Pal et al. 2003; Curray 2005; Awasthi and Ray 2019), in spite of fact that there exists no outcrop which shows them to occur directly (stratigraphically) below the Mithakhari rocks. All the groups are distinctly separated by unconformities, except for the boundary between the Andaman Flysch and Mithakhari groups that remains indistinguishable (Awasthi and Ray 2019). The Eocene Mithakhari Group consists of trench-slope deposits of conglomerate, sandstone, shale and volcaniclastics. This group of rocks is believed to have been deposited prior to 40 Ma based on their foraminiferal assemblages and detrital mineral ages (Allen et al. 2008). The basal conglomerate of the Mithakhari Group contains clasts/pebbles of basalt, gabbro and serpentinite that apparently belong to the Ophiolite Group. Such clasts are also found in the ejecta of mud volcanoes that occur at the top of the accretionary prism, which points to the fact that the ophiolitic rocks form part of the basement complex of the Andaman accretionary wedge. The Oligocene Andaman Flysch Group is made up of submarine fan deposits consisting of unfossiliferous siliciclastic turbidites in the form of sandstone-shale-rhythmite and was deposited during 35–20 Ma (Awasthi and Ray 2019). Apatite fission track and U–Th/He age data suggest exhumation of the sequence at ~20 Ma (Allen et al. 2008). The Mio-Pliocene Archipelago Group represents shallow marine shelf deposits comprising carbonates, with minor siliciclastic and volcaniclastic sediments/tephra (Bandopadhyay and Ghosh 2015). This unit experienced folding along N-S axis with dips 30–40° (Pedersen et al. 2010). Sr-isotope stratigraphic dates put the depositional age of the group in the bracket of 18–6 Ma (Awasthi and Ray 2019). Pleistocene limestone, beach deposits and tuffs represent the topmost unnamed group, sometimes referred to as the Nicobar Group (Curray 2005; Awasthi and Ray 2019).

The ophiolite sections are exposed sporadically primarily along the east coast of the main chain of the Andaman Islands and the Rutland Island (Fig. 1b). They have also been reported from Teressa, Tilanchong and Bompoka islands of the Nicobars (e.g., Jafri and Sheikh 2013). The contact relations between various members of the ophiolite are sparsely observed, except in a few localities. Field evidence suggests it to be a dismembered ophiolite sequence. Most of the sections contain massive pillow lavas and a few have plagiogranites, gabbros and serpentinized peridotites. No outcrop of sheeted dikes has so far been reported. Figure 2 presents geological maps of the South and the Middle Andaman Islands prepared based on information given in Acharyya (2007), Bandopadhyay and Carter (2017), Pal et al. (2003), Pal (2011) and our own field observations. Since most parts of the islands remain inaccessible or restricted, the structural details remain scanty and debatable. It is, however, generally accepted that the AAP is made up of easterly dipping imbricate thrust slices (Fig. 2) and the dip of these slices increases eastward with the last slice being almost vertical (Pal et al. 2003). There exist numerous field evidences for these thrusting events and associated thermal events, the most prominent ones being the tectonites and psuedotachylite veins/dikes found at several places along the east coast (Pal et al. 2003; Ghosh et al. 2009). Two examples of such features that occur on the hanging wall of the easternmost thrust slice are shown in Fig. 3.

Eleven whole rock samples were collected from various crustal and mantle sections exposed on thrusted/faulted blocks along the east coast of Rutland, South Andaman and Middle Andaman Islands for geochemical and isotopic studies (Table 1). Nine samples were dated using ^{147}Sm–^{143}Nd method and four were selected for ^{40}Ar–^{39}Ar dating. The locations of the four ^{40}Ar–^{39}Ar samples are marked on the geological map in Figs. 1 and 2. Most of the exposed sections of the Andaman Ophiolite have undergone various degrees of alteration. We selected relatively less altered portions of the samples for geochemical/isotopic and dating studies after a thorough petrographic investigation. AND-9-60 is a relatively less altered peridotite and it contains olivine, clinopyroxene, orthopyroxene and spinel (Fig. 4a); hence, is classified as a spinel-lherzolite. AND-9-32 is a heavily altered peridotite that has serpentine (relict olivine), altered clinopyroxene and spinel (Fig. 4b). The basalts generally show low degree of alteration, display polygonal arrangement of plagioclase laths and flow structures (Fig. 4c, d). Interstitial spaces are filled with subhedral grains of clinopyroxene (sometime altered to chlorite), opaque minerals and glass. The plagiogranites are medium to fine grained, consisting predominantly of quartz and plagioclase with minor hornblende (Fig. 4e, f). The rock type, mineralogy of the samples and the investigation done on them are provided in Table 1.

3 Analytical Methods

Trace element and Sm–Nd isotopic analyses were carried out on powdered whole-rock samples at the Physical Research Laboratory, Ahmedabad. Trace elements were analyzed in sample solutions which were prepared by dissolving about 60 mg of powdered sample using conventional HF–HNO_3 acid dissolution protocol. Concentrations of trace elements were measured on a Thermo Q-ICPMS using BHVO-2 as a calibration standard. BHVO-2 was also run as an unknown for accuracy and precision checks. The details of analytical procedure followed can be found in Chatterjee and Ray (2017). For Nd isotopic ratio analyses, sample powders were dissolved using

Fig. 3 Two examples of field evidence for tectonothermal events along the east coast of the main chain of the Andaman Islands: **a** tectonite at Panchwati, Middle Andaman; **b** pseudotachylite in gabbro at Rangachang, South Andaman

HF–HNO_3–HCl dissolution procedure for silicate rocks (Awasthi et al. 2010). Separation of rare earth elements (REE) was done by conventional cation exchange chemistry and Nd was separated from other REEs using Ln-specific resin from Eichrom with dilute HCl as elutant. The Nd isotopic ratios of $^{143}Nd/^{144}Nd$ were measured in static multi-collection mode on an ISOPROBE-T mass spectrometer and the ratios were corrected for mass fractionation using $^{146}Nd/^{144}Nd = 0.7219$ (Ray et al. 2013;

Table 1 Details of samples from the Andaman Ophiolite studied in this work

Sample	Location	Rock type	Mineralogy	Methods of study
PB-07-05A	N11.5308; E92.7242	Plagiogranite	Plagioclase + quartz + hornblende	Chem; Sm–Nd
AND-09-32	N13.2704; E92.9625	Peridotite	Serpentinized olivine + relict clino pyroxene + spinel	Chem; Sm–Nd
AND-09-50	N11.5921; E92.6722	Gabbro	Altered plagioclase + clino pyroxene	Chem; Sm–Nd
AND-09-60	N11.5211; E92.7174	Peridotite	Olivine (partially serpentinized) + clino pyroxene + ortho pyroxene + spinel	Chem; Sm–Nd; Ar–Ar
AND-11-1	N11.4914; E92.6650	Basalt (altered)	Plagioclase + clino pyroxene + opaques	Chem; Sm–Nd
AND-11-2	N11.4914; E92.6671	Basalt (altered)	Plagioclase + clino pyroxene + opaques	Chem; Sm–Nd
AND-11-3	N11.4876; E92.6689	Basalt	Plagioclase + minor clino pyroxene + chlorite + glassy ground mass	Chem; Sm–Nd; Ar–Ar
AND-11-8	N11.4900; E92.7096	Basalt	Plagioclase + clino pyroxene	Chem; Sm–Nd
AND-11-10	N11.4889; E92.7094	Plagiogranite	Plagioclase + quartz + hornblende + opaques	Chem; Sm–Nd; Ar–Ar
AND-11-19	N11.5667; E92.7409	Diabase	Plagioclase + clino pyroxene	Chem; Sm–Nd
AND-11-45	N11.5689; E92.9650	Basalt	Plagioclase + clino pyroxene	Chem; Ar–Ar

Note Chem, Sm–Nd and Ar–Ar in the last column stand for trace element geochemistry (Table 2), ^{147}Sm-^{143}Nd dating (Table 3) and ^{40}Ar-^{39}Ar dating (Table 4), respectively

Awasthi et al. 2010). The average value for JNdi-1 standard was determined to be 0.512104 ± 0.000004 (2σ) during the course of our measurements.

About 200 mg of pre-cleaned whole-rock powder (grain size >250 mm), cleared off carbonates by 1% HCl leaching, was packed in aluminium capsules and irradiated in the DHRUVA reactor at BARC, Mumbai, for ~120 h. The 17.3 ± 0.2 Ma (2σ) B4B

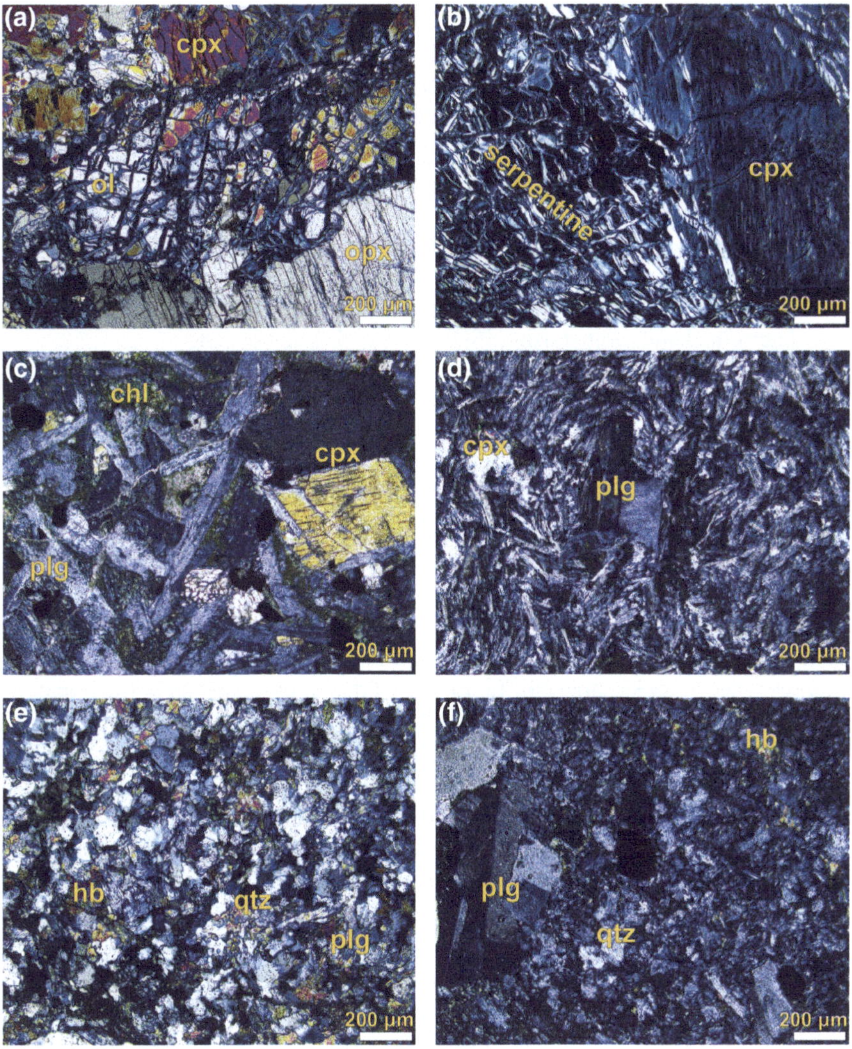

Fig. 4 Photomicrographs (crossed polars) of whole rocks of some of the samples listed in Table 1. The rocks types shown are: **a** fresh Peridotite (AND-9-60); **b** serpentinized Peridotite (AND-9-32); **c** basalt (AND-11-03); **d** basalt (AND-11-45); **e** plagiogranite (PB-07-05A); **f** plagiogranite (AND-11-10)

biotite (Flisch 1982) was used as the flux monitor and high-purity CaF_2 and K_2SO_4 salts for interference corrections arising from the production of Ar from Ca and K isotopes. Argon was extracted by incremental heating between 500 and 1350 °C, at steps of 50 °C and isotopic ratios were measured in a Thermo Fisher ARGUS-VI multi-collector mass spectrometer at the Department of Earth Sciences, IIT Bombay, India (Ray et al. 2015; Pande et al. 2017). ^{40}Ar blank contributions were <2% for all temperature steps. The irradiation parameter J for each sample was corrected for neutron flux variation using the activity of nickel wires irradiated with each sample. Values of fluence-corrected J values ($\pm 2\sigma$) for our samples are: AND-11-03, 0.001151 $\pm$ 0.000035; AND-11-10, 0.001390 $\pm$ 0.000042; AND-11-45, 0.001496 $\pm$ 0.000046; and AND-9-60, 0.001392 $\pm$ 0.000042. The mean values for the correction factors $(^{36}Ar/^{37}Ar)_{Ca}$, $(^{39}Ar/^{37}Ar)_{Ca}$, and $(^{40}Ar/^{39}Ar)_K$ are 0.000461, 0.001175, and 0.004529, respectively. We define a plateau age as weighted mean of ages of contiguous and concordant step ages comprising >50% of the total ^{39}Ar released. Ages were calculated using the decay constant of Steiger and Jäger (1977), and the value used for $^{40}Ar/^{36}Ar$ of atmospheric argon was 295.5. Plateau and isochron ages were determined and plotted using ISOPLOT 4.15 (Ludwig 2012).

4 Results

Trace element concentrations in ophiolitic samples are presented in Table 2. Normal mid-oceanic ridge basalt (N-MORB) normalized data are plotted in Fig. 5. The mantle rocks (AND-09-32; AND-09-60) and the cumulate gabbro (AND-09-50) show depleted light rare earth element (LREE) patterns. The wehrlite sample (AND-09-32) has the lowest trace element contents, with most elements being less than 0.2 ppm. All three types of rocks show Nb, Ta, Zr and Hf depletions, and Nb/Ta fractionation. The gabbro sample has a similar trace element pattern as that of the lherzolite sample (AND-09-60) except that it has much higher contents of Ba, Pb and Sr. Basalts and plagiogranites that make up the bulk of the crustal section show enriched patterns for most of the large ion lithophile elements (LILEs) and LREEs (Fig. 5a). These samples show Nb and Ta anomalies and most depict Nb/Ta fractionation, although at a smaller magnitude compared to that observed in the peridotites and gabbros (Fig. 5a). Pb contents of these rocks are highly variable (0.18–3.26 ppm) that results in both positive and negative anomalies. A plagiogranite sample (PB07-05A) and a basalt sample (AND-11-10) display negative Pb anomalies and also show prominent negative Zr and Hf anomalies (Fig. 5a). In chondrite normalized plots (not shown) basalt and plagiogranites samples show LREE enriched patterns with a few showing minor Eu anomalies, whereas gabbro and peridotite samples display depleted LREE patterns.

The ^{147}Sm–^{144}Nd data for basalt, gabbro and peridotites samples are presented in Table 3. These data yield a 9 points isochron (MSWD = 1.12) giving an age of 98.1 $\pm$ 8.2 Ma (2σ, Fig. 6). Considering that the isochron involves an entire sequence of ophiolite, peridotite-gabbro-plagiogranite-basalt, the Sm–Nd date can be considered as the age of crystallization/formation of the Andaman ophiolite. The y-intercept of the isochron gives an initial $^{143}Nd/^{144}Nd$ value of 0.512913 (ε_{Nd} (t) = + 7.7), which suggests derivation of these rocks from an LREE depleted mantle source.

The summary of ^{40}Ar–^{39}Ar dating results are presented in Table 4. The apparent age spectra for all the four samples are plotted in Fig. 7. These data suggest a clear two stage evolution of the K–Ar system, a high temperature (1400–1000 °C) and a

Table 2 Trace element data for Andaman ophiolites

	PB-07-05A	AND-09-32	AND-09-50	AND-09-60	AND11-01	AND11-02	AND-11-03	AND11-08	AND11-10	AND11-19	AND-11-45	BHVO-2	BHVO-2
	Plagiogranite	Peridotite	Gabbro	Peridotite	Basalt	Basalt	Basalt	Basalt	Plagiogranite	Basalt	Basalt	Measured	Reported
Rb	2.14	0.03	0.07	0.18	3.87	2.65	0.80	5.34	1.37	2.84	0.13	9.58	9.11
Ba	31.40	0.19	30.51	1.47	44.39	63.42	24.35	40.51	6.32	36.46	9.60	132	131
Th	0.35	bdl	bdl	0.01	0.82	0.52	0.39	0.07	0.39	0.29	0.56	1.16	1.22
U	0.12	0.01	bdl	bdl	0.32	0.18	0.16	0.01	0.11	0.11	0.20	0.38	0.40
Nb	1.3	bdl	0.1	0.1	6.8	1.9	3.0	1.4	1.1	1.2	1.7	18.7	18.1
Ta	0.07	0.02	0.03	0.02	0.48	0.18	0.31	0.17	0.16	0.12	0.17	1.10	1.14
La	6.24	0.003	0.13	0.18	13.71	4.81	7.25	3.85	5.41	4.35	7.43	15.4	15.2
Ce	16.34	0.02	0.41	0.67	35.38	11.82	18.52	11.52	12.70	11.48	18.73	37.7	37.5
Pr	2.57	0.01	0.09	0.15	5.29	1.79	2.94	2.04	2.01	1.83	2.82	5.38	5.35
Pb	0.19	bdl	0.13	0.00	0.52	0.46	0.62	0.18	3.26	0.61	1.08	1.6	1.6
Sr	109.90	1.95	91.65	11.84	129.10	164.00	81.41	138.60	161.75	147.30	62.54	404	396
Nd	12.33	0.06	0.61	0.98	23.98	8.43	14.33	11.19	9.43	9.06	13.29	24.5	24.5
Hf	0.48	0.02	0.10	0.15	4.07	1.70	2.20	0.37	1.27	1.43	1.93	4.22	4.36
Zr	12.27	0.43	2.89	3.64	121.60	68.20	80.01	7.81	59.53	45.92	86.02	177	172
Sm	3.30	0.05	0.32	0.43	5.94	2.19	3.93	3.43	2.59	2.58	3.46	6.05	6.07
Eu	0.77	0.02	0.20	0.16	1.58	0.69	1.40	1.11	0.77	0.98	1.04	2.05	2.07
Gd	3.58	0.10	0.51	0.56	6.19	2.41	4.27	4.05	2.93	2.84	3.61	6.16	6.24
Tb	0.57	0.03	0.10	0.10	0.94	0.39	0.67	0.66	0.48	0.47	0.56	0.91	0.92
Dy	3.61	0.22	0.74	0.70	5.67	2.47	4.24	4.24	3.14	3.02	3.48	5.25	5.31
Y	20.12	1.52	4.83	3.85	31.28	13.17	25.20	23.63	19.60	18.06	20.32	27	26

(continued)

Table 2 (continued)

	PB-07-05A	AND-09-32	AND-09-50	AND-09-60	AND11-01	AND11-02	AND-11-03	AND11-08	AND11-10	AND11-19	AND-11-45	BHVO-2	BHVO-2
	Plagiogranite	Peridotite	Gabbro	Peridotite	Basalt	Basalt	Basalt	Basalt	Plagiogranite	Basalt	Basalt	Measured	Reported
Ho	0.72	0.05	0.16	0.14	1.10	0.50	0.87	0.87	0.65	0.61	0.69	0.97	0.98
Er	1.94	0.16	0.46	0.38	3.05	1.41	2.43	2.35	1.87	1.73	1.89	2.5	2.54
Tm	0.26	0.02	0.06	0.05	0.40	0.19	0.34	0.32	0.27	0.24	0.26	0.32	0.33
Yb	1.6	0.2	0.4	0.3	2.6	1.3	2.2	2.0	1.7	1.5	1.6	1.9	2.0
Lu	0.19	0.03	0.06	0.05	0.36	0.18	0.32	0.27	0.24	0.21	0.22	0.267	0.274

Reproducibility (2σ) for REE is <3% and for others <5%. A comparison of measured (mean of 5 analyses) and recommended values for standard BHVO-2 (Jochum et al. 2005) are given as an indication of accuracy. bdl = below detection limit

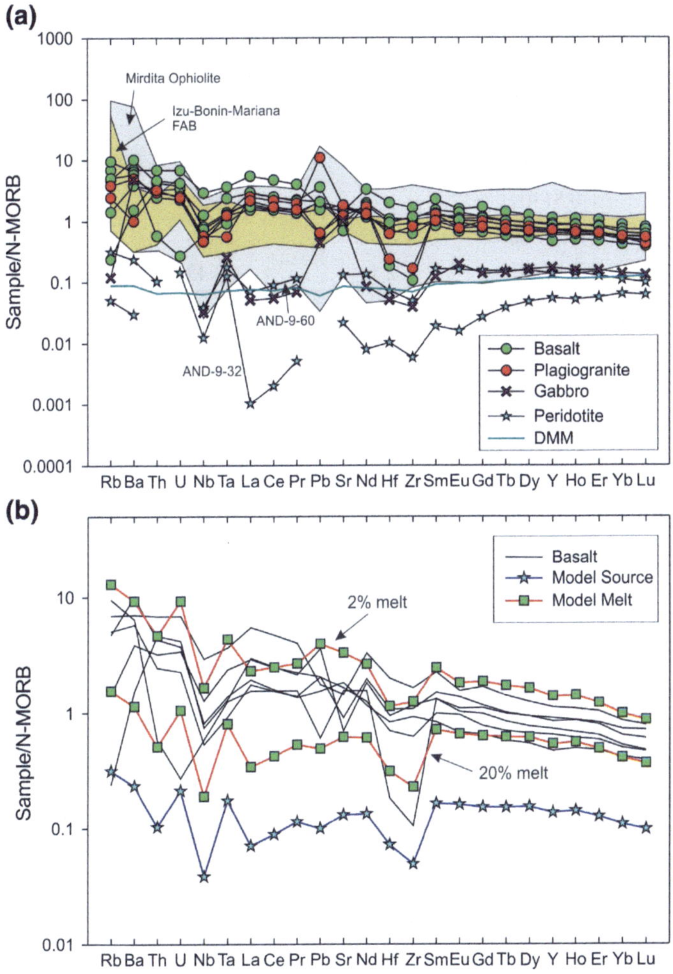

Fig. 5 **a** Normal-MORB normalized trace element patterns for samples of Andaman ophiolites studied in this work (Table 1). Samples are categorized into different rocks types. The two peridotites samples are marked: AND-9-32: wehrlite; AND-9-60: lherzolite. DMM: Depleted MORB mantle (Workman and Hart 2005). The shaded regions encompass data for volcanic and subvolcanic rocks, except rhyolites, from the Mirdita ophiolite from Albania (Dilek et al. 2008) and that for forearc basalts (FAB) from Izu-Bonin-Marian subduction system (Reagan et al. 2010); data from both regions possess SSZ signatures. Normalizing values are from Sun and McDonough (1989). **b** N-MORB normalized patterns for only the basalts of figure (**a**) compared with model generated melts representing 20 and 2% non-modal batch melting of a hypothetical lherzolitic source. The model parameters used are the following. Source mineralogy (%): olivine (ol): clino-pyroxene (cpx): ortho-pyroxene (opx): spinel (sp) = 45:20:30:5; proportion in the melt (%): ol: cpx: opx: sp = 20:40:10:30; source trace element contents: AND-9-60 (Table 2) + U and Pb are 0.01 and 0.03 ppm, respectively; mineral-melt partition coefficients for all elements except Eu for cpx-melt are from Bedard (2006) and that for Eu (cpx-melt) is from McKenzie and O'Nions (1991)

Table 3 Sm-Nd analytical data for ophiolite samples

Sample	$^{143}Nd/^{144}Nd$	Sm (ppm)	Nd (ppm)	$^{147}Sm/^{144}Nd$
PB-07-05A	0.513017	3.30	12.33	0.1619
AND-9-32	0.513240	0.05	0.06	0.5233
AND-9-50	0.513123	0.32	0.61	0.3179
AND-9-60	0.513091	0.43	0.98	0.2687
AND-11-1	0.513003	5.94	23.98	0.1497
AND-11-2	0.513024	2.19	8.43	0.1572
AND-11-8	0.513028	3.43	11.19	0.1855
AND-11-10	0.513023	2.59	9.43	0.1659
AND-11-19	0.513017	2.58	9.06	0.1720

Note Refer to Table 1 for sample descriptions. The $^{143}Nd/^{144}Nd$ for JNdi-1 measured is 0.512104 ± 0.000004 (n = 10; 2σ). Internal precision (2σ) for $^{143}Nd/^{144}Nd$ ratio is 0.00001. Total procedural blank for Nd was <500 pg. Sm and Nd contents are Q-ICPMS data

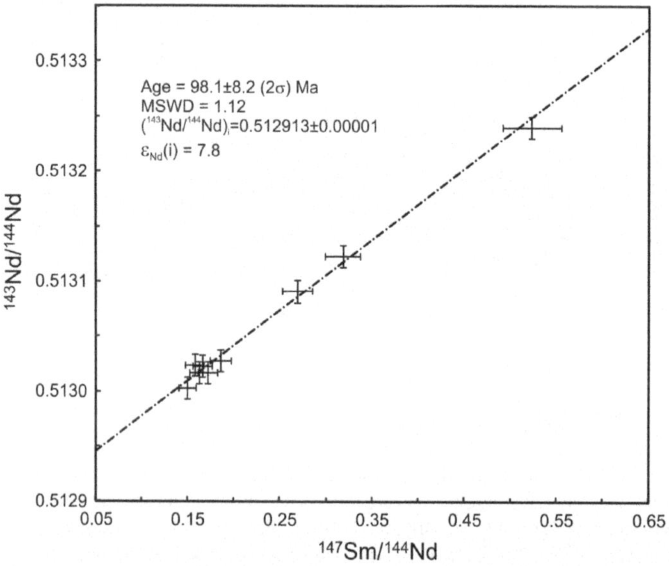

Fig. 6 Plot of $^{143}Nd/^{144}Nd$ versus $^{147}Sm/^{144}Nd$ for Andaman Ophiolite. Samples and data used for the plot are listed in Table 3. Isoplot 4.15 (Ludwig 2012) has been used for the isochron fitting and age calculations. A uniform 6% error (at 2σ) is assumed for $^{147}Sm/^{144}Nd$ in the isochron fitting

low-medium temperature (960–400 °C). The former stage shows staircase Ar-release patterns in all samples, whereas low temperature steps display plateaus (Fig. 7). The plateau steps yield indistinguishable ages though with large errors. The $^{40}Ar/^{36}Ar$ versus $^{39}Ar/^{36}Ar$ data for the plateau steps are plotted in isochron diagrams in Fig. 8. The plateau and normal/inverse isochron ages are indistinguishable and intercepts of

Table 4 Summary of $^{40}Ar/^{39}Ar$ data for Andaman Ophiolite

Sample	Plateau				Inverse isochron			Integrated
	Steps	^{39}Ar (%)	Age (Ma)	MSWD	Age (Ma)	$(^{40}Ar/^{36}Ar)_i$	MSWD	Age (Ma)
AND-11-03	9	77.1	0.7 ± 0.5	0.03	0.7 ± 0.5	295.2 ± 2.3	0.015	11.4 ± 0.8
AND-11-10	7	55.4	0.9 ± 1.2	0.001	0.9 ± 1.2	295.5 ± 3.8	0.001	31.5 ± 1.2
AND-11-45	6	81.0	1.0 ± 0.5	0.01	1.0 ± 0.6	295.6 ± 6.9	0.014	12.4 ± 0.3
AND-9-60	5	70.1	0.9 ± 1.1	0.004	0.9 ± 1.2	295.4 ± 2.8	0.003	12.9 ± 1.2

Note Sample details are given in Table 1. Ages are weighted averages and errors are 2σ. Errors on ages are without error on J

MSWD Mean Square Weighted Deviate

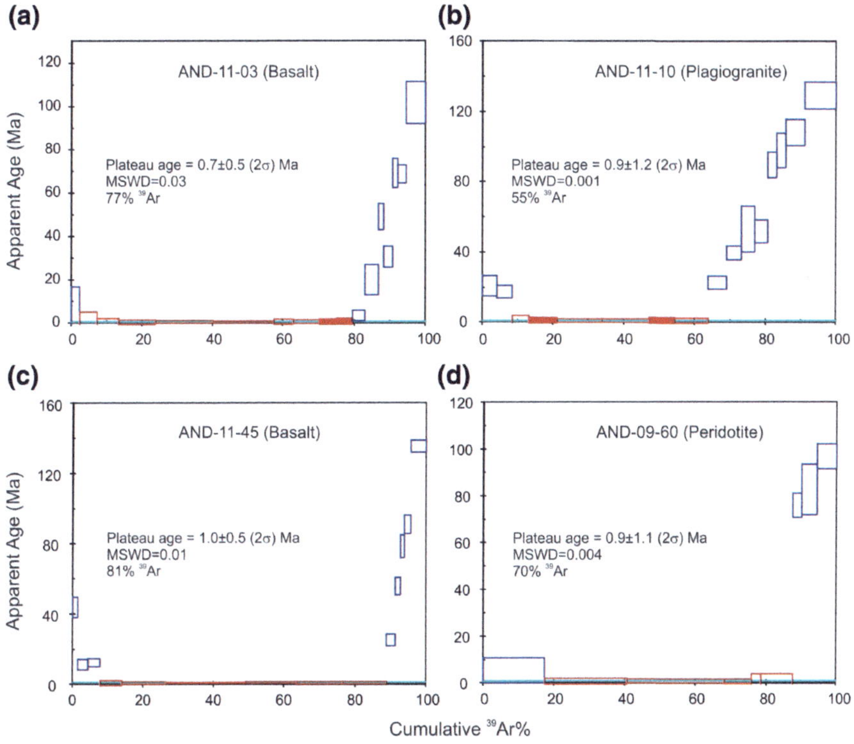

Fig. 7 Step-heating ^{40}Ar–^{39}Ar apparent age spectra for whole rock samples from Andaman Ophiolite. Plateau (steps marked in red) ages, defined as weighted means of contiguous and concordant step ages comprising >50% of the total gas released, are indicated. Age spectra were plotted using ISOPLOT 4.15 (Ludwig 2012)

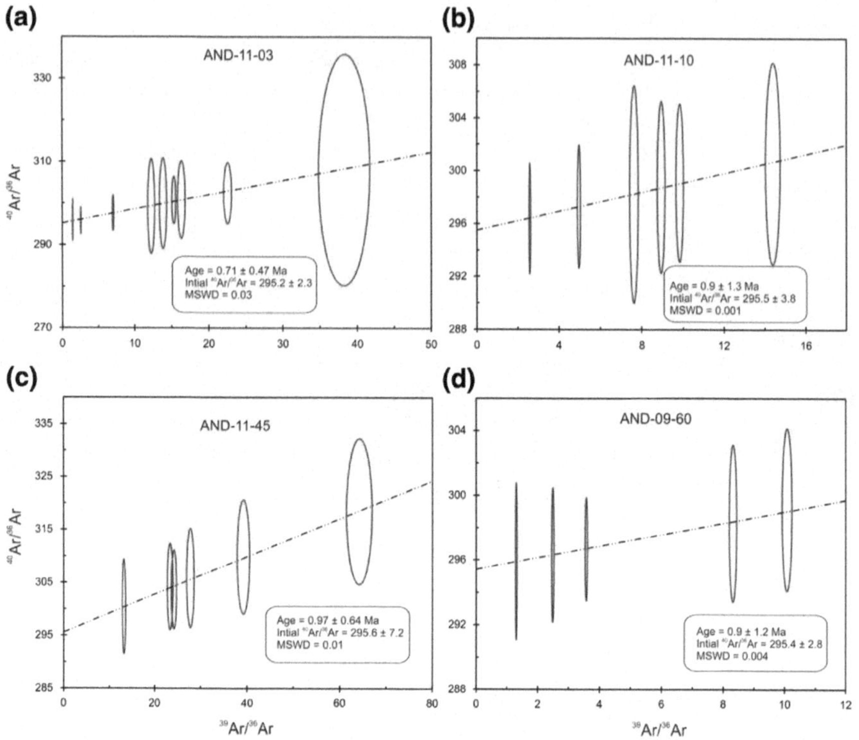

Fig. 8 $^{40}Ar/^{36}Ar$ versus $^{39}Ar/^{36}Ar$ isochron plots for plateau steps for the samples in Fig. 6. Error ellipses are 2σ. Isochrons were plotted ISOPLOT 4.15 (Ludwig 2012)

isochrons display atmospheric composition (Fig. 8; Table 4). Therefore, the weighted mean of all four plateau ages, 0.9 ± 0.3 Ma (2σ), can be considered as the timing of closure of the K–Ar system in these rocks. It should be noted that the laboratory temperature for Argon extraction from samples is not directly correlatable to the temperature the rock has seen in the natural environment. The apparent ages of the highest temperature steps, in all four samples, fall in the range of 136–97 Ma, albeit with very high errors (Fig. 7). This observation points to the fact that the crystallization/formation age of these rocks is older than 100 Ma, which is consistent with the Sm–Nd isochron age.

5 Discussion

Most exposures of the Andaman Ophiolite are crustal sections that contain pillow basalt (lavas), diabase and plagiogranite (dikes) and gabbros (intrusives). Outcrops of mantle sections are limited and mostly serpentinized. Our geochemical data reveal that the lavas and intrusives possess much higher contents of LILE compared to N-MORB (Fig. 5a). Higher contents of LILE and REE, and lower contents of Nb, Hf and Zr in the lherzolite sample, compared to that of an average depleted MORB mantle (DMM; Fig. 5a), are typical signatures of a metasomatized mantle wedge that occurs above a subduction zone (e.g., Prouteau et al. 2001; Munker et al. 2004; Kimura et al. 2016 and references therein). The volcanic/subvolcanic rocks of the ophiolite possess Th/Yb ratios of 0.04–0.35 for very low ratios of Ta/Yb (0.004–0.015; Table 2), which can be considered as typical signatures of SSZ lavas (Dilek and Furnes 2014). The N-MORB normalized patterns for the basalts and plagiogranites of the Andaman ophiolite (Fig. 5a) overlap with that of SSZ signature bearing volcanic rocks of the Mirdita ophiolite (Dilek et al. 2008) and forearc basalts from the Izu-Bonin-Mariana subduction system (Reagan et al. 2010). Thus, the Andaman ophiolite can be considered as an SSZ ophiolite, corroborating some of the earlier suggestions (e.g., Ghosh et al. 2017; Saha et al. 2018). This inference is further substantiated by Nb/Ta fractionation observed in both mantle and crustal rocks (Fig. 5a), with averages of 2.8 and 10.0 (Table 2), respectively. Such low ratios for mantle rocks have also been reported from the Manipur ophiolite, the northern counterpart of the Andaman ophiolite along the same convergent margin (Kingson et al. 2019). The Nb/Ta fractionation trend (i.e., Nb/Ta < 3) in mantle wedges is believed to have been produced by interaction of slab derived fluids that carry the signal from a rutile bearing slab (e.g., Huang et al. 2012). These signals get reduced in the partial melts, which end up as crustal sections of the ophiolites. To demonstrate this, we compared the N-MORB trace element patterns of the basalt samples with that of the model generated partial melts from a metasomatized mantle wedge (AND-9-60). As shown in Fig. 5b, we observed that melts generated through a non-modal batch/equilibrium melting process could explain the data for the basalts with the degree of melting in the range of 2–20%, the latter melt almost replicating the pattern of the source. The existence of chemical equilibrium between the lherzolitic source and the basaltic melt prior to the latter's extraction can be established by the fact that both source rock and the products fall on the Sm–Nd isochron (Fig. 6). The model melts could generate Nb, Zr and Hf anomalies, Nb/Ta fractionation and LILE-REE enrichments as observed in the basalts, thus confirming the SSZ nature of the entire ophiolite section studied in this work (Table 1).

The ^{147}Sm–^{143}Nd whole rock isochron age of 98.1 ± 8.2 (2σ) Ma for samples from both the mantle and crustal sections of the Andaman Ophiolite (Fig. 6) is indistinguishable from the U–Pb zircon ages of 93.6 ± 1.3 (2σ) Ma and 95 ± 2 (2σ) Ma for plagiogranites of the formation, determined earlier by Sarma et al. (2010) and Pedersen et al. (2010), respectively. This observation establishes that the eastern belt of ophiolite (Fig. 1) belongs to a single magmatic event (cogenetic) and the weighted mean of all the three ages, 94 ± 1 (2σ) Ma, can be considered as the age of formation of this ophiolite belt. In a geochemical and isotopic study of crustal xenoliths in Barren Island lavas, Ray et al. (2015) had inferred that the crust beneath the current volcanic arc is part of the same lithosphere that makes up the Andaman ophiolite at the forearc, located ~100 km to the west. If so, then the forearc crust likely to have developed as an SSZ ophiolite during $\geq$106 and 94 Ma, where 106 Ma is the age of the crust beneath the volcanic arc (Ray et al. 2015). Thus, it is likely that the Andaman ophiolite, which represents the newly created SSZ oceanic lithosphere on an overriding plate, formed during 110–90 Ma as part of the Woyla-Incertus subduction system (Hall 2012). The age of the Manipur ophiolite (117–118 Ma; Singh et al. 2017), which represents the northern counterpart of the Andaman ophiolite, is consistent with this hypothesis. The age of the Andaman ophiolite is also strikingly similar to that of the Semail ophiolite, Oman (93.5–97.9 Ma, Tilton et al. 1981; Warren et al. 2005) and the Troodos ophiolite, Cyprus (91.6 ± 1.4 Ma; Mukasa and Ludden 1987) and they all display signatures of SSZ origin. These data confirm that the Andaman ophiolite is part of the Tethyan ophiolite belt that occurs around the Indian plate margin from west to east through north. However, its further extension to south up to Sumatra and Java remains equivocal (e.g., Agard et al. 2007; Dilek and Furnes 2009; Nicolas and Boudier 2011).

The resetting of K–Ar system in our samples at ~0.9 Ma points to a thermal metamorphic event. The fact that these samples came from the hanging wall of the easternmost thrust slice and are located close to fault planes (Fig. 2) links the thermal event to a reactivation event of the thrust. This event possibly was responsible for the exhumation of the eastern belt ophiolite. Because the age of formation of the eastern belt ophiolite is ~94 Ma and parts of the ophiolite form the basement of the accretionary prism and contributed sediments to the younger trench-forearc deposits, in particular to the Mithakhari Group (Awasthi and Ray 2019), the 0.9 million year old exhumation event appears to represent the second such event in the evolutionary history of the AAP. The first ophiolite exhumation event is likely to be older than 55 Ma, the maximum age of the Mithakhari Group (Awasthi and Ray 2019). The NW-SE pull-apart extension that initiated the backarc spreading and formation of the Burma plate during the Mio-Pliocene (Khan and Chakraborty 2005) could have resulted in net compression parallel to the accretionary wedge (e.g., Allen et al. 2008). We speculate that this process was responsible for the ~0.9 Ma ophiolite exhumations onto the accretionary prism. Interestingly, this age is similar to the ^{40}Ar–^{39}Ar age of the Mile Tilek felsic tuff of the South Andaman (0.7 ± 0.2 Ma; Awasthi et al. 2015). This could mean that either the tuff deposit has preserved records of the exhumation event or represents a very large scale, explosive, volcanic eruption somewhere along the Burma-Andaman-Indonesia volcanic arc at this time. In any case, this age and events associated with the exhumation suggest a very large scale tectonic reconfiguration in the Burma-Andaman-Sumatra subduction system at ~0.9 Ma.

Using results of our study, tectonic reconstructions of Hall (2012) for the Indonesian region and the Indian Ocean, tectonic models of Curray (2005), Morley (2017) and Morley and Searle (2017) for the Andaman Sea, and the existing geophysical and geochronological data for the region (Curray 2005) we propose a tectonic evolutionary model for the Andaman subduction zone (Fig. 9). Since the time of the breakup of the Gondwanaland, the north-eastward movement of the Greater India was facilitated by at least two subduction systems located to the southwest of Southeast Asia (Hall 2012). We hypothesize that the Andaman ophiolite formed as a new oceanic lithosphere at the time of initiation of subduction along the leading edge of the Greater

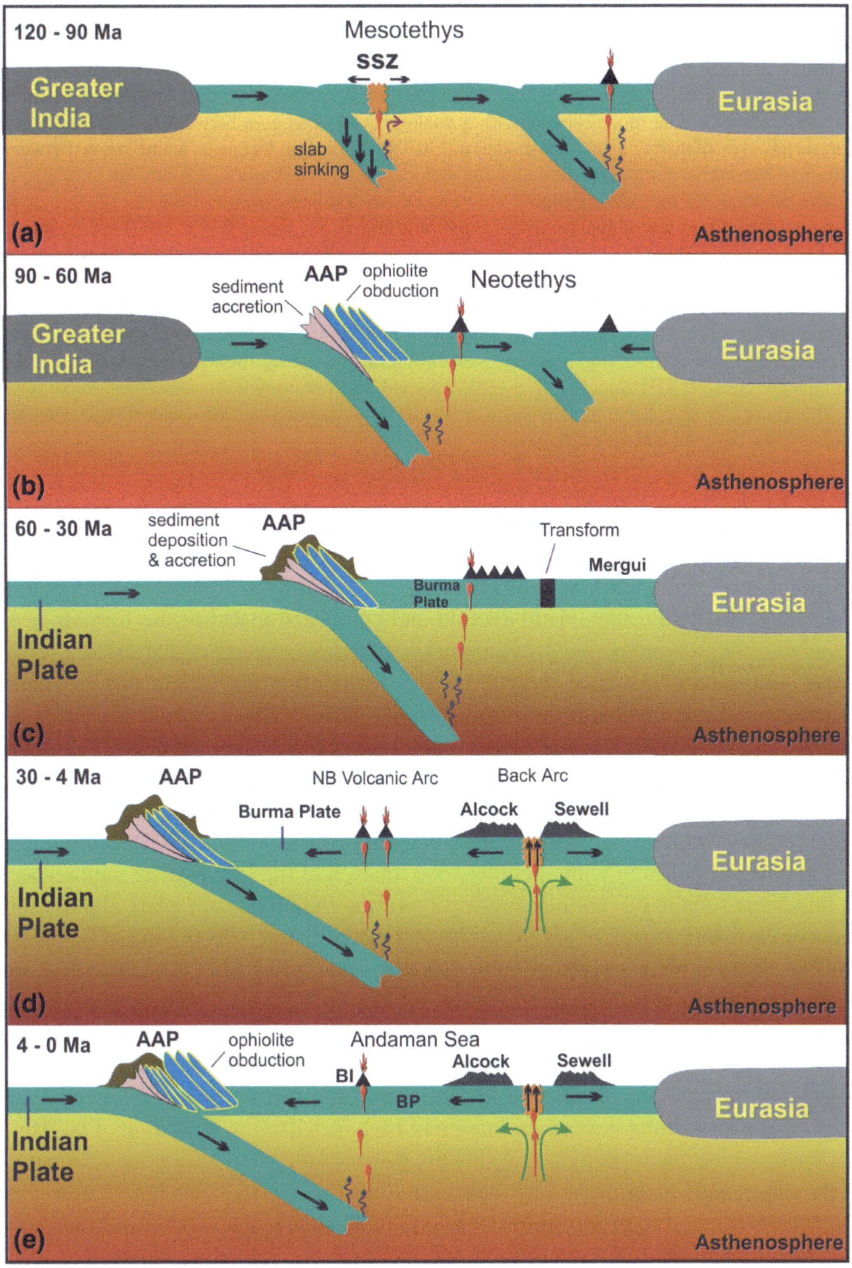
120 - 90 Ma
Mesotethys
SSZ
Greater India
Eurasia
slab sinking
(a)
Asthenosphere
90 - 60 Ma
AAP
ophiolite obduction
sediment accretion
Neotethys
Greater India
Eurasia
(b)
Asthenosphere
60 - 30 Ma
sediment deposition & accretion
AAP
Transform
Mergui
Burma Plate
Eurasia
Indian Plate
(c)
Asthenosphere
30 - 4 Ma
AAP
NB Volcanic Arc
Back Arc
Burma Plate
Alcock
Sewell
Eurasia
Indian Plate
(d)
Asthenosphere
4 - 0 Ma
AAP
ophiolite obduction
Andaman Sea
Alcock
Sewell
BI
BP
Eurasia
Indian Plate
(e)
Asthenosphere

◂**Fig. 9** The figure shows a schematic of our proposed tectonic evolutionary model for the Andaman Ophiolite as part of the Andaman Accretionary Prism (AAP). The profiles are not specific to a particular transect rather they represent a generalized configuration of subduction along the eastern margin of Eurasia. The model begins with a dual subduction system between Greater India and Eurasia based on plate reconstruction model of Hall (2012). Our model covers a time period between 120 and 0 Ma, and encompasses the following major events: (1) formation of Andaman Ophiolite in a supra subduction zone (SSZ) setting during 110–90 Ma, far away from Eurasia (**a**); (2) formation of the AAP and the first obduction of SSZ ophiolite sometime during 90–60 Ma (**b**); (3) counter clockwise rotation of the Indian plate that resulted in movement of the Greater India away from the subduction zone, continued growth of the AAP with exhumation of trench sediments (55–38 Ma) and formation of the Burma plate and Mergui basin (~32 Ma) (**c**); (4) further growth of AAP with addition of forearc turbidites (30–5 Ma), initiation of backarc spreading at ~4 Ma with separation of Alcock and Sewell rises (the older arc system) and formation of the modern volcanic arc, the Narcondam-Barren (NB) arc, at ~3.5 Ma (**d**); a second SSZ ophiolite obduction event on the AAP at ~0.9 Ma (**e**). *Data sources* This work; Curray (2005), Allen et al. (2008), Hall (2012), Bhutani et al. (2014), Awasthi and Ray (2019)

Indian plate, sometime during Early Cretaceous (Fig. 9a). This possibly caused the formation of the northern sector of the Woyla subduction zone, called the Incertus subduction zone, marking the beginning of the northward journey of the Indian plate (Hall 2012). Prolonged subduction and compression led to the formation of an accretionary complex, the AAP, along the Incertus convergent margin as a result of accumulation of material scraped off the downgoing plate. Onto this complex, the SSZ ophiolite got obducted during Late Cretaceous through multiple thrusting events (Fig. 9b). The presence of ophiolitic pebbles in the trench-slope sediments of the Mithakhari Group (55–38 Ma; Awasthi and Ray 2019) bears the testimony of this first obduction process. This type of accretionary emplacement is commonly observed in Cordilleran type ophiolites (Wakabayashi and Dilek 2003). Subsequent tectonic reconfiguration with anticlockwise rotation of the Indian plate with respect to Southeast Asia led to oblique subduction of the Indian oceanic plate and its subsequent collision, during 60–30 Ma. The oblique collision between the Indian plate and the Eurasia plate caused an anticlockwise rotation of Southeast Asia leading to development of a transform boundary in place of the convergent margin west of Eurasia (Fig. 9c). The sliver plate formed as result of this is the Burma plate (Fig. 9c). During this period the AAP continued to grow with deposition and uplift of forearc sediments, the nascent volcanic arc developed into a mature arc (Alcock-Sewell Arc), and compression along the eastern margin of Eurasia led to the formation of the Mergui basin. Subsequent periods saw slowing down of the subduction due to rotation of the plates and cessation of magmatic activity in the Alcock-Sewell Arc; however, the AAP continued to grow. Towards the end of Miocene, a new but minor volcanic arc (Narcondam-Barren Arc) formed and the back-arc spreading began (Fig. 9d). The back arc spreading divided the older volcanic arc into two; the Alcock Rise and the Sewell Rise. The compressional stresses resulting from the spreading possibly caused the second phase of obduction of the SSZ ophiolite, which was present as the oceanic lithosphere of the leading edge of the overriding plate, onto the AAP at ~0.9 Ma (Fig. 9e).

6 Conclusions

Based on our current study of geochemistry and geochronology of the eastern belt ophiolite of the Andaman Islands, we make the following inferences about the origin and evolution of these rocks vis-à-vis that of the accretionary prism on which they occur.

1. The geochemical properties of the crustal and mantle sections of the eastern belt ophiolite point to their formation in an SSZ setting. The crustal sections appear to have been produced by equilibrium melting (2–20%) of a metasomatized lherzolitic mantle wedge.
2. The ^{147}Sm–^{143}Nd whole rock isochron dating of basalts, plagiogranites and peridotites yields a crystallization age of 98.1 ± 8.2 (2σ) Ma for the ophiolite. This age is concordant with previously determined U–Pb zircon ages of the plagiogranites, and therefore, suggests an age of 94 ± 1 (2σ) Ma, the weighted mean of all age data, for the formation of the eastern belt ophiolite. Considered together with other available age data for geochemically similar rocks in the region, our result suggests that the SSZ ophiolite of the ASZ formed during 110–90 Ma.
3. The ^{40}Ar–^{39}Ar dating of both crustal and mantle rock samples indicates thermal resetting of the K–Ar system at 0.9 ± 0.3 Ma (2σ), the mean of four plateau ages. Since these rocks represent the hanging wall of the easternmost thrust imbricate, the age of the thermal event most likely corresponds to the age of exhumation of the ophiolite that happened as a result of reactivation of this thrust slice.
4. Our results, stratigraphy of sediments exposed on the AAP, and data from one of the most widely accepted tectonic reconstruction models for the region (Hall 2012) clearly advocate for a two-stage obduction process for the Andaman ophiolite, one sometime prior to 55 Ma and the other at ~0.9 Ma.

Acknowledgements We thank Neeraj Awasthi and Gaurav Sharma for help during field work. Bivin George helped with trace element measurements on Q-ICPMS. Critical reviews by Rajneesh Bhutani and Biswajit Ghosh are gratefully acknowledged.

References

Acharyya SK (2007) Collisional emplacement history of the Naga-Andaman ophiolites and the position of the eastern Indian suture. J Asian Earth Sci 29:229–242. https://doi.org/10.1016/j.jseaes.2006.03.003

Agard P, Jolivet L, Vrielynck B, Burov E, Monié P (2007) Plate acceleration: the obduction trigger. Earth Planet Sci Lett 258:428–441. https://doi.org/10.1016/j.epsl.2007.04.002

Allen R, Carter A, Najman Y, Bandopadhyay PC, Chapman HJ, Bickle MJ, Garzanti E, Vezzoli G, Andò S, Foster GL, Gerring C (2008) New constraints on the sedimentation and uplift history of the Andaman-Nicobar accretionary prism, South Andaman Island. Geol Soc Am 223–255. https://doi.org/10.1130/2008.2436(11)

Awasthi N, Ray JS (2019) The Palaeogene record of Himalayan erosion in the Andaman Basin. J Earth Syst Sci. https://doi.org/10.1007/s12040-019-1266-7

Awasthi N, Ray JS, Laskar AH, Kumar A, Sudhakar M, Bhutani R, Sheth HC, Yadava MG (2010) Major ash eruptions of Barren Island volcano (Andaman Sea) during the past 72 kyr: clues from a sediment core record. Bull Volcanol 72:1131–1136. https://doi.org/10.1007/s00445-010-0408-1

Awasthi N, Ray JS, Pande K (2015) Origin of the Mile Tilek Tuff, South Andaman: evidence from ^{40}Ar–^{39}Ar chronology and geochemistry. Curr Sci 108:205–210

Bandopadhyay PC, Carter A (2017) Geological framework of the Andaman-Nicobar Islands. Geol Soc Lond Mem 47:75–93. https://doi.org/10.1144/M47.6

Bandopadhyay PC, Ghosh B (2015) Provenance analysis of the Oligocene turbidites (Andaman Flysch), South Andaman Island: a geochemical approach. J Earth Syst Sci 124:1019–1037. https://doi.org/10.1007/s12040-015-0586-5

Bedard JH (2006) A catalytic delamination-driven model for coupled genesis of Archaean crust and sub-continental lithospheric mantle. Geochim Cosmochim Acta 70:1188–1214. https://doi.org/10.1016/j.gca.2005.11.008

Bhutani R, Pande K, Ray JS, Smitha RS, Awasthi N, Kumar A (2014) ^{40}Ar-^{39}Ar geochronology of Narcondam island volcano, Andaman-Indonesian island arc, Andaman Sea: voluminous andesitic eruption at 0.56 Ma and preservation of > 1.8 Ma old plagioclase xenocrysts. GSA Annu Meet Vancouver Br Columbia. Abstract No. 268-9

Chatterjee A, Ray JS (2017) Sources and depositional pathways of mid-Holocene sediments in the Great Rann of Kachchh, India: implications for fluvial scenario during the Harappan Culture. Q Int 443:177–187. https://doi.org/10.1016/j.quaint.2017.06.008

Chew DM, Daly JS, Magna T, Page LM, Kirkland CL, Whitehouse MJ, Lam R (2010) Timing of ophiolite obduction in the Grampian orogen. Geol Soc Am Bull 122:1787–1799. https://doi.org/10.1130/B30139.1

Curray JR (2005) Tectonics and history of the Andaman Sea region. J Asian Earth Sci 25:187–232. https://doi.org/10.1016/j.jseaes.2004.09.001

Dilek Y, Furnes H (2009) Structure and geochemistry of Tethyan ophiolites and their petrogenesis in subduction rollback systems. Lithos 113:1–20. https://doi.org/10.1016/j.lithos.2009.04.022

Dilek Y, Furnes H (2014) Ophiolites and their origins. Elements 10:93–100. https://doi.org/10.2113/gselements.10.2.93

Dilek Y, Furnes H, Shallo M (2008) Geochemistry of the Jurassic Mirdita Ophiolite (Albania) and the MORB to SSZ evolution of a marginal basin oceanic crust. Lithos 100:174–209. https://doi.org/10.1016/j.lithos.2007.06.026

Flisch M (1982) Potassium-argon analysis. In: Odin GS (ed) Numerical dating in stratigraphy. Wiley, Chichester, pp 151–158

Ghosh B, Pal T, Bhattacharya A, Das D (2009) Petrogenetic implications of ophiolitic chromite from Rutland Island, Andaman—a boninitic parentage in supra-subduction setting. Miner Petrol 96:59–70. https://doi.org/10.1007/s00710-008-0039-9

Ghosh B, Bandyopadhyay D, Morishita T (2017) Andaman-Nicobar Ophiolites, India: origin, evolution and emplacement. Geol Soc Lond Mem 47:95–110. https://doi.org/10.1144/M47.7

Hall R (2012) Late Jurassic–Cenozoic reconstructions of the Indonesian region and the Indian Ocean. Tectonophysics 570–571:1–41. https://doi.org/10.1016/j.tecto.2012.04.021

Huang J, Xiao Y, Gao Y, Hou Z, Wu W (2012) Nb-Ta fractionation induced by fluid-rock interaction in subduction-zones: constraints from UHP eclogite- and vein-hosted rutile from the Dabie orogen, Central-Eastern China. J Metamorph Geol 30:821–842. https://doi.org/10.1111/j.1525-1314.2012.01000.x

Jafri SH, Sheikh JM (2013) Geochemistry of pillow basalts from Bompoka, Andaman–Nicobar Islands, Bay of Bengal, India. J Asian Earth Sci 64:27-37. https://doi.org/10.1016/j.jseaes.2012.11.035

Jochum KP, Pfänder J, Woodhead JD, Willbold M, Stoll B, Herwig K, Hofmann AW (2005) MPI-DING glasses: new geological reference materials for in situ Pb isotope analysis. Geochem Geophys Geosyst 6:1–15

Khan PK, Chakraborty PP (2005) Two-phase opening of Andaman Sea: a new seismotectonic insight. Earth Planet Sci Lett 229:259–271. https://doi.org/10.1016/j.epsl.2004.11.010

Kimura J-I, Gill JB, Skora S, van Keken PE, Kawabata H (2016) Origin of geochemical mantle components: role of subduction filter: origin of earth's geochemical components. Geochem Geophys Geosyst 17:3289–3325. https://doi.org/10.1002/2016GC006343

Kingson O, Bhutani R, Balakrishnan S, Dash JK, Shukla AD (2019) Subduction-related Manipur Ophiolite Complex, Indo-Myanmar Ranges: elemental and isotopic record of mantle metasomatism. Geol Soc Lond Spec Publ SP481.9. https://doi.org/10.1144/SP481.9

Ludwig KR (2012) Isoplot-3.75, a geochronological toolkit for Microsoft Excel. Berkeley Geochronol Cent Spec Publ 5:75 (covers versions of Isoplot from 3.75 to 4.15)

McKenzie D, O'Nions RK (1991) Partial melt distributions from inversion of rare earth element concentrations. J Petrol 32:1021–1091. https://doi.org/10.1093/petrology/32.5.1021

Morley CK (2017) Cenozoic rifting, passive margin development and strike-slip faulting in the Andaman Sea: a discussion of established v. new tectonic models. Geol Soc Lond Mem 47:27–50. https://doi.org/10.1144/M47.4

Morley CK, Searle M (2017) Regional tectonics, structure and evolution of the Andaman-Nicobar Islands from ophiolite formation and obduction to collision and back-arc spreading. Geol Soc Lond Mem 47:51–74. https://doi.org/10.1144/M47.5

Mukasa SB, Ludden JN (1987) Uranium-lead isotopic ages of plagiogranites from the Troodos ophiolite, Cyprus, and their tectonic significance. Geology 15:825. https://doi.org/10.1130/0091-7613

Munker C, Wörner G, Yogodzinski G, Churikova T (2004) Behaviour of high field strength elements in subduction zones: constraints from Kamchatka–Aleutian arc lavas. Earth Planet Sci Lett 224:275–293. https://doi.org/10.1016/j.epsl.2004.05.030

Nicolas A, Boudier F (2011) Structure and dynamics of ridge axial melt lenses in the Oman ophiolite. J Geophys Res 116. https://doi.org/10.1029/2010JB007934

Pal T (2011) Petrology and geochemistry of the Andaman ophiolite: melt–rock interaction in a suprasubduction-zone setting. J Geol Soc Lond 168:1031–1045. https://doi.org/10.1144/0016-76492009-152

Pal T, Chakraborty PP, Gupta TD, Singh CD (2003) Geodynamic evolution of the outer-arc forearc belt in the Andaman Islands, the central part of the BurmaJava subduction complex. Geol Mag 140:289–307. https://doi.org/10.1017/S0016756803007805

Pande K, Cucciniello C, Sheth H, Vijayan A, Sharma KK, Purohit R, Jagadeesan KC, Shinde S (2017) Polychronous (Early Cretaceous to Palaeogene) emplacement of the Mundwara alkaline complex, Rajasthan, India: 40Ar/39Ar geochronology, petrochemistry and geodynamics. Int J Earth Sci 106:1487–1504. https://doi.org/10.1007/s00531-016-1362-8

Pedersen RB, Searle MP, Carter A, Bandopadhyay PC (2010) U-Pb zircon age of the Andaman ophiolite: implications for the beginning of subduction beneath the Andaman-Sumatra arc. J Geol Soc Lond 167:1105–1112. https://doi.org/10.1144/0016-76492009-151

Prouteau G, Scaillet B, Pichavant M, Maury R (2001) Evidence for mantle metasomatism by hydrous silicic melts derived from subducted oceanic crust. Nature 410:197–200. https://doi.org/10.1038/35065583

Ray JS, Pande K, Bhutani R, Shukla AD, Rai VK, Kumar A, Awasthi N, Smitha RS, Panda DK (2013) Age and geochemistry of the Newania dolomite carbonatites, India: implications for the source of primary carbonatite magma. Contrib Miner Petrol 166:1613–1632. https://doi.org/10.1007/s00410-013-0945-7

Ray JS, Pande K, Bhutani R (2015) $^{40}Ar/^{39}Ar$ geochronology of subaerial lava flows of Barren Island volcano and the deep crust beneath the Andaman Island Arc, Burma Microplate. Bull Volcanol 77. https://doi.org/10.1007/s00445-015-0944-9

Reagan MK, Ishizuka O, Stern RJ, Kelley KA, Ohara Y, Blichert-Toft J, Bloomer SH, Cash J, Fryer P, Hanan BB, Hickey-Vargas R, Ishii T, Kimura J-I, Peate DW, Rowe MC, Woods M (2010) Fore-arc basalts and subduction initiation in the Izu-Bonin-Mariana system. Geochem Geophys Geosyst 11:1–17. https://doi.org/10.1029/2009GC002871

Saha A, Santosh M, Ganguly S, Manikyamba C, Ray J, Dutta J (2018) Geochemical cycling during subduction initiation: evidence from serpentinized mantle wedge peridotite in the south Andaman ophiolite suite. Geosci Front 9:1755–1775. https://doi.org/10.1016/j.gsf.2017.12.017

Sarma DS, Jafri SH, Fletcher IR, McNaughton NJ (2010) Constraints on the tectonic setting of the Andaman ophiolites, Bay of Bengal, India, from SHRIMP U-Pb zircon geochronology of plagiogranite. J Geol 118:691–697. https://doi.org/10.1086/656354

Singh AK, Chung S-L, Bikramaditya RK, Lee HY (2017) New U-Pb zircon ages of plagiogranites from the Nagaland-Manipur ophiolites, Indo-Myanmar orogenic belt, NE India. J Geol Soc Lond 174:170–179. https://doi.org/10.1144/jgs2016-048

Steiger RH, Jäger E (1977) Subcommission on geochronology: convention on the use of decay constants in geo- and cosmochronology. Earth Planet Sci Lett 36:359–362. https://doi.org/10.1016/0012-821X(77)90060-7

Sun SS, McDonough WF (1989) Chemical and isotopic systematics of oceanic basalts: implications for mantle composition and processes. Geol Soc Lond Spec Publ 42:313–345. https://doi.org/10.1144/GSL.SP.1989.042.01.19

Tilton GR, Hopson CA, Wright JE (1981) Uranium-lead isotopic ages of the Semail Ophiolite, Oman, with applications to Tethyan ocean ridge tectonics. J Geophys Res Solid Earth 86:2763–2775. https://doi.org/10.1029/JB086iB04p02763

Vaughan APM, Scarrow JH (2003) Ophiolite obduction pulses as a proxy indicator of super plume events? Earth Planet Sci Lett 213:407–416. https://doi.org/10.1016/s0012-821x(03)00330-3

Wakabayashi J, Dilek Y (2003) What constitutes 'emplacement' of an ophiolite? Mechanisms and relationship to subduction initiation and formation of metamorphic soles. Geol Soc Lond Spec Publ 218:427–447. https://doi.org/10.1144/gsl.sp.2003.218.01.22

Warren CJ, Parrish RR, Waters DJ, Searle MP (2005) Dating the geologic history of Oman's Semail ophiolite: insights from U–Pb geochronology. Contrib Mineral Petrol 150:403–422. https://doi.org/10.1007/s00410-005-0028-5

Workman RK, Hart SR (2005) Major and trace element composition of the depleted MORB mantle (DMM). Earth Planet Sci Lett 231:53–72. https://doi.org/10.1016/j.epsl.2004.12.005

Facies Stacking of a Deep Sea Depositional Lobe: Case Study from Palaeogene Andaman Flysch, South Andaman Islands, India

Sandip K. Roy and Santanu Banerjee

Abstract This study presents an architectural element analysis of submarine fan deposits based on facies analysis in excellent outcrops of the Andaman Flysch in South Point section at the eastern margin of South Andaman Islands. The 292 m thick section of the Andaman Flysch consists of four facies viz. thick-bedded sandstone, thin-bedded sandstone, heterolithics and shale. Greater part of the section, particularly towards the top and bottom is dominated by amalgamated, laterally persistent thick-bedded sandstone. A detailed analysis of 78 m thick section at the central part of the section reveals vertical stacking of dominantly thin-bedded sandstones, heterolithics and shale encased within amalgamated thick-bedded sandstone. The vertical stacking of facies corresponds to dominantly lobe fringe and inter-lobe elements at the central part and lobe axis and lobe flank elements at the top and bottom of the section. The occurrence of coarse-grained clastic equivalents of Andaman Flysch with erosional bases of sandstones in north and middle Andaman Islands and very fine-grained clastics in Great Nicobar Island, along with present investigation in South Andaman Islands reveals an arc-parallel transformation of architectural elements from channel, channel-lobe transition zone and lobes from north to south.

Keywords Andaman Flysch · Architectural elements · Facies analysis · Channel-lobe transition · Turbidites · Palaeogene

1 Introduction

Deep-water clastic depositional system has witnessed a chain of studies on various aspects such as variable turbidite thickness (Talling 2001; Prélat and Hodgeson 2013), defining submarine fan and channel-levee complex (Hickson and Lowe 2002), positioning of channel-lobe transition complexes (Wynn et al. 2002), subaqueous sediment density flows (Talling et al. 2012), distinction between sheet and channel strata (Shanmugam and Moiola 1988), architectural elements (Baas et al. 2005) and

S. K. Roy · S. Banerjee (✉)
Department of Earth Sciences, Indian Institute of Technology Bombay, Powai, Mumbai 400076, India
e-mail: santanu@iitb.ac.in

J. S. Ray and M. Radhakrishna (eds.), *The Andaman Islands and Adjoining Offshore: Geology, Tectonics and Palaeoclimate*, Society of Earth Scientists Series,
https://doi.org/10.1007/978-3-030-39843-9_3

facies architecture of basin plain (Sumner et al. 2012). The thick-bedded turbidite has received significant attention as an architectural element (Normark 1970, 1978; Mutti and Lucchi 1975; Pickering et al. 1986, 1989, 1995; Mulder 2011). However, the thin-bedded clastic facies has received lesser attention, though being expressed by some detailed studies (Bouma 1992, 2000; Basu and Bouma 2000; Prélat and Hodgson 2013; Hansen et al. 2017; Stow and Omoniyi 2018; Omoniyi et al. 2018). The Andaman Flysch has been investigated for thick-bedded gravity flow deposits (Chakraborty and Pal 2001; Mukhopadhay et al. 2003; Pal et al. 2003; Bandopadhyay 2012; Roy and Banerjee 2016, 2017; Roy 2017). The lesser dominating facies in Andaman Flysch, the thin-bedded sandstone has not received much attention in previous works.

The South point section in south Andaman Islands has been interpreted as a sand-rich depositional lobe with an intervening section of thin-bedded turbidite facies (Roy 2017; Roy and Banerjee 2017). This study aims to characterize the stacking pattern of the Andaman Flysch along a 292 m-thick section at South Point, South Andaman Islands. Within this 292 m section, a 78 m-thick section has been chosen for detailed facies analysis highlighting the thin-bedded turbidite facies of the Andaman Flysch, sandwiched between thicker sandstone. Finally, we have interpreted the facies stacking patterns in terms of architectural elements of submarine fan deposits.

2 Geologic Setting and Tectonic Background

The Andaman Island Arc system was formed by the northward and anticlockwise movement of the Indian plate and its subduction beneath the Eurasian plate, since Late Cretaceous. The Andaman and Nicobar Islands represent a trench-slope break or an uplifted accretionary prism flanked by the fore-arc to the east. The major tectonic elements are oriented N-S (arc-parallel) with major faults being dextral (Curray et al. 1979; Roy 1983; Misra and Roy 1984; Roy and Das Sharma 1993; Curray 2005). The stratigraphy of the basin reveals a siliciclastic-dominated Palaeogene succession passing upward to a carbonate-dominated Neogene sequence (Roy 1983; Roy and Das Sharma 1993; Pal et al. 2003). Earlier sedimentological studies, particularly on outcrops of south Andaman Islands invoked a fan model (Chakraborty and Pal 2001; Pal et al. 2003; Mukhopadhyay et al. 2003; Bandopadhyay 2012) and an arc-parallel, channel-lobe transition zone (Roy and Banerjee 2017; Roy 2017).

The Andaman Flysch of Late Eocene-Oligocene age overlies the Late Cretaceous-Early Eocene Formation, known as the Mithakhari Formation (Roy 1983). The Andaman Flysch is increasingly drawing attention for hydrocarbon exploration in view of its distinguished presence as a major reservoir all along the island arc chain and in the adjoining fore-arc.

3 Methods and Terminology

This study involves detailed logging and field observations of the South Point section with emphasis on bed boundaries, sedimentary structures, grain size variations, facies relationship and cyclicity. The definition of thick- and thin-bedded turbidite facies has been debated in the past. Basu and Bouma (2000) defined Thin-Bedded Turbidite (TBT) varying in thickness from 5 to 60 cm. Nichols (2009) defined various classes as: very thin beds (<1 cm), thin beds (1–10 cm), medium beds (10–30 cm), thick beds (30–100 cm) and very thick beds (>100 cm). Hansen et al.'s (2017) definition of thin beds (<10 cm) and medium-thick beds (10–100 cm) differed from Omoniyi et al. (2018), who distinguished very thin-bedded turbidite/VTBT (1–3 cm) and thin-bedded turbidite/TBT (3–10 cm).

Although the detailed logging for this study was carried out in the South Point section, the exposures of Andaman Flysch were investigated in Karmatang, Parlob Jig No-15 (23kms before Rangat) and Kaushalya nagar Ward No-1(12° 31′ and 11″ N, 92° 49′ 14″ E) up to Bakultala in Middle Andaman Islands and Durgapur, Kalipur, Diglipur, Sitanagar, Smith Island, Ross Island, Craggy and Cory Island in North Andaman Islands. Based on field observations of Andaman Flysch at various outcrops, following terms are used in this study for different thicknesses of sandstones: very thin beds (1–5 cm), thin beds (5–30 cm), medium beds (30–60 cm), thick beds (60–100 cm) and very thick beds (>100 cm).

4 Facies Analysis of Andaman Flysch at South Point

Logging of Andaman Flysch in north, middle and south Andaman Islands and Great Nicobar Islands, in multiple outcrops has resulted in identification of five facies and several sub-facies on the basis of bed definition, bed contacts, bed relationship, sedimentary structure, sorting and grain size (Table 1). The five constituent facies viz. conglomerate/pebbly sandstones, thick-bedded sandstone, thin-bedded sandstone, heterolithics and shale have been coded with A, B, C, D, E respectively in Table 1. Out of the five facies, the last four are most abundant within the Andaman Flysch across Andaman Islands. Each facies consists of a number of sub-facies which are suffixed with a number in Table 1. The constituent sub-facies of north and middle Andaman Islands are generally coarse- to medium-grained, and those from south Andaman Islands and Great Nicobar Islands are fine- to very fine-grained. Therefore, sub-facies B1, B2, B3, B4 and B5 occur exclusively in north and middle Andaman Islands, while these are absent in south Andaman Islands. These sub-facies are replaced by B6, B7, B8, B9 and B10 in south Andaman Islands (Table 1).

Detailed bed-by bed logging of the Andaman Flysch reveals the broad facies distribution in the South Point Section (Fig. 1). Table 2 summarizes description of each sub-facies and their process interpretation in south Andaman Islands. Following

Table 1 Spatial distribution of different sub-facies of Andaman Flysch across the Andaman-Nicobar Islands (colours representing relative abundance)

Facies Name	Facies	Sub-facies	Coarse-grained sandstone		Medium/fine-grained sandstone	
			North Andaman Islands	Middle Andaman Islands	South Andaman Islands	Great Nicobar Island
A	Conglomerate/ pebbly sandstone					
B	Thick-bedded sandstone	Thick, massive sandstone	**B1**	**B1**	**B6**	**B6**
		Thick, massive to faintly laminated sandstone	**B2**	**B2**	**B7**	
		Thick, massive sandstone with floating clasts	**B3**	**B3**	**B8**	
		Thick sandstone with planar laminae, with clasts	**B4**	**B4**	**B9**	
		Thick, convolute laminated sandstone	**B5**	**B5**	**B10**	
C	Thin-bedded sandstone	Thin sandstone sheets	**C1**	**C1**	**C2**	**C2**
D	Heterolithics	Sand-shale alternations, planar and ripple laminate sandstone	**D1**	**D1**	**D1**	**D1**
		Sand-shale alternations, convolute laminated sandstone			**D2**	
E	Shale	Planar laminated shale	**E1**	**E1**	**E1**	**E1**
		Shale with thin siltstone			**E2**	

	Most abundant
	selective occurrence
	limited
	not present

Fig. 1 Partial view of the South Point section showing vertical dipping beds. Note the alternations between laterally persistent thick and thin sandstone beds (men standing on rocks at upper left corner of the photo for scale)

discussion provides a brief description of constituent facies of the Andaman Flysch across the Andaman Islands (Fig. 2).

Facies A: Conglomerate/pebbly sandstone

These polygmictic Conglomerates and gritty sandstones occur exclusively in exposures of North and Middle Andaman Islands. The thickness of the facies varies from 0.2 m to a few m and it occurs at the contact between Andaman Flysch and Mithakhari Formation. It contains more than 15% matrix. The clasts consist of quartz, chert, feldspar, rock fragments calcite and mica.

Facies B: Thick-bedded sandstone

The facies consists of light grey to khaki, well-sorted, fine- to medium-grained sandstone, with bed thickness exceeding a metre. The sandstone beds have sharp basal contacts. The upper contact of the facies is usually gradational with heterolithics. The bases of the sandstone beds exhibit profuse flute casts, groove casts and prod marks (Fig. 3). The vertical sections of the beds exhibit soft sediment deformation structures like water escapes, dish and pillar structures and convolute lamination. The width of the flute casts may be as big as ~25 mm and relief varying between a few mm and a few cm. The beds occasionally exhibit bioturbation. Sub-facies include: Massive sandstone (B6), faintly laminated sandstone (B7), planar laminated sandstone with mud clasts and concretions (B8), massive sandstones with random clasts (B9) and massive sandstones with slumps (B10). The coarse-grained equivalents of these facies (B1–B5) are absent in South Andaman Islands (Table 1).

Table 2 Brief description and interpretation of the constituent facies and sub-facies of the Andaman Flysch

Facies	Sub-facies	Sub-facies description	Sedimentary structures	Inferred depositional process
Thick- bedded sandstone (B)	Massive sandstone (B6)	Khaki-coloured, massive, fine- to medium-grained, poor to moderately sorted sandstone. Scoured base, and gradational top; thickness = 1–2 m	Flute, groove and prod casts at bed base, no apparent grading, abundant dish and pillars	High-density turbidity current with rapid fallout from sediment suspension. Dish and pillar structures indicates water escape
	Faintly laminated sandstone (B7)	Khaki coloured fine- to medium-grained, poorly-sorted, erosional basal contact, upper contact gradational, sandstone thickness = 1–2 m	Crude or faint planar laminae with dish and pillars. Flute, groove and prod casts at bed soles	Deposition from a waning sediment gravity flow/high-density turbidity currents Planar laminae and climbing ripple laminae traction deposits
	Massive sandstone with random mud clasts (B8)	Fine-grained massive sandstone with conspicuous floating mud clasts. The base is often sharp and top is gradational, sandstone thickness = 1–2 m	Structures absent	High density turbidity current deposits with sudden freezing of flow. Floating clasts indicate matrix strength in debris flows
	Planar laminated sandstone with mud clasts (B9)	Graded, fine-grained sandstones containing clasts. Sheet-like sandstone beds with sharp basal and gradational upper contacts, thickness = 1–2 m	Planar laminae, Flute casts abundant at the bed soles	High-density turbidity current deposits. Amalgamated beds indicate recurrent high-density turbidity currents flow

(continued)

Table 2 (continued)

Facies	Sub-facies	Sub-facies description	Sedimentary structures	Inferred depositional process
	Massive sandstone with slumps (B10)	Light grey, fine-grained, massive sandstone, thickness = 1–2 m	Slump fold	Slumped product in a sandy debris flow
Thin- bedded sandstone (C)		Sheet-like, khaki- coloured sandstone beds, fine grained and moderately sorted. Scoured base and gradational top. Thickness of sandstone = 10–30 cm	Faint to distinct planar laminae near the base, overlain by ripple laminae. Sole marks at base	Traction related planar and ripple laminae relates to turbidity currents. Products of a low-density turbidite
Heterolithics (D)	Heterolithics with planar and ripple laminae (D1)	Cm-scale alternations of sand—shale, overall fining-upward, Sheet-like geometry of sandstone beds. Average thickness of shale and sandstone varies from 5 to 10 cm and from 8 to 20 cm respectively	Planar and climbing ripple laminae	Plane parallel lamination with occasional ripples indicates traction currents. Products of low-density turbidity currents

(continued)

Table 2 (continued)

Facies	Sub-facies	Sub-facies description	Sedimentary structures	Inferred depositional process
	Heterolithics with convolute laminae (D2)	Alternations of sandstone and shale with convolute laminae. Fine-grained and poorly sorted. Sharp basal and gradational upper contacts. Thickness of sandstone up to 10 cm	Convolute laminae, dish and pillars	Products of muddy debris flows
Shale	Plane laminated shale (E1)	Dark grey to black shale with occasional siltstone interbeds, thickness = 0.5–2 m	Planar laminae	Heavy suspension sedimentation in a low-density turbidity flow regime. Common deposits in deep- water environment
	Convolute laminated shale with siltstone (E2)	Highly contorted, convolute laminated m-thick shale with siltstone interbeds, thickness = 0.5–2 m	Convolute laminae	Slumping in a muddy system

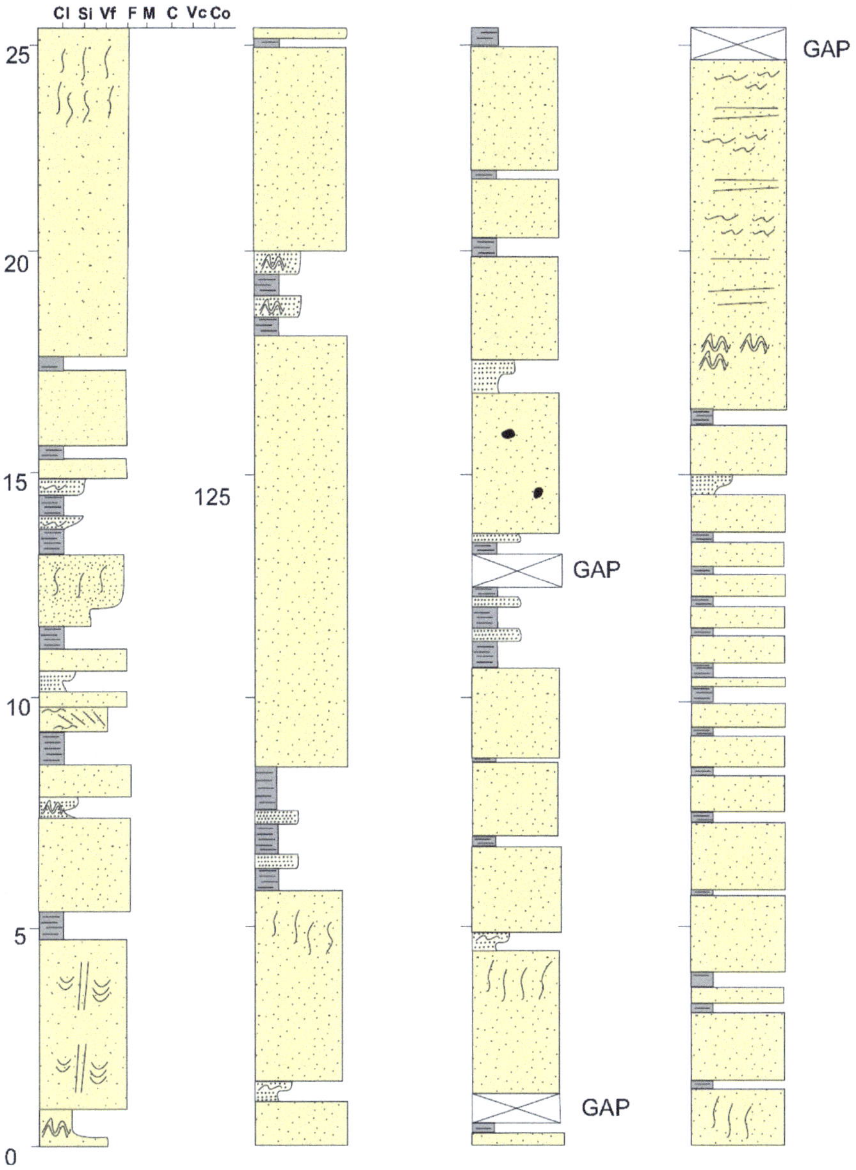

Fig. 2 Graphic log showing bed-by bed logging of Andaman Flysch at South Point in south Andaman Islands. Architecture of the Flysch shows vertical stacking of four repetitive facies viz. thick-bedded sandstone, thin-bedded sandstone, heterolithics and shale. Height above the base of the measured section has been indicated to the left of the log. The stratigraphic position of the detailed log presented in Fig. 7 has been indicated (between 122 and 200 m)

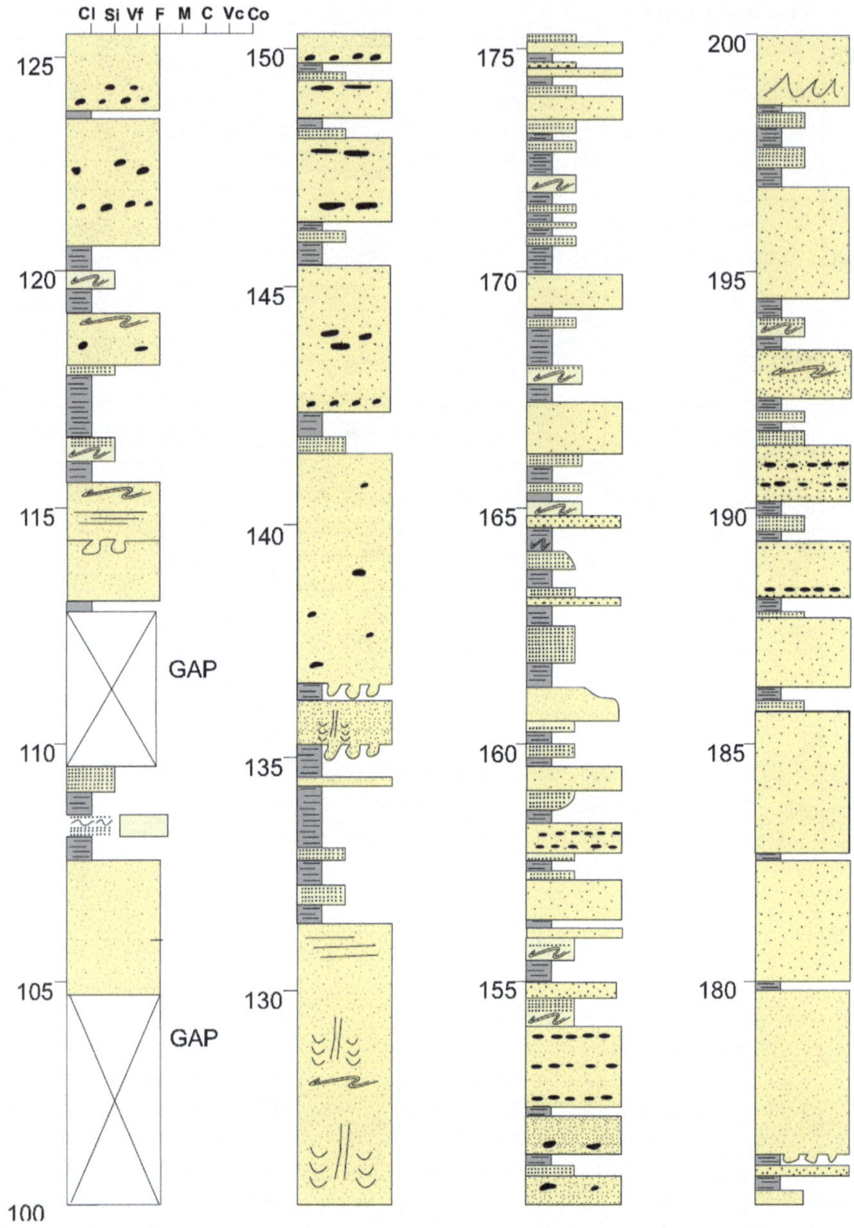

Fig. 2 (continued)

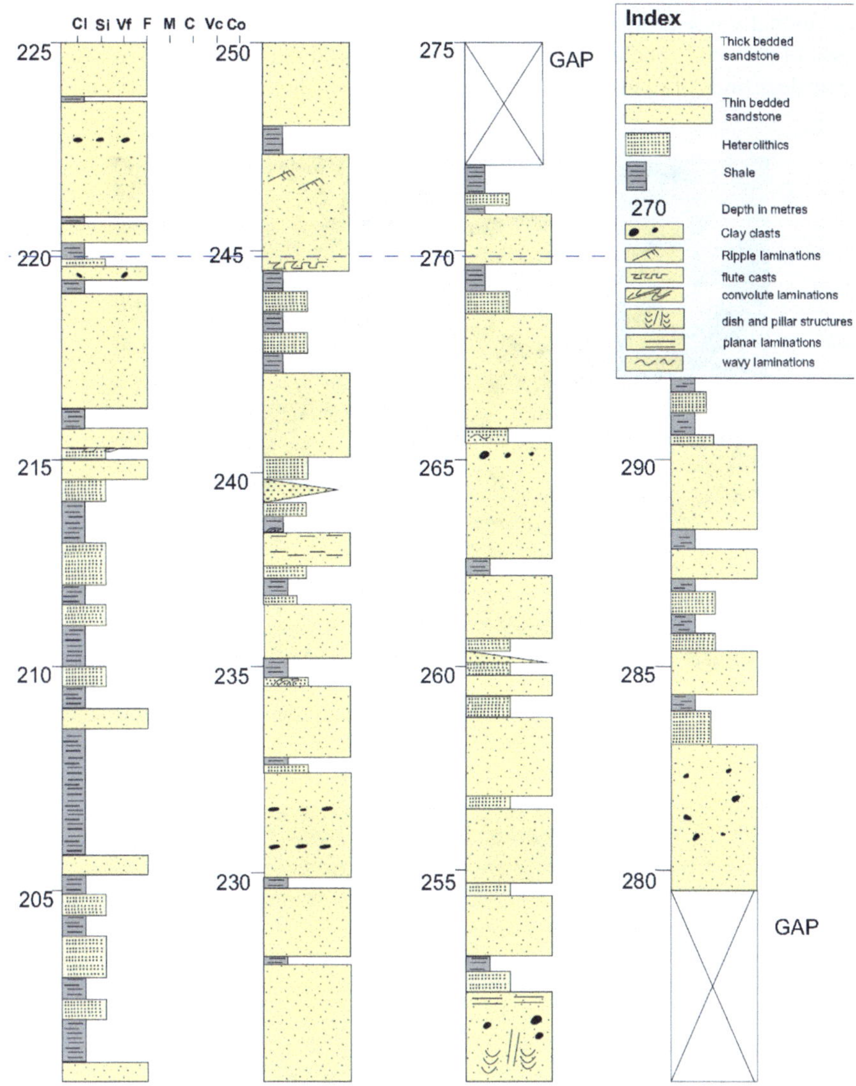

Fig. 2 (continued)

Facies C: Thin-bedded sandstone

The facies is characterized by tabular, sheet-like, khaki-coloured sandstone beds consisting of fine-grained and moderately sorted sandstones. The average thickness of the sandstone beds varies from 10 to 30 cm. The sandstone beds exhibit faint to distinct planar laminae near the base, overlain by ripple laminae. These beds usually have sharp basal contacts with underlying shale. The thin-bedded sandstone gradationally passes upward to heterolithic facies with incorporation of shale. The bases of the sandstone beds exhibit flute marks, prod marks and groove casts. The vertical sections of sandstone beds exhibit abundant water escape structures (Fig. 4).

Fig. 3 Large-scale flute casts at the base of a steeply dipping, thick-bedded sandstone (hammer length = 38 cm)

Fig. 4 Vertical section of a thin-bedded sandstone showing water escape structures (lens cap for scale)

Facies D: Heterolithics facies

This facies characteristically exhibits planar- and climbing ripple-laminated sandstone of mm- to cm-scale, intercalated with shale of similar thickness (Fig. 5). The facies occurs above the thick-bedded sandstone with a planar contact and is capped by a shale facies or the thick-bedded sandstone facies. It is the most frequently occurring facies of the Andaman Flysch. Occasionally the vertical section of the facies show convolute laminae (Fig. 6).

Fig. 5 Vertical section of a sandstone bed showing climbing ripple laminae

Fig. 6 Vertical section showing convolute laminae in heterolithic facies (pen length = 14 cm)

Facies E: Shale
This facies consists of dark grey shale with interlayered horizontal silt laminae. Thickness of the facies varies from 5 cm to 1 m.

5 Architectural Elements

The 292 m-thick South Point outcrop comprises sand-rich stratal packages alternating with clay-rich intervals including thin-bedded turbidites. The consideration of bed thickness of sandstone and proportion of sandstone: shale differentiates four architectural elements of a few m to several metre thick, viz. lobe axis, lobe flank, lobe fringe and inter-lobe (cf. Łapcik 2017) (Table 3). A 78 m thick section of the Andaman Flysch in the central part of the South Point section (between 122 and 200 m in Fig. 2), South Andaman Island, reflects the alternations of these facies associations in the vertical sequence (Figs. 7 and 8).

Lobe axis
It is predominantly sandy, consisting of a few m-thick, vertically stacked, thick-bedded sandstone at the bottom and top parts of the 74 m-thick section (Table 2). The thickness of the amalgamated beds varies from 5 to 12 m. The sandstone is fine- to medium-grained, moderately sorted beds with sheet-like geometry. Thin (<0.05 m) shale partings occasionally occur between thick-bedded sandstones. This architectural element makes up the major part of the South Point outcrop. This predominantly sandy unit with amalgamation of thick-bedded sandstone represents the axial part of the lobe (cf. Łapcik 2017; Mutti and Normark 1987; Prélat and Hodgson 2013; Mulder 2011). The erosive bases of some of the facies association possibly indicate channelized lobes in mid-fan environment (cf. Mutti and Nomark 1987). High sand: mud ratio and amalgamated, thick bedded sandstone represents deposition from high density turbidity flows (cf. Lowe 1982; Mutti 1992).

Lobe flank
This element consists of thick-bedded sandstone, containing sub-equal amounts of the heterolithics and the shale. The thickness of this element varies from 8 to 14 m. The individual bed thickness of the thick sandstone facies varies from 0.65 to 2.3 m. The proportion of thick-bedded sandstone reduces to 50% of the total constituents (Table 3). The thinness and fineness of sandstone beds reflects flanks of lobes (cf. Shanmugam and Moiola 1991, Deptuck et al. 2008). The component with predominant heterolithics and shale facies reflects the distal end of lobe flank.

Lobe fringe
The heterolithics facies is the most prominent constituent in this element, besides thick-bedded sandstone, shale and thin-bedded sandstone. The thickness of this element varies from 5 to 6 m (Table 3). Thin-bedded sandstone frequently alternates with shale within this element. The increase in proportion of heterolithic facies,

Table 3 Relative abundance of each facies within different architectural elements of the Andaman Flysch of the 78 m-thick section

Facies	Relative % of constituent facies	Individual bed thickness (m)	Remarks
Lobe axis element (stratal package thickness = 10.3–10.7 m)			
Thick-bedded sandstone	93	2.7–7.9 (amalgamated)	Average thickness of amalgamated beds ~3 m. Average thickness of individual beds ~1 m
Heterolithics	3	0.05–0.8	
Shale	4	0.05–0.48	
Thin-bedded sandstone	0	Nil	
Lobe flank element (stratal package thickness = 8.1–14.0 m)			
Thick-bedded sandstone	56	0.65–2.3	
Heterolithics	18	0.15–1.6	
Shale	25	0.05–0.3	
Thin-bedded sandstone	1	0.3	
Lobe fringe element (stratal package thickness = 5.05–6.07 m)			
Thick-bedded sandstone	24	0.55–0.80	
Heterolithics	55	0.15–0.65	Most conspicuous
Shale	13	0.05–0.30	
Thin-bedded sandstone	8	0.22–035	Conspicuous
Inter-lobe element (stratal package thickness = 4.6–11.8 m)			
Thick-bedded sandstone	12	0.7–1.1	
Heterolithics	48	0.05–0.55	Most conspicuous
Shale	34	0.05–025	
Thin-bedded sandstone	6	0.22–0.23	Conspicuous

reduced content of thick-bedded sandstone facies suggests deposition within the lobe fringe zone (cf. Deptuck et al. 2008; Prélat and Hodgson 2013). The typical presence of Bouma type turbidite sandstones (TBT) with fining-upward sequences indicate deposition from waning turbulent flow from low-density turbidity currents.

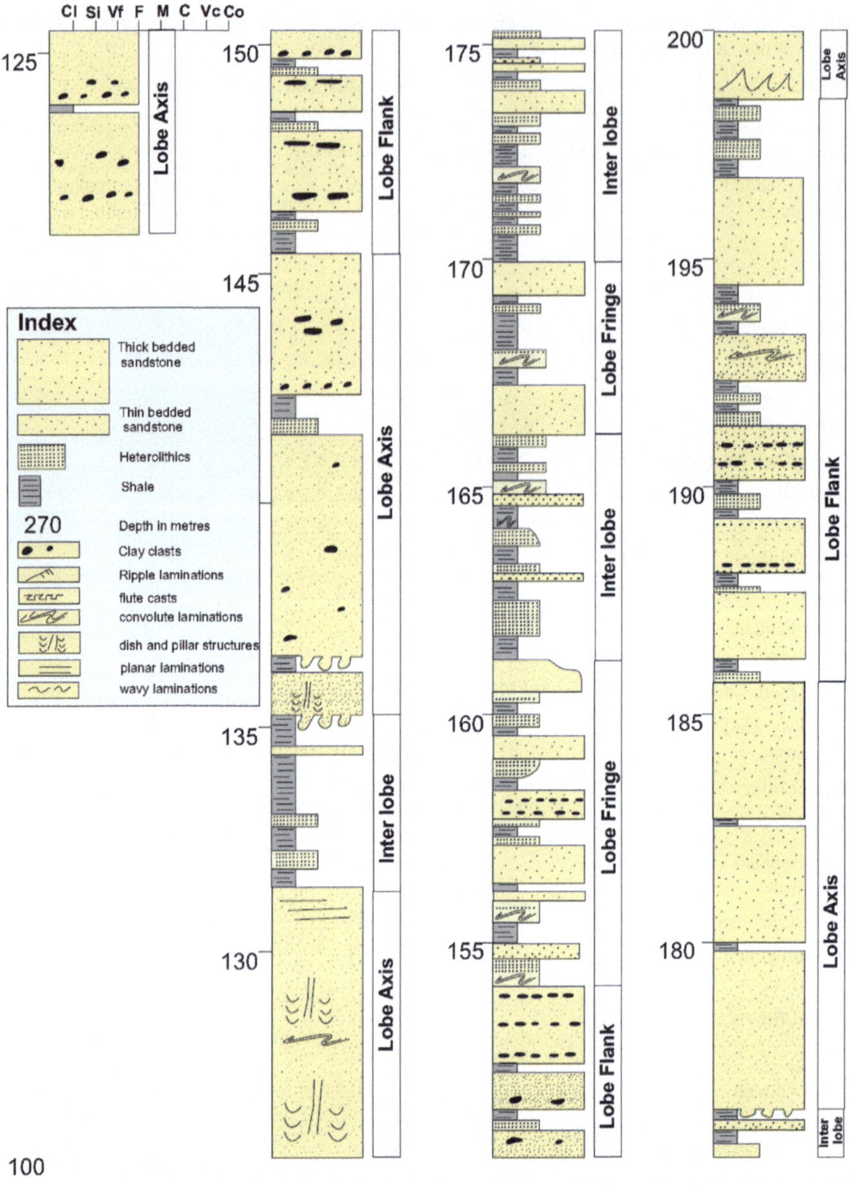

Fig. 7 Log showing vertical dispostion of architectural elements in the central part (78 m) of exposures at South Point section

Fig. 8 Outcrop photo showing steeply dipping exposures of the Andaman Flysch depicting the four principal facies viz thick-bedded sandstone, heterolithics, shale and thin-bedded sandstone. The lobe axis and lobe flank occurs towards the right while the lobe fringe and inter lobe towards the left of the photo. The thick-bedded sandstone is ~3.5 m thick

Inter-lobe

More than 90% of this element comprises heterolthics and shale (Table 3). Both lobe fringe as well this element characteristically exhibits a few cm-thick, sheet-like, planar laminated, fine-grained, moderate to well sorted, thin-bedded turbidite (TBT). The presence of abundant heterolithics and shale and the absence of thick-bedded sandstone attributes this element to an interlobe (cf. Pickering et al. 1995; Grundvåg et al. 2014). The dominance of hemipelagic and pelagic mud and heterolithics suggest deposition from low-density turbidity currents and suspension fallouts.

6 Discussion: Dynamic Stratigraphy and Stacking Patterns

This study highlights the vertical stacking of lobes and other architectural elements within the overall depositional milieu of the Andaman Flysch (Figs. 7 and 8). The thick-bedded sandstone facies is the most conspicuous type in the South Point section. The m-scale, typical fining-upward cycle consists of a thick-bedded sandstone, often with a scoured base (flute and groove, and prod casts) that gradually passes upward to heterolithics, and is capped by shale. In places, the thick-bedded sandstone is amalgamated, with occasional thin veneer of shale. Thickness of many of these thick-bedded sandstone facies exceed 5 m, and may be up to 12 m because of amalgamation. The sheet-like geometry of beds and absence of pinch out relates the thick-bedded sandstone to lobe axis and lobe flank elements (Fig. 8). The thin-bedded sandstone facies, usually consists of 20–30 cm thick, planar laminated, and tabular beds. The increase in mud content relates thin-bedded sandstone to lobe fringe and inter-lobe elements (Figs. 7, 8 and 9). The central part of the South Point section exhibits thin-bedded sandstone, heterolithics and shale and overlain and underlain by amalgamated thick-bedded sandstone facies (Fig. 7). This part displays classical turbidites (thin-bedded sandstone), and is capped by heterolithics and shale in each sedimentation cycle. From the standpoint of hydrocarbon exploration, these lobe fringe and inter lobe elements are similar to proximal levees.

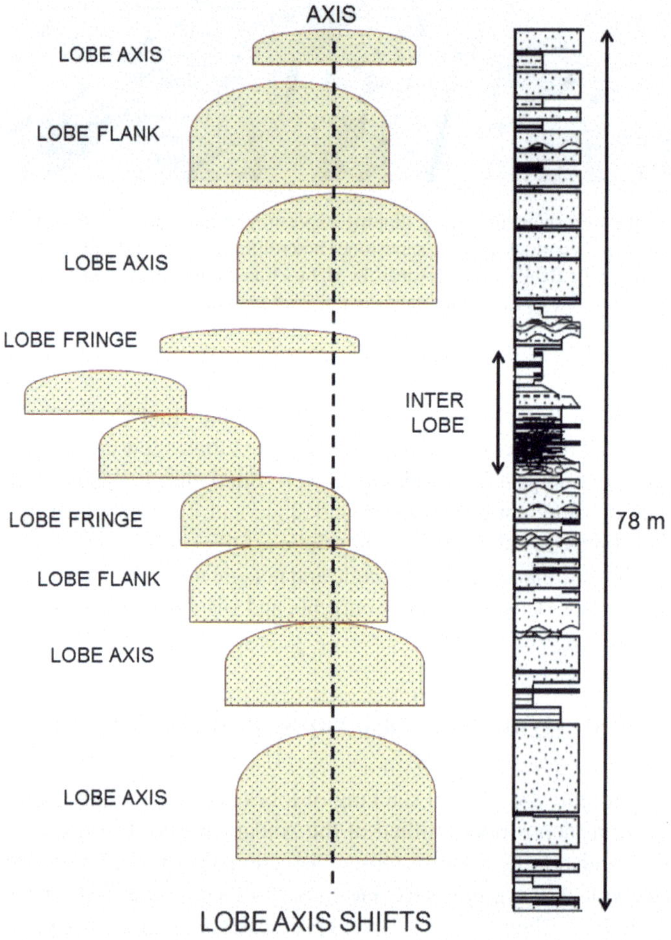

Fig. 9 Conceptual diagram showing the shifting of lobes and other architectural elements at the central part of the Andaman Flysch exposures at South Point. The log at the right indicate lithological variations across the section (stippled for sandstone, black for shale). Note the dominance of muddy sediments at the central part. The vertical dashed line indicates the initial position of lobe axis. While the lobe axis and lobe flank elements occur at the top and bottom part of the section, lobe fringe and inter lobe characterize the central part of the section

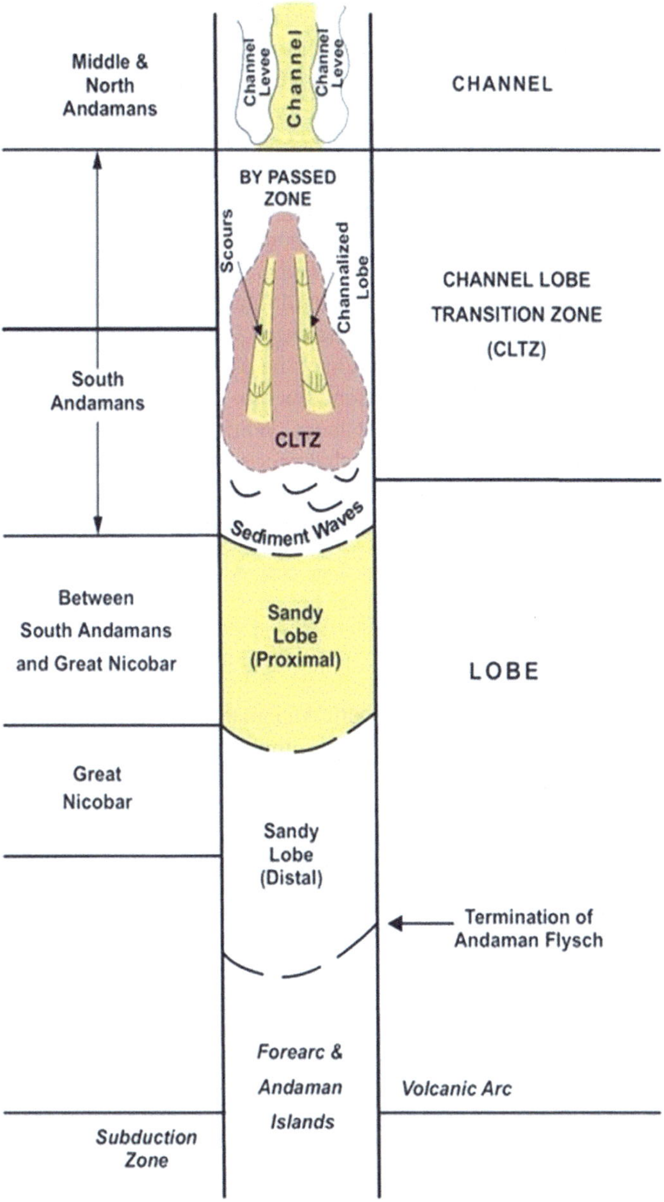

Fig. 10 Spatial distribution of channel, channel-lobe transition complex and lobes of Andaman Flysch. The Flysch at South Andaman Islands, reflect a channel-lobe transition zone and channelized and non-channelized lobes at South Point Adapted from Roy and Banerjee (2017)

In the South Point section, the amalgamated thick-bedded sandstone facies corresponds to lobe axis and lobe flank elements the central mud-rich portion corresponds lobe fringe and inter lobe facies associations (Figs. 7, 8 and 9; Table 3). The thick-bedded sandstone displays sheet-like geometry of beds with no lateral pinch outs, and with occasional erosive bases with flute casts and it corresponds to channelized and non-channelized lobes. In North and Middle Andaman Islands, the Flysch consist dominantly of coarse- to medium-grained sandstone, with prominently scoured bases of beds and lateral pinch-outs, corresponding to channel elements in basin slope (Roy et al 2011). While in Great Nicobar Islands, the flysch is even more finer-grained and with predominance of heterolithics and shale facies and paucity of sheet-like thick sandstone beds (Roy and Banerjee 2017). Considering the facies and facies associations and over all distribution of the Andaman Flysch on the Andaman Islands, along the edge of the subduction zone, we propose a dominantly channel regime in north and middle Andaman Islands. While a channel-lobe transition and channelized and non-channelized lobes characterizes the Andaman Flysch in south Andaman Islands. The Andaman Flysch in the Great Nicobar Islands represent a distal lobe (cf. Wynn et al. 2002, Roy and Banerjee 2017) (Fig. 10).

7 Conclusions

(a) The entire Andaman Flysch in South Andaman consists of four major facies, which include thick-bedded sandstones, thin-bedded sandstones, heterolithics and shale.
(b) Fining-upward cycles and occasional erosive bases in the thick-bedded sandstone facies indicates channelized and non-channelized lobes of a sand-rich deepwater clastic deposit on a basin floor, separated by levees.
(c) While the predominance of thick-bedded sandstone facies represents the lobe axis, the thin-bedded sandstone, heterolithics and shale corresponds to lobe fringe and inter lobe elements.
(d) An N-S, arc-parallel architecture of channel, channel lobe transition zone and lobes is envisaged for the Andaman Flysch on the subduction zone.

Acknowledgements Authors are thankful to Indian institute of technology Bombay for infrastructure support. Authors are thankful to anonymous reviewers for comments and suggestions for an earlier version of the manuscript.

References

Baas JH, McCaffrey WD, Knipe RJ (2005) The deep-water architecture knowledge base: towards an objective comparison of deep-marine sedimentary systems. Pet Geosci 11:309–320

Bandopadhyay PC (2012) Reinterpretation of age and environment of deposition of Paleogene turbidites in the Andaman and Nicobar Islands, Western Sunda Arc. J Asian Earth Sci 45:126–137

Basu D, Bouma AH (2000) Thin-bedded turbidites of the Tanqua Karoo: physical and depositional characteristics. In: Bouma AH, Stone CG (eds) Fine-grained turbidite systems, SEPM Special Publication, vol 68, pp 263–278

Bouma AH (1992) Locations and characteristics of thin bedded turbidites in passive margin setting submarine fans. Gulf Coast Assoc Geol Soc Trans 42:425–432

Bouma AH (2000) Coarse grained and fine grained turbidite systems as end member models: applicability and dangers. Mar Pet Geol 17:137–143

Chakraborty PP, Pal T (2001) Anatomy of a fore arc submarine fan: upper Eocene-Oligocene Andaman Flysch Group, Andaman Islands, India. Gondwana Res 4:477–486

Curray JR (2005) Tectonics and history of the Andaman Sea region. J Asian Earth Sci 25:187–232

Curray JR, Moore DG, Lawver LA, Emmel FJ, Raitt RW, Henry M, Kieckhefer R (1979) Tectonics of the Andaman Sea and Burma. In: Geological and geophysical investigations of continental margins. In: American association of petroleum geologists memoir vol 29, pp 189–198

Deptuck ME, Piper DJW, Savoye B, Gervais A (2008) Dimensions and architecture of Late Pleistocene submarine lobes off the northern margin of East Corsica. Sedimentology 55:869–898

Grundvåg SA, Johannessen EP, Hansen WH, Plink- Björklund P (2014) Depositional architecture and evolution of progradationally stacked lobe complexes in the Eocene Central Basin of Spitsbergen. Sedimentology 61:535–569

Hansen L, Callow R, Kane I, Kneller B (2017) Differentiating submarine channel-related thin-bedded turbidite facies: outcrop examples from the Rosario Formation, Mexico. Sediment Geol 358:19–34

Hickson TA, Lowe DR (2002) Facies architecture of a submarine fan channel–levee complex: the Juniper Ridge Conglomerate, Coalinga, California. Sedimentology 49:335–362

Łapcik P (2017) Facies heterogeneity of a deep sea depositional lobe complex: case study from the Slonne section of Skole Nappe, Polish Outer Carpethians. In: Annales Societatis Geologorum Poloniae, vol 87, pp 301–324

Lowe DR (1982) Sedimentary gravity flows: II. Depositional models with special reference to the deposits of high-density turbidity currents. J Sediment Res 52:279–297

Misra PC, Roy TK (1984) Exploration in Andaman forearc basin: its evaluation, facies trend and prospects- a review. In: Proceedings of 5th Offshore South East Asia Conference (OFFSEA 84), Singapore, South East Asia Petroleum Exploration Society, pp 466–483

Mukhopadhay B, Chakraborty PP, Paul S (2003) Facies clustering in turbidite successions: case study from Andaman Flysch Group, Andaman Islands, India. Gondwana Res 6:918–925

Mulder T (2011) Gravity processes and deposits on continental slope, rise and abyssal plains. In: Hüneke H, Mulder, T (eds) Deep-sea Sediments. Developments in Sedimentology, vol 63, pp 25–148

Mutti E (1992) Turbidite sandstones. Agip, San Donato Milanese

Mutti E, Normark WR (1987) Comparing examples of modern and ancient turbidite systems: problems and concepts. In: Leggett JK, Zuffa GG (eds) Marine clastic sedimentology. Graham and Trotman, London

Mutti E, Lucchi KF (1975) Turbidite facies and facies associations. In: Mutti E, Parea GC, Ricci Lucchi F, Sagri M, Zanzucchi G (eds) Examples of turbidite facies and facies associations from selected formations of Northern Appennines, pp 21–36

Nichols G (2009) Sedimentology and stratigraphy, 2nd edn. Wiley-Blackwell, US

Normark WR (1970) Growth patterns of deep sea fans. Am Assoc Pet Geol Bull 54:2170–2195

Normark WR (1978) Fan valleys, channels and depositional lobes on modern submarine fans: characters for recognition of sandy turbidite environments. Am Assoc Pet Geol Bull 62:912–931

Omoniyi BA, Stow DAV, Gardiner AR (2018) Characterization of thin-bedded turbidites in the North Brae Field, South Viking Graben, North Sea. In: Turner CC, Cronin BT (ed) Rift-related Coarse-grained Submarine Fan Reservoirs; the Brae Play, South Viking Graben, North Sea. AAPG Memoir, vol. 115, pp 257–281

Pal T, Chakraborty PP, Gupta TD, Singh CD (2003) Geodynamic evolution of the outer arc forearc belt in the Andaman islands, the central part of Burma-Java subduction complex. Geol Mag 140:289–307

Pickering K, Stow DAV, Watson M, Hiscott R (1986) Deep-water facies, processes and models: a review and classification scheme for modern and ancient sediments. Earth Sci Rev 23:75–174

Pickering KT, Hiscott RN, Hein FJ (1989) Deep marine environments: clastic sedimentation and tectonics. Unwin Hyman, London

Pickering KT, Hiscott RN, Kenyon NH, Ricci Lucchi F, Smith RDA (1995) Atlas of deep water environments: architectural style in turbidite systems. Chapman and Hall, London

Prélat A, Hodgson DM (2013) The full range of turbidite bed thickness patterns in submarine lobes: controls and implications Journal of the Geological Society. London 170:209–214

Roy TK (1983) Geology and hydrocarbon prospects of Andaman-Nicobar. Petrol Asia J 1:37–53 KDMIPE, ONGC, Dehradun

Roy SK (2017) Facies architecture of the Andaman Flysch: implications for deep water sedimentation and depositional model, Andaman basin. Unpublished Ph.D. thesis, Indian Institute of Technology Bombay

Roy SK, Banerjee S (2016) Thick bedded sandstone facies of Andaman Flysch: variability and process controls along the Andaman Island Arc. Abstracts. 33rd Convention, Indian Association of Sedimentologists, BHU, Varanasi, India Nov 2016

Roy SK, Banerjee S (2017) Commonality of Sedimentary facies of Andaman Flysch in South Andamans and Great Nicobar Island: implications for depositional model in southern part of Andaman basin Jour. Indian Assoc Sedimentol 34:75–92

Roy SK, Das Sharma S (1993) Evolution of Andaman Fore arc Basin and its hydrocarbon potential. In: Proceedings second seminar on Petroliferous Basins of India 1, 407–435

Roy SK, Paul S, Sadhu R, Banerjee S (2011) Provenance and tectonic setting of a coarse grained Flysch: case study from the Andaman Flysch at trench slope break, North Andaman Islands. GSA Minneapolis Annual Meeting, Oct 11th (Abstract, Poster session). Geological Society of America Abstracts with Programs. 43, 5, 432

Shanmugam G, Moiola RJ (1988) Submarine fans: characteristics, models, classification, and reservoir potential. Earth Sci Rev 24:383–428

Shanmugam G, Moiola RJ (1991) Types of submarine fan lobes: models and implications. AAPG Bulletin 75:156–179

Stow DAV, Omoniyi BA (2018) Thin-Bedded turbidites: overview and petroleum perspective. In: Turner CC, Cronin BT (ed) Rift-related coarse-grained submarine fan reservoirs; the Brae Play, South Viking Graben, North Sea. AAPG Memoir, vol 115, pp 97–117

Sumner EJ, Talling PJ, Amy LA, Wynn RB, Stevenson C, Frenz M (2012) Facies architecture of individual basin-plain turbidites: comparison with existing models and implications for flow processes. Sedimentology 59:1850–1857

Talling PJ (2001) On the frequency distribution of turbidite thickness. Sedimentology 48:1297–1329

Talling PJ, Masson DG, Sumner EJ, Malgesini G (2012) Subaqueous sediment density flows: depositional processes and deposit types. Sedimentology 59:1937–2003

Wynn RB, Kenyon NH, Masson DG, Stow DAV, Weave R, Philip PE (2002) Characterization and recognition of deep-water channel-lobe transition zones. AAPG Bulletin 86:1441–1462

Spatial Variation in the Composition of Andaman Flysch Across the Andaman Islands in Relation to Source of Sediments and Tectonics

Sandip K. Roy, Angana Chaudhuri and Santanu Banerjee

Abstract Detailed petrographical and geochemical investigations of the Andaman Flysch provide distinct yet contrasting provenance and tectonic setting in different parts of the arc-parallel, uplifted subduction complex. Petrographical and geochemical data from multiple sections in North, Middle and South Andaman Islands highlight the spatial variation in the composition of sediments from north to south. While lithic wacke dominates North and Middle Andaman Islands, the composition of sandstone varies from lithic wacke to quartz wacke in the South Andaman Islands. The QFR data corresponds to the dissected arc origin of samples. Flysch sediments of North Andaman Islands exhibit a moderate to high Al_2O_3/SiO_2, variable K_2O/Na_2O ratios and high Fe_2O_3+MgO contents, while those in South Andaman Islands show a low to moderate Al_2O_3/SiO_2, high K_2O/Na_2O ratios and high TiO_2. Flysch sediments of South and Middle Andaman Islands bear the signature of active continental margin to continental island arc system while those of the North Andaman Islands bear oceanic island signature. The REE data indicates the predominance of felsic source of the South Andaman Island samples. The combined petrographical and geochemical study suggests the derivation of sediments in the North Andaman Islands possibly from the Irrawaddy delta, while Andaman Flysch in Middle and South Andaman Islands possibly sourced from rocks occurring to the east and northeast direction.

Keywords Provenance · Tectonic setting · Andaman Flysch · Irrawaddy delta · Subduction zone · Fore-arc

1 Introduction

Geochemical perspectives, constraining the provenance in convergent plate margins in subduction zones and fore-arcs have always remained an enigma because of the mixing of arc-derived material with old upper crustal sources (McLennan et al. 1990; Guo et al. 2012; Armstrong-Altrin et al. 2015). Provenance study in convergent plate

S. K. Roy · A. Chaudhuri · S. Banerjee (✉)
Department of Earth Sciences, Indian Institute of Technology Bombay, Powai, Mumbai 400076, India
e-mail: santanu@iitb.ac.in

J. S. Ray and M. Radhakrishna (eds.), *The Andaman Islands and Adjoining Offshore: Geology, Tectonics and Palaeoclimate*, Society of Earth Scientists Series,
https://doi.org/10.1007/978-3-030-39843-9_4

margin setting is particularly challenging as the mineralogical and chemical composition of clastic sedimentary rocks are affected by a multitude of factors, including variable source rock composition, weathering, erosion, transportation, burial and diagenesis (Dickinson and Suczek 1979; Dickinson et al. 1983; Bhatia 1983, 1985; Roser and Korsch 1986; McLennan et al. 1993; Fedo et al. 1996; Paikaray et al. 2008; Saha et al. 2010, 2018; Armstrong-Altrin et al. 2015; Garzanti et al. 2016; Chaudhuri et al. 2018, 2020). Provenance analysis for the Andaman Flysch, elucidating sediment derivation to the basin has remained controversial. The mineralogical and chemical composition of Andaman Flysch and Mithakari Formation has been addressed in a few studies, primarily from South Andaman Islands (Allen et al. 2008; Garzanti et al. 2013; Bandopadhyay and Ghosh 2015), and additionally from Rutland Islands and Great Nicobar (Garzanti et al. 2016). Roy et al. (2011) provide the initial results of major and trace element geochemistry for the Andaman Flysch of the North Andaman Islands. However, the temporal variation in mineralogical and geochemical composition of sediments across North, Middle, and South Andaman Islands has not been presented in previous studies.

This study focuses on the understanding of textural, mineralogical and geochemical characteristics of samples of the Andaman Flysch along the arc-chain for the assessment of the provenance and tectonic setting of the source area, covering multiple outcrops in the Andaman Islands. The objectives of this study are to provide (a) the provenance and tectonic setting covering North, Middle, and South Andaman Islands, and (b) to infer sediment dispersal pattern along the arc-chain. We have presented field, petrographical and geochemical data to achieve the objectives.

2 Geological Setting and Tectonic Background

The Andaman Basin is an island arc system, initiated by northward motion, anticlockwise rotation and impingement of the Indian plate beneath the Eurasian plate since Late Cretaceous time (Roy 1983; Roy and Das Sharma 1993; Curray et al. 1979; Curray 2005). The major tectonic elements include outer arc, Andaman trench, accretionary prism, trench-slope break or structural high, fore-arc, volcanic arc and back-arc with arc-parallel, and dextral strike-slip faults (Roy and Das Sharma 1993; Pal et al. 2003; Roy and Banerjee 2016). The Andaman and Nicobar Islands represent an uplifted accretionary prism on the edge of a fore-arc. A standardized stratigraphy of the basin on the basis of outcrop studies in the Andaman and Nicobar group of Islands reveal domination of clastics in the Cretaceous to Palaeogene and carbonates in the Neogene sections (Roy 1983; Roy and Das Sharma 1993; Pal et al. 2003) (Table 1). The Andaman Flysch, an Oligocene deep marine siliciclastic sequence is increasingly gaining credence for hydrocarbon exploration because of its distinguished presence as a major reservoir facies all along the island arc chain on the Andaman Islands, and adjoining fore-arc and subduction zone (Roy and Banerjee 2017). Published sedimentological studies on the Andaman Flysch of the main Andaman Group of Islands used the Bouma sequence and classical submarine fan model in the South

Table 1 Stratigraphy of Andaman Islands

Age	Stratigraphic Unit	Lithology
Pleistocene to recent	Archipelago Group	Beach, tidal flat deposits, mudstones, coral reefs, raised beaches
Pliocene		Foraminiferal mudstones, calcareous sandstones, siltstones
Early-Mid Miocene		White nano-foram chalk, volcanic ash, calcareous sandstones, mudstones, siltstones
Early Miocene		Calcareous sandstones, conglomerates, marl, siltstones
Late Eocene Oligocene	Andaman Flysch Group	Massive to plane laminated, buff to light grey colourec fine to coarse grained sandstones with or without clay clast and concretions. Sand-shale alternations, shale and conglomerates
Late Cretaceous Palaeocene	Mithakari Group	Dark grey, compact shales. Coarse grained ophiolite derived greenish sandstones and polymictic conglomerates with extraneous ophiolite, mudstone, sandstone and limestone clasts
Late Cretaceous		Pink radiolarian cherts, jaspers, quartzites, white limestones and marbles
	Ophiolite Group	Oceanic basement- ophiolite suite

Andaman Islands (Chakraborty and Pal 2001; Pal et al. 2003; Mukhopadhyay et al. 2003). The Andaman Flysch exposures are observed all along the arc-chain in most of the Andaman Islands (cf. Roy and Banerjee 2017). This study uses outcrop samples of Andaman Flysch in North, Middle, and South Andaman Islands. The Andaman Flysch comprises interbedded thick- and thin-bedded sandstones, heterolithics and shales, which are coarse-grained in North and Middle Andaman Islands exposures, and fine-grained in the South Andaman and Great Nicobar Islands. The underlying Mithakari Formation comprises clastics of a more heterogenous nature consisting of primarily ophiolite-derived greywacke, pebbly sandstone, conglomerate with extraneous older basement fragments, andesite, glass shards, andesite fragments, broken and fresh plagioclase (Roy 1983; Bandopadhyay 2005; Curray 2005). However, the identification of formation boundaries remain contentious (cf. GSI 2012).

3 Samples and Methods

Outcrop samples were collected from different Islands of North (Smith Island, Ross Island, Craggy Island, Kalipur, Durgapur), Middle (Rampur–Karmatang) and South Andaman (Baratang Island, South Point, North Bay, Collinpur and Wandoor) Islands. Petrographic investigations were carried out on 65 rock samples using Leica DM 4500P polarizing microscope attached with Leica DFC420 camera in the Department of Earth Sciences, Indian Institute of Technology Bombay.

Major and trace element analysis was carried out for 52 samples. All samples were crushed and powdered to <75 micrometers in an agate mortar. 0.25 g of sample powder was added to 0.75 g lithium metaborate ($LiBO_2$) and 0.50 g of lithium tetraborate (LiB_4O_7) in a platinum crucible, and fused in a muffle furnace at 1050 °C for 10 min. The crucible was immersed in 75 ml of 1 N HCl in a 100-ml glass beaker after cooling, followed by magnetic stirring for 1 h until the fusion bead had dissolved completely. The sample volume was made up to 100 ml in a volumetric flask. 5 ml of this solution was diluted to 50 ml with distilled water and analyzed using inductively coupled plasma atomic emission spectrometry (ICP-AES) facility (ARCOS by Spectro®, Germany) at the Sophisticated and Analytical Instrument Facility (SAIF), IIT Bombay. USGS standards, MAG-1, SCo-1 and GSJ standards, JG-1a and JB-1b were used as standards for the analysis. Loss on ignition (LOI) values for samples were determined by calculating the difference in weight of samples before and after heating at 950 °C.

Rare earth elements (REEs) were extracted for 8 samples using column chromatography. For this, chromatographic glass columns (internal diameter—2.0 cm and length—25 cm) fitted with sintered glass disks and PTFE (polytetrafluoroethylene) stopcocks were charged with 30 g of 200–400 mesh AG5OW-X8® cation resin obtained from Bio-Rad Laboratories. This gave a settled height of ~10 cm in water, equivalent to a working volume of ~30 cm^3. After equilibrating the ion exchange resin with 100 ml of 1 N HCl, the sample solution was loaded onto the column. The sample container was washed with 1 N HCl and the washings (~30 ml) added to the column. First, the unwanted elements were eluted with 300 ml 1.75 N HC1 and discarded. Then the REE, Y, Sc and Hf were eluted with 200 ml 6 N HCl. The resin was cleaned with 50 ml 8 N HCl and washed immediately with 200 ml distilled water, and the column was then left in 1 N HCl. The final elute of 200 ml containing the REE, Y, Sc and Hf was carefully evaporated to a volume of ~25 ml and transferred to a 50 ml glass beaker. This solution was evaporated to dryness. The residue was re-dissolved in 10 ml of 2 N HNO_3 immediately prior to analysis using ICP-AES facility (Jobin Yvon Ultima-2), Department of Earth Sciences, Indian Institute of Technology Bombay. Pure single element synthetic standard solutions were used as standards for the analysis.

4 Description of Lithotypes

Field and petrographic studies have outlined five broad lithotypes. Detailed descriptions of each of the lithotype are as follows. The grain size of the flysch progressively decreases from north to south of the Andaman Islands.

4.1 Conglomerate

The polymictic conglomerate, varying in thickness from 0.2 to 5 m occurs primarily in the North Andaman Islands, at the contact between Andaman Flysch and Mithakari Formation. It contains more than 15% matrix. The detrital framework grains are sub-angular to sub-rounded, consisting of quartz (15%), feldspar (20%), rock fragments (50–60%), calcite and mica (total 5%). The rock fragments include chert and lithic fragments including volcanic rock (trachy andesite, pumice, glass, lapilli and mafic and ophiolite fragments), schist and sedimentary rock such as shale, limestone and sandstone (Fig. 1).

Fig. 1 Photomicrograph in plane polarized light showing the matrix in conglomerate of Smith Island, North Andaman with clasts of various composition viz., shale, mudstone, ophiolites, trachy andesite

4.2 *Pebbly Sandstone*

This lithotype consists of poorly sorted, coarse-grained, pebbly sandstone, and it occurs near the base of Andaman Flysch in Middle and North Andaman Islands. Grains are angular to sub-rounded. The content of matrix exceeds 15%. The framework grains include quartz, feldspar, rock fragments and basement-derived ophiolites. The content of feldspar exceeds quartz in most sandstones. The content of polycrystalline quartz is higher in the North Andaman Islands compared to those in the Middle Andaman Islands. The amount of rock fragment is high, varying from 40 to 50%. Rock fragments include andesite, ophiolite, mud clast and chert.

4.3 *Medium- to Coarse-Grained Sandstone*

This lithotype, consisting of moderately-sorted, medium- to coarse-grained lithic wacke, crops out mostly in the North Andaman islands and occasionally in the Middle Andaman Islands. Grains are angular to sub-rounded with the content of matrix exceeding 15%. The framework grains contain quartz, feldspar, rock fragments and opaque minerals (Fig. 2). Monocrystalline quartz with undulatory extinction dominates, although polycrystalline variety also occurs. The plagioclase variety dominates

Fig. 2 Photomicrograph under cross-polarized light showing medium-grained, poorly sorted lithic wacke with sub-angular grains from Rampur-Karmatang, middle Andaman Islands with monocrystalline quartz, polycrystalline quartz, plagioclase and rock fragments containing chert, schist and mafic minerals

over the k-feldspar. Rock fragments include volcanics, chert and schist. The sandstone is weakly cemented by calcite. This lithotype is rare in the South Andaman Island.

4.4 Coarse Siltstone to Medium-Grained Sandstone

This forms one of the most common lithotypes of the Andaman Flysch in South Andaman Islands. Samples from South Point in South Andaman Islands include brown to light brown, moderate to well sorted, very fine-grained sandstone and siltstone. Occasionally they exhibit cm-scale mudclast. Petrographic investigation reveals the lithic wacke composition of sandstone consisting of detrital quartz (30–40%), feldspar (30–35%), rock fragment (13–15%), opaque mineral (10–12%), pyroxene (ca 3%) with very little detrital mica (2%) as framework particles, and a brownish-yellow matrix (20–30%) (Fig. 3). Grains are sub-angular to sub-rounded with poor sorting. Quartz grains are mostly monocrystalline, while the polycrystalline variety makes up ~10% of the total. Lithic fragments are dominantly sedimentary (chert and ortho-quartzite) and metamorphic (gneiss). Chlorite and other clay minerals constitute the matrix. Lithic fragments and feldspar grains are frequently squashed to form pseudo-matrix. Quartz overgrowth and calcite cement are rare.

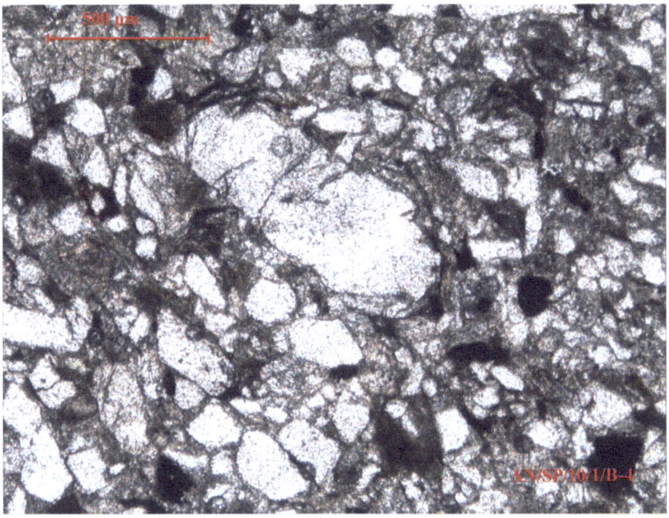

Fig. 3 Photomicrograph in plane polarized light showing fine-grained, lithic wacke with sub angular grains

4.5 Siltstone to Fine-Grained Sandstone

This lithotype occurs in the clay-rich part of each exposure of the Andaman Flysch across the Andaman Islands but predominantly in the South. Typically, they are tabular with planar top and bottom contacts. The matrix-rch sandstone belongs to lithic to feldspathic wacke. The content of matrix varies from 15 to 40%. The sandstone consists of quartz (both monocrystalline and polycrystalline), feldspar, rock fragments, mafic minerals. The content of polycrystalline quartz is higher in samples of the North Andaman Islands than those of South.

4.6 Heterolithics

This lithotype consists of mm- to cm-scale, planar and ripple laminated sandstone intercalated with shale of comparable thickness. The lithotype usually occurs above the medium- to coarse-grained sandstone with a planar contact. The sandstone gradationally passes into pelagic or hemipelagic shale. It is the most common lithotype across the arc-chain, and is increasingly abundant in the South Andaman Islands.

4.7 Shale

This lithotype comprises dark grey to black, massive to planar laminated shale with occasional silt interbeds. The thickness of the shale bed ranges from a few cm to a few m. The lithotype is marked by a planar basal contact with underlying sand-shale alternations and an erosive top, overlain by younger sandy channel sediments. The thickness of the shale varies from 15 cm to 2 m. A 6 to 10 m thick, black shale bed, occurring at the top part of the Andaman Flysch serves as a marker bed in the South Andaman Islands. Occasionally the shale is rich in organic carbon.

5 Ternary Q-F-L Plot

The detailed petrographic investigation, including point-counting, was carried out on fifteen sandstone samples of Andaman Flysch. Most of the samples from the North Andaman Islands occupy the dissected arc regime on the QFL plot (Fig. 4). While samples of the North Andaman Islands bear abundant altered feldspar and rock fragments, those in the South Andaman Islands are devoid of this component.

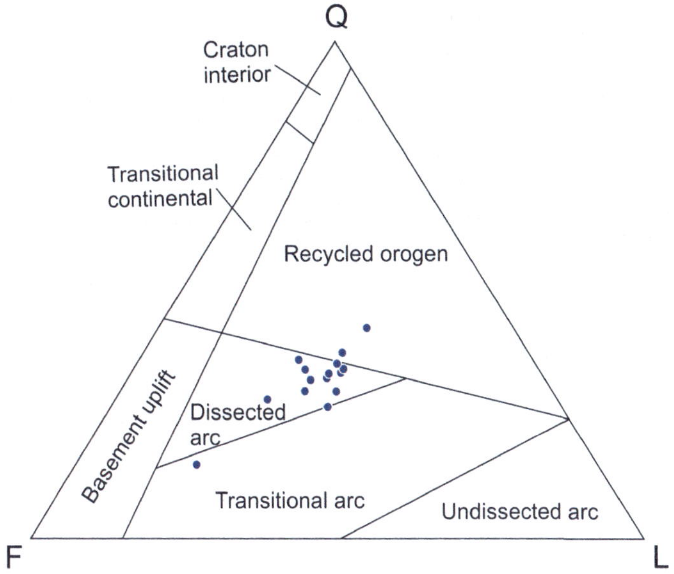

Fig. 4 QFL plot showing the compositions of Andaman Flysch showing the dominance of dissected arc signature (Q = total quartzose grains including monocrystalline (Qm) and polycrystalline (Qp) varieties, F = feldspar grains, L = lithic fragments)

6 Distinction Between Mithakari Formation and Andaman Flysch

We investigate the samples of Andaman Flysch and Mitakhari Formation in Kalipur and Ross Island, respectively, in the North Andaman islands. In Kalipur, three sets of conglomerates of 20–50 cm thickness alternate with pebbly sandstone and shale, mark the gradational contact between the Andaman Flysch and the Mithakari Formation. In Ross island, similarly, three conglomerate beds demarcate the contact between two formations. The conglomerates in Kalipur and Ross Islands in the North Andaman Islands bear clasts of polycrystalline quartz, shale, chert, andesite, limestone, metamorphic rock and pieces of wood (Fig. 5a). The Mithakari Formation (Ross Island, Kalipur in the North Andaman Islands and Namunagarh, Hope town in the South Andaman Islands) consists primarily of distinct green and coarse sandstones, possibly offscrapped from the trench (Fig. 5b). In sharp contrast, the Andaman Flysch in the South Andaman Island is coarse- to fine-grained, khaki in colour (Fig. 5c). It also exhibits increased quartz content, completely dominated by monocrystalline quartz with subequal amounts of feldspar and rock fragment including mud clast, and rock fragments, which is primarily chert with and occasional mica. Published geological maps show outcrops of the Mithakari Formation in the North

Fig. 5 Field photographs showing conglomerate at the boundary of Andaman Flysch and Mithakari Formation (**a**), Mithakari sandstone with characteristic greenish colour of sands (**b**) and thick-bedded sanstone of Andaman Flysch (**c**)

Andaman Islands and eastern half of Middle Andaman and Baratang Islands. However, Andaman Flysch crops out along the Andaman Trunk road, as well as on the east coast. The Baratang island also exposes Andaman Flysch along with Mithakari Formation in places.

7 Geochemical Characteristics of Flysch Samples

7.1 Major Elements

The major element oxides (wt%) in samples of North, Middle and South Andaman Islands are provided in Table 2. The SiO_2 content of samples varies considerably from north to south of the Andaman Islands. The content of SiO_2 ranges from 44.40% to 71.78% (av. 58.69%) in samples of the North Andaman Islands. The average SiO_2 content in flysch samples is, therefore, lower than that of PAAS (62.80%) (cf. Taylor and McLennan 1985). The content of SiO_2 ranges from 66.6% to 80.20% (av. 73.76%) and 52.00% to 86.30% (av. 68.73%) in samples of Middle and South Andaman Islands, respectively. Generally, cratonic and recycled sediments along passive margins have higher silica content compared to arc-derived sediments, associated with active orogens (Dickinson et al. 1983). Pebbly sandstone and conglomerate samples (Mithakari Formation) of the North Andaman Islands show lower SiO_2 content than those in flysch. The Al_2O_3 content of samples is lower than those of PAAS (18.90%), with average concentrations 13.02%, 11.03% and 12.63% for North, Middle and South Andaman samples, respectively. Samples of the Middle Andaman islands exhibit the least content of Fe_2O_3 (av. 3.66%), followed by those of the North Andaman Islands (av. 5.01%). The Fe_2O_3 contents in both cases are lower than that of PAAS (6.50%). Most samples are enriched in Na_2O compared to K_2O. However, The Na_2O content of samples is higher in conglomerate and pebbly sandstone (Mithakari Formation) in the North Andaman Islands compared to the flysch sediments in the South Andaman Islands. The CaO content of samples exhibits

Table 2 Major oxide concentrations in analyzed samples

Major oxides	SiO_2	TiO_2	Al_2O_3	Fe_2O_3	MnO	MgO	CaO	Na_2O	K_2O	P_2O_5	LOI	Total
North Andaman samples												
N1	49.4	0.8	11.5	4.9	0.16	4.1	12	3.5	0.8	0.18	11.6	98.94
N2	63.3	0.7	14.3	6	0.06	4.2	1.8	3.4	1.6	0.14	6.1	101.6
N3	49.4	0.6	10.5	3	0.12	1.9	16.3	3.3	0.9	0.13	14.4	100.55
N4	63.8	0.6	14.4	6.5	0.06	3.1	2.4	3.3	1.5	0.12	5.4	101.18
N5	46.3	0.6	11.4	4.7	0.31	2.8	16.3	3	0.9	0.21	14.9	101.42
N6	63.2	0.8	14.5	5.3	0.04	4.3	2.4	4.1	0.9	0.16	5.5	101.2
N7	55.9	1.1	15.4	7.9	0.08	5.9	2	4.6	0.8	0.28	7.2	101.16
N8	47.7	1.1	15.2	11.1	0.13	9.9	1.3	4.6	0.7	0.15	9.7	101.58
N9	46.3	0.6	11.3	2.4	0.7	1.4	17.5	3.2	1.3	0.14	15.7	100.54
N10	44.4	0.8	11.4	5.6	0.3	3.9	15.7	3.5	0.8	0.24	14.7	101.34
N11	61.7	1	14.84	6.9	0.07	3.7	1.9	4.3	0.9	0.21	5.2	100.72
N12	77.89	0.54	9.74	4.25	0.14	1.55	0.37	1.47	1.34	0.05	4.37	101.71
N13	60.94	0.82	10.4	3.11	0.26	1.6	9.38	3.47	0.63	0.13	10.92	101.67
N14	70.4	0.67	12.86	4.46	0.02	1.66	0.24	4.49	0.81	0.15	5.31	101.06
N15	69.98	0.6	13.08	3.28	0.02	2.43	1.65	3.9	1.54	0.14	4.43	101.05
N16	70.62	1.16	12.59	4.26	0.03	2.07	1.54	3.39	1.48	0.16	3.6	100.89
N17	58.09	0.67	11.74	4.26	0.28	2.24	7.51	4.02	1.4	0.16	9.03	99.4
N18	61.14	0.86	14.22	5.86	0.13	3.69	2.47	4.49	1.39	0.26	4.66	99.18
N19	59.88	0.86	17.36	6.86	0.02	2.18	0.25	2.14	1.85	0.08	9.81	101.29
N20	46.72	0.58	10.54	2.54	0.87	1.86	17.01	3.18	1.1	0.17	16.13	100.69

(continued)

Table 2 (continued)

Major oxides	SiO_2	TiO_2	Al_2O_3	Fe_2O_3	MnO	MgO	CaO	Na_2O	K_2O	P_2O_5	LOI	Total
N21	49.18	0.55	13.94	9.66	0.07	4.26	0.62	7.05	0.58	0.13	13.75	99.78
N22	60.33	0.4	9.66	2.87	0.17	1.99	10.06	2.81	1.02	0.13	11.58	101
N23	59.45	0.71	11.94	5.8	0.12	1.87	6.87	2.48	0.82	0.07	9.58	99.7
N24	65.46	0.69	12.98	3.86	0.06	1.97	3.91	4.94	1.65	0.17	5.93	101.62
N25	57.75	0.53	14.67	4.77	0.13	2.63	6.74	3.86	1.41	0.09	8.89	101.47
N26	70.32	0.74	13.83	2.83	0.04	1.47	1.84	4.14	1.32	0.15	4.79	101.46
N27	55.08	0.52	16.7	4.43	0.1	2.72	6.22	4.96	1.31	0.08	9.38	101.51
N28	66.77	0.62	14.03	3.71	0.03	2.48	1.62	4.12	1.31	0.13	5.34	100.16
N29	60.95	0.7	11.67	6.56	0.16	2.61	5.06	4.19	0.93	0.02	8.59	101.44
N30	71.78	0.82	13.99	2.5	0.02	1.15	1.46	4.78	1.7	0.16	3.37	101.73
Middle Andaman samples												
M1	69.95	0.73	14.04	2.58	0.04	2.23	2.33	4.13	1.26	0.19	3.88	101.36
M2	69.2	0.83	13.42	3.17	0.04	2.95	1.85	4.24	1.27	0.36	3.9	101.23
M3	66.66	0.8	12.69	5.52	0.08	3.47	3.17	3.3	1.32	0.42	3.47	100.89
M4	69.29	0.69	13.23	3.66	0.04	1.83	1.68	3.96	0.93	0.17	5.2	100.69
M5	80.2	0.3	6.7	5.13	0.02	0.6	0.14	1.8	0.4	0.08	2.9	98.27
M6	78.8	0.4	9.7	2.9	0.04	1.5	1	2.2	1.4	0.09	3.3	101.33
M7	78.1	0.4	9.2	3.2	0.06	1.6	1.07	2.04	1.3	0.13	4.4	101.5
M8	77.9	0.4	9.3	3.1	0.04	1.4	0.95	2.1	1.3	0.09	4.3	100.88
South Andaman samples												
S1	65.3	0.7	15.5	6.1	0.03	2.6	0.7	3.4	0.9	0.06	5.8	101.09

(continued)

Table 2 (continued)

Major oxides	SiO_2	TiO_2	Al_2O_3	Fe_2O_3	MnO	MgO	CaO	Na_2O	K_2O	P_2O_5	LOI	Total
S2	86.3	0.2	6.1	2.4	0.06	1.2	1	1.2	0.7	0.08	2.3	101.54
S3	63.5	0.7	10.6	3.9	0.32	1.4	7.1	1.9	1.7	0.12	9.5	100.74
S4	75.2	0.5	11	5.3	0.09	2	0.9	2.9	0.6	0.1	3.4	101.99
S5	68.9	0.8	13.7	6.3	0.05	1.3	0.3	1.6	2.1	0.14	4.7	99.89
S6	70.6	1.4	13.7	6.2	0.07	1.2	0.1	1.4	2.1	0.07	4.4	101.24
S7	70.1	1.2	12.2	6.3	0.06	1.9	0.4	2.4	2.3	0.36	4.2	101.42
S8	70.1	1.1	13.1	6.3	0.03	1.3	0.1	1.4	2.1	0.16	5.7	101.39
S9	70.6	1.2	11.5	6.7	0.04	1.7	0.4	2.2	2.2	0.32	4.9	101.76
S10	67.3	0.9	13.2	8.3	0.11	2.3	0.6	2.1	2.1	0.25	4.1	101.26
S11	64.9	0.6	14.7	6.2	0.09	3.3	1.8	4.7	0.8	0.14	3.9	101.13
S12	52	2.2	16.2	14.5	0.13	3.4	0.5	5.4	0.8	0.31	6.3	101.74

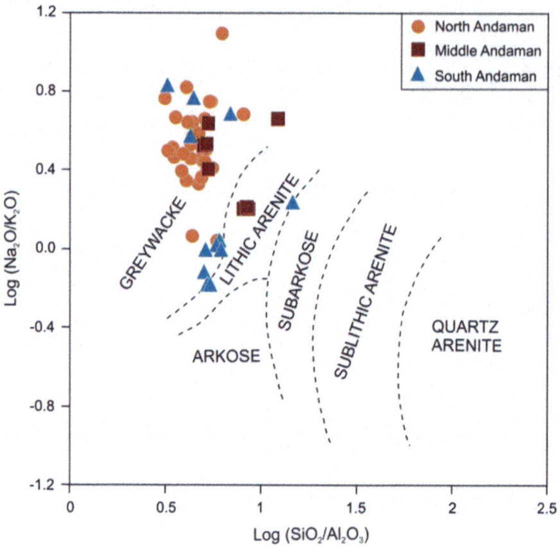

Fig. 6 Geochemical classification of terrigenous sands. Adapted from Pettijohn et al. (1972)

wide variations. The varying content of calcite cement is one of the reasons for the variable CaO content of samples. The MgO content in the North Andaman Islands samples is higher than those of Middle and South Andaman Islands.

The relationship between log (SiO_2/Al_2O_3) vs log (Na_2O/K_2O) (after Pettijohn et al. 1972) distinguishes between mature and immature sediments (Fig. 6). All samples of the North Andaman Islands plot within the greywacke field. While those of the South Andaman Islands occupy both greywacke and lithic arenite fields. Most samples of the Middle Andaman Islands occupy the greywacke field and a few plot in lithic arenite field.

A cross plot between log (SiO_2/Al_2O_3) and log (Fe_2O_3/K_2O) distinguishes different sediment composition (Herron 1988). Most samples of the North Andaman Islands occupy shale, Fe-shale, Fe-sand and wacke fields, while those in the South Andaman Islands plot mostly within the fields of wacke and Fe-shale (Fig. 7). Most samples of the Middle Andaman Islands occupy litharenite, wacke and Fe-sand fields.

7.2 *Trace Elements*

The trace element concentrations (ppm) of all the samples are provided in Table 3. The contents of Ba, Sc, V, Zr and Sr are lower in Andaman samples compared to the PAAS. However, the concentration of Cr in most samples are higher than that of PAAS (110 ppm). The total REE (ΣREE) concentration of the siltstone or fine-grained sandstones (S1, S3, S5, S6, S7, S8) from the South Andaman islands

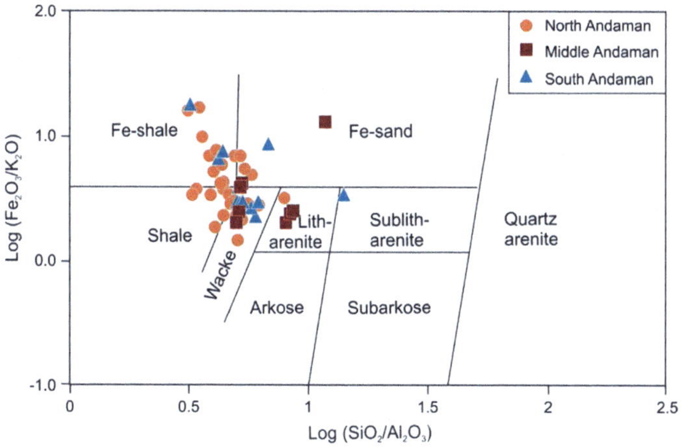

Fig. 7 Geochemical classification of terrigenous sands and shales. Adapted from Herron (1988) for samples of Andaman Flysch

ranges from 131 to 289 ppm. Concentrations of REE in samples are higher compared to the chondrite. However, they exhibit chondrite-normalized patterns similar to post-Archean Australian shales (PAAS) (Fig. 8; Table 4). They also show LREE enrichment (high $(La/Yb)_N$, $(La/Sm)_N$ and ΣLREE/ΣHREE), negative Eu anomaly (Eu/Eu* = 0.61–0.69), near-flat HREE ((Gd/Yb)n = 1.6–2.08) and a low Ce anomaly (Ce/Ce* = 1.19–1.3) (Fig. 8). The coarse-grained flysch sandstones (S1, M5) show low ΣREE (56.1 and 39.2 ppm respectively) concentration with LREE enrichment (high (La/Yb)n, (La/Sm)n, and ΣLREE/ΣHREE), flat HREE (Gd/Yb)n = 1.06–1.4), low Eu anomaly (Eu/Eu* = 0.88–0.84) and positive Ce anomaly (Ce/Ce* = 1.29–1.46). The conglomerate sample (N8) from Ross island in the North Andaman Islands shows a more or less flat REE pattern ((La/Yb)n = 2.15 and ΣLREE/ΣHREE = 2.94), low Eu anomaly (Eu/Eu* = 0.779) and a positive Ce anomaly (Ce/Ce* = 1.52).

8 Interpretations of Major and Trace Elements Data

Since, Zr and Ti behave as immobile phases during weathering, transportation, deposition and diagenesis, the ratio of these two elements is considered as a reliable indicator of the source rock (Hayashi et al. 1997). This plot distinguishes between felsic, intermediate and mafic igneous rocks (Guo et al. 2012). Most sandstone samples of Middle and North Andaman Islands reveal the intermediate sources (Fig. 9). While most samples of the South Andaman Islands scatter between felsic and intermediate fields.

The REE data was normalized with respect to chondrite meteorite values (cf. Boynton 1983) and compared with chondrite-normalized REE concentrations of

Table 3 Trace element concentrations in the analyzed samples from Andaman

Elements	Ba	Cr	Sr	Zr	Sc	V	Zn	La	Y
North Andaman									
N1	148.3	806.4	200.7	109.9	16.5	115	46.9	17.4	31
N2	158	191.2	168.1	162.5	15	111	45.5	18.5	27.2
N3	213.5	318.4	229.4	75.4	12.7	78.8	32.2	16.1	22.5
N4	192.8	311.7	176.3	128.7	11.2	101.7	88.5	16.5	21.2
N5	198.8	638.5	213.8	98	18	121.1	48.9	11.4	25.2
N6	181.5	540.1	209.7	103.6	18.4	123.8	49.7	14.5	27.3
N7	195.3	544.8	190.3	149.9	25.2	179.8	72	18.8	8.6
N8	184.3	366.7	129.9	141.2	26.8	199.7	103	6.2	25
N9	320.6	174	254.3	88.8	10.3	69.2	26.7	19.6	29.7
N10	254.1	1333.8	244.5	97.9	18.2	132.5	52.5	11.5	26.1
N11	299.9	1065	193	111.7	22.3	155	66.9	12.2	28.2
N12	249.73	651.95	58.05	205.56	11.04	95.06	49.98	16	20.43
N13	84.74	346.52	305.54	170.37	10.51	82.49	44.92	10.41	18.86
N14	115.58	275.76	66.35	135.55	13.37	87.99	52.11	11.88	19.85
N15	351.8	262.67	174.23	110.12	12.06	81.42	43.31	16.77	20.66
N16	323.09	460.05	163.43	163.06	25.47	146.81	55.92	15.75	29.27
N17	418.62	370.44	195.18	111.74	12.69	87.23	37.24	19.94	31.04
N18	503.18	539.28	185.34	120.79	17.31	118.72	50.14	18.6	27.65
N19	130.27	110.08	77.17	152.56	19.2	140.78	74.28	18.8	19.9
N20	309.72	224.71	268.2	93.39	10.94	77.62	25.44	15.56	26.41

(continued)

Table 3 (continued)

Elements	Ba	Cr	Sr	Zr	Sc	V	Zn	La	Y
N21	98.04	17.28	65.19	59.52	11.27	73.93	24.73	47.16	32.34
N22	164.02	64.86	146.48	80.49	9.01	61.44	22.88	20.86	22.02
N23	78.9	164.89	186.5	145.6	16.86	106	49.83	11.24	23.93
N24	406.02	133.15	202.49	135.59	12.93	93.14	41.63	20.09	27.53
N25	227.14	163.04	222.5	79.24	13.39	98.18	32.11	8.29	18.19
N26	246.5	155.77	206.54	124.76	13.13	87.52	31.3	14.71	24.67
N27	205.8	192.36	245.06	63.47	14.4	92.97	36.86	9.16	18
N28	315.13	155.75	181.73	107.31	11.21	81.51	31.33	17.67	22.37
N29	158.85	310.08	256.6	234.89	16.5	154.37	36.79	9.6	14.86
N30	384.07	247.53	226.41	127.87	10.86	87.74	63.64	18.14	23.23
Middle Andaman									
M1	306.47	225.46	236.25	120.68	12.6	85.09	31.87	17.17	26.15
M2	371.86	285.92	196.39	150.41	15.89	118.76	51.23	21.6	29.77
M3	354.6	598.39	195.84	110.17	17.39	125.08	74.75	27.65	45.36
M4	268.83	357.4	186.68	123.85	12.36	105.82	46.18	21.46	24.4
M5	81.34	133.29	41.1	64.23	5.67	50.96	21.49	8.1	8.98
M6	142.22	78.35	80.38	239.17	7.52	58.3	31.04	21.86	19.96
M7	132.4	90.52	74.23	90.24	7.23	57.37	41.78	20.51	21.94
M8	138.19	81.47	72.5	144.77	7.08	55.86	60.28	20.49	17.69

(continued)

Table 3 (continued)

Elements	Ba	Cr	Sr	Zr	Sc	V	Zn	La	Y
S1	62.1	130.4	79.36	118.11	14.23	115.69	49.93	13.81	14.02
South Andaman									
S2	96.21	246.34	28.7	201.11	4.51	34.36	23.4	11.5	9.28
S3	324.23	164.78	183.14	54.94	10.41	97.19	58.45	31.48	27.95
S4	96.8	492.9	69.2	145.44	9.62	83.14	70.03	14.11	14.69
S5	316.44	112.3	91.78	42.11	10.98	98.29	97.35	29.84	30.35
S6	342.25	237.4	81.42	504.05	13.13	146.17	131.12	62.2	49.94
S7	335.79	239.09	109.57	226.65	15.39	144.65	130.88	51.4	46.39
S8	395.51	205.67	106.85	178.33	13.92	132.88	145.72	80.4	43.81
S9	346.72	281.99	123.53	389.48	15.2	147.46	174.52	77.36	70.11
S10	310.26	177.31	119.61	49.06	13.59	117.13	126.94	33.05	32.06

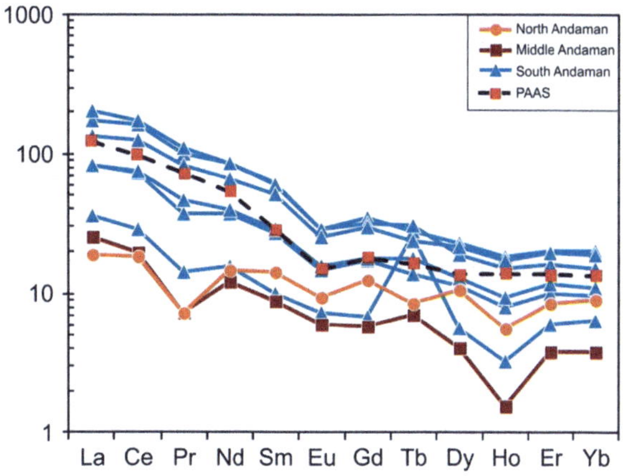

Fig. 8 Chondrite-normalized patterns for samples of South Andaman (chondrite meteorite concentrations from Boynton 1983)

Table 4 Rare earth element concentrations in the analysed samples from Andaman

Elements	North	Middle	South					
	N8	M5	S1	S3	S5	S6	S7	S8
La	6.0	8.0	11.1	25.8	26.1	53.3	42.1	63.5
Ce	15.1	16.0	23.5	58.4	60.3	130.3	102.1	142.0
Pr	0.9	0.9	1.7	4.5	5.7	12.4	10.2	13.5
Nd	8.9	7.2	9.3	22.2	23.6	51.0	40.5	51.8
Sm	2.8	1.7	2.0	5.2	5.6	12.2	9.9	11.8
Eu	0.7	0.4	0.5	1.1	1.2	2.1	1.9	2.1
Gd	3.2	1.5	1.8	4.5	4.6	9.0	7.8	8.3
Tb	0.4	0.3	1.3	0.7	0.8	1.3	1.1	1.4
Dy	3.4	1.3	1.8	3.7	4.2	7.4	7.0	6.1
Ho	0.4	0.1	0.2	0.6	0.7	1.3	1.2	1.1
Er	1.8	0.8	1.3	2.1	2.5	4.3	4.1	3.4
Yb	1.9	0.8	1.3	2.0	2.3	4.2	3.9	3.2
Lu	0.3	0.2	0.2	0.4	0.4	0.7	0.6	0.5
∑LREE	45.8	39.3	56.1	131.2	137.9	289.6	232.5	308.8
La/Yb	0.2	0.1	0.2	0.3	0.3	0.3	0.3	0.3
(La/Yb)n	13.4	0.7	0.6	1.0	0.8	0.9	0.8	1.5
(La/Sm)n	3.4	0.7	0.8	0.7	0.7	0.6	0.6	0.8
(Gd/Yb)n	2.1	1.1	0.8	1.4	1.2	1.3	1.2	1.6
Eu/Eu*	0.6	1.3	1.4	1.1	1.1	0.9	1.0	1.0
Ce/Ce*	1.2	1.4	1.2	1.2	1.1	1.2	1.1	1.1

Post-Archean average Australian Shale (PAAS) and Mid Oceanic Ridge Basalt (MORB) (cf. Nance and Taylor 1976; Sun and McDonough 1989) (Fig. 10; Table 4). The LREEs show relative enrichment in comparison to HREEs. The samples exhibit negative Eu anomalies. This negative Eu anomaly is related to the presence of feldspar in the samples. Therefore, the analysed samples indicate felsic igneous origin.

Roser and Korsch (1986) used the relationship between K_2O/Na_2O and SiO_2 to distinguish between tectonic settings viz. island arc (ARC), active continental margin (ACM) and passive margins (PM). The samples of the North Andaman Islands plot exclusively within the ACM field. While samples of the Middle Andaman Islands plot within ARC and ACM fields, those of the south Andaman islands exhibit scatter (Fig. 10).

A cross plot between TiO_2 and $Fe_2O_3 + MgO$ distinguishes between active continental margin(ACM), passive continental margin (PCM), continental island arc (CIA) and oceanic island arc (OIA) (Bhatia 1983) (Fig. 11). While the data points scatter widely, many of the samples of North and South Andaman Islands plot within the fields of continental island arc and oceanic island arc. Considerable overlap exists between data points of samples from different Islands. Several workers reported similar scatter of data points for this particular plot (cf. Kassi et al. 2015; Verma and Armstrong-Altrin 2016).

The relationship between $Fe_2O_3 + MgO$ and Al_2O_3/SiO_2 distinguishes between oceanic island arc, continental island arc, active continental margin and passive margin. The samples of North and Middle Andaman Islands mostly plot within the fields of oceanic island arc and continental island arc while those in the South Andaman Islands scattered around most fields (Fig. 12).

9 Discussion

Sediments in convergent margins, like Andaman Island Arc, possibly derives from multiple sources (Garzanti et al. 2013). The east of the Andaman Basin comprises the Indian plate with oceanic settings. The south of the Andaman Basin merges with the Sumatra island arc and the Java trench. The Mergui terrace, with a barrier of volcanic arc in between the fore-arc and back-arc, occupies the west of Andaman basin. This leaves the northern side of the Andaman Basin as the major source of sediments. Petrographical and geochemical data suggest that the nearest cratonic platform, the Myanmar shelf on the north, provided the detritus for Andaman Flysch in the eastern part of the North Andaman Islands because of their similarity with modern Irawaddy sediments (cf. Garzanti et al. 2013). The rising subduction complex in the form of Andaman Islands on the west contributed the magmatic basement derived sediments. Intermediate igneous rock fragments were possibly sourced by the volcanic arc in the east. Upper Eocene–lower Oligocene rocks are absent in the western Irrawaddy delta (Wandrey 2006). An erosional unconformity marks the post-Eocene succession in the Myanmar shelf in Irrawaddy delta area and the Indo Burman Range (Wandrey 2006). This unconformity indicates that sediments in the

Andaman Flysch possibly derived through eroded shelf sediments of the Myanmar shelf and transported through the submarine canyons before depositing as coarse-grained, texturally and mineralogically immature, channel-filled medium- to coarse-grained sandstones in the North Andaman Islands. The rising subduction complex contributed sediments simultaneously, in the form of accreted ophiolite as observed in flysch samples of the North Andaman Islands, along with volcanic effusives from the volcanic arc.

Earlier geochemical studies indicated dominant recycled orogenic sources, with subordinate arc-derived sediment input for the Andaman Flysch of the South Andaman Islands (Allen et al. 2008). Allen et al. (op. cit.) suggested the derivation of sediments from Myanmar due to the ongoing India Eurasia collision. Bandopadhyay and Ghosh (2015) considered an active continental margin and associated magmatic arc for the Andaman Flysch on the basis of major and trace element concentrations in the South Andaman Islands. The sediments in the Flysch is likely to be derived from granitoid rocks with minor sedimentary and metamorphic rocks and arc massif. They further indicated that sediments Andaman Flysch in the South Andaman Islands primarily derived from plutonic-metamorphic provenance with subordinate contributions from sedimentary/metamorphic province, matching with the Shan-Thai Block of north-eastern Myanmar. On the basis of petrographical, mineralogical and detrital zircon data Limonta et al. (2017) considered a significant derivation of Andaman Flysch from the continental crust signal with only a minor input from volcanic arc. Awasthi (2017) analyzed whole rock major, trace, and rare earth element compositions of Palaeogene and Neogene rocks of Andaman Islands and attributed supply of sediments to the Andaman Flysch from first–cycle active continental margin sources, probably located in the Tibetan and eastern Myanmar region and recycled quartzose sedimentary sources within the nascent Himalayas. Awasthi (2017) related uniformly high CIA and PIA values in the Andaman Flysch to high precipitation and strong chemical weathering associated to initiation of Indian monsoon.

Our study, reveals the compositional differences between North, Middle and South Andaman Islands. Geochemical study indicates that active continental margin to continental island arc system for flysch in South and Middle Andaman Islands. The coarse-grained sandstones of the north Andaman flysch samples bear the signature of oceanic island arc setting (Figs. 9, 10, 11 and 12). Trace element concentrations of sediments of South Andaman and Middle Andaman Islands indicate felsic source while flysch samples of the North Andaman, conglomerate and Mithakari sandstone bear intermediate rock signature. The Indian plate has moved westward since the initiation of subduction (Roy 1992; Pal et al. 2003; Curray 2005). Envisaged sediment dispersal pattern of Andaman Flysch superimposed on a Bouguer gravity map of the Andaman Basin reflects the spreading centre in Plio-Pleistocene time (Fig. 13) (cf. Kamesh Raju et al. 2004). This data suggests the occurrence of the subduction zone and the fore-arc closer to the Sagaing Fault during Late Eocene-Early Oligocene with respect to their present position. The Andaman trench has shifted westward with time (Roy 1992; Pal et al. 2003; Curray 2005). Considering the arcuate nature of the plate

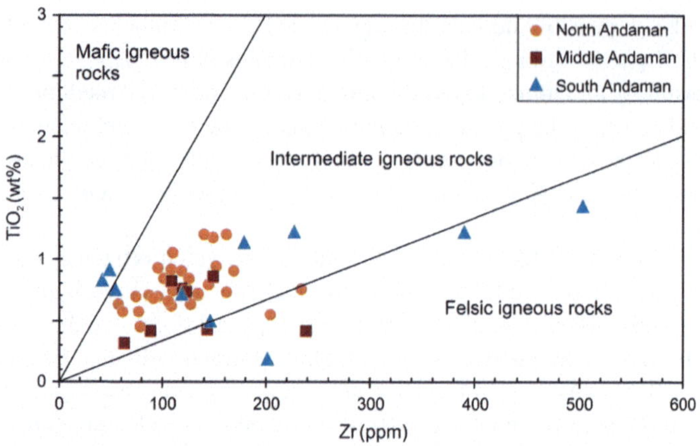

Fig. 9 Source rock discrimination diagram based on TiO_2 and Zr. Adapted from Hayashi et al. (1997) for samples of Andaman Flysch

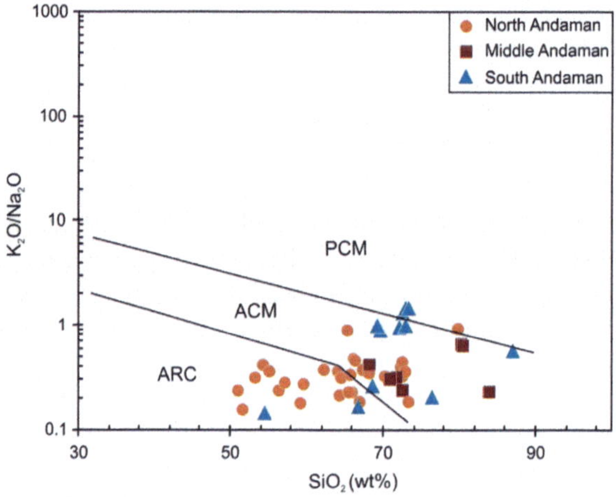

Fig. 10 Tectonic setting discrimination diagram based on K_2O/Na_2O and SiO_2. Adapted from Roser and Korsch (1986) for samples of Andaman Flysch (PCM = passive continental margin, ACM = Active Continental Margin and ARC = Oceanic island arc)

boundary, the South Andaman was closer to the N-S oriented Sagaing Fault. Therefore, the Irrawaddy delta possibly supplied sediments in the North Andaman Islands while Sagaing Fault derivatives accumulated in South Andaman islands (Fig. 13). The global sea level fall during the Oligocene led to the subaerial exposure of the Irrawaddy shelf. The missing Oligocene section on the Irrawaddy shelf records the global sea level fall. The eroded shelfal sediments were transported to deepwater as Andaman Flysch, and entrapped in the rising subduction zone. The volcanic arc and the rising subduction complex possibly contributed sediments to the Andaman

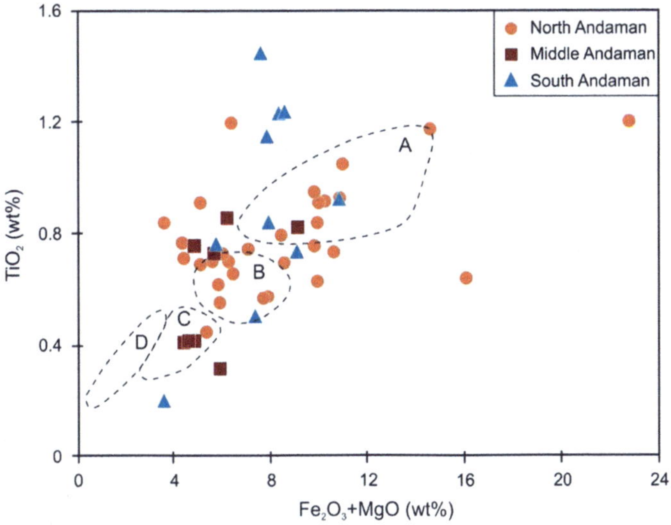

Fig. 11 Tectonic setting discrimination diagram based on TiO_2 and (Fe_2O_3 + MgO). Adapted from Bhatia (1983) for samples of Andaman Flysch (fields A—oceanic island arc; B—continental island arc; C—active continental margin and D—passive margin)

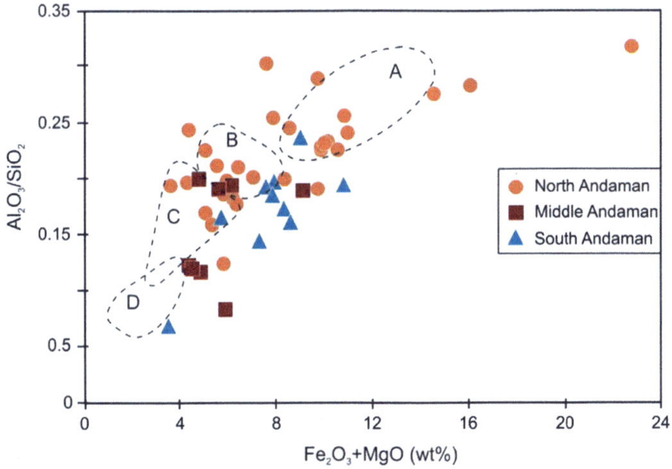

Fig. 12 Tectonic setting discrimination diagram based on Al_2O_3/SiO_2 and (FeO + MgO). Adapted from Bhatia (1983) for samples of Andaman Flysch (fields A—oceanic island arc; B—continental island arc; C—active continental margin and D—passive margin)

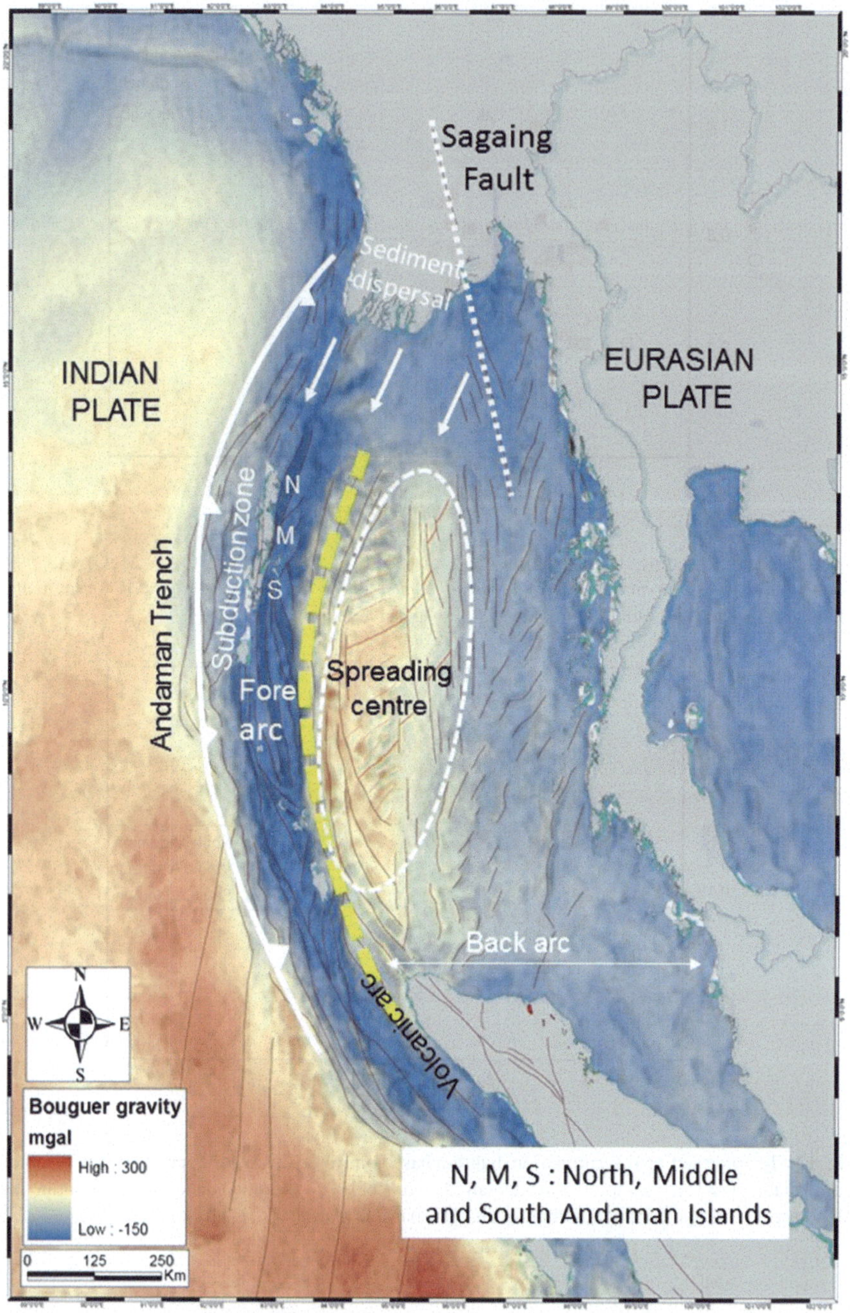

Fig. 13 Envisaged sediment dispersal pattern of Andaman Flysch shown on a Bouguer gravity map of Andaman Basin

Flysch in North Andaman Islands, while the Flysch is nearly devoid of volcanic fragments in the South Andaman Islands. While our data support the view of sediment supply for Andaman Flysch from continental sources like eastern Myanmar (Awasthi 2017), our data are insufficient to indicate the sediment input from the Himalayas.

10 Conclusions

Petrographical and geochemical investigations of Andaman Flysch lead to the following conclusions.

(a) The Andaman Flysch comprises various lithotypes viz. medium- to coarse-grained sandstones, coarse siltstone and medium-grained sandstone, siltstone to fine-grained sandstone, heterolithics and shale. The underlying Mithakari Formation consists of conglomerate and coarse-grained sandstones.
(b) Major and trace element compositions of the Andaman Flysch of South and Middle Andaman Islands corresponds to active continental margin to continental island arc settings, with inputs from the east and northeast. The flysch samples of the North Andaman Islands bear the oceanic island arc signature and indicate sediment derivation from the north. The sediment derivation from the Andaman trench decreases progressively from the north to the south.
(c) The Mithakari samples show lower SiO_2 and higher Na_2O compared to those in Andaman Flysch, and are more enriched in mafic minerals.
(d) The Andaman Flysch in North Andaman Islands received sediments possibly from the Irawaddy delta. While the Andaman Flysch in the Middle and South Andaman Islands received sediments from the east. Therefore, the Andaman Flysch received sediments from multiple sources.

Acknowledgements Authors thank Indian Institute of Technology Bombay for infrastructure facilities. Authors are thankful to Swagata Paul and Rajib Sadhu for chemical analysis of samples.

References

Allen RC, Najman AY, Bandhopadhay PC, Chapman HJ, Bickle MJ, Garzanti E, Vezzoli G, Ando S, Foster GL, Gerring C (2008) New constraints on the sedimentation and uplift history of the Andaman-Nicobar accretionary prism, South Andaman Island. Geol Soc America Special Publ 436:223–255

Armstrong-Altrin JS, Machain-Castillo ML, Rosales-Hoz L, Carranza-Edwards A, Sanchez-Cabeza JA, Ruiz-Fernandez AC (2015) Provenance and depositional history of continental slope sediments in the Southwestern Gulf of Mexico unraveled by geochemical analysis. Conti Shelf Res 95:15–26

Awasthi N (2017) Provenance and paleo–weathering of Tertiary accretionary prism–forearc sedimentary deposits of the Andaman Archipelago, India. J Asian Earth Sci 150:45–62

Bandopadhyay PC (2005) Discovery of abundant pyroclasts in the Namunagarh Grit, South Andaman: evidence for arc volcanism and active subduction during the Paleogene in the Andaman area. J Asian Earth Sci 25:95–107
Bandopadhyay PC, Ghosh PC (2015) Provenance analysis of the Oligocene turbidites (Andaman Flysch), South Andaman Island: a geochemical approach. J Earth Sys Sci 124:1019–1037
Bhatia MR (1983) Plate tectonics and geochemical composition of sandstones. J Geol 91:611–627
Bhatia MR (1985) Composition and classification of flysch mudrocks of eastern Australia: implication in provenance and tectonic setting interpretation. Sed Geol 41:249–268
Boynton WV (1983) Cosmochemistry of the rare earth elements. In: Hendersen P (ed) Rare earth element geochemistry, pp 63–114
Chaudhuri A, Banerjee S, Le Pera E (2018) Petrography of Middle Jurassic to Early Cretaceous sandstones of the Kutch Rift Basin: implications on provenance and basin evolution. J Palaeogeo 7:2–14
Chaudhuri A, Banerjee S, Banerjee S, Chauhan G (2020) Compositional evolution of sediments recording the increasing tectonic stability of a pericratonic rift: Mesozoic Kutch Basin, western India. Mar Pet Geol 111:475–476
Chakraborty PP, Pal T (2001) Anatomy of a fore arc submarine fan: upper Eocene-Oligocene Andaman Flysch Group, Andaman Islands, India. Gond Res 4:477–486
Curray JR, Moore DG, Lawver LA, Emmel FJ, Raitt RW, Henry M, Kieckhefer R (1979) Tectonics of the Andaman Sea and Burma. In: Geological and geophysical investigations of continental margins. AAPG Mem, vol 29, pp 189–198
Curray JR (2005) Tectonics and history of the Andaman Sea region. J Asian Earth Sci 25:187–232
Dickinson WR, Suczek CA (1979) Plate tectonics and sandstone composition. AAPG Bull 63:2164–2182
Dickinson WR, Beard LS, Brackenridge GR, Erjavec JL, Ferguson RC, Inman KF, Knepp RA, Lindberg FA, Ryberg PT (1983) Provenance of North American Phanerozoic sandstones in relation to tectonic setting. Geol Soc America Bull 94:222–235
Fedo CM, Eriksson KA, Krogstad EJ (1996) Geochemistry of shales from the Archean (3.0 Ga) Buhwa Greenstone Belt, Zimbabwe: implications for provenance and source-area weathering. Geochim Cosmochim Acta 60:1751–1763
Garzanti E, Limonta M, Resentini A, Bandopadhyay PC, Najman Y, Sergio VG (2013) Sediment recycling at convergent plate margins (Indo-Burman Ranges and Andaman-Nicobar Ridge). Earth Sci Rev 123:113–132
Garzanti E, Wanf JG, Vezzoli G, Limonta M (2016) Tracing provenance and sediment fluxes in the Irrrawaddy river basin (Myanmar). Chem Geol 440:73–90
Geological Survey of India (2012) Geology and mineral resources of Andaman and Nicobar islands. Misc Publ Geol Surv India 30:1–40
Guo Q, Xiao W, Windley BF, Mao Q, Han C, Qu J, Ao S, Li J, Song D, Yong Y (2012) Provenance and tectonic settings of Permian turbidites from the Beishan Mountains, NW China: implications for the Late Paleozoic accretionary tectonics of the southern Altaids. J Asian Earth Sci 49:54–68
Hayashi KI, Fujisawa H, Holland HD, Ohmoto H (1997) Geochemistry of ~1.9 Ga sedimentary rocks from northeastern Labrador. Canada Geochim Cosmochim Acta 61:4115–4137
Herron MM (1988) Geochemical classifications of terrigenous sands and shalesfrom core or log data. J Sed Pet 58:820–829
Kamesh Raju KA, Ramprasad T, Rao PS, Ramalingeswara Rao B, Varghese J (2004) New insights into the tectonic evolution of the Andaman basin, northeast Indian Ocean. Earth Planet Sci Lett 221:145–162 (National Institute of Oceanography, Dona Paula, Goa)
Kassi AM, Grigsby JD, Khan AS, Kasi AK (2015) Sandstone petrology and geochemistry of the Oligocene-Early Miocene Panjgur Formation, Makran accretionary wedge, southwest Pakistan: Implications for provenance, weathering and tectonic setting. J Asian Earth Sci 105:192–207
Limonta M, Resentini A, Carter A, Bandopadhyay PC, Garzanti E (2017) Chapter 10 Provenance of Oligocene Andaman sandstones (Andaman-Nicobar Islands): Ganga-Brahmaputra or Irrawaddy derived? Geol Soc London Mem 47:141–152

McLennan SM, Taylor SR, McCullogh MT, Maynard JB (1990) Geochemical and Nd–Sr isotope composition of deep-sea turbidites: crustal evolution and plate tectonic associations. Geochim Cosmochim Acta 54:2015–2050
McLennan SM, Hemming S, McDaniel DK, Hanson GN (1993) Geochemical approaches to sedimentation, provenance, and tectonics. In: Johnson MJ, Basu A (eds) Processes controlling the composition of clastic sediments. Geological Society of America Special Paper, vol 284, pp 21–40
Mukhopadhyay B, Chakraborty PP, Paul S (2003) Facies clustering in turbidite successions: case study from Andaman Flysch Group, Andaman Islands, India. Gond Res 6:918–925
Nance WB, Taylor SR (1976) Rare earth element patterns and crustal evolution-I: Australian post Archean sedimentary rocks. Geochim Cosmochim Acta 40:1539–1551
Paikaray S, Banerjee S, Mukherji S (2008) Geochemistry of shales from the Paleoproterozoic to Neoproterozoic Vindhyan Supergroup: Implications on provenance, tectonics and paleoweathering. J Asian Earth Sci 32:34–48
Pal T, Chakraborty PP, Gupta TD, Singh CD (2003) Geodynamic evolution of the outer arc forearc belt in the Andaman islands, the central part of Burma-Java subduction complex. Geol Mag 140:289–307
Petijohn FJ, Potter PE, Siever R (1972) Sand and sandstone. Springer, New York, p 618
Roser BP, Korsch RJ (1986) Determination of tectonic setting of sandstone–mudstone suites using SiO_2 content and K_2O/Na_2O ratio. J Geol 94:635–650
Roy TK (1983) Geology and hydrocarbon prospects of Andaman–Nicobar Basin. In: Bhandari LL (ed) Petroliferous Basins of India. Pet Asia J, pp 37–50
Roy SK (1992) Accretionary prism in Andaman Forearc. Geol Surv India Spec Publ 29:273–278
Roy SK, Paul S, Sadhu R, Banerjee S (2011) Provenance and tectonic setting of a coarse grained Flysch: case study from the Andaman Flysch at trench slope break, North Andaman islands. In: Geological Society of America Abstracts with Programs, vol 43, pp 432
Roy SK, Das Sharma S (1993) Evolution of Andaman Forearc Basin and its hydrocarbon potential. In: Proceedings of the Second Seminar on Petroliferous Basins of India. Indian Petroleum Publishers, Dehradun, vol 1, pp 407–435
Roy SK, Banerjee S (2016) Soft Sediment Deformation Structures in Andaman Flysch, Andaman Basin: Evidence for Palaeogene Seismic Activity in the Island Arc. Berita Sed 35:55–64
Roy SK, Banerjee S (2017) Commonality of Sedimentary facies of Andaman Flysch in South Andamans and Great Nicobar Island: Implications for depositional model in southern part of Andaman basin. J Indian Asso Sedi 34:75–92
Saha S, Banerjee S, Burley SD, Ghosh A, Saraswati PK (2010) The influence of flood basaltic source terrains on the efficiency of tectonic setting discrimination diagrams: an example from the Gulf of Khambhat, western India. Sed Geol 228:1–13
Saha S, Burley S, Banerjee S (2018) Mixing processes in modern estuarine sediments from the Gulf of Khambhat, western India. Mar Pet Geol 91:599–621
Sun SS, McDonough WF (1989) Chemical and isotopic systematics of oceanic basalts; implications for mantle composition and processes. In: Saunders AD, Norry MJ (eds) Magmatism in the Ocean Basins. Geological Society London, vol 42, pp 313–345
Taylor SR, McLennan SM (1985) The continental crust: its composition and evolution. Blackwell, Oxford, UK
Verma SP, Armstrong-Altrin JS (2016) Geochemical discrimination of siliciclastic sediments from active and passive margin settings. Sed Geology 332:1–12
Wandrey BC (2006) Eocene to Miocene Composite Total Petroleum System, Irrawaddy-Andaman and North Burma Geologic Provinces, Myanmar. In: U.S. Geological Survey Bull, vol 2208, p 26

Tracing the Sources and Depositional Pathways for the Oligocene Sediments in the Andaman Forearc

Neeraj Awasthi, Bivin G. George and Jyotiranjan S. Ray

Abstract Finding provenance of the siliciclastic turbidites deposited in the Andaman Trench-Forearc Basin had remained contentious with most indicators pointing towards major sediment sourcing either from eastern Myanmar or from nascent Himalayan mountain belt. Whereas the palaeo-Irrawaddy River was considered as the main transporting agent for sediments originating in Myanmar, the Himalayan sediments were believed to have been recycled from the earliest Bengal Fan. Here, we examine existing Sr–Nd isotopic data for Oligocene sediments deposited in the basin and exposed today on the Andaman Accretionary Prism (AAP) to determine their provenance and transport pathways. We also present new data for sediments in the Barail Group, Bengal Basin and use them to test the sediment recycling hypothesis. Results of our study confirm that material from the Himalayan sources, transported through the submarine Bengal Fan, contributed significantly to the Andaman basin during Oligocene. Sr–Nd isotopic compositions considered along with published U–Pb detrital zircon age data reveal that both the Tethys Himalaya and Higher Himalaya were the major sediment suppliers to the basin, with a minor contribution coming from juvenile source(s) in the Indo-Myanmar ophiolite belt and/or the central Myanmar volcanic arc. We propose that the large Himalayan contribution was a result of rapid exhumation and high rate of erosion in the north-eastern Himalaya and Tibet during Late Eocene and Oligocene and sediment transport by the palaeo-channels of the Yarlung-Siang-Brahmaputra river system via the Bengal Basin/Fan.

Keywords Andaman forearc · Andaman flysch · Turbidites · Sr–Nd isotopic ratios · Palaeo-Irrawaddy · Palaeo-Brahmaputra

---

N. Awasthi (✉)
Department of Earth and Planetary Sciences, Prof. Rajendra Singh (Rajju Bhaiya) Institute of Physical Sciences for Study and Research, VBS Purvanchal University, Jaunpur 222003, India
e-mail: aneeraj.geology@gmail.com

B. G. George · J. S. Ray
Physical Research Laboratory, Navrangpura, Ahmedabad 380009, India

J. S. Ray and M. Radhakrishna (eds.), *The Andaman Islands and Adjoining Offshore: Geology, Tectonics and Palaeoclimate*, Society of Earth Scientists Series,
https://doi.org/10.1007/978-3-030-39843-9_5

1 Introduction

The Andaman and Nicobar Islands in the north-eastern Indian Ocean is an archipelago comprising 572 islands that run from north to south between Myanmar and Indonesia. All these islands, except the two volcanic islands of Barren and Narcondam, represent subaerially exposed parts of the accretionary prism of the Andaman Subduction Zone (ASZ), which is the northward extension of the Sunda-Banda subduction zone (Curray 2005; McCaffrey 2009). At the ASZ, the Indian Plate is subducting obliquely beneath the Burma Plate along the Andaman Trench (Fig. 1a). The current subduction along this convergent margin is believed to have begun during late Cretaceous (Bhattacharya et al., this book) and as a result several tectono-stratigraphic features including the Andaman accretionary complex formed between Myanmar and Sumatra (Allen et al. 2007). The accretionary complex is primarily made up of ophiolite thrust sheets and trench-forearc siliciclastic and carbonate turbidites. Earlier studies have revealed that the complex had undergone multiple episodes of folding, thrusting, subsidence, and uplift (e.g., Allen et al. 2007 and references therein).

The provenance of the Oligocene siliciclastic turbidite sequence of the Andaman accretionary prism, called the Andaman Flysch Group, remains controversial because of opposing views about the locations of the primary sources of sediments and modes

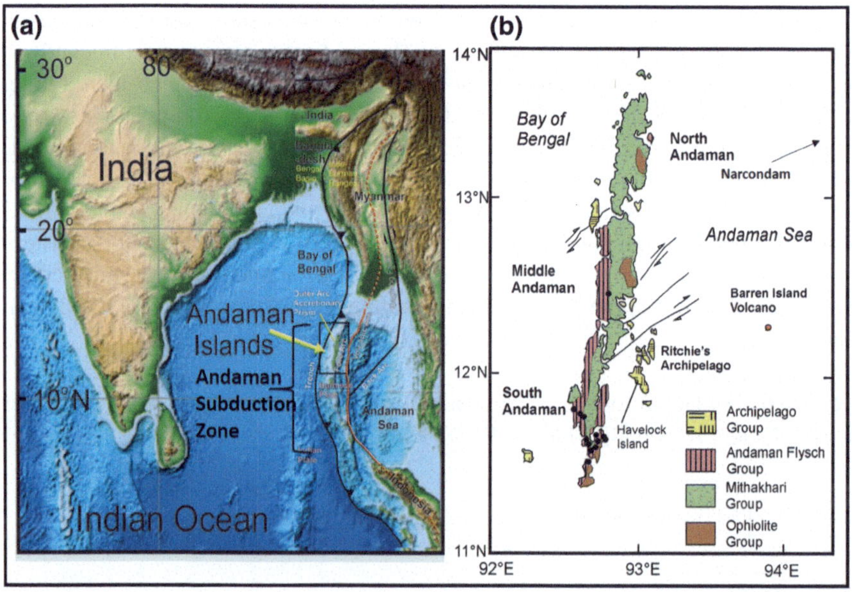

Fig. 1 **a** A geotectonic map of South Asia showing the Andaman Subduction Zone and its components. The boxed area shows the location of the Andaman Islands, which form part of the accretionary prism. Locations of the Bengal Basin and the Indo-Burman Ranges are marked. **b** A geological map of the Andaman Islands (modified from Pal et al. 2003) showing major stratigraphic units. Our sampling locations for the Andaman Flysch Group sediments are marked as solid circles

of sediment transport into the Andaman basin (e.g., Pal et al. 2003; Curray 2005; Allen et al. 2007, 2008; Garzanti et al. 2013; Bandopadhyay and Ghosh 2015; Limonta et al. 2017; Awasthi and Ray 2019). All these studies were based either on traditional methods such as petrography and heavy-mineral abundances or modern methods such as detrital zircon geochronology and radiogenic isotope (Sr–Nd) finger printing. Many favour the idea that sediments that make up the bulk of the Andaman Flysch came from Myanmar through palaeo-Irrawaddy river system, whereas only a few believe that they represent off-scraped material from the Bengal submarine fan. Although, there is a consensus on the presence of a strong continent crust signal in these sediments, none of the studies categorically indicate their derivation from the nascent Himalayas. Interestingly, the nascent Himalayas have been considered as sources of sediments to some of the Palaeogene and most of the Neogene rocks of the Indo-Burman ranges (IBR) located to the north of the Andaman basin (Allen et al. 2008). In this work, we make an attempt to resolve this conundrum by comparing Sr–Nd isotopic data from the Andaman Flysch Group with the newly generated data from the Barail Group of the Bengal Basin. We also make use of the published U–Pb detrital zircon data from the Andaman Flysch and that from the potential source regions to supplement our inferences.

2 Geology and Earlier Work

Due to the complex structure of the accretionary prism, interpretation of the geology of the Andaman and Nicobar Islands has always been challenging. The earliest proposed stratigraphy of the Andaman Islands was based on lithological mapping done by Oldham (1885), Rink (1847), Hochstetter (1869) and Ball (1870). Oldham (1885) first divided the Andaman lithology into an older Port Blair Series and a younger Archipelago Series. Palaeontological constraints have been used since 1960s to place the lithostratigraphic units within a temporal framework (Guha and Mohan 1965; Karunakaran et al. 1968). The seismic, drilling, and outcrop stratigraphic studies carried in the islands during the mid-1970s by the Oil and Natural Gas Commission (ONGC) have helped understanding the geology of the Andaman Islands in the context of an accretionary setting. During the last 150 years of study, the stratigraphy of the Andamans has been modified and names of formations have changed/redefined multiple times.

The Andaman and Nicobar Islands form the outer-arc ridge of the Andaman fore-arc and is made up of imbricate stack of east-dipping thrust slices and folds that expose sections of oceanic lithosphere, pelagic sediments and turbidites (Pal et al. 2003; Bandopadhyay and Ghosh 2015). Stratigraphically, the entire sedimentary succession is divided into four major groups (Curray 2005). The Late Cretaceous Ophiolite Group that contains ocean floor volcanic-plutonic rocks and pelagic sediments forms the basement (~95 Ma; Pedersen et al. 2010; Fig. 1), followed by the Eocene Mithakhari Group (>40 Ma; Allen et al. 2007). The Mithakhari Group is further subdivided into the Hope Town Conglomerate and the Namunagarh Grit formations. The former is made up of conglomeratic beds, having clasts of mafic and ultramafic igneous rocks with subordinate fragments of limestone, jasper and radiolarian chert, interbedded with greenish-grey volcanic rich sandstone, whereas the latter is composed of greenish-grey volcaniclastic sandstone and andesitic tuff deposits (Bandopadhyay 2005). The late Eocene-early Miocene Andaman Flysch Group (~35–20 Ma; Awasthi and Ray 2019) overlies the Mithakhari Group. It contains folded, faulted and sheared turbiditic sandstones and mudrocks deposited in a submarine fan environment (Chakraborty and Pal 2001; Bandopadhyay and Ghosh 2015). Petrographic and mineralogical studies of these rocks have shown that it is made up of very fine to fine grained sand, dominantly feldspatho-quartzose to quartzose in composition with abundant mica (Allen et al. 2007; Limonta et al. 2017). It also contains lithic fragments of both sedimentary and metamorphic origin, and subordinate felsic volcanic to ultramafic grains. However, burial diagenesis and/or recycling have led to selective dissolution of less stable minerals and enrichment of quartz (Allen et al. 2007; Limonta et al. 2017). Allen et al. (2007) suggested that the Andaman Flysch sediments were mostly derived from recycled orogenic sources in Myanmar affected by the India-Asia collision (e.g., Shan-Thai Terrane; Mogok Metamorphic Belt) with minor arc-derived material. Limonta et al. (2017) on the other hand, suggested these turbidites have a provenance similar to the sediments of the Palaeogene Bengal Fan and received very little contribution from the Myanmar sources. The Mio-Pliocene Archipelago Group overlies the Andaman Flysch Group. This unit, which is made up of shallow to deep water carbonates with minor siliciclastics and volcaniclastics, is considered as the youngest stratigraphic unit by most workers (Pal et al. 2005; Bandopadhyay and Ghosh 2015). However, Curray (2005) and later Awasthi and Ray (2019) suggested a fifth group, named as the Nicobar Group, to place the spatially less extensive Pleistocene limestone, beach deposits, unclassified volcanic rocks and tuff in the Andaman stratigraphy.

Bandopadhyay (2012) divided the exposed part of the AAP into two tectonic domains. The first domain is named as the ‘chaotic’ terrane that comprises the eastern side of the South and Middle Andaman Islands, and the entire North Andaman Island. The second domain is named as the ‘coherent’ terrane that extends along the western margin of Middle Andaman Island and alternate with rocks of the ‘chaotic’ terrane in the South Andaman. The Jarwa Thrust passing through the middle of the islands separates these two terranes (Pandey et al. 1992). The thrust sheets and olistoliths of pre-ophiolitic rocks, ophiolites and associated pelagic/hemipelagic sediments are included in the chaotic terrane. The Mithakhari Group rocks contain both the coherent and chaotic units whereas the Andaman Flysch Group rocks are entirely included in the ‘coherent’ terrane.

3 Samples and Methods

Thirteen sandstone and mudstone samples from the Andaman Flysch Group, collected from the outcrops across the Andaman Islands, were studied in this work.

Seven sandstone samples from the Oligocene Barail Group, southern Assam, were also collected and analysed in this work. The details of the samples are provided in Table 1 and the Andaman sampling locations are marked in Fig. 1. Each sample was crushed into small pieces; chips were handpicked and powdered into a homogenized mixture. The powdered sample was decarbonated with 2 N HCl and combusted at 600 °C to remove organic matter. Samples were dissolved using standard HF-HNO_3-HCl dissolution procedure, and Sr and Nd were separated from other elements by column chemistry using cation exchange and Ln-specific resins, respectively (Awasthi et al. 2014). Sr–Nd isotopic ratio measurements were carried out on an Isoprobe-T TIMS at Physical Research Laboratory, Ahmedabad (Ray et al. 2013). The ratios were corrected for fractionation using $^{86}Sr/^{88}Sr = 0.1194$ and $^{146}Nd/^{144}Nd = 0.7219$. The average values for international standards NBS 987 and JNdi-1 measured over a period of 4 years were $^{87}Sr/^{86}Sr = 0.710234 \pm 0.000008$ (2σ) and $^{143}Nd/^{144}Nd = 0.512104 \pm 0.000004$ (2σ), respectively (Ray et al. 2013). The U–Pb ages of zircons compiled from the available literature have been utilized to differentiate between various sources contributing sediments to the Andaman basin. The complete methodological details on U–Pb dating of zircons are given in the respective publications. Following the procedure given in Gehrels et al. (2011), $^{206}Pb/^{238}U$ ratios were used to determine best ages for samples younger than 1000 Ma and $^{207}Pb/^{206}Pb$ ratios were used for samples older than 1000 Ma. The U–Pb zircon ages were plotted on a probability density diagram using Isoplot 4.15 (Ludwig 2012).

4 Results and Discussion

4.1 Provenance of Andaman Flysch Sediments

The ε_{Nd} and $^{87}Sr/^{86}Sr$ data for samples from the Andaman Flysch Group are given in Table 1 and plotted in Fig. 2. Isotopic data from the Oligocene formations of the Barail Group, Central Myanmar Basin (CMB) and Subathu sub-basin (Dagshai and Kasauli formations, Himachal), and those from the modern Irrawaddy and Ganga-Brahmaputra sediments, Naga-Manipur-Andaman ophiolites, Myanmar volcanics, and Himalayan sources are also plotted in Fig. 2 for comparison. The $^{87}Sr/^{86}Sr$ and ε_{Nd} of the Andaman Flysch vary between 0.71548 and 0.73049, and 12.7 and −9.3, respectively. In the ε_{Nd} vs. $^{87}Sr/^{86}Sr$ plot (Fig. 2), the data for Andaman Flysch lie in-between the variations shown by the mantle derived rocks (ophiolites and volcanics) and Himalayan sources. As compared to the Oligocene sediments of the CMB, the Andaman Flysch sediments contain a significant continental (crustal) component. However, their $^{87}Sr/^{86}Sr$ is lower and ε_{Nd} is higher than those of the Himalayan foreland sediments (e.g., Dagshai and Kasauli formations, Himachal). The sediments in the Barail Group have overlapping isotopic compositions, with $^{87}Sr/^{86}Sr$ and ε_{Nd} varying in the ranges of 0.71772 to 0.75636, and −20.1 to −11.5, respectively (Table 1; Fig. 2). This overlap points to the possibility that the sediments

Table 1 Geochemical and isotopic compositions of Andaman flysch and Barail groups (Assam) siliciclastic sediments

Sample	Latitude/longitude	$^{87}Sr/^{86}Sr$	$^{143}Nd/^{144}Nd$	$\varepsilon_{Nd}(0)$	Sm/Nd	T_{DM}	$f_{Sm/Nd}$
Andaman flysch group							
AND-09-12	N 12° 31.692′, E 92° 49.890′	0.73049	0.511987	−12.7	0.193	1.62	−0.41
Collinpur	–	0.72395	0.512050	−11.5	0.168	1.35	−0.48
AND-11-22	N 11° 41.327′, E 92° 36.468′	0.72823	0.512004	−12.4	0.178	1.47	−0.45
AND-11-23	N 11° 41.394′, E 92° 36.309′	0.72902	0.512033	−11.8	0.192	1.56	−0.41
AND-09-43	N 11° 36.674′, E 92° 42.544′	0.71605	0.512059	−11.3	0.191	1.51	−0.41
AND-09-42	N 11° 36.674′, E 92° 42.544′	0.72136	0.512085	−10.8	0.207	1.62	−0.36
AND-11-6a	N 11° 29.190′, E 92° 40.144′	0.71695	0.512066	−11.2	0.197	1.55	−0.39
AND-09-61	N 11° 37.246′, E 92° 43.567′	0.72516	0.512091	−10.7	0.220	1.75	−0.32
AND-09-61A	N 11° 37.246′, E 92° 43.567′	0.72323	0.512089	−10.7	0.220	1.75	−0.32
PB-08-09	N 11° 35.423′, E 92° 37.139′	0.72105	0.512007	−12.3	0.180	1.49	−0.45
AND-09-04	N 11° 38.058′, E 92° 43.180′	0.71548	0.512142	−9.7	0.205	1.51	−0.37
PB-08-07	N 11° 39.700′, E 92° 45.336′	0.72031	0.512163	−9.3	0.197	1.40	−0.40
AND-09-36	N 11° 37.985′, E 92° 42.180′	0.71837	0.512157	−9.4	0.204	1.47	−0.37
Barail group (Assam)							
NEI-08	N 25° 01′ 45.64″, E 92° 25′ 37.59″	0.74550	0.511945	−13.5	0.195	1.71	−0.40
NEI-10	N 25° 01′ 8.52″, E 92° 25′ 38.03″	0.71926	0.511951	−13.4	0.205	1.80	−0.37
NEI-11	N 25° 1′ 38.64″, E 93° 10′ 49.82″	0.71772	0.511963	−13.2	0.191	1.64	−0.41
NEI-23	N 25° 06′ 25.67″, E 92° 58′ 27.62″	0.71941	0.512005	−12.4	0.212	1.80	−0.35
NEI-24	N 25° 07′ 17.01″, E 93° 1′ 21.77″	0.72247	0.511974	−13.0	0.208	1.80	−0.36
NEI-25	N 25° 08′ 33.13″, E 93° 01′ 20.19″	0.72642	0.512048	−11.5	0.204	1.64	−0.37
NEI-29	N 25° 02′ 38.39″, E 92° 47′ 09.05″	0.75636	0.511610	−20.1	0.215	2.48	−0.34

Note Sr, Nd isotopic data are given as ratios and T_{DM} (Nd-model age) in Ga. $\varepsilon_{Nd}(0) = \{[(^{143}Nd/^{144}Nd)_S - (^{143}Nd/^{144}Nd)_{CHUR}]/(^{143}Nd/^{144}Nd)_{CHUR}\} \times 10^4$ where subscript 's' stands for sample and 'CHUR' stands for Chondrite Uniform Reservoir. The present day $^{143}Nd/^{144}Nd$ value of CHUR is 0.512638. $T_{DM} = 1/\lambda\ \{1 + [(^{143}Nd/^{144}Nd)_S - (^{143}Nd/^{144}Nd)_{DM}]/[(^{147}Sm/^{144}Nd)_S - (^{147}Sm/^{144}Nd)_{DM}]\}$ where subscripts 'DM' stands for depleted mantle, $(^{143}Nd/^{144}Nd)_{DM} = 0.513114$ and $(^{147}Sm/^{144}Nd)_{DM} = 0.222$ (Michard et al. 1985), $f_{Sm/Nd} = [(^{147}Sm/^{144}Nd/0.1967) - 1]$

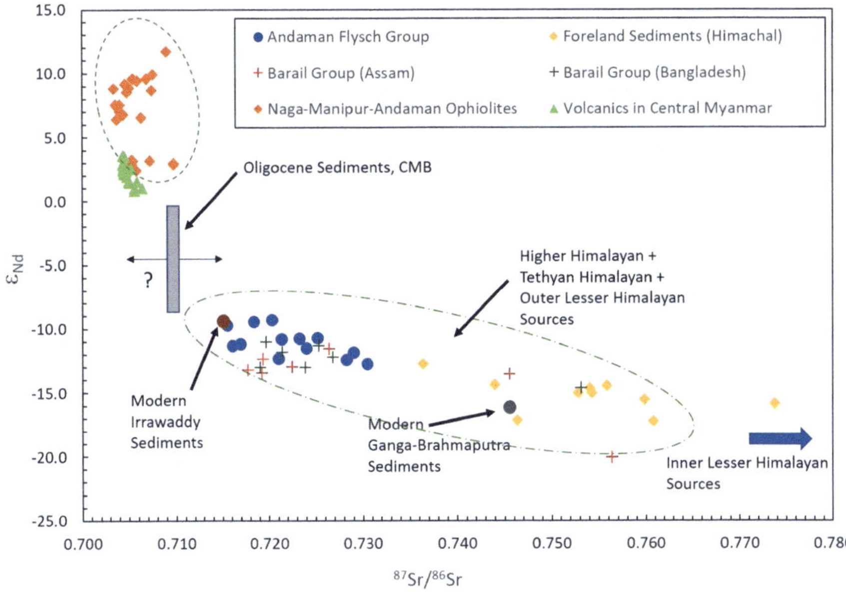

Fig. 2 ε_{Nd} versus $^{87}Sr/^{86}Sr$ plot for sediment samples from the Andaman Flysch Group compared with data of contemporaneous deposits in the Bengal Basin (Barail Group), Subathu Sub-basin (Dagshai and Kasauli Formations) and Central Myanmar Basin (Shwezetaw, Padaung, Okhmintaung Formations). Also shown are average compositions of modern Irrawaddy and Ganga-Brahmaputra sediments and those for Naga-Manipur-Andaman ophiolite belt and the volcanic arc. Data Sources: Andaman Flysch-this work and Awasthi and Ray (2019); Barail Group—this work, Najman et al. (2008); Bracciali et al. (2015); Myanmar Volcanics-Lee et al. (2016); Ophiolites-Kumar (2011), Kingson et al. (2019); Subathu Sub-basin-Najman et al. (2000); Central Myanmar Basin-Licht et al. (2016); Average Irrawaddy (n = 3)—Allen et al. (2008), Colin et al. (1999), Damodararao et al. (2016); Average Ganga-Brahmaputra (n = 2)—Galy and France-Lanord (2001)

in both the groups (i.e., Barail and Andaman Flysch) share a common parentage. The modern Irrawaddy sediments contain more radiogenic in Nd and less radiogenic in Sr compared to sediments in the above groups and appear to have a significant component from the ophiolitic-volcanic sources of Myanmar (Fig. 2). In contrast, the average ε_{Nd} and $^{87}Sr/^{86}Sr$ for the Ganga-Brahmaputra sediments, as expected, plot within the field of the Himalayan rocks (Fig. 2).

From the ε_{Nd} vs. $^{87}Sr/^{86}Sr$ plot (Fig. 2), the Andaman Flysch sediments appear to be a mixture of sediments derived from juvenile (mantle) sources and continental (crustal) sources, with the latter being the major component. The continental component here is represented by the Himalayan rocks, as inferred by Awasthi and Ray (2019). The juvenile sources most likely were proximal to the Andaman Basin and were located in the forearc (ophiolite belts) and/or in the volcanic arcs of the India-Burma convergent margin. There may have been some contribution from similar sources located in the distal Trans Himalayas, as observed in Palaeogene sediments in the IBR and Bengal Basin (e.g., Allen et al. 2008; Najman et al. 2008; Vadlamani et al. 2015). However, it is not possible to differentiate these two juvenile sources; i.e., Indo-Myanmar versus Trans Himalayan, in the Andaman Flysch sediments, because they have overlapping Sr–Nd isotopic compositions (Fig. 2). Locating the sediment sources having the continental crustal signatures in the Himalayas is difficult, because these could be exposed in any of the three main Himalayan lithounits; the Tethys Himalaya (TH), the Higher Himalaya (HH), and the Lesser Himalaya (LH) and its subunits (inner and outer). In addition, the isotopic compositions of all these sources, except the rocks of the inner LH, overlap (Fig. 2), thus identification and quantification of contributions from these sources to both the Bengal and Andaman basins become difficult. Interestingly, however, a few samples from the Barail Group (and Dagshai and Kasauli formations of Himachal) possess Sr–Nd isotopic compositions which could be attributed to contributions from the inner LH rocks (Fig. 2). This observation is consistent with the findings of Najman et al. (2000, 2008). The Andaman Flysch sediments; though have overlapping composition with the rocks of the TH, HH and outer LH rocks, were likely sourced from the first two domains since the rocks of the outer LH had not been exhumed and were not available for weathering/erosion (age of exhumation is <16 Ma; Myrow et al. 2015) when the Andaman Flysch was getting deposited.

A probability density plot, Fig. 3, for published U–Pb ages of detrital zircons throw more light on the differences between the probable sediment sources. The plot compares zircon ages from the Andaman Flysch Group, Barail Group, Padaung Formation of CMB, Palaeogene IBR, modern Irrawaddy, Myanmar volcanics, and Himalayan sources. The U–Pb ages of the detrital zircons in the Andaman Flysch Group vary from Proterozoic to Neogene, with prominent peaks in the distribution at 2500, 1600 and 1000 Ma (Fig. 3a). Besides these, the data also show appreciable number of zircons with Cambro-Ordovician (540–440 Ma), Carboniferous (350–300 Ma) and Cretaceous-Palaeocene (150–50 Ma) ages. This age distribution is very similar to that of the Barail Group (Fig. 3b). Both the groups show maximum abundance of zircons with ages at about 1000 and 500 Ma (Fig. 3b). The HH rocks (Fig. 3c) have zircons in age groups of 540–750 Ma (orthogneisses), 800–1200 Ma, 1600–1900, and 2400–2700 Ma; however, they show maximum abundance of zircons with ages between 800 and 1200 Ma. The detrital zircons in the inner LH rocks (Fig. 3e) are older than 1400 Ma, with a major mode in-between 1700 and 2000 Ma and minor mode in-between 2400 and 2800 Ma, whereas zircons ages in the outer LH rocks (Fig. 3f) show multiple peaks in the age ranges of 480–570, 750–1200, 1700–2000, and 2430–2560 Ma. The TH rocks (Fig. 3d), which are stratigraphically correlatable with the outer LH rocks, also show detrital zircon age peaks at similar intervals, however, the modes of 480–570 and 750–1200 Ma are prominent, which is likely a result of their derivation from the HH (Gehrels et al. 2011). Compared to the Himalayan sources, the Trans-Himalayan and Myanmar volcanic arc sources generally possess zircons with ages <200 Ma (Gehrels et al. 2011).

There exist significant differences between the detrital zircon age distributions for the Andaman Flysch and the Palaeogene rocks of Myanmar (e.g., Padaung Formation of CMB) and IBR (Fig. 3g, h). The zircons in the CMB rocks are predominantly of late Cretaceous-Palaeogene antiquity (Fig. 3h) and the same is observed in central

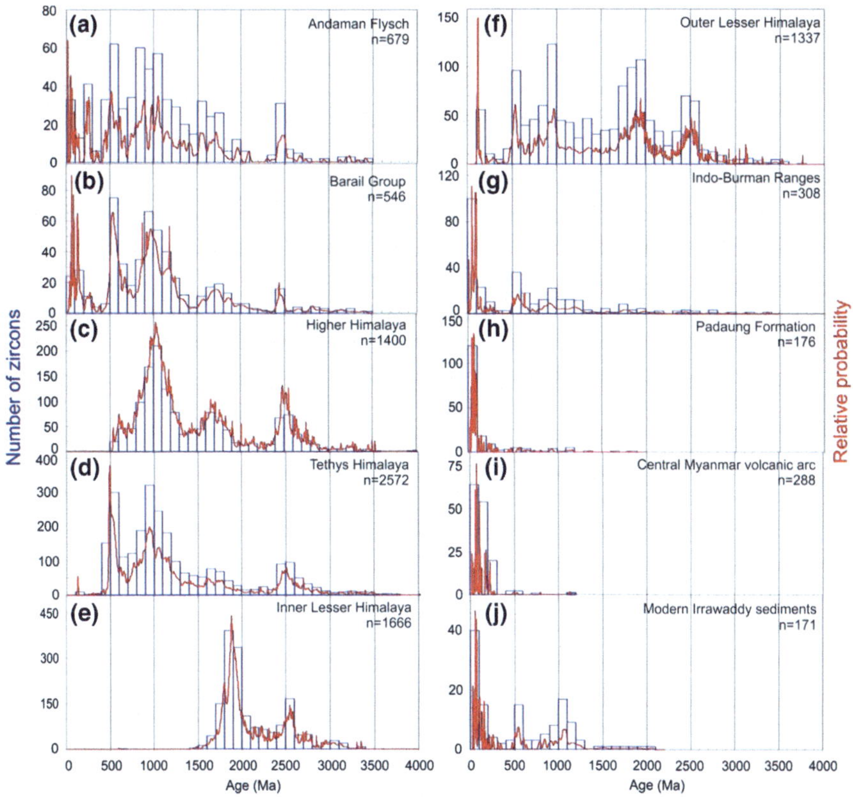

Fig. 3 Histogram/Probability Density plot of detrital zircon U–Pb ages for the Andaman Flysch Group (Allen et al. 2007; Limonta et al. 2017), the Barail Group (Najman et al. 2008; Bracciali et al. 2015; Vadlamani et al. 2015), the Padaung Formation, CMB (Robinson et al. 2014), the Palaeogene rocks of IBR (Allen et al. 2008), sediments of the Irrawaddy (Bodet and Scharer 2000; Limonta et al. 2017), the Himalayan lithounits (Gehrels et al. 2011), and Myanmar volcanic arc (Lee et al. 2016)

Myanmar volcanic arc (Fig. 3i), which suggest that the sediments in the CMB is likely sourced from the latter. Similar zircon populations are also observed in the Palaeogene rocks of the IBR; in addition, these rocks contain Proterozoic zircons as well (Fig. 3g). Based on zircon age data and other proxies it has been suggested that sediments in these rocks (of IBR) have provenances in both the Myanmar volcanic arc and the Himalayas (Allen et al. 2008). The detrital zircon age data for the modern Irrawaddy river sediments (Fig. 3j) show prominent peaks for Cretaceous-Palaeogene and late Mesoproterozoic-Neoproterozoic time periods suggesting that they were primarily derived from magmatic arc sources in central Myanmar (directly or through recycling) and from Proterozoic sources. From the detrital zircon age spectra, it is apparent that arc sources in Myanmar was not a significant sediment contributor to the Andaman trench-forearc basin through the modern or palaeo-Irrawaddy river system during the Palaeogene, since the sediments in the latter primarily contain Cretaceous-Palaeogene zircons, whereas there is a dominance of Proterozoic zircons from crustal sources in the Andaman Flysch sediments. The zircon age distribution for the Andaman Flysch matches closely with those for the HH and TH (Fig. 3) suggesting possible sediment sourcing from these units into the Andaman Basin in Palaeogene. This could mean that the less radiogenic Sr (more radiogenic Nd) observed in the Andaman Flysch (Fig. 2) might not have been generated by sediment contributions from juvenile sources. The major peaks of zircon age spectrum of the Barail Group are very similar to that of TH sources (Fig. 3). This suggests that the rapidly exhuming TH acted as a prominent source of sediments in the Bengal Basin during Palaeogene. Since the TH rocks themselves have some sediment component derived from the HH, separation of both the signals is difficult.

The similarity of the Sr–Nd isotopic compositions and detrital zircon age distributions of the Barail and Andaman Flysch Groups clearly advocates for identical sediment sources for both. Additional confirmation comes from the cross plot $f^{Sm/Nd}$ versus ε_{Nd} (Fig. 4), where sediments in both the groups not only overlap with each other but also suggest their derivation from Precambrian (continental) crustal sources older than 1.6 Ga. The presence of significant number of zircons older than 1.6 Ga in both the TH and HH units make them most likely sources.

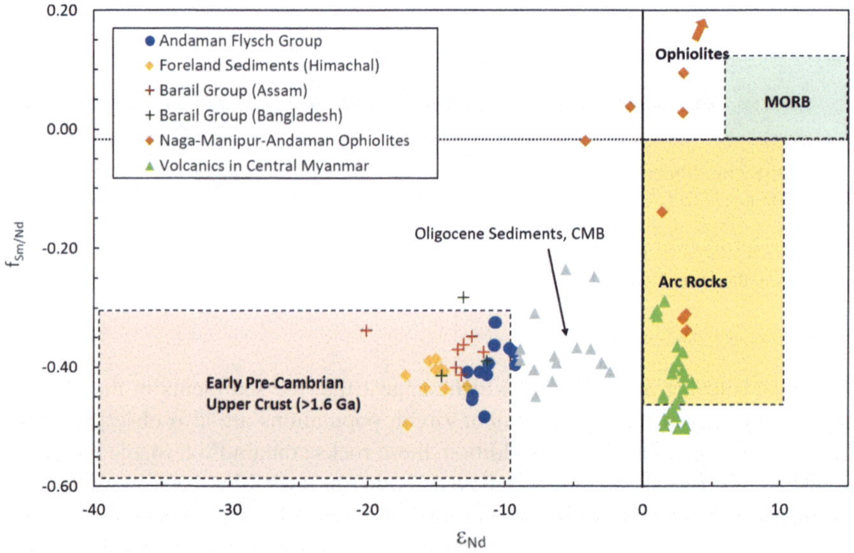

Fig. 4 $f^{Sm/Nd}$ versus ε_{Nd} plot for the Andaman Flysch Group, the Barail Group (Assam and Bangladesh), the Dagshai and Kasauli Formations of Himachal, the Oligocene rocks of the CMB (Shwezetaw, Padaung and Okhmintaung formations), the Naga-Manipur-Andaman Ophiolites, and volcanics in Central Myanmar. Various fields are after McLennan et al. (1993). Data sources are same as in Fig. 2

4.2 Sediment Transport to the Andaman Basin

Having determined that the Andaman Trench-Forearc basin received sediments from the Tethys and Higher Himalayan sources it becomes necessary that we discover the transport/depositional pathway(s) to substantiate the provenance analysis. A palaeo-drainage evolution model for the source regions can possibly help in this regard, however, it would be highly speculative considering that we have access to only limited petrological, geochemical and palaeogeographical information. Despite this, there have been several proposals which reported major changes in the drainage network (sediment-routing) south-east of Tibet in response to the Himalayan Orogeny (e.g., Clift et al. 2006; Liang et al. 2008; Cina et al. 2009; Robinson et al. 2014; Zhang et al. 2019). Although these proposals differ in the proposed timings of the changes to the pre-Himalayan fluvial systems, many suggest that precursors to the modern Irrawaddy/Brahmaputra existed even before Miocene and were connected to Yarlung-Tsangpo of southern Tibet through the eastern Himalayan syntaxis (e.g., Robinson et al. 2014; Lang and Huntington 2014, Limonta et al. 2017). The Himalayan Orogen attained its topography after multiple episodes of exhumations during the Cenozoic (An et al. 2001). We propose that the increase in continental sediment input to the Andaman Basin as well as to other peri-Himalayan basins was a result of progressive uplift and rapid erosion of the north-eastern Himalaya and Tibet. The Andaman Flysch group was likely deposited following one such exhumation event during the Late Eocene-Oligocene. The Yarlung-Siang-Brahmaputra river system draining across the eastern Himalayan syntaxis probably delivered sediments to the Andaman Trench-Forearc Basin via the Bengal Basin and Bengal Fan. Our model is consistent with the post-Gondwana palaeogeographic models, proposed by Hall (2012), Boonchaisuk et al. (2013) and Replumaz et al. (2014), which describe the position of Indian sub-continent during the Oligocene with respect to other tectonic blocks of the Eurasian Plate, and the palaeocurrent directions given for the Andaman sediments (Chakraborty and Pal 2001).

5 Conclusions

We made an attempt to determine the sources and transport pathways for sediments in the Andaman Flysch Group, Andaman Islands and Barail Group, Assam, which were deposited primarily during the Oligocene. Provenance analysis using newly generated Sr–Nd isotope ratios and existing detrital zircon U–Pb age data reveals that these sediments have overlapping isotopic compositions and similar zircon age distributions, which point to derivation from the same or similar sources. It further reveals that rocks of the Tethys Himalaya and Higher Himalaya could have been the primary sediment sources to the Andaman Trench-Forearc Basin and the Bengal Basin during the Oligocene. Minor contributions from juvenile source(s) have also been recognized. These sources could have been either the Myanmar volcanic arc or the Naga-Manipur-Andaman ophiolite belt or the Trans-Himalayan volcanics, or more than one of these. The location of the primary sources in relations to the depositional basins advocates for sediment transport through palaeo-channels of the Yarlung-Siang-Brahmaputra river system via Bengal Basin and Bengal Fan. Significant Himalayan contribution points to a rapid exhumation and high erosion in the north-eastern Himalaya and Tibet during the Late Eocene to Oligocene.

Acknowledgements NA acknowledges Science and Engineering Research Board, Government of India, for research funding. Alok Kumar, Gaurav Sharma and Basudev Dutta are thanked for their help during field trips. We thank Debajyoti Paul for his comments and suggestions.

References

Allen R, Carter A, Najman Y, Bandopadhyay PC, Chapman HJ, Bickle MJ, Garzanti E, Vezzoli G, Ando' S, Foster GL (2007) New constraints on the sedimentation and uplift history of the Andaman-Nicobar accretionary prism, South Andaman Island. Geol Soc Am Spec Pap 436:223–255

Allen R, Najman Y, Carter A, Barfod D, Bickle MJ, Chapman HJ, Garzanti E, Vezzoli G, Ando S, Parrish RR (2008) Provenance of the Tertiary sedimentary rocks of the Indo-Burman Ranges, Burma (Myanmar): Burman arc or Himalayan-derived? J Geol Soc 165:1045–1057

An Z, Kutzbach JE, Prell WL, Porter SC (2001) Evolution of Asian monsoons and phased uplift of the Himalaya-Tibetan plateau since late Miocene times. Nature 411:62–66

Awasthi N, Ray JS, Singh AK, Band ST, Rai VK (2014) Provenance of the late quaternary sediments in the Andaman Sea: implications for monsoon variability and ocean circulation. Geochem Geophys Geosyst 15:3890–3906

Awasthi N, Ray JS (2019) The Palaeogene record of Himalayan erosion in the Andaman Basin. J Earth Syst Sci (https://doi.org/10.1007/s12040-019-1266-7)

Ball V (1870) Notes on the geology of the vicinity of Port Blair, Andaman Islands', foot-note on p. 231. J Asiatic Soc Bengal 39

Bandopadhyay PC (2005) Discovery of abundant pyroclasts in the Namunagarh Grit, South Andaman: evidence for arc volcanism and active subduction during the Palaeogene in the Andaman area. J Asian Earth Sci 25:95–107

Bandopadhyay PC (2012) Re-interpretation of the age and environment of deposition of Paleogene turbidites in the Andaman and Nicobar Islands, Western Sunda Arc. J Asian Earth Sci 45:126–137

Bandopadhyay PC, Ghosh B (2015) Provenance analysis of the Oligocene turbidites (Andaman Flysch), South Andaman Island: a geochemical approach. J Earth Syst Sci 124:1019–1037

Bodet F, Scharer U (2000) Evolution of the SE-Asian continent from U–Pb and Hf isotopes in single grains of zircon and baddeleyite from large rivers. Geochim Cosmochim Acta 64:2067–2091

Boonchaisuk S, Siripunvaraporn W, Ogawa Y (2013) Evidence for middle Triassic to Miocene dual subduction zones beneath the Shan-Thai terrane, western Thailand from magnetotelluric data. Gondwana Res 23:1607–1616

Bracciali L, Najman Y, Parrish RR, Akhter SH, Millar I (2015) The Brahmaputra tale of tectonics and erosion: early Miocene river capture in the Eastern Himalaya. Earth Planet Sci Lett 415:25–37

Clift PD, Blusztajn J, Nguyen AD (2006) Large-scale drainage capture and surface uplift in eastern Tibet-SW China before 24 Ma inferred from sediments of the Hanoi Basin, Vietnam. Geophys Res Lett 33:L19403

Chakraborty PP, Pal T (2001) Anatomy of a forearc submarine fan: upper Eocene-Oligocene Andaman Flysch Group, Andaman Islands, India. Gondwana Res 4:477–486

Cina SE, Yin A, Grove M, Dubey CS, Shukla DP, Lovera OM, Kelty TK, Gehrels GE, Foster DA (2009) Gangdese arc detritus within the eastern Himalayan Neogene foreland basin: implications for the Neogene evolution of the Yalu-Brahmaputra River system. Earth Planet Sci Lett 285:150–162
Colin C, Turpin L, Bertaux J, Desprairies A, Kissel C (1999) Erosional history of the Himalayan and Burman ranges during the last two glacial-interglacial cycles. Earth Planet Sci Lett 171:647–660
Curray JR (2005) Tectonics and history of the Andaman Sea region. J Asian Earth Sci 25:187–232
Damodararao K, Singh SK, Rai VK, Ramaswamy V, Rao PS (2016) Lithology, monsoon and sea-surface current control on provenance, dispersal and deposition of sediments over the Andaman continental shelf. Frontiers Mar Sci 3:118
Galy A, France-Lanord C (2001) Higher erosion rates in the Himalaya: geochemical constraints on riverine fluxes. Geology 29:23–26
Garzanti E, Limonta M, Resentini A, Bandopadhyay PC, Najman Y, Andò S, Vezzoli G (2013) Sediment recycling at convergent plate margins (Indo-Burman ranges and Andaman–Nicobar Ridge). Earth-Sci Rev 123:113–132
Gehrels G, Kapp P, DeCelles P, Pullen A, Blakey R, Weislogel A, Ding L, Guynn J, Martin A, McQuarrie N (2011) Detrital zircon geochronology of pre-tertiary strata in the Tibetan-Himalayan orogen. Tectonics 30: TC5016
Guha DK, Mohan M (1965) A note on upper cretaceous microfauna from the middle Andaman Island. Bull Geol Min Metall Soc India 33:73
Hall R (2012) Late Jurassic-Cenozoic reconstructions of the Indonesian region and the Indian Ocean. Tectonophysics 570:1–41
Hochstetter FV (1869) In contribution to the study of physical geology of the Nicobar Islands in voyage of the Austrian frigate novara around the world in 1857–1859 (Geology, Part 2, vol 2, pp 83–112, Vienna) (trans: Stoliczka F). Records of Geological Survey of India, vol 203, pp 59–73
Karunakaran C, Ray KK, Saha SS (1968) Tertiary sedimentation in the Andaman-Nicobar geosyncline. J Geol Soc India 9:32–39
Kingson O, Bhutani R, Balakrishnan S, Dash JK, Shukla AD (2019) Subduction-related Manipur ophiolite complex, Indo-Myanmar ranges: elemental and isotopic record of mantle metasomatism. Geol Soc Lond Spec Publ 481:195–210
Kumar A (2011) Geochemical and isotopic studies of rocks from the Barren Island Volcano and Andaman subduction zone, India. Ph.D. thesis, The M.S. University of Baroda, Vadodara, India
Lang KA, Huntington KW (2014) Antecedence of the Yarlung-Siang-Brahmaputra River, eastern Himalaya. Earth Planet Sci Lett 397:145–158
Lee H-Y, Chung S-L, Yang H-M (2016) Late Cenozoic volcanism in central Myanmar: geochemical characteristics and geodynamic significance. Lithos 245:174–190
Liang Y-H, Chung S-L, Liu D, Xu Y, Wu F-Y, Yang J-H, Wang Y, Lo C-H (2008) Detrital zircon evidence from Burma for reorganization of the eastern Himalayan river system. Am J Sci 308:618–638
Licht A, Reisberg L, France-Lanord C, Naing Soe A, Jaeger J-J (2016) Cenozoic evolution of the central Myanmar drainage system: insights from sediment provenance in the Minbu Sub-Basin. Basin Res 28:237–251
Limonta M, Resentini A, Carter A, Bandopadhyay PC, Garzanti E (2017) Provenance of Oligocene Andaman sandstones (Andaman-Nicobar Islands): Ganga-Brahmaputra or Irrawaddy derived? Geol Soc London Mem 47:141–152
Ludwig KR (2012) Isoplot v3.75. A Geochronological Toolkit for Excel. Berkeley Geochronology Center, Special Publication Number 5, Berkeley Geochronological Centre, Berkeley, California, pp 75
McCaffrey R (2009) The tectonic framework of the Sumatran subduction zone. Annu Rev Earth Planet Sci 37:345–366
McLennan SM, Hemming S, McDaniel DK, Hanson GN (1993) Geochemical approaches to sedimentation, provenance, and tectonics. Geol Soc Am Spec Pap 284:21–40

Michard A, Gurriet P, Soudant M, Albarede F (1985) Nd isotopes in French Phanerozoic shales: external vs internal aspects of crustal evolution. Geochimica et Cosmochimica Acta 49:601–610

Myrow PM, Hughes NC, Derry LA, McKenzie NR, Jiang G, Webb AAG, Banerjee DM, Paulsen TS, Singh BP (2015) Neogene marine isotopic evolution and the erosion of lesser Himalayan strata: implications for cenozoic tectonic history. Earth Planet Sci Lett 417:142–150

Najman Y, Bickle M, Chapman H (2000) Early Himalayan exhumation: isotopic constraints from the Indian foreland basin. Terra Nova 12:28–34

Najman Y, Bickle M, BouDagher-Fadel M, Carter A, Garzanti E, Paul M, Wijbrans J, Willett E, Oliver G, Parrish R (2008) The Paleogene record of Himalayan erosion: Bengal Basin, Bangladesh. Earth Planet Sci Lett 273:1–14

Oldham RD (1885) Notes on the geology of the Andaman Islands. Rec Geol Surv India 18:135–145

Pandey J, Agarwal RP, Dave A, Maithani A, Trivedi KB, Srivastava AK, Singh DN (1992) Geology of Andaman. Bull ONGC 29:19–103

Pal T, Chakraborty PP, Gupta TD, Singh CD (2003) Geodynamic evolution of the outer-arc-forearc belt in the Andaman Islands, the central part of the Burma-Java subduction complex. Geol Mag 140:289–307

Pal T, Gupta TD, Chakraborty PP, Gupta SCD (2005) Pyroclastic deposits of Mio-Pliocene age in the Arakan Yoma-Andaman-Java subduction complex, Andaman Islands, Bay of Bengal, India. Geochem J 39:69–82

Pedersen RB, Searle MP, Carter A, Bandopadhyay PC (2010) U–Pb zircon age of the Andaman ophiolite: implications for the beginning of subduction beneath the Andaman-Sumatra arc. J Geol Soc 167:1105–1112

Ray JS, Pande K, Bhutani R, Shukla AD, Rai VK, Kumar A, Awasthi N, Smitha RS, Panda DK (2013) Age and geochemistry of the Newania dolomite carbonatites, India: implications for the source of primary carbonatite magma. Contrib Miner Petrol 166:1613–1632

Replumaz A, Capitanio FA, Guillot S, Negredo AM, Villasenor A (2014) The coupling of Indian subduction and Asian continental tectonics. Gondwana Res 26:608–626

Rink PH (1847) Die Nikobaren Insel, Kobenhagen, translated selections. Rec Gov India 77(1870):105–153

Robinson R, Brezina CA, Parrish RR, Horstwood MSA, Oo NW, Bird MI, Thein M, Walters AS, Oliver GJH, Zaw K (2014) Large rivers and orogens: the evolution of the Yarlung Tsangpo-Irrawaddy system and the eastern Himalayan syntaxis. Gondwana Res 26:112–121

Vadlamani R, Wu F-Y, Ji W-Q (2015) Detrital zircon U–Pb age and Hf isotopic composition from foreland sediments of the Assam Basin, NE India: constraints on sediment provenance and tectonics of the Eastern Himalaya. J Asian Earth Sci 111:254–267

Zhang P, Najman Y, Mei L, Millar I, Sobel E, Carter A, Barfod D, Dhuime B, Garzanti E, Govin G (2019) Palaeodrainage evolution of the large rivers of East Asia, and Himalayan-Tibet tectonics. Earth Sci Rev 192:601–632

Seismicity, Lithospheric Structure and Mantle Deformation in the Andaman Nicobar Subduction Zone

G. Srijayanthi and M. Ravi Kumar

Abstract In this paper, we synthesize the seismicity, seismic structure and deformation of the hitherto less investigated Andaman–Nicobar subduction zone (ANSZ) utilizing results obtained from new data accrued from a broadband seismological network. Parameterization of the subsurface through joint inversion for source mechanisms and velocity structure indicates a Moho at 30 km depth. Hypocenters of local earthquakes located using this model reveal a N-S trend to the east of the trench. The northern part of the archipelago is seismically very active compared to the middle and southern parts. Interestingly, high attenuation is also observed in the northern Andaman. Further, the attenuation characteristics of ANSZ, akin to those in the eastern Himalaya and southern Tibet, reveal an active role of intrinsic attenuation attributed to the presence of fluids associated with subduction. Collation of results from various seismological studies indicates that the geometry of the subducting Indian plate controls the rupture characteristics of major earthquakes and deformation pattern along the arc. Seismic images constructed using Receiver functions reveal variations in the crustal and lithospheric thickness along the ANSZ and tearing in the subducting Indian plate between 7° N and 8° N latitude. Clustering of earthquake swarms, their magmatic affinity and shear wave anisotropy in the mantle wedge seem to be governed by this tear. Also, the subslab anisotropy is controlled by the dip of the subducting plate, as evidenced by trench-parallel and perpendicular fast axis azimuths in steeper (>40°) and shallower segments, respectively. This observation is consistent along the whole Andaman–Sumatra–Java arc.

Keywords Andaman subduction zone · Seismicity and attenuation · Receiver functions · Slab tear · Shear wave splitting · Trench parallel anisotropy

---

Present Address:
G. Srijayanthi · M. Ravi Kumar
Institute of Seismological Research, Raisan, Gandhinagar, Gujarat, India

M. Ravi Kumar (✉)
CSIR-National Geophysical Research Institute, Hyderabad, India
e-mail: mravingri@gmail.com

J. S. Ray and M. Radhakrishna (eds.), *The Andaman Islands and Adjoining Offshore: Geology, Tectonics and Palaeoclimate*, Society of Earth Scientists Series,
https://doi.org/10.1007/978-3-030-39843-9_6

1 Introduction

The Andaman Nicobar subduction zone (ANSZ) occupies the central part of the 5000 km long eastern convergence margin of the Indo-Australian plate, starting from Burma in the north to Andaman in the south, extending further south-east into Sumatra–Java. The unique characteristics associated with this subduction zone are, the variation in the nature of convergence from a continental type near Burma to an oceanic type near Andaman, oblique subduction owing to varying convergence rates; ~5.7 cm/year near Sumatra to ~3.7 cm/year in the Andaman, distinct variation in the age of the subducting lithosphere; 110 Ma near Java to 55 Ma offshore Sumatra and again 95 Ma in the northern Andaman, arc parallel volcanism, an ill-defined spreading ridge between 10 and 12° N in the Andaman Sea and the Andaman Sea rift basin lying to the east of the Andaman Islands being a pull-apart basin rather than a typical backarc basin (Fig. 1) (Curray 1989). The rate of spreading at the ridge is similar to that at the active ones, evidenced by a 118 km wide opening in the central Andaman Basin in the past ~4 Ma (Kamesh Raju et al. 2004). The region is host to many prominent transform faults like the north-south trending West Andaman Fault (WAF), the Sumatra fault system and also the Sagaing fault situated further east, which demarcates the eastern boundary of the Burma microplate. The Sagaing fault is an extensive fault parallel to the convergence boundary that divides the central low lands from the eastern high lands of Burma and extends to the Andaman Sea rift zone which is characterized by seafloor spreading (Curray et al. 1979). The inner volcanic arc is marked by discontinuous submarine ridges of volcanic seamounts (Sewell) and two volcanoes namely the Barren and Narcondam that are ~150 km east of the main Islands within the Andaman Sea (Hamilton 1979). The two volcanoes are distinctly unalike; Barren being active and Narcondam being extinct, with no recent scripted activity (Pal et al. 2007). After lying dormant for ~200 years, the Barren volcano erupted in March 1991 and subsequent eruptions were observed in 1994–1995, 2005–2006, 2008–2009 (Sheth et al. 2010) and 2017. The volcanic arc extends further northwards to Burma (central molasses basin) and southwards to the Barisan range in Sumatra (Curray et al. 1982). The region between the Andaman Islands and the volcanic arc forms the foredeep sedimentary trough (Dickinson and Seely 1979) and the central Andaman basin constitutes the back-arc region. Therefore, an EW geo-tectonic profile along the ANSZ typically replicates a matured subduction zone identified with a concave shaped trench, followed by a mineral-rich sedimentary accretionary prism, an uplifted structural high that constitutes the islands (roughly 300 m in the Andaman and Nicobar), the forearc sub-basin, the vibrant volcanic arc and the back-arc sub-basin that hosts the neo-extensional margin in the Andaman Sea and finally the Mergui Terrance.

The P wave velocity models along the Burma–Andaman–Sumatra arc obtained from joint inversion of source and structure using non-linear methods provided a broad picture of the subducting Indo-Australian slab (Pesicek et al. 2010). It is postulated that the 1300 km long rupture of the 2004 megathrust earthquake terminated at a region between the Andaman Islands and Burma where a near-vertical

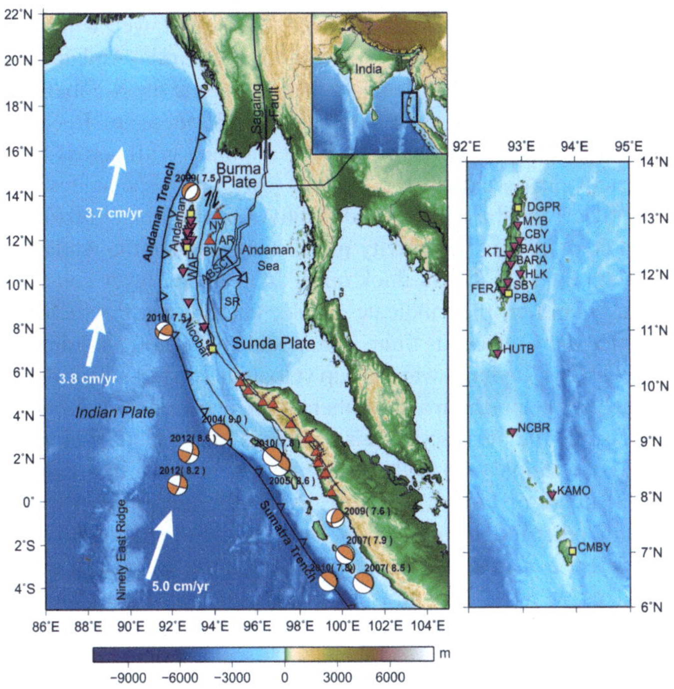

Fig. 1 Seismotectonic scenario of the Andaman Subduction zone along with focal mechanisms of major earthquakes (M > 7.5), their year of occurrence and magnitude. The pink inverted triangles indicate locations of 11 broadband seismological stations of the National Geophysical Research Institute and yellow squares correspond to 3 stations installed by the India Meteorological Department. The red triangles denote the volcanic arc. ABSC: Andaman Back-Arc Spreading Centre, AR: Alcock Rise, BV: Barren Volcano, NV: Narcondam Volcano, SR: Sewell Rise, WAF: West Andaman fault, white arrows: relative plate motion between India and Sunda (Subarya et al. 2006). The scale at the bottom indicates the topography and bathymetry in meters (GEBCO 2003). Inset: The Andaman subduction zone (study region), the central part of the eastern plate boundary of Indian plate, is demarcated with a black rectangle. Detailed map of the stations is indicated in the right panel

lithospheric tear was proposed by Richards et al. (2007). Images constructed from seismic tomography reveal evidence for absence of a subducted slab in this region, which progressively manifests with an east dipping trend further south towards the Andaman and Sunda arcs (Fig. 7; Pesicek et al. 2010). Further, a clear variation in the dip and depth of the slab is evident from the P wave velocity tomograms along the arc. A steeply dipping slab is traced to depths of 660 km in the upper mantle in the Andaman region. Dasgupta and Mukhopadhyay (1993) also proposed a 40°–55° east dipping Benioff zone beneath the Andaman arc by studying the trend of earthquake hypocenters down to 200 km depth. The gravity anomalies show a characteristic variation along and across the subduction zone (Mukhopadhyay 1988). The region of minimum gravity corresponds to the trench and the sedimentary islands (Andaman and Nicobar), whereas the segment coinciding with the maximum gravity field is associated with the volcanic arc located further east, well within the overriding Burma microplate. The gravity anomalies along the Burma–Andaman subduction zone reveal a high anomaly of ~100 mGal in the south near the Nicobar Islands and a very low anomaly of ~20 mGal in the north near the Burmese arc. Recent modelling studies using GEOSAT gravity anomaly data (2-minute grid) across the Andaman subduction system clearly indicated two broad lows, one associated with the trench (−60 to −100 mGal) and the other related to the Andaman arc (−150 to −240 mGal) (Radhakrishna et al. 2008). The gravity models suggest that the Andaman Nicobar ridge is underlain by a 40–47 km thick upper lithospheric layer consisting mainly of oceanic crust. Further, they indicate a thick oceanic crust, which is attributed to accretion since the Cretaceous time. The models derived from joint inversion of receiver functions and Rayleigh wave group velocities suggest a continent like crust, ~24–32 km thick, in the northern and southern Andaman regions (Gupta et al. 2016).

The ANSZ is one the most active seismic zones of the globe which is classified as zone V (i.e. the region with highest seismic potential) in the seismic zonation map of India (Bureau of Indian Standards 2002). Further, it is also important to understand many other interesting tectonic aspects that the region inherits, such as, the solitary existence of an active volcano in the Barren Island and the absence of major earthquakes (M > 8.0) in contrast to the contiguous Sumatra arc further south. Also, it is important to understand the role of subduction and transform faulting, due to the sliver plate movement, in partitioning the release of seismic energy along the Andaman arc. However, this archipelago was seismologically ill-instrumented. The solitary station at Port Blair installed by the Indian Meteorological Department (IMD) was the only source of seismological waveform data, which was later upgraded to a three-component digital broadband seismometer in 2004 (Mishra et al. 2007). Subsequently, two more broadband seismological stations were established by IMD in 2008 (Fig. 1). Although the Geological Survey of India (GSI) was the first organisation to instrument this region, the installations were limited to short-period sensors in the main Andaman Island, operated for a very short duration, specifically to monitor the aftershocks of the 13 September 2002 M_w 6.5 Andaman earthquake (Kayal et al. 2004) and 26 December 2004 M_w 9.1 Sumatra earthquake (Mishra et al. 2007). Therefore, detailed knowledge of the seismotectonics, seismic structure and deformation was scanty due to paucity of data from local seismological networks. In view of this, the CSIR-NGRI (National Geophysical Research Institute) installed a 11 station broadband seismological network to investigate the seismicity, lithospheric structure and deformation of the region, under a project supported by the Indian National Center for Ocean Information Services (INCOIS). In this work, we present an overview of the hitherto less studied subduction system, based on the analysis of seismological waveform data from 11 NGRI stations together with those from the 3 stations installed by the IMD, along the Andaman subduction arc, between 6 and 14° N, spanning ~800 km in the NS direction. In a first ever study using waveform data from the newly established network, Rao et al. (2011) determined the fault plane solutions of the 2009 Digilipur earthquake (M_w 7.5) and its 4 aftershocks. They performed a full waveform inversion to simultaneously determine the velocity structure of the Andaman region and the moment tensor solutions. The results reveal a high bulk crustal V_p/V_s of ~1.81, akin to that found in an oceanic crust, with the Moho at 30 km. The latter was also ascertained by the preliminary results from receiver functions analysis at 5 stations along the Andaman arc (Rao et al. 2011). This velocity model was adopted to locate ~1214 local earthquakes during the period April 2009 to February 2011, since this model resulted in the lowest root mean square residual error (Srijayanthi et al. 2012). These local earthquake waveforms were further utilized for the estimation of attenuation and seismic anisotropy character of this subduction system.

2 Seismic Attenuation and Crustal Heterogeneities

The study of seismic attenuation is based on the well-established fact that the amplitudes of seismic waves decrease with increasing epicentral distance. The rate at which the amplitude decreases and its dependence on the frequency of seismic waves is important because it not only yields important information about the subsurface structure but is also useful in evaluating the seismic hazard, especially in regions such as ANSZ, that are host to many earthquakes. Seismic attenuation (Q factor) has been widely investigated using different seismic waves like the P, S, Coda and Lg waves, denoted with Q subscripted with the wave type like Q_p, Q_s, Q_c and Q_{Lg}. The dimensionless quantity simply called Q or quality factor defines the attenuation character of a medium which can be defined as the ratio of the energy of a seismic wave to the energy dissipated during one cycle of a wave (per wave oscillation) (Knopoff and Hudson 1964; Aki and Chouet 1975). When a seismic wave propagates through the earth, some of its energy (amplitude) decays due to factors such as geometric spreading (ever increasing size of the spherical wavefront surface originating from a point source), scattering attenuation and intrinsic attenuation. In scattering attenuation, the energy of a wavefield is scattered incoherently whenever it encounters heterogeneities in the medium or a change in material properties, which leads to differential amplitude decay pattern. The scattering attenuation is defined mathematically as the ratio of the inhomogeneity of the medium to the wavelength. The seismic energy dissipated in intrinsic or inelastic attenuation is due to transformation of mechanical energy to heat energy at interfaces and grain boundaries, which decreases the amplitude and broadens the pulse of the seismic wave. The factor Q assumes a pivotal role in understanding the subsurface characteristics, localizing inhomogeneities in the medium, characterise the seismotectonic activity and also in exploration seismology (for oil and gas). It is also established that the value of Q is higher in stable continental regions compared to seismically active regions (Sato and Fehler 1998). Further, the inverse of the attenuation factor (Q_0 or Q^{-1}) at unit frequency (1 Hz) can be used to quantify the seismicity associated with the degree of fractures in a region. The coda Q or Q_c is the most widely studied factor globally (Ibanez et al. 1990; Ugalde et al. 2002; Mukhopadhyay and Tyagi 2007; Singh et al. 2012a, b; Dasović et al. 2013; Hazarika et al. 2013; Singh et al. 2015, 2017; Dobrynina et al. 2016; Irandoust et al. 2016; Nazemi et al. 2017; Biswas and Singh 2019) to evaluate seismic hazard with respect to the seismicity and tectonics of a region. Thus, the variation of factor Q in the Andaman region is also examined to locate the crustal deformation zones by estimating the attenuation characteristics (Singh et al. 2017, 2019), which is discussed in this section.

To determine the coda wave attenuation factor, its frequency dependency and spatial variation, Singh et al. (2017, 2019) utilised good quality waveform data from 399 local earthquakes within 100 km epicentral distance. In this study, they estimated the Q_c values using the single scattering model of Sato (1977). Subsequently, the Multiple Lapse Time Window Analysis (MLTWA) technique (Hoshiba et al. 1991; Fehler et al. 1992; Mayeda et al. 1992) was also adapted by them to isolate attenuation due to scattering (Q_{sc}) and intrinsic absorption (Q_i). Further, the spatial distribution of Q_c was correlated with the tectonics and seismicity of the Andaman Islands.

The Q_c values are found to increase largely from 101 to 951 with an increase in frequency from 1.5 to 18 Hz, clearly establishing its frequency dependency. This is an expression of the heterogeneities present in the medium (Aki 1980a; Mayeda et al. 1992) and indicates a tectonically active lithosphere. The frequency dependence of Q_c is generally expressed as a product of Q_0 (value of Q_c at 1 Hz) and a term f^n, where n is a frequency relation parameter. The observed low value of Q_0 in the Andaman region indicates high attenuation, attributed to the complex structure of the subduction zone. Further, the MLTWA analysis indicates an active role of intrinsic attenuation compared to scattering attenuation ($Q_i > Q_{sc}$) in the study region. Both the intrinsic absorption and scattering attenuation factors influence coda Q in any given scenario. However, the major factor contributing to coda wave attenuation is debated. Few studies reported that the scattering phenomenon significantly affects the amplitude of the coda wave (Aki 1980a, b) which is also evidenced by the observations from Garwhal–Kumaun Himalaya (Mukhopadhyay et al. 2010). On the contrary, Frankel and Wennerberg (1987) argued that though both the phenomena (scattering and intrinsic) control the attenuation of coda waves, with the latter being more dominant (Del-Pezzo et al. 1995; Bianco et al. 2002; Giampiccolo et al. 2004; Singh et al. 2016, 2017). The Q_{sc} and Q_i values obtained for the ANSZ are comparable with those in the eastern Himalaya and southern Tibet (Padhy and Subhadra 2013), rather than the values from other subduction zones (Mukhopadhyay et al. 2010; Ugalde et al. 1998, 1999; Badi et al. 2009; Hoshiba et al. 2001). Akin to the observations from Nepal Himalaya (Singh et al. 2016), the dominance of intrinsic Q_c in the study region is evidenced by the higher values of Q_i than Q_{sc} at all frequencies. The high intrinsic coda Q corroborates the presence of fluids, probably generated by the metamorphism of the subducting plate.

The analysis of spatial distribution of Q_c at different frequencies, Q_0 and n in the Andaman region, by Singh et al. (2019) reveals interesting attenuation characteristics of the crust. The Q_c values calculated using small lapse times, signifying the subsurface properties of the crust, suggest that the entire north-south segment of the Andaman region is equally tectonically active because of the strong frequency dependence of Q_c. However, it was observed that the Q_c pattern distinctly varies along the arc, vindicating differences in the attenuation characteristics due to various structural heterogeneities. Since the region is categorized as a high intrinsic attenuation zone, the distinct spatial Q_c values can be attributed to the intrinsic mechanisms. At lower frequencies (1.5 and 3 Hz), low Q_c values are observed in the near vicinity of the Narcondam volcanic Island and the Andaman trench, correlating well with the low velocity zones demarcated from seismic tomography studies (Mishra et al. 2011). The low velocity feature mapped near Narcondam could be associated with a concealed magma chamber in the subsurface, that could be a potential source to initiate dehydration of the subducting plate, thereby lowering the velocities in the vicinity. The change in medium properties or the seismotectonic characteristics of a region can be assessed by the spatial distribution of Q_0 and n values. A high attenuation is dominantly observed in the entire Andaman region with highest values observed in the northernmost segment of the Andaman Island and pockets of low attenuation near the southern part of the Andaman Island. Further, depending on the spatial distribution of the n value, it is suggested that the coda Q values in the western, eastern and northern segments of the Andaman Island are highly frequency dependent and the Q_0 values reveal contrasting differences in the attenuation characteristics not only along the arc (from south to north) but also across the Andaman arc (from west to east). The lowest and highest Q_0 values are observed in the northern and southern Andaman regions respectively. Thus, the seismic waves are highly attenuated due to intrinsic absorption in e regions near the northern Andaman than those in the southern part. An inverse relation is established between the Q_0 and n values in this region. It is observed that the regions with low V_p (Mishra et al. 2011) and the segments with low Q_0 and high n values are very well correlated. A prominent low coda Q_0 zone is observed to the west of the north Andaman region, where majority of the local earthquakes including the large magnitude earthquakes ($M > 5$) were reported by Srijayanthi et al. (2012). Further, previous global seismicity studies (Malik et al. 2006) also indicated that the majority of earthquakes in the ANSZ are concentrated either along the trench or along the spreading centre, while being sparsely distributed in the main Islands. The spatial distribution of Q_0, almost mimics this seismicity pattern; low-Q_0-high seismicity and high Q_0-low seismicity. Therefore, it was concluded that the regions with low Q_0 could probably indicate potential zones of future earthquakes or rupture zones. An active role of crustal heterogeneities in earthquake genesis, was also suggested.

3 Seismic Image of the ANSZ Using Receiver Functions

A new era in delineating the crustal seismic discontinuities began when the ratio of vertical to the horizontal component spectral responses of P waveforms were modelled to infer the structure beneath the station (Phinney 1964). This was regarded as the pioneering version of the Receiver Function (RF) technique. Later, Burdick and Langston (1977) and Vinnik (1977) initiated the P receiver function (PRF) technique in its present form, where the converted phases from discontinuities beneath a seismic station form the primary source of information. This method emerged as an efficient technique to investigate the crustal and upper mantle seismic discontinuities. To map the crustal and uppermost mantle seismic structure in the ANSZ, both the body wave converted phase techniques namely, the P-to-s (P receiver functions: PRF) (Ammon 1991) and S-to-p (S receiver function: SRF) (Farra and Vinnik 2000), were adopted utilising a large number of teleseismic waveforms recorded at the 14 three-component broadband seismic stations (Srijayanthi 2016; Kumar et al. 2016). For the P receiver functions (PRF), the waveforms of earthquakes in the epicentral distance range of 30°–95° and for the S receiver functions (SRF), waveforms of earthquakes in the epicentral distance range of 60°–85° are utilized. Waveforms of teleseismic earthquakes having magnitude >5.5, with a high signal to noise ratio and a clear parent phase (P wave for PRF and S wave for SRF) are only considered to compute the RF(s). The results and the inferences drawn from this study are discussed below.

3.1 Variable Crustal Character Along the Andaman Arc

To investigate the crustal structure beneath the Andaman arc, a total of 3017 good quality PRFs that are moveout corrected to a reference slowness of 6.4 s/° (corresponding to the mean epicentral distance of 67°) are utilised (Srijayanthi 2016). The primary results of P receiver functions clearly indicate a variable crustal character along the arc, evident from the dissimilar responses at stations sampling different regions (Fig. 2). At a few stations (like HUTB and PBA), two positive pulses are observed within the first 4 s after the direct P wave. The first positive amplitudes close to ~1 s correspond to a very shallow and strong intra crustal layer (probably the top of the subducting Indian plate). A positive conversion prominently seen close to 2–3 s at most of the stations (DGPR, BAKU, BARA, KTL, MYB, CBY, SBY and PBA) corresponds to the Moho (marked as M in Fig. 2). These conversion times translate to depths in the range of 16–24 km. The variation in conversion times even at stations in close proximity reveal contrasting crustal configurations suggesting a complex structure along the Andaman subduction zone. Interestingly, the first positive conversion for the stations sampling the Nicobar Islands (HUTB, KAMO) is delayed to ~4.0 s, indicating a sudden increase in the depth to the Moho (~30 km). The P receiver functions from four stations DGPR, BAKU, PBA and KAMO sampling different portions of the Andaman–Nicobar arc are shown in Fig. 2. The RF wiggle plots clearly indicate a sudden shift in the times of the arrival of the first positive conversion marked as M at the southern stations (like KAMO). Also, a strong negative conversion is observed at ~6–7 s (marked as L in Fig. 2).

3.2 Evidence for Tearing in the Subducting Slab

The motion of a subducting plate could be interrupted by various complex phenomena leading to thinning and tearing of the downgoing oceanic plate. Evidence for such phenomena is documented from many subduction zones globally, like the

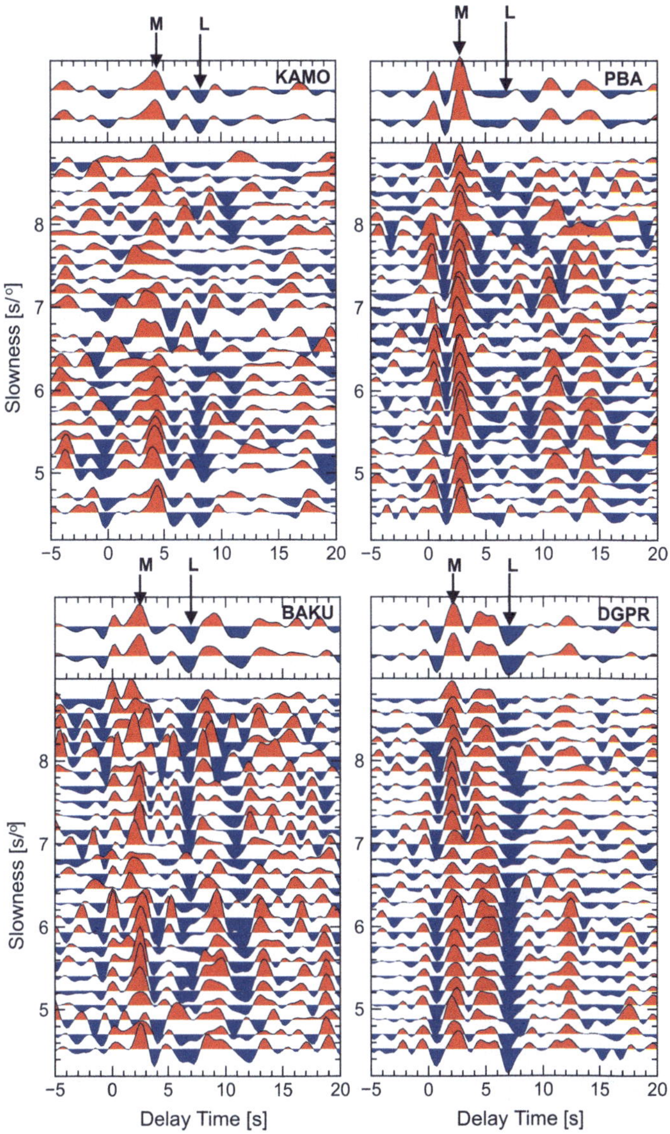

Fig. 2 Q (SV) components of P wave receiver functions stacked in narrow slowness bins, at stations KAMO, PBA, BAKU and DGPR in the Andaman and Nicobar Islands. Stations are arranged from south to north. Summation traces moveout corrected for converted (bottom) and multiple (top) phases are shown in the top panel of RF section at each station. The labels M and L correspond to the Moho and LAB discontinuities

South Sandwich trench (Barker 2001), southern part of the Izu–Bonin arc (Miller et al. 2005), Tonga trench (Millen and Hamburger 1998), Lesser Antilles (Clark et al. 2008), Gibraltar arc (Meighan et al. 2013), and north Apennines (Rosenbaum and Piana Agostinetti 2015). Common conversion point images of both the PRF(s) and SRF(s) constructed using waveform data from the seismological network in the Andaman–Nicobar Island arc provided a new understanding on the nature and geometry of the subducting Indian plate (Kumar et al. 2016). The time to depth migrated images along two profiles AB, CD across the arc and profile EF parallel to the arc (Figs. 3 and 4) clearly show a complex lithospheric architecture beneath the Andaman Islands. A close examination of the top 20–30 km depth image along profiles AB and CD reveals the plate dipping from the Indian Ocean side towards the Andaman back arc side (Fig. 3). Akin to the observations mentioned in the previous section (PRF study), the RF image along the Andaman arc (profile EF) also establishes variations at the crustal level and in the subducting oceanic Indian plate (Fig. 4). The base of the lithospheric plate can be traced at a depth of about 50 km for an 80–90 Ma old Indian oceanic plate subducting below the Andaman–Nicobar region, except between 7 and

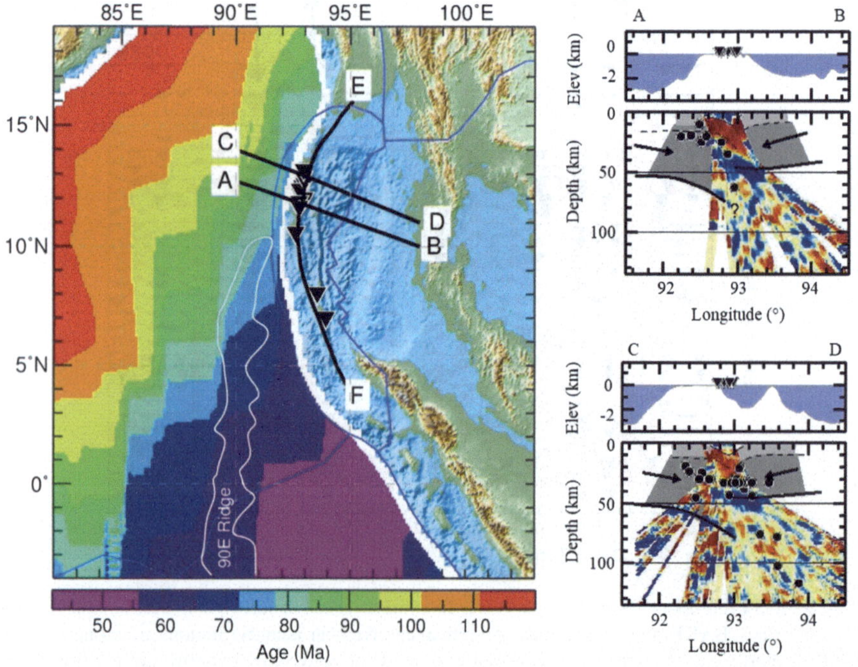

Fig. 3 Topography map of the Andaman-Nicobar Islands along with the sea floor age (Müller et al. 2008), broadband seismic stations (black inverted triangles). Time to depth migrated receiver function images along the two profiles AB and CD are shown in the right panel. The bathymetry with station locations (black inverted triangles) is also shown at the top of each profile. Modified after Kumar et al. (2016)

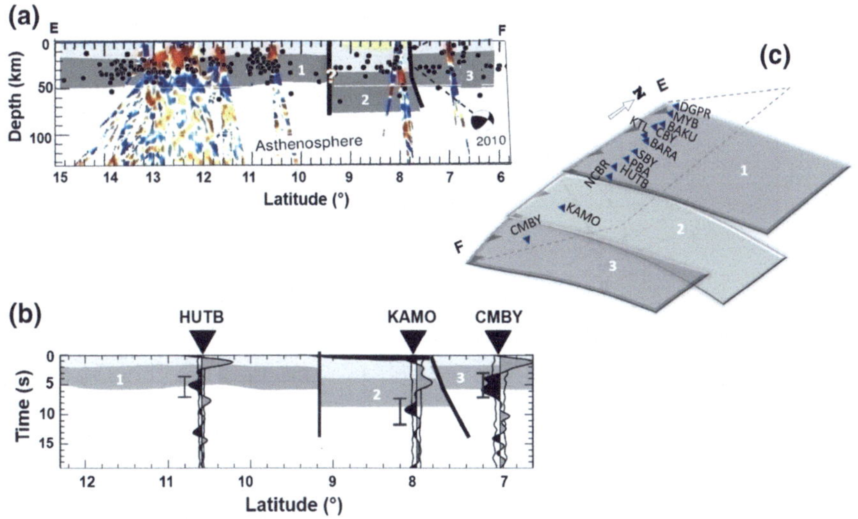

Fig. 4 **a** Time to depth migrated images of Receiver functions along the north-south profile EF (shown in Fig. 3). The earthquakes ($M_b > 5$) are plotted with black dots. Focal mechanism of the strike-slip earthquake of 2010 is also shown. The numerals 1, 2 and 3 denote the three distinct lithospheric segments observed beneath our seismic sections. **b** The stacked RF traces from stations sampling these three different zones clearly show a shift in the positive and negative amplitudes corresponding to MOHO and LAB respectively. **c** The conceptual model of slab tearing. The station locations are shown with blue inverted triangles. The figure is modified after Kumar et al. (2016)

8° N latitude, where a sharp increase in depth down to ~70 km is observed. Interestingly, a similar trend is observed in the positive phase representing the Moho, which suddenly deepens to ~40 km, where the base of the lithosphere is deepest. Therefore, the north-south section clearly reveals three distinct segments of the subducting Indian plate along the trench (i) to the north of the intersection of the 90E ridge with the Andaman trench, (ii) between the Andaman and Nicobar islands (iii) on the Nicobar side. An abrupt offset of ~20 km with respect to the adjacent segments is noticed between 7 and 8° N latitude (Fig. 4). The two slabs, one on the northern side and the other on the southern side show gentler dips, whereas the middle segment shows a steeper subduction angle (Figs. 4 and 5), with an offset of ~20 km relative to the other two. This distinguishing offset provides a direct seismological evidence for tearing of the Indian oceanic plate at shallow depths along the Andaman arc. Further, the presence of a tear gains support from the hypocentral distribution of the earthquakes catalogued in the ISC bulletin, along the profiles that are roughly perpendicular to the trench (Fig. 5). The seismicity trends that mimic the downgoing slabs indicate varying slopes along the arc. The slopes traced by the hypocentres along the profiles in the neighbourhood where the slab-tear is proposed are higher compared to other regions along the arc, akin to the differential dips in the Mariana slab that are attributed to tearing (van der Hilst and Seno 1993). The abrupt break in seismicity (Reyners et al. 1991; Millen and Hamburger 1998) or a sudden change in

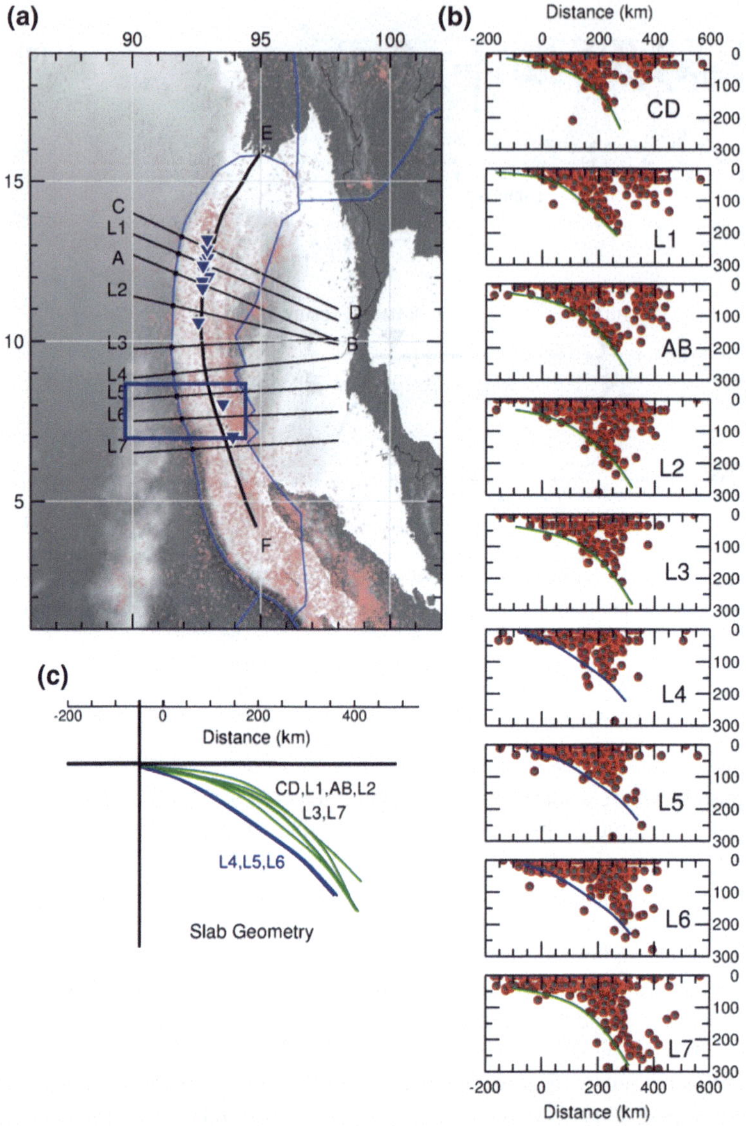

Fig. 5 Seismicity in the ANSZ (left). **a** The pink circles represent the epicentres of earthquakes from 1960 to 2016, catalogued in the Bulletin of the International Seismological Centre (ISC). The blue inverted triangles are the stations, the blue lines are the major tectonic features and the black lines are the profiles perpendicular to the trench. The blue box demarcates the region where tearing in the subducting oceanic Indian slab is observed. **b** The depth sections along these profiles include earthquakes within ±50 km. The curved lines represent the subducting oceanic Indian lithosphere. **c** The slab geometry along all the profiles is plotted together to show the variation in slab dips along the trench; steeper slopes are observed along the profiles L4, L5 and L6 (blue color), where tearing is observed from RF studies. Modified after Kumar et al. (2016)

the velocity anomalies in the tomographic models (Lucente et al. 1999; Benoit et al. 2011; Giacomuzzi et al. 2012) are considered as direct indicators for tear in the slab. Previously, the evidence of tearing in the subducting plates was either based on low resolution tomographic images or hypocentral distribution. However, the RF imaging using high resolution body waves is a pioneering documentation of a tear in the subducting oceanic Indian plate beneath the Andaman–Nicobar Islands. Further, the lithospheric tears are believed to be direct pathways for 3D mantle flow around the slab. The possibility of such phenomena in the study region is also explored by studying the anisotropy in the mantle wedge, which will be discussed in the next section. Some of the interesting and supporting observations associated with the proposed region of slab-tear are (i) the largest horizontal coseismic offset of the 2004 Sumatra earthquake inferred from GPS studies (Gahalaut et al. 2006; Subarya et al. 2006; Yadav et al. 2013) coincides exactly with the area where tearing of the lithosphere is imaged. It is observed that the torn slab slips more than the adjacent ones during a large earthquake, such as the 2004 Sumatra–Andaman earthquake. (ii) the largest swarm activity (2005 swarm) observed globally after approximately a month after the 2004 megathrust earthquake also coincides with the region (same latitude). (iii) the source characteristics of the 2012 earthquake swarms in the proximity are linked to magmatic activity (Srijayanthi et al. 2017).

4 Deformation Scenario in the ANSZ

Subduction marks the most dynamic phase in the plate recycling process, initiating many complex deformation phenomena. Studying the deformation is of prime importance, since the causative mechanisms are linked to the dynamics of a subduction zone. Deformation in the mantle rocks is generally imprinted in the preferred orientation of the minerals or partial melt, which can be probed using seismic anisotropy inferred from shear wave splitting (SWS). Seismic anisotropy, an intrinsic property of the Earth materials, signifies the directional dependence of seismic wave velocities. The two parameters that are used to describe seismic anisotropy are, delay time (δt) between the fast and slow waves and the orientation of the fast axis (ϕ). The basic underlying principle in shear wave splitting measurement methods is to invert the splitting, using a grid search, to find the optimal combination of fast axis and delay time, which best removes the splitting. Here, we review the deformation pattern along the Andaman arc in the crust, mantle wedge and subslab using the shear wave splitting analysis results from Srijayanthi and Kumar (2019).

4.1 Seismic Anisotropy in Subduction Zones

In a subduction system like the Andaman, the contribution to seismic anisotropy comes from four different sources (zones), namely, the overriding crust (continental/oceanic), the mantle wedge, the subducting slab and the subslab mantle. Therefore, to understand the seismic anisotropy scenario of the ANSZ, shear wave splitting from a combination of different shear waves that sample different zones, due to the difference in the travel paths, are utilised. For example, the splitting measurements from both the local S and teleseismic SK(K)S phases are analysed to constrain the depth distribution of anisotropy.

The genesis of seismic anisotropy in each zone of the subduction system (slab, subslab, overriding crust and wedge) is addressed separately since the mechanisms governing the orientation of anisotropy are different for each of these. The subslab mantle is the most extensively explored zones of a subduction system, in terms of seismic anisotropy. The plethora of studies on subslab mantle anisotropy indicate large variability in the anisotropy parameters (both ϕ and δt) among various subduction zones. The initial global compilation of the fast polarisation directions from different subduction zones (South America, Tonga, Izu–Bonin, Mariana) revealed trench parallel FPA, a deviation from the classical 2D entrained flow model, which is based on the general assumption that ϕ coincides with the average principal compressional stress direction, which is parallel to the collision or subduction direction, meaning trench-perpendicular (Russo and Silver 1994; Long and Silver 2008, 2009; Long and Becker 2010). On the other hand, new results from subduction zones like northern Chile, Cascadia (Juan de Fuca), Sumatra and Central America show trench-normal fast polarization azimuths (FPA) (Long and Becker 2010; Lynner and Long 2014a, b). Thus, to interpret the distinct variations of anisotropy observed at various subduction zones, three theories were proposed. The first theory is based on the 3D flow concept that jointly addresses the subslab (Russo and Silver 1994) and the mantle wedge anisotropy invoking a mantle flow system parallel to the strike of the trench. This is essentially governed by the rates of trench migration and convergence (Long and Silver 2008). A recent hypothesis proposed by Song and Kawakatsu (2012) suggests dominance of radial anisotropy from the oceanic asthenosphere in the subslab as a major contributor to the orientations of fast axis azimuths. They demonstrated how splitting varies from trench-parallel to sub-parallel or trench-normal in shallow to steep subduction zones with varying incidence angles of the sampling wave. Later, Lynner and Long (2014a, b) presented a direct relation between the subslab anisotropy and age of the subducting lithosphere after extensively studying the source side anisotropy utilising the shear waves that essentially sample only the subslab mantle. They proposed that in the older subducting lithospheres (>95 Ma), the fast axis azimuths are trench-parallel while in the younger lithospheres (<95 Ma), they are parallel to absolute plate motion. Walpole et al. (2017) in a recent study concluded that trench-parallel measurements observed globally are confined to shallow depths (<50 km), possibly due to trench-parallel faults associated with slab bending. Several alternative theories that involve hydrated faults within the slab (Faccenda et al. 2008) and change in the olivine slip system (Jung et al. 2006; Karato et al. 2008) due to change in temperature and water content are suggested to explain the orientation of fast axis azimuths.

Conventionally, the a-axis of an anisotropic mineral like olivine is oriented in the direction of deformation, which prevails at conditions where high temperatures, low stresses and water content exist (Karato et al. 2008). However, in a subduction zone where the possibilities like presence of water (since oceanic slabs contain huge amounts of water), 2D and 3D flow and temperature and pressure changes exist, the LPO geometry can be complicated. The LPO transforms from A-type to B-type under water-deprived and water-rich conditions accompanied by flow-parallel and flow-perpendicular fast axis azimuths (Ohuchi et al. 2012). Also, the B-type alignment of the olivine mineral is observed perpendicular to the active mantle flow or deformation direction (Jung et al. 2006; Karato et al. 2008). Thus, anisotropy in the mantle wedge is more complex than in the subslab mantle.

In this section, we discuss the contribution of seismic anisotropy from different zones of the Andaman Nicobar subduction system using results from local and teleseismic shear wave splitting analysis (Srijayanthi and Kumar 2019). The local and teleseismic (SK(K)S) shear waves carry signatures of seismic anisotropy from the regions they encounter during their travel from the source to the receiver. Considering the station location, geometry of subduction and also the wavelengths of the sampling waves, Srijayanthi and Kumar (2019) advocated that the local and teleseismic shear waves provide information on the deformation character of the overriding crust, mantle wedge and subslab respectively (Fig. 6). Further, it is observed that the delay times estimated from local SWS are mostly $\leq$0.5 s and those from teleseismic SWS are $\geq$0.8 s, which is attributed to the difference in travel paths, since the SK(K)S waves primarily sample the subslab mantle and the local shear waves sample the slab, mantle wedge and the overriding crust (Fig. 6). The fast axis orientations derived from the teleseismic shear waves are found to be primarily in the NS direction (trench-parallel), and varying from -25 to 10° N (Fig. 7a). Unlike the teleseismic splitting results, the local SWS reveals a complex scenario with scattered fast polarisation azimuths (Fig. 7b). However, ~70% of the measured ϕ are mainly parallel to the trench or approximately in the NS direction. Each scenario is explained below.

4.2 Dominance of Arc Parallel Anisotropy in the Overriding Crust

In the previous section, it was discussed how the Moho depth varies along the arc from 20 to 40 km. Thus, to decipher the contribution of seismic anisotropy from the overriding crust, only the SWS measurements from shallow local earthquakes having depth $\leq$30 km were considered. Generally, the delay time due to crustal anisotropy measures up to ~0.3 s (Savage 1999) and ~96% of the delay times measured from

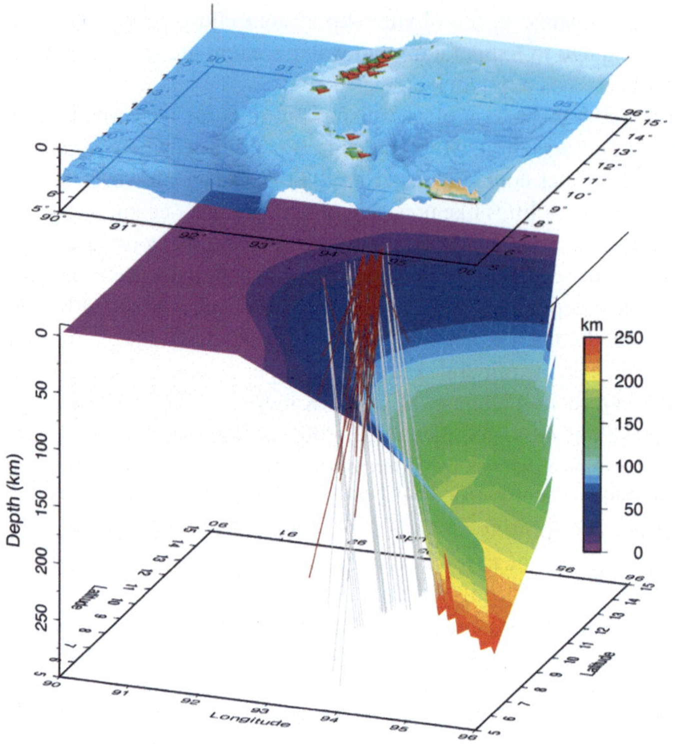

Fig. 6 Travel-paths of the local (maroon) and core refracted shear waves (SK(K)S) (gray) plotted along with the geometry of the Andaman subducting slab. Colour code indicates the slab depth extracted from slab depth contour lines at 50 km interval (Gudmundsson and Sambridge 1998)

local SWS analysis are ≤0.5 s. The shear waves from shallow earthquakes (≤30 km) essentially travel through the overriding crust and show complex fast axis orientations. The large scatter in the fast axis azimuths in the crust is basically attributed to complex structures in the Andaman region that formed during its tectonic evolution. Though a large variation in fast axis azimuths is observed, a predominance of trench-parallel or sub-parallel fast axis azimuths in the crust is still evident (70% of the observed ϕ are trench-parallel) (Fig. 7b). This is related to the presence of arc-parallel strike-slip fault systems like the West Andaman Fault. Thus, anisotropy in the crust can be attributed to the SPO (shape preferred orientation) mechanism associated with the orientation of cracks, fractures, partially filled melt intrusions and fluid filled pores that are aligned in the direction of maximum stress (Crampin 1984, 1994). Previous studies by Fitch (1972) and Shapiro et al. (2008) proposed a series of arc parallel faults like the West Andaman Fault that developed due to complex strain partitioning along the Andaman arc. Further, this scenario is validated by arc parallel fault planes of normal earthquakes at shallow depths (Fig. 8) within the Burmese micro plate and strike-slip earthquakes with NE–SW and NW–SE oriented

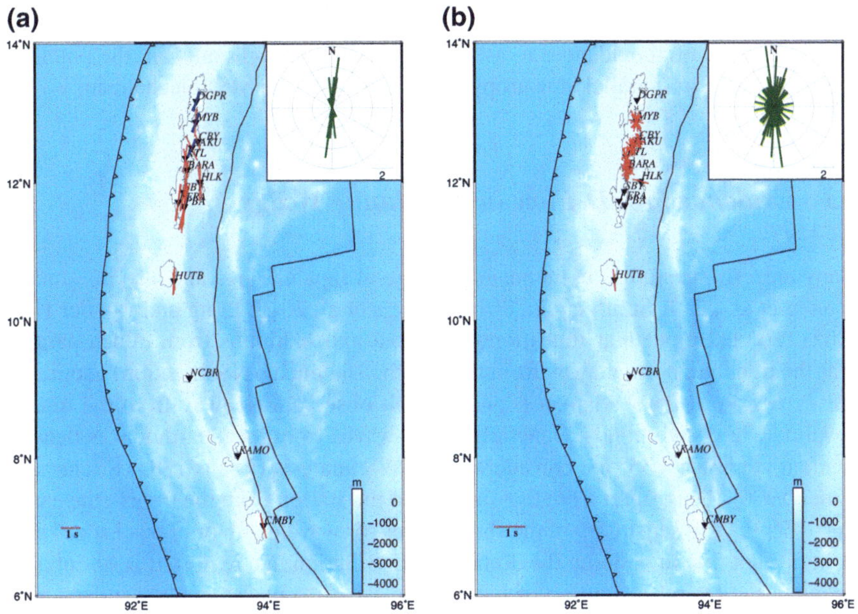

Fig. 7 **a** Teleseismic shear wave (SK(K)S) and **b** Local shear wave splitting measurements at each station. The blue bars at stations DGPR, MYB and BAKU indicate the probable fast axis azimuths inferred from null measurements. Inset shows the frequency distribution of the fast axis azimuths indicating a predominant NS orientation. The length of the bar (red) is proportional to the delay time and the orientation of the bar indicates the fast axis azimuth at respective stations

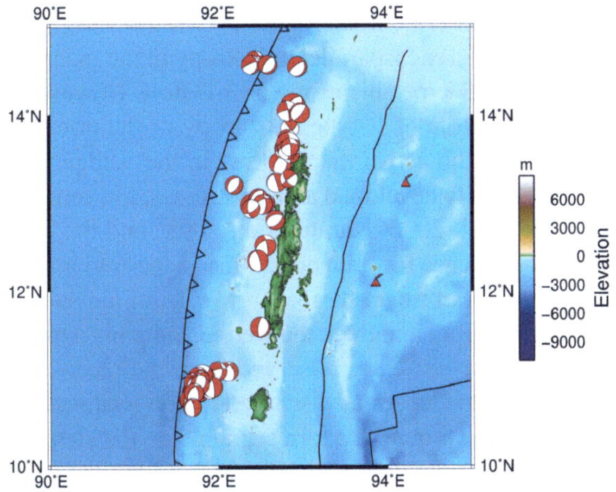

Fig. 8 The normal fault earthquake mechanisms in the Andaman region extracted from the Global CMT database indicate arc parallel fault planes

fault planes below 20 km in the subducting Indian Plate (Rao et al. 2011). Thus, these are substantial evidences to ascertain that arc-parallel faults are the potential sources for the arc-parallel anisotropy, associated with the SPO mechanism.

4.3 Evidence for 3D-Flow in the Mantle Wedge

Many anisotropic studies in the mantle wedge in Japan, Tonga and also in Sumatra (Huang et al. 2011; Smith et al. 2001; Collings et al. 2013; Long and van der Hilst 2005) reported a peculiar transition of the fast axis azimuths from trench-parallel near the trench to normal away from the trench. The linear (NS direction) geographical extension of the Andaman Islands inhibit observations away from the trench, to validate any such trends. Generally, the S-waves from local earthquakes that are incident at angles $\leq 35°$ are considered for SWS analysis (Evans 1984; Fischer and Wines 1996). Thus, apart from the distance of the station from the trench, the earthquake depths and sensitivity of the shear waves (or wavelength or Fresnel zone) play an important role in defining the dominant contribution in a region. It is essential to calculate the lateral resolution of local shear waves to investigate the influence of mantle wedge on the shear wave paths and their corresponding splitting parameters. Therefore, the anisotropy parameters inferred from deeper earthquakes (≥ 70 km) having Fresnel zones >13 km were only considered to characterise the mantle wedge anisotropy. Majority of the fast axis azimuths in this category are either parallel or sub-parallel to the trench. Further, a good correlation between the observed fast axis azimuths and the delay times is observed, where the fast axis azimuths associated with larger delay times (>0.4 s) are trench-parallel (within $-20°$ to $+20°$) (Srijayanthi and Kumar 2019). Such, trench-parallel anisotropy in the mantle wedge could be hypothesised either due to B-type anisotropy or 3-D flow. However, Srijayanthi and Kumar (2019) negated the role of B-type anisotropy or the orientation of fast axis azimuths perpendicular to the active mantle flow in the wedge of the ANSZ. The B-type anisotropy is generally developed at high stresses, low temperatures and high water conditions (Jung et al. 2006; Karato et al. 2008) and the current knowledge provides no supporting evidence for such possibility in the Andaman wedge. Further, they advocated trench-parallel flow due to 3D flow from the subslab mantle. This can strongly influence the supra mantle in the wedge, inducing a trench-parallel flow. The presence of slab-tear at ~8° N, as observed from the receiver function images along the ANSZ and other evidences as discussed in previous sections, ascertain the possibility of such a 3D flow around the torn segments of the slab. These tears in the subducting plates could be the potential outlets from where the subslab mantle can flow into the mantle wedge (Russo and Silver 1994). Therefore, the trench-parallel anisotropy in the mantle wedge of the ANSZ is attributed to the 3D flow mechanism.

4.4 Correlation Between Slab Dip and Subslab Anisotropy

The fast axis azimuths obtained by analysing the SK(K)S are found to be unambiguously oriented in the NS direction; parallel to the Andaman trench, unalike the azimuths inferred from analysis of splitting of local S-waves. The most intuitive explanation for the coincidence of fast axis azimuths with the relative plate motion in the Andaman segment (Fig. 9), is oblique subduction. The observed trench-parallel anisotropy can also be interpreted in terms of the geometry or the dip angle of the subducting plate in the ANSZ (dip 40°–55°), invoking the model of Song and Kawakatsu (2012). Further, akin to the SK(K)S splitting results (Srijayanthi and Kumar 2019), a trench-parallel anisotropy was also deduced from source side anisotropy studies (Lynner and Long 2014a), which was associated with the age of the subducting lithosphere. The previous studies hitherto focussed on a particular segment and the global studies considered only a single segment to model and interpret the subslab anisotropy of the whole arc. Therefore, considering that the ANSZ is only the northern part of a long subduction system, Srijayanthi and Kumar (2019) synthesised the source side anisotropy and the SK(K)S results from previous studies along the whole arc starting from Andaman in the North followed by Sumatra in the centre and Java to the extreme southeast, to understand the subslab anisotropy of the arc as a whole. This unwrapped an intriguing observation and indicated a sudden change in the fast axis azimuths, from trench-normal in Sumatra to trench-parallel in both Andaman and Java (Fig. 9). Further, a transition in the fast polarisation azimuths from trench-normal near to the trench to trench-parallel away from the trench was observed from previous studies (Hammond et al. 2010; Collings et al. 2013). To recap the characteristics of the subduction as discussed in the introduction section (Figs. 1 and 5), a change in age of the subducting plate, variation in convergence rate and change in slab depth and dip are evident along the Andaman–Sumatra–Java arc, which could influence the mantle flow in the subslab. The dynamics of subslab mantle in most of the subduction zones can be directly linked to the age of subducting slab. The older slabs are denser, colder and hence have steeper dips resulting in trench-parallel fast axis azimuths. The younger subducting slabs are lighter, hotter and relatively buoyant that resist subduction. Therefore, these are associated with gentle dips that result in trench-normal fast axis azimuths. The Andaman and Java have older subducting slabs, steeper dips and deeper slabs and trench-parallel fast axes. The central part (Sumatra) is younger having a gentle dip and hence trench-normal fast axes. Further, akin to the Nazca subduction zone, the fast axis azimuths observed along the Andaman–Sumatra–Java arc are trench-normal at the centre of the plate and parallel at the plate edges (indicated by broad red arrows in Fig. 9). This can be clearly explained by invoking the concept of 3D flow, where the mantle flow is perpendicular at the centre of the plate where the younger Sumatra segment is present and changes its course laterally or horizontally along the plate edges (Andaman and Java). Though the theories based on 3D flow and age of the subducting lithosphere seem to explain the anisotropy observations along the arc, both of them fail to explain the variation of ϕ within the Sumatra arc. However, a clear correlation is observed

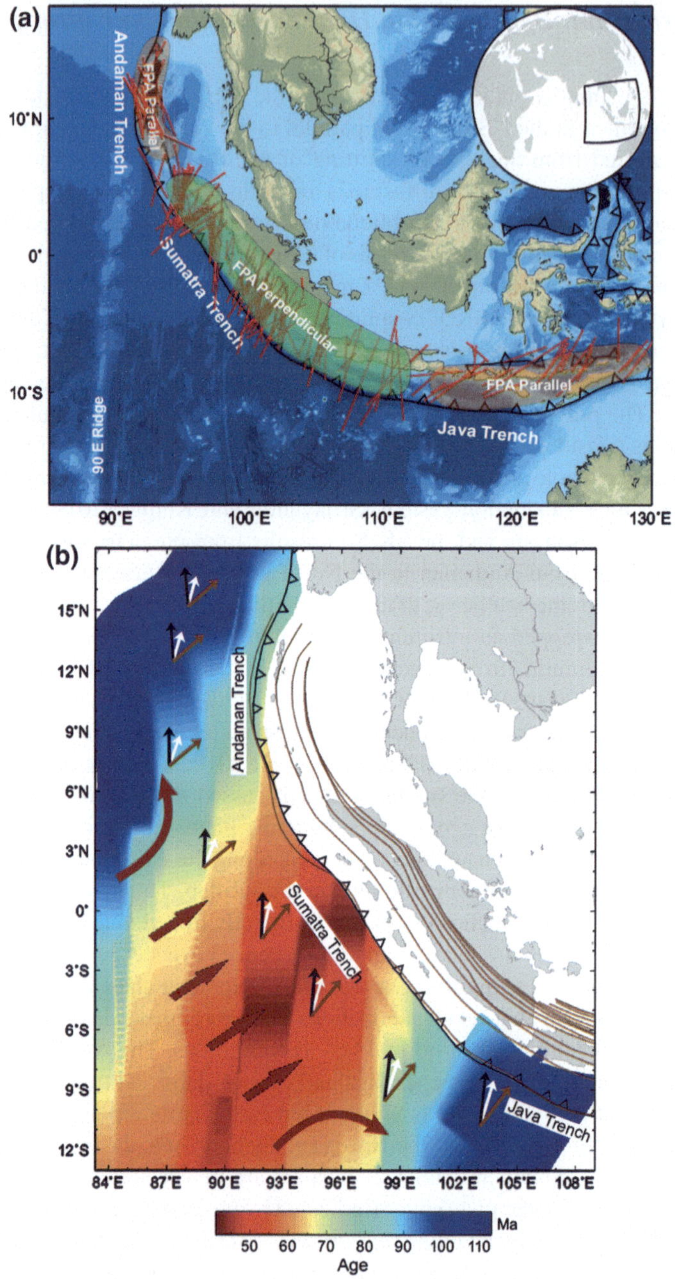
(a)
Andaman Trench
FPA Parallel
Sumatra Trench
FPA Perpendicular
FPA Parallel
Java Trench
90 E Ridge
10°N
0°
10°S
90°E
100°E
110°E
120°E
130°E
(b)
Andaman Trench
Sumatra Trench
Java Trench
15°N
12°N
9°N
6°N
3°N
0°
3°S
6°S
9°S
12°S
84°E
87°E
90°E
93°E
96°E
99°E
102°E
105°E
108°E
Ma
50 60 70 80 90 100 110
Age

◂**Fig. 9** **a** Overview of the subslab anisotropy along the whole Andaman–Sumatra–Java arc. The subslab anisotropy (FPA: Fast Polarization Azimuth (ϕ)) (Srijayanthi and Kumar 2019; Lynner and Long 2014a) in the Andaman and Java is trench-parallel (red shaded region) and trench-normal in Sumatra (green shaded region), **b** Brown arrows represent the absolute plate motion of the oceanic Indian plate in a no net rotation frame and white arrows indicate the relative plate motion with respect to the overriding Sunda plate, predicted by the MORVEL plate motion model (De Mets et al. 2010). Black arrows represent the relative plate motion of the Indian plate in a hotspot reference frame (HS3-Nuvel-1A; (Gripp and Gordon 2002)). Dotted lines parallel to the trench are the slab depth contour lines at 50 km interval (Gudmundsson and Sambridge 1998). The bold red arrows indicate the probable sub-mantle flow along the arc

between the fast axis azimuths in a ray coordinate system (ϕ^*) and slab dip, which could unequivocally explain the results of anisotropy. A compilation of all the results brings out a strong correlation between the dip of the subducting Indian plate and the observed fast axis orientations along the whole arc. Also, it is demonstrated that when the slab dip is shallow, the fast polarization azimuths in a ray coordinate system (ϕ^*) vary from trench-normal to trench-oblique with the variation in incidence angle and ray propagation direction (Song and Kawakatsu 2012) and for slab dips >40° the fast polarization azimuths are clearly trench-parallel.

5 Overview

The Andaman-Nicobar subduction zone involving the Indian and Burmese plates is an exceptional natural laboratory to investigate many interesting geodynamic problems related to subduction and earthquake genesis, crustal deformation, active tectonics, and volcanism. However, this subduction zone is perhaps the least studied in comparison to the other ones globally, and also its contiguous segments that constitute a ~5000 km long arc. For instance, many studies were carried out in the central Sumatran segment, immediately after the region was jolted by the M_w 9.1 megathrust earthquake on 26 December 2004. In the recent past, new data from the broadband seismological stations operated by NGRI and IMD provided many insights into the dynamics of the subducting plate and the source processes. In this review, important inferences from various seismological studies in the ANSZ are recapitulated and a comprehensive overview is presented.

The seismological studies in the ANSZ started with creation of a local earthquake database (Srijayanthi et al. 2012). The northern Andaman region adjoining the Narcondam islands is observed to be seismically more active. The seismicity is dominant to the east of the Andaman trench, mostly clustered between 92 and 93° E longitude. Interestingly, earthquakes from different regions in the ANSZ show different characteristic features. Full waveform inversion of the waveforms of 10 August 2009 Andaman earthquake of M_w 7.5 and its aftershocks (Rao et al. 2011) shows a predominance of normal and strike-slip earthquakes deviating from the thrust type that are generally expected in a subduction system. Further, the location of the main shock and its aftershocks coincides with the region where a vertical tear and lack of evidence for active subduction are proposed (Pesicek et al. 2010; Kumar et al. 1996). The distinct vertical distribution of hypocentres lends credence to this scenario. The local 1D velocity model derived jointly with the moment tensor solutions indicates an oceanic crust having a high crustal V_p/V_s of 1.81 and a Moho at 30 km depth. In contrast to the conventional tectonic earthquakes, an earthquake swarm was observed for nine days during 16 April to 25 April 2012, in the vicinity of Nicobar (9° N, 94° E) (Srijayanthi et al. 2017). Interestingly, this swarm preceded the 11 April 2012 twin strike-slip earthquakes of M_W 8.6 and M_W 8.2, in the Indian Ocean (Fig. 10). These earthquakes also show normal and strike-slip earthquake mechanisms, but seem to be associated with magmatic activity, since they show a high non-double-couple component (70%). Similarly, in the near vicinity of the 2012 earthquake swarm, one of the most energetic earthquake swarm activity was observed in 2005, which was again preceded by the M_W 9.1 2004 Sumatra earthquake (Kundu et al. 2012). The earthquakes of 2005 swarm also show a combination of magmatic and tectonic sources. Although, a 10 year periodicity is observed in the earthquake swarms from the Andaman sea region, the latest ones (2005 and 2012) appear to be triggered by largest magnitude earthquakes in the south. Therefore, seismicity in the ANSZ is a complex manifestation of both tectonic (north Andaman region) and magmatic (Andaman Sea region) activities in response to subduction and spreading.

The seismic attenuation (Q factor values) from the ANSZ are comparable with the eastern Himalaya and southern Tibet and show a dominance of intrinsic absorption rather than seismic wave scattering, which is ascribed to the large scale heterogeneities in the crust. Generally, the intrinsic attenuation annihilates the high frequency component of the waves due to the conversion of mechanical energy generated by the elastic waves into heat energy. Higher seismic wave attenuation is observed both in the vicinity of the trench and the volcanic island, consistent with the low velocities imaged from seismic tomography. In the central part of Andaman, low values of seismic wave attenuation coincide with the region having low seismicity.

The body wave receiver functions (PRF and SRF) have been very effective in delineating the lithospheric structure of the ANSZ. Although the north-south geographical extension of the Andaman archipelago restricts sampling of regions away from the trench, the along arc crustal and lithospheric variations reveal interesting features. The RF images across the Andaman arc clearly indicate an east dipping feature that correlates with the subducting Indian plate. Interestingly, the PRF and SRF images along the arc show that the Moho at depths of ~20 km suddenly deepens to ~40 km, where the base of the lithosphere is deepest (~70 km) between 7 and 8° N latitude. This offset (~20 km) in both the Moho and LAB is a direct evidence for tearing of the subducting oceanic Indian plate. This also gains support from seismicity trends across the Andaman arc which clearly indicate a steeply dipping seismicity pattern in the vicinity of the tear, compared to the adjacent blocks (depth sections in Fig. 5). It is observed that variation in the convergent rates, similar to the scenario in the ANSZ, with varying degree of obliquity and rate of subduction (Sieh and Natawidjaja 2000), tends to migrate the slab backwards, resulting in physical deformation such as bending and/or tearing (Jarrard 1986; Schellart 2007). This

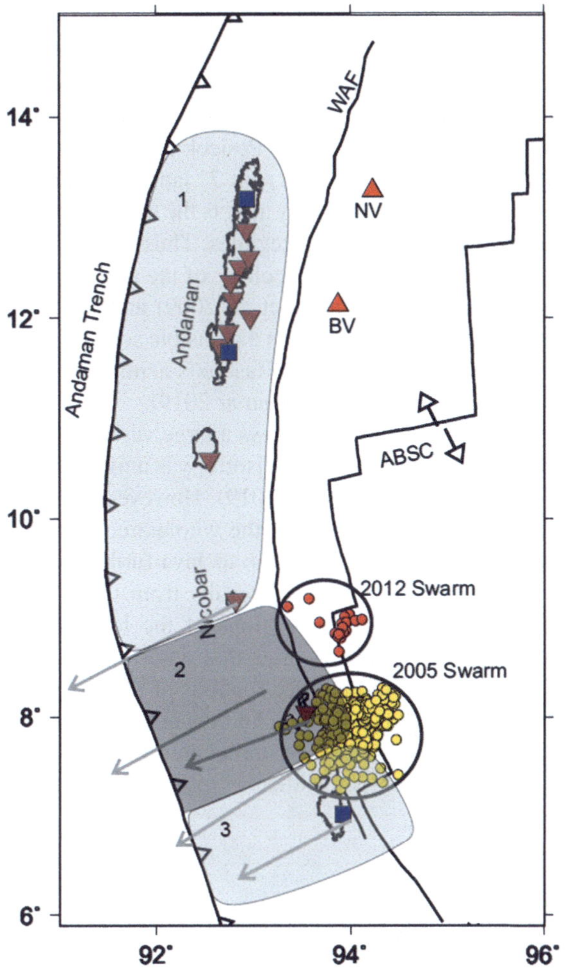

Fig. 10 The region shaded with dark gray colour represents the location of the imaged slab tear in the subducting Indian plate between 7 and 8° N latitude, where a ~20 km offset in both the Moho and LAB depths is observed compared to stations within the light gray coloured shaded regions. The 2005 (yellow circles) and 2012 (red circles) earthquake swarms coincide with the proposed region of tearing. The largest coseismic displacement during the 2004 Sumatra earthquake indicated by gray arrows (Gahalaut et al. 2006) is also observed in this region

reflects as a break or discontinuity in the geometry of the subducting slab along the trench, evidenced by an offset between adjacent segments of the plate (Spence 1977; Aki 1979). The location of two earthquake swarms in 2005 and 2012 (as described above) (Fig. 10) is possibly linked to a volcanic source. Also, a new twin cratered seamount classified as submarine dormant stratovolcano is discovered exactly at the center of the 2005 Nicobar swarm (Kamesh Raju et al. 2012), coinciding with the latitude where tearing is observed. Interestingly, GPS studies related to the 2004

Sumatran earthquake (Gahalaut et al. 2006; Subarya et al. 2006; Yadav et al. 2013) indicate the largest horizontal coseismic offset exactly in the area where tearing is proposed (Fig. 10), demonstrating that the torn slab slips more than the adjacent one. Further, the 1200 km long rupture of this earthquake halted to the north of Andaman Island, where a vertical tear was proposed (Pesicek et al. 2010). Therefore, tearing in the subducting lithosphere, in the south near 7° latitude (Kumar et al. 2016) and north at ~14° (Pesicek et al. 2010), not only affects the source characteristics of local earthquakes but also modulates the stress regimes. Thus, these slab tears play a vital role in controlling the geodynamics and geology of the region (Gvirtzman and Nur 1999; Rosenbaum et al. 2008; Gasparon et al. 2009) and act as vents to outpour magma. It is further observed that the tear is a plausible source for 3D flow from the subslab mantle, resulting in trench-parallel fast axis azimuths in the mantle wedge near the trench (Fig. 11) (Srijayanthi and Kumar 2019).

Unlike the lithospheric structure that shows a large variation from north to south along the trench of ANSZ, trench-parallel anisotropy is uniformly observed at all the stations (Fig. 7a) (Srijayanthi and Kumar 2019). However, a drastic change in the splitting directions is clearly evident along the whole arc, starting from Andaman in the north to Sumatra in the central portion to Java further south-east. Although both the source side and SK(K)S splitting results from the northern (Andaman), central (Sumatra) and southern (Java) sections of the long subduction zone are interpreted using different available theories that describe the subslab anisotropy, a synthesis of the results from all the three sections of the arc establishes a unique relation between slab dip and the fast axis azimuths in a ray coordinate system. The regions having slab dip >40° show trench-parallel anisotropy and the gently dipping segments show oblique to trench-perpendicular splitting directions. Therefore, it is

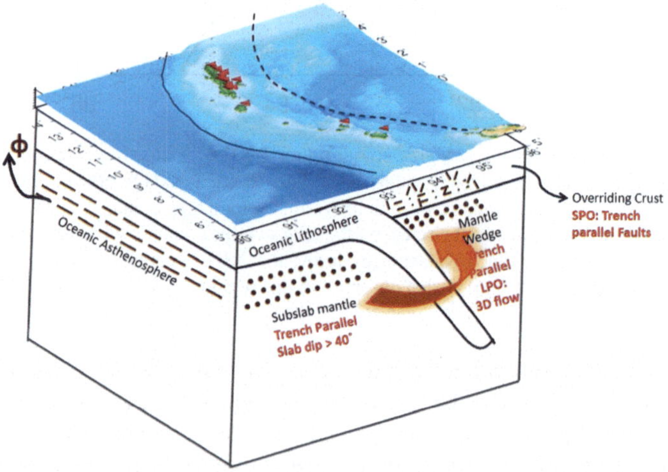

Fig. 11 A pictorial summary of the seismic anisotropy in the ANSZ (depth not to be scale). The broad arrow (orange) indicates the influence of 3D mantle flow from the subslab to the mantle wedge, evidenced by trench-parallel anisotropy in the mantle wedge

inferred that the geometry of the subducting slab significantly influences the sub-slab anisotropy along the whole Andaman–Sumatra–Java arc. The local shear wave splitting directions associated with the overriding crust are complex and are predominantly trench-parallel (Fig. 11). This can be attributed to the SPO mechanism owing to the trench-parallel faults like the WAF.

Acknowledgements The Andaman experiment was supported by INCOIS, Ministry of Earth Sciences, Government of India. All the figures in this paper are prepared using GMT (Generic Mapping Tools), an invaluable software system from Wessel and Smith (1998). We thank Prof. M. Radhakrishna, for inviting us to write this review article.

References

Aki K (1979) Characterization of barriers on an earthquake fault. J Geophys Res 84:6140–6148. https://doi.org/10.1029/JB084iB11p06140

Aki K (1980a) Attenuation of shear-waves in the lithosphere for frequencies from 0.05 to 25 Hz. Phys Earth Planet Inter 21:50–60

Aki K (1980b) Scattering and attenuation of shear waves in the lithosphere. J Geophys Res 85:6496–6504

Aki K, Chouet B (1975) Origin of coda waves: source attenuation and scattering effects. J Geophys Res 80:3322–3342

Ammon CJ (1991) The isolation of receiver effects from teleseismic P waveforms. Bull Seismol Soc Am 81(6):2504–2510

Badi G, Del Pezzo E, Ibanez JM, Bianco F, Sabbione N, Araujo M (2009) Depth dependent seismic scattering attenuation in the Nuevo Cuyo region (southern central Andes). Geophys Res Lett 36(24)

Barker PF (2001) Scotia Sea regional tectonic evolution: implications for mantle flow and palaeocirculation. Earth Sci Rev 55(1–2):1–39

Benoit MH, Torpey M, Liszewski K, Levin V, Park J (2011) P and S wave upper mantle seismic velocity structure beneath the northern Apennines: new evidence for the end of subduction. Geochem Geophys Geosyst 12:Q06004. https://doi.org/10.1029/2010GC003428

Bianco F, Del-Pezzo E, Castellano M, Ibanez J, di Luccio F (2002) Separation of intrinsic and scattering seismic attenuation in the Southern Apennine zone Italy. Geophys J Int 150(1):10–22

Burdick LJ, Langston C (1977) Modeling crustal structure through the use of converted phases in teleseismic body-wave forms. Bull Seismol Soc Am 67:677–691

Bureau of Indian Standards (2002) IS 1893 (Part I): Indian standard criteria for earthquake resistant design of structures (5th revision). Bureau of Indian Standards, New Delhi. http://www.mausam.gov.in/WEBIMD/seismic_zone_map.jsp

Clark SA, Sobiesiak M, Zelt CA, Magnani MB, Miller MS, Bezada MJ, Levander A (2008) Identification and tectonic implications of a tear in the South American plate at the southern end of the Lesser Antilles. Geochem Geophys Geosyst 9:Q11004. https://doi.org/10.1029/2008GC002084

Collings R, Rietbrock A, Lange D, Tilmann F, Nippress S, Natawidjaja D (2013) Seismic anisotropy in the Sumatra subduction zone. J Geophys Res Solid Earth 118:5372–5390. https://doi.org/10.1002/jgrb.50157

Crampin S (1984) Effective anisotropic elastic constants for wave propagation through cracked solids. Geophys J R Astron Soc 76:135145

Crampin S (1994) The fracture criticality of crustal rocks. Geophys J Int 118(2):428–438. https://doi.org/10.1111/j.1365-246X.1994.tb03974.x

Curray JR (1989) The Sunda arc: a model for oblique plate convergence. Neth J Sea Res 24:131–140

Curray JR, Moore DG, Lawver LA, Emmel FJ, Raitt RW, Henry M, Kieckhefer R (1979) Tectonics of the Andaman Sea and Burma in geological and geophysical investigations of continental margins. Am Assoc Petrochem Geol Memoir 29:189–198
Curray JR, Emmel FJ, Moore DG, Raitt RW (1982) Structure tectonics and geological history of the northeastern Indian Ocean. In: Nairn AEM, Stehli FG (eds) The ocean basins and margins. The Indian Ocean, vol 6. Plenum Press, New York, pp 399–450
Dasgupta S, Mukhopadhyay M (1993) Seismicity and plate deformation below the Andaman arc northeastern Indian ocean. Tectonophysics 225:529–542
Dasović I, Herak M, Herak D (2013) Coda-Q and its lapse time dependence analysis in the interaction zone of the Dinarides, the Alps and the Pannonian basin. Phy Chem Earth, Parts A/B/C 1(63):47–54
De Mets C, Gordon RG, Argus DF (2010) Geologically current plate motions. Geophys J Int 181:180. https://doi.org/10.1111/j.1365-246X.2009.04491.x
Del Pezzo E, Ibanez J, Morales J, Akinci A, Maresca R (1995) Measurements of intrinsic and scattering seismic attenuation in the crust. Bull Seismol Soc Am 85:1373–1380
Dickinson WR, Seely DR (1979) Structure and stratigraphy of forearc regions. Am Asso Petrol Geol Bull 63:2–31
Evans R (1984) Anisotropy: pervasive feature fault zones? Geophys J R Astron Soc 76:157–163
Faccenda M, Burlini L, Gerya TV, Mainprice D (2008) Fault-induced seismic anisotropy by hydration in subducting oceanic plates. Nature 455:1097–1100
Farra V, Vinnik L (2000) Upper mantle stratification by P and S receiver functions. Geophys J Int 1;141(3):699–712
Farra V, Vinnik LP, Romanowicz B, Kosarev GL, Kind R (1991) Inversion of teleseismic S particle motion for azimuthal anisotropy in the upper mantle: a feasibility study. Geophys J Int 106(2):421–431
Fehler M, Hoshiba M, Sato H, Obara K (1992) Separation of scattering and intrinsic attenuation for the Kanto–Tokai region Japan using measurements of S-wave energy versus hypocentral distance. Geophys J Int 108:787–800
Fischer KM, Wiens DA (1996) The depth distribution of mantle anisotropy beneath the Tonga subduction zone. Earth Planet Sci Lett 142(1–2):253–260. https://doi.org/10.1016/0012-821X(96)00084-2
Fitch TJ (1972) Plate convergence transcurrent faults and internal deformation adjacent to SE Asia and Western Pacific. J Geophys Res 77:4432–4460
Frankel A, Wennerberg L (1987) Energy-flux model of seismic coda: separation of scattering and intrinsic attenuation. Bull Seismol Soc Am 77:1223–1251
Gahalaut VK, Nagarajan B, Catherine JK, Kumar S (2006) Constraints on 2004 Sumatra–Andaman earthquake rupture from GPS measurements in Andaman–Nicobar Islands. Earth Planet Sci Lett 242:365–374
Gasparon M, Rosenbaum G, Wijbrans J, Manetti P (2009) The transition from subduction arc to slab tearing: Evidence from Capraia Island, northern Tyrrhenian Sea, J Geodyn 47:30–38
GEBCO (2003) 1-minute Global Bathymetry Grid. http://www.ngdc.noaa.gov/mgg/gebco/gebco.html
Giacomuzzi G, De Gori Civalleri MP, Chiarabba C (2012) A 3D Vs model of the upper mantle beneath Italy: insight on the geodynamics of central Mediterranean. Earth Planet Sci Lett 35:105–120
Giampiccolo E, Gresta S, Rasconà F (2004) Intrinsic and scattering attenuation from observed seismic codas in Southeastern Sicily (Italy). Phys Earth Planet Inter 145(1–4):55–66
Gripp AE, Gordon RG (2002) Young tracks of hotspots and current plate velocities. Geophys J Int 150:321–361
Gudmundsson O, Sambridge M (1998) A regionalized upper mantle (RUM) seismic model. J Geophys Res 103:7121–7136. https://doi.org/10.1029/97JB02488

Gupta S, Borah K, Saha G (2016) Continental like crust beneath the Andaman Island through joint inversion of receiver function and surface wave from ambient seismic noise. Tectonophysics 687:129–138

Gvirtzman Z, Nur A (1999) The formation of Mount Etna as the consequence of slab rollback. Nature 401:782–785

Hamilton W (1979) Tectonics of the Indonesian region. United States Geological Survey Professional Paper 1078

Hammond JOS, Wookey J, Kaneshima S, Inoue H, Yamashina T, Harjadi P (2010) Systematic variation in anisotropy beneath the mantle wedge in the Java-Sumatra subduction system from shear wave splitting. Phys Earth Planet Inter 178:189–201

Hazarika P, Kumar MR, Kumar D (2013) Attenuation character of seismic waves in Sikkim Himalaya. Geophys J Int 195(1):544–557

Hoshiba M, Sato H, Fehler M (1991) Numerical basis of the separation of scattering and intrinsic absorption from full seismogram envelope: a Monte-Carlo simulation of multiple isotropic scattering. Papers Meteorol Geophys 42:65–91

Hoshiba M, Rietbrock A, Scherbaum F, Nakahara H, Haberland C (2001) Scattering attenuation and intrinsic absorption using uniform and depth dependent model—application to full seismogram envelope recorded in Northern Chile. J Seismol 5:157–179

Ibanez J, Del Pezzo E, De Miguel F, Herraiz M, Alguacil G, Morales J (1990) Depth-dependent seismic attenuation in the Granada zone (Southern Spain). Bull Seismol Soc Am 80(5):1232–1244

Jarrard RD (1986) Relations among subduction parameters. Rev Geophys 24(2):217–284

Jung H, Katayama I, Jiang Z, Hiraga T, Karato S (2006) Effect of water and stress on the lattice preferred orientation (LPO) of olivine. Tectonophysics 421:1–22

Kamesh Raju KA, Ramprasad T, Rao PS, Rao BR, Varghese J (2004) New insights into the tectonic evolution of the Andaman basin northeast Indian Ocean. Earth Planet Sci Lett 221(1–4):145–162. https://doi.org/10.1016/S0012-821X(04)00075-5

Kamesh Raju KA, Ray D, Mudholkar A, Murty GPS, Gahalaut VK, Samudrala K, Paropkari AL, Ramachandran R, Prakash LS (2012) Tectonic and volcanic implications of a cratered seamount off Nicobar Island Andaman Sea. J Asian Earth Sci 56:42–53

Karato S-I, Jung H, Katayama I, Skemer P (2008) Geodynamic significance of seismic anisotropy of the upper mantle: new insights from laboratory studies. Ann Rev Earth Planet Sci 36:59–95

Kayal JR, Gaonkar SG, Chakraborty GK, Singh OP (2004) Aftershocks and Seismotectonic Implication of the 13 September 2002 Earthquake (M_W 6.5) in the Andaman Sea Basin. Bull Seismol Soc Am 94:326–333

Knopoff L, Hudson JA (1964) Scattering of elastic waves by small inhomogeneities. J Acoust Soc Am 36(2):338–343

Kumar MR, Rao NP, Chalam SV (1996) A seismotectonic study of the Burma and Andaman arc regions using centroid moment tensor data. Tectonophysics 253:155–165

Kumar P, Srijayanthi G, Kumar MR (2016) Seismic evidence for tearing in the subducting Indian slab beneath the Andaman arc. Geophys Res Lett 43(10): 489–4904. https://doi.org/10.1002/2016GL068590

Kundu B, Legrand D, Gahalaut K, Gahalaut VK, Mahesh P, Kamesh Raju KA, Catherine JK, Ambikapthy A, Chadha R. K. (2012) The 2005 volcano-tectonic earthquake swarm in the Andaman Sea: triggered by the 2004 great Sumatra–Andaman earthquake. Tectonics 31(5). https://doi.org/10.1029/2012TC003138

Long MD, Becker TW (2010) Mantle dynamics and seismic anisotropy. Earth Planet Sci Lett 297:341–354

Long MD, Silver PG (2008) The subduction zone flow field from seismic anisotropy: a global view. Science 319(5861):315–318

Long MD, Silver PG (2009) Mantle flow in subduction systems: the subslab flow field and implications for mantle dynamics. J Geophys Res 114 (B10)

Lucente FP, Chiarabba C, Cimini GB, Giardini D (1999) Tomographic constraints on the geodynamic evolution of the Italian region. J Geophys Res 104(B9):20307–20327. https://doi.org/10.1029/1999JB900147

Lynner C, Long MD (2014a) Sub-slab anisotropy beneath the Sumatra and circum-Pacific subduction zones from source-side shear wave splitting observations. Geochem Geophys Geosyst 15:2262–2281. https://doi.org/10.1002/2014GC005239

Lynner C, Long MD (2014b) Testing models of sub-slab anisotropy using a global compilation of source-side shear wave splitting data. J Geophys Res 119(9):7226–7244

Malik JN, Murty C, Rai DC (2006) Landscape changes in the Andaman and Nicobar Islands (India) after the December 2004 great Sumatra earthquake and Indian Ocean tsunami. Earthq. Spectra 22(S3):43–66

Mayeda K, Koyanagi S, Hoshiba M, Aki K, Zeng Y (1992) A comparative study of scattering intrinsic and coda Q for Hawaii Long Valley and central California between 1.5 and 15.0 Hz. J Geophys Res 97:6643–6659

Meighan HE, ten Brink U, Pulliam J (2013) Slab tears and intermediate-depth seismicity. Geophys Res Lett 40:1–5. https://doi.org/10.1002/grl.50830

Millen DW, Hamburger MW (1998) Seismological evidence for tearing of the Pacific plate at the northern termination of the Tonga subduction zone. Geology 26(7):659–662

Miller MS, Gorbatov A, Kennett BLN (2005) Heterogeneity within the subducting Pacific plate beneath the Izu–Bonin–Mariana arc: evidence from tomography using 3D ray-tracing inversion techniques. Earth Planet Sci Lett 235:331–342

Mishra OP, Kayal JR, Chakrabortty GK, Singh OP, Ghosh D (2007) Aftershock investigation in the Andaman–Nicobar Islands of India and its seismotectonic implications. Bull Seismol Soc Am 97:71–85

Mishra OP, Zhao D, Ghosh C, Wang Z, Singh OP, Ghosh B, Mukherjee KK, Saha DK, Chakrabortty GK, Gaonkar SG (2011) Role of crustal heterogeneity beneath Andaman–Nicobar Islands and its implications for coastal hazard. Nat Hazards 57(1):51–64

Mukhopadhyay M (1988) Gravity anomalies and deep structure of the Andaman-Arc. Marine Geophysical Research 9:197–211

Mukhopadhyay S, Tyagi C (2007) Lapse time and frequency-dependent attenuation characteristics of coda waves in the Northwestern Himalayas. J Seismol 11(2):149–158

Mukhopadhyay S, Sharma J, Del-Pezzo E, Kumar N (2010) Study of attenuation mechanism for Garwhal–Kumaun Himalayas from analysis of coda of local earthquakes. Phys Earth Planet Inter 180:7–15

Müller RD, Sdrolias M, Gaina C, Roest WR (2008) Age spreading rates and spreading asymmetry of the world's ocean crust. Geochem Geophys Geosyst 9:Q04006. https://doi.org/10.1029/2007GC001743

Ohuchi T, Kawazoe T, Nishihara Y, Irifune T (2012) Change of olivine a-axis alignment induced by water: origin of seismic anisotropy in subduction zones. Earth Planet Sci Lett 317:111–119

Padhy S, Subhadra N (2013) Separation of intrinsic and scattering seismic wave attenuation in Northeast India. Geophys J Int 195:1892–1903

Padhy S, Subhadra N, Kayal JR (2011) Frequency-dependent attenuation of body and coda waves in the Andaman Sea Basin. Bull Seismol Soc Am 101:109–125

Pal T, Bandopadhyay PC, Mitra SK, Raghav SR (2007) Recent eruption (2005) of Barren volcano: an explosive inner arc volcanism in Andaman Sea. J Geol Soc India 69:1195–1202

Pesicek JD, Thurber CH, Zhang H, De Shon HR, Engdahl ER, Widiyantoro S (2010) Teleseismic double-difference relocation of earthquakes along the Sumatra–Andaman subduction zone using a 3D model. J Geophys Res 115(B10) https://doi.org/10.1029/2010jb007443

Phinney RA (1964) Structure of the Earth's crust from spectral behavior of long period body waves. J Geophys Res 69(14):2997–3017

Radhakrishna M, Lasitha S, Mukhopadhyay M (2008) Seismicity gravity anomalies and lithospheric structure of the Andaman arc NE Indian Ocean. Tectonophysics 460(1):248–262. https://doi.org/10.1016/j.tecto.2008.08.021

Rao NP, Rao ChN, Hazarika P, Tiwari VM, Kumar MR, Singh A (2011) Structure and tectonics of the Andaman Subduction Zone from modeling of seismological and gravity data. In: Sharkov EV (ed) New Frontiers in tectonic research-general problems sedimentary basins and island arcs. Intech Publishers, Rijeka, pp 249–268
Reyners M, Gledhill K, Waters D (1991) Tearing of the subducted Australian plate during the Te Anau New Zealand earthquake of 1988 June. Geophys J Int 104:105–115
Richards S, Lister G, Kennett B (2007) A slab in depth: three-dimensional geometry and evolution of the Indo-Australian plate. Geochem Geophys Geosyst 8:Q12003. https://doi.org/10.1029/2007GC001657
Rosenbaum G, Piana Agostinetti N (2015) Crustal and upper mantle responses to lithospheric segmentation in the northern Apennines. Tectonics 34:648–661. https://doi.org/10.1002/2013TC0034981
Rosenbaum G, Gasparon M, Lucente FP, Peccerillo A, Miller MS (2008) Kinematics of slab tear faults during subduction segmentation and implications for Italian magmatism. Tectonics 27(2)
Russo RM, Silver PG (1994) Trench-parallel flow beneath the Nazca plate from seismic anisotropy. Science 263(5150):1105–1111
Sato H (1977) Energy propagation including scattering effects single isotropic scattering approximation. J Phys Earth 25:27–41
Savage MK (1999) Seismic anisotropy and mantle deformation: what have we learned from shear wave splitting? Rev Geophys 37(1):65–106
Schellart WP (2007) The potential influence of subduction zone polarity on overriding plate deformation, trench migration and slab dip angle. Tectonophysics 20;445(3–4):363–372
Shapiro NM, Ritzwoller MH, Engdahl ER (2008) Structural context of the great Sumatra–Andaman Islands earthquake. Geophys Res Lett 35(5). https://doi.org/10.1029/2008GL033381
Sheth HC, Ray JS, Bhutani R, Kumar A, Awasthi N (2010) The latest (2008–09) eruption of Barren Island volcano and some thoughts on its hazards logistics and geotourism aspects. Curr Sci 98:620–626
Sieh K, Natawidjaja D (2000) Neotectonics of the sumatran fault, Indonesia. J Geophys Res 105:28295–28326
Singh C, Basha S, Shekar M, Chadha RK (2012a) Spatial variation of coda wave attenuation in the Southern Indian shield and its implications. Geol Acta 10:309–318
Singh C, Bharathi VS, Chadha RK (2012b) Lapse time and frequency dependent attenuation characteristics of Kumaun Himalaya. J Asian Earth Sci 54:64–71
Singh C, Mukhopadhyay S, Singh S, Chakraborty P, Kayal JR (2015) Study of lapse time dependence coda Q in the Andaman Islands using the aftershocks of the 2002 earthquake (M_w 6.5). Natural Hazards 75(1):779–793
Singh C, Biswas R, Srijayanthi G, Kumar MR (2017) Relative role of intrinsic and scattering attenuation beneath the Andaman Islands India and tectonic implications. Phys Earth Planet Inter 271:19–28
Singh C, Jaiswal N, Mukhopadhyay S (2019) Pg Q model for Nepal Himalaya. Phys Earth Planet Inter 286:13–20
Song T-RA, Kawakatsu H (2012) Subduction of oceanic asthenosphere: evidence from sub-slab seismic anisotropy. Geophys Res Lett 39(17)
Spence W (1977) The aleutian are: Tectonic blocks, episodic subduction, strain diffusion, and magma generation. J Geophys Res 82:213–230. https://doi.org/10.1029/JB082i002p00213
Srijayanthi G (2016) Investigation of seismicity and lithospheric structure beneath the Andaman-Nicobar subduction zone, Ph.D. Thesis, Osmania University, Hyderabad, India
Srijayanthi G, Kumar MR (2019) Subslab anisotropy in the Andaman subduction zone controlled by slab dip? Phys Earth Planet Inter 286:21–28
Srijayanthi G, Kumar MR, Sirisha T, Sushini K, Prasad GS, Raju PS, Singh A, Rao NP (2012) The ISLANDS network in the Andaman–Nicobar subduction zone. Seismol Res Lett 83:686–696
Srijayanthi G, Kumar MR, Prasanna S, Rao NP (2017) Source characteristics of the 2012 earthquake swarm activity in the Andaman spreading ridge. J Ind Geophys Union 21(1):25–33

Subarya C, Chlieh M, Prawirodirdjo L, Avouac J-P, Bock Y, Sieh K, Meltzner AJ, Natawidjaja DH, McCaffrey R (2006) Plateboundary deformation associated with the great Sumatra–Andaman earthquake. Nature 440:46–51

Ugalde A, Vargas CA, Pujades LG, Canas JA (2002) Seismic coda attenuation after the $M_W = 6.2$ Armenia Colombia earthquake of 25 January 1999. J Geophys Res 107(B6):1–12

van der Hilst R, Seno T (1993) Effects of relative plate motion on the deep structure and penetration depth of slabs below the Izu–Bonin and Mariana island arcs. Earth Planet Sci Lett 120(3–4):395–407

Vinnik LP (1977) Detection of waves converted from P to SV in the mantle. Phys Earth Planet Inter 15:39–45

Walpole J, Wookey J, Kendall JM, Masters TG (2017) Seismic anisotropy and mantle flow below subducting slabs. Earth Planet Sci Lett 465:155–167

Wessel P, Smith WHF (1998) New improved version of generic mapping tools released. Eos Trans Am Geophys Union 79(47):579. https://doi.org/10.1029/98EO00426

Yadav RK, Kundu B, Gahalaut K, Catherine J, Gahalaut VK, Ambikapthy A, Naidu MS (2013) Coseismic offsets due to the 11 April 2012 Indian Ocean earthquakes (Mw 8.6 and 8.2) derived from GPS measurements. Geophys Res Lett 40:3389–3393.https://doi.org/10.1002/grl.50601

Lithospheric Framework of Sumatra–Andaman Subduction Zone—A Review

R. Yadav and V. M. Tiwari

Abstract Sumatra–Andaman subduction zone, where the Indo-Australian plate subducts obliquely beneath Burmese–Sumatra plate, is one of the most seismically active zones and produced two great megathrust earthquakes (Mw 9.1, 2004 Sumatra and Mw 8.6, 2005 Nicobar) in the recent past. Our understanding of this subduction system is considerably advanced due to several studies carried out after the 2004 great Sumatra earthquake. This article presents an overview of the prominent results on lithospheric structure obtained from the analyses of seismological, gravity, geoid and other data sets. The geometry of subduction zone, structure of accretionary prism and Benioff zone show extensive diversity along the subduction axis viz: broad accretionary prism area in the central region (~ 0–10° N) compared to its north and south; variations in lithospheric thickness of Indo-Australian plate; steeper subducting slab in the Andaman region compared to the Sumatra; missing subducting slab in the northern Andaman region; trench parallel seismic anisotropy in Andaman-Nicobar in contrast to trench perpendicular in Sumatra region and deeper Moho under the Ninety-degree east (NER), Alcock Rise (AR) and Sewell Rise (SR). These observations suggest along axis variations in the subduction geometry, dynamics and their relations with earthquake rupture dynamics.

Keywords Gravity · Crustal thickness · Lithosphere density structure · Andaman–Sumatra subduction zone

1 Introduction

The subduction zones are convergent plate boundaries, which are prone to seismic hazard and host almost 80% of worldwide earthquakes (particularly Mw > 8.0). The 90% of the convergent boundary are subduction zones formed by the underthrusting of the lithospheric plate (usually ocean) underneath the overriding plate (oceanic/continental). The subduction process gives rise to different features like trenches, accretionary prisms and volcanic/island arcs. The trenches are very narrow

R. Yadav · V. M. Tiwari (✉)
CSIR-National Geophysical Research Institute, Uppal Road, Hyderabad 500007, India
e-mail: virendra.m.tiwari@gmail.com

J. S. Ray and M. Radhakrishna (eds.), *The Andaman Islands and Adjoining Offshore: Geology, Tectonics and Palaeoclimate*, Society of Earth Scientists Series,
https://doi.org/10.1007/978-3-030-39843-9_7

and deepest features of the Earth's ocean surface, formed when the subducting plate pushes the overriding plate, causing it to bend and form a steep, V-shaped depression. The trench marks the line where subduction begins, and it moves backwards/forward relative to the motion of the subducting plate. The accretionary prism is folded and faulted piles of the material that have been scraped off from the tops of the subducting lithospheric plates. The sediment of the prism is usually marine since the downgoing plate is oceanic. The melange (mixture) is material of prism and a mass of chaotically mixed brecciated blocks in a highly sheared matrix. The volcanic arc formed when oceanic lithospheric subducts into the warmer asthenosphere, and its temperature rises, resulting in loss of water and other volatile material from the descending slab. The volatile material rises to the overlying mantle wedge and forms a volcanic belt at the surface. A series of volcanoes over the overriding plate, parallel to the trench and perpendicular to the direction of plate motion is produced by the subduction process. The subduction process is responsible for mixing surface material into the deep earth mantle, plate driving force, mountain building, and growth of the new continental crust (Christensen and Hofmann 1994; Davidson and Arculus 2006; Forsyth and Ueda 1975). These complex convergence processes lead to megathrust earthquakes like Chile 1960, 2010 (Mw 9.5, 8.8), Sumatra–Andaman 2004 (Mw 9.1) and Tohoku 2011 (Mw 9.0).

The Sumatra–Andaman subduction zone formed by underthrusting of the Indo-Australian plate (IAP) below the Burmese-Sumatra plate (BSP), has received global attention after the megathrust earthquake of 26th December 2004 and the devastating Tsunami. The nature of convergence varieties along axis: (1) continent-continent type near Burma (2) ocean-ocean in Andaman and again ocean-continental in Sumatra (Fig. 1). This is characterised by oblique subduction, variation in the rate of convergence (~3.7 to ~5.7 cm/year), variation in the age of the subducting lithosphere (~20–100 Ma) and spreading ridge in the overriding plate (Fig. 1). The free-air gravity anomaly map (Fig. 2) of the Andaman–Sumatra arc region reveals an excellent correlation with all known structural elements. The convergence is orthogonal to the south of Java region, and it becomes progressively oblique to the west of the northern Andaman and Myanmar region (Curray et al. 1979; McCaffrey 1992; Socquet et al. 2006) (Fig. 1). The obliquity leads to slip partitioning into a trench normal thrust component along with plate interface and predominantly strike-slip motion in the Andaman Sea, the ridge-transform system in the back-arc region and the Sumatra fault system in the south (Fitch 1972; Curray 2005; McCaffrey 1992; Sieh and Natwidjaja 2000). Such type of motion causes shearing in the overriding plate. This shearing is accommodated on a variety of structures, which varies along the convergence zone and referred to as a sliver plate (Jarrard 1986; McCaffrey 2009). In Fig. 3, we present the block diagram of such a sliver plate (after McCaffrey 2009). The sliver plate was formed due to increasing obliquity of IAP during the middle to late Eocene (around 44 Ma). Several faults such as Sagaing Fault (SF) in Myanmar, West Andaman Fault (WAF), Sumatran Fault System (SFS), and Andaman Spreading Centre (ASC) have been identified in it (Curray 2005). Several other faults in this area/plate are hidden because of a thick pile of Bengal Fan sediments of the Himalayan-source (Curray et al. 2002). These sediments have been accreted

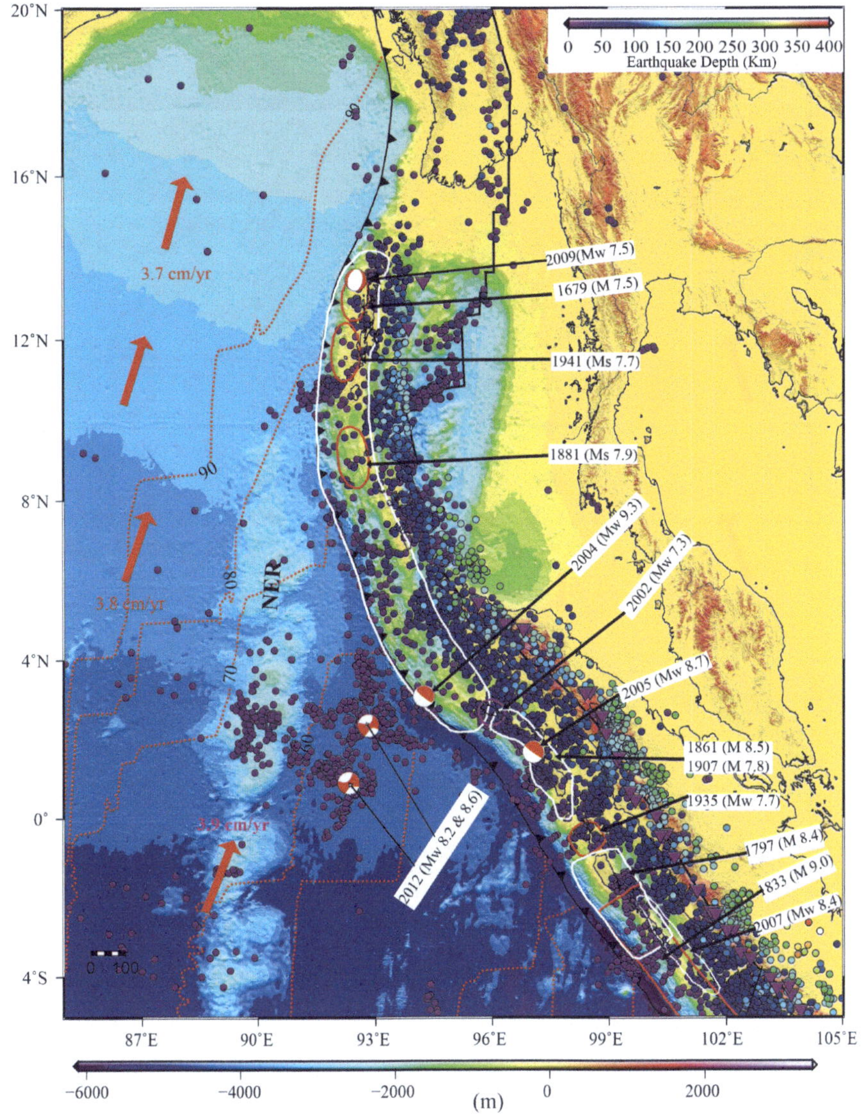

Fig. 1 Bathymetry-topography of the Sumatra–Andaman region. The age of the Indo-Australian plate is superimposed with red dotted contour. The red arrows illustrate the rates and direction of convergence between Indo-Australian and Eurasian Plates. The bounded areas with different colour lines are ruptures regions of the different earthquakes ruptures. The solid circles and red triangles are locations of earthquakes and volcanoes

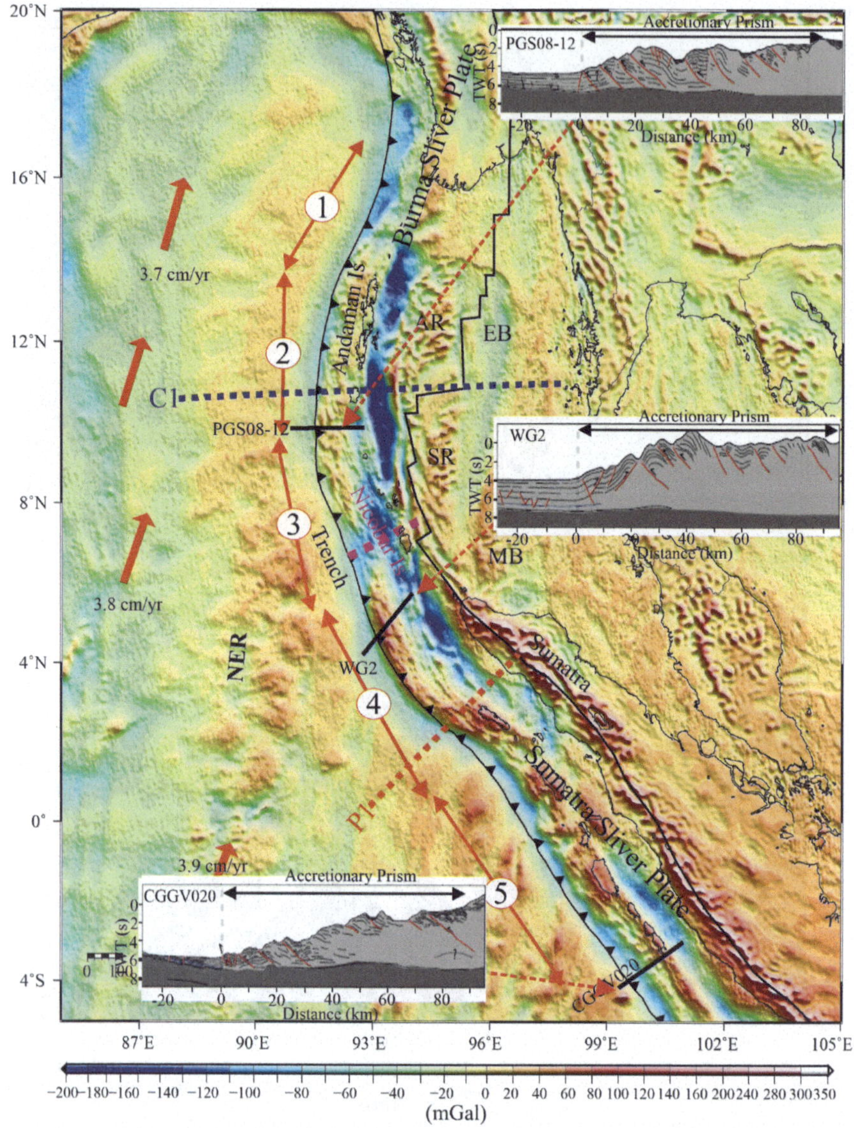

Fig. 2 Free air gravity anomaly map, which shows high over NER, low over the trench, a gradual increase over accretionary prism and a decreasing trend with minimum values over the forearc region. The red duble ended arrows mark the segments of accretionary prism: (1) Narrow, steep prism and seaward vergence, (2). Widening, steep toe-plateau prism and mixed vergence (3) Wide, steep toe-plateau to normal prism landward vergence (4) wide, steep toe-plateau prism landward vergence (5) narrow, steep prism and mixed vergence. The dotted blue and red lines show the 2D Profile (P1) and a cross-section of 3D density (C1) model, which are illustrated in Figs. 4 and 6

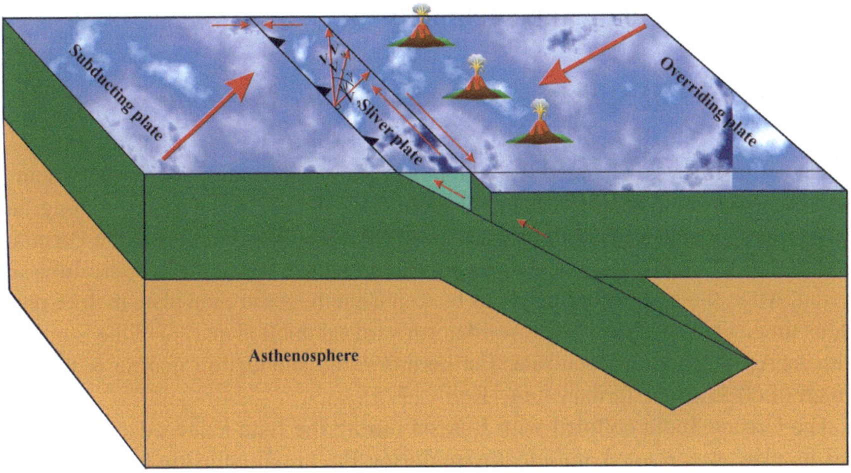

Fig. 3 Block diagram of sliver plate (after McCaffrey 2009). The V, V_{ss} and V_t are total relative, strike-slip and thrust vector motions. The ψ is the slip vector obliquity, γ is the convergence obliquity, and V_n is the velocity normal to the trench

and uplifted into the form of Indo-Burma ranges, Andaman-Nicobar Ridge, and the outer-arc ridge of Sumatra (Fig. 1). However, Ninety East Ridge (NER) is the most extended linear feature (6000 km) in the Indian Ocean extending from 30° S at the Broken Ridge to 18° N, follows the 90° E Meridian is interpreted as a trace of the hotspot (Curray et al. 1982). The variation of bathymetry at the trench extends from the Java to northward where it is overwhelmed by sediments of the Bengal Fan. The Global Positioning System (GPS) study (Gahalaut and Gahalaut 2007) reported that the frontal arc accommodates a movement of about 32 mm/year through oblique motion while the back-arc region accommodates about 21 mm/year mostly through the opening of the Andaman Sea and strike-slip motion on the SFS. Such movements affect the shallow to deeper structures on a large scale. The WAF and the SFS in the Andaman Sea separate the shallow fore-arc and deep back-arc regions. The WAF intersects the south-west end of the Andaman spreading centre near at 10° 15′ N, 93° 40′ E (Curray et al. 1979).

In the present paper, we review the findings of various seismic, seismological, geological, and other geophysical studies and discuss current initiatives to develop a comprehensive understanding of the Sumatra–Andaman subduction system.

2 Geology, Tectonic and Seismicity

2.1 Tectonics of Sumatra–Andaman Subduction Zone

Indian continent was separated from Gondwanaland, situated in the southern hemisphere, moved into the northern hemisphere. The subduction took place along the Sunda, and Himalayan margin dipping toward the NE direction during the Permian time (Katili 1973) and later moved towards the Indian Ocean in the Triassic-Jurassic period. After, the Indian continent was broken from the Australian plate in the Cretaceous time, was started moving toward northward subduction margin of the Eurasian with increasing convergence rate. Furthermore, the subduction margin reached a length of 6000 km in Tertiary time (Katili 1975).

The Greater India collided with Eurasia during the Late Palaeocene (~59 Ma), and its plate was rotated anti-clockwise during the initial collision (up to 55 Ma) (Klootwijk et al. 1992). Later during 55 and 45 Ma, Indian plate rotated in a clockwise direction, which caused the bending of the northwestern Sunda subduction zone. This rotation also caused increases in the obliquity of convergence (Klootwijk et al. 1992; Curray 2005) resulting in the generation of SF (a strike-slip Sliver type fault) and WAF on the east (Peltzer and Tapponnier 1988) during Eocene (~44 Ma). The increased obliquity with time resulted in the opening of several extensional basins in which the combination of back-arc extension and the strike-slip motion has taken place. Many features have been developed in such extensional basins viz. Mergui Basin (MB) (~32 Ma), Alcock Rise (AR) and Sewell Rise (SR) (~23 Ma) and East Basin (EB) separating the rises from the foot of the continental slope ~15 Ma. Also, the AR and SR were separated by the formation of the Central Andaman Basin (CAB) ~4 Ma (Curray 2005; Kamesh Raju et al. 2004).

2.2 Ninety East Ridge (NER)

The NER is a longest (~6000 km) linear volcanic feature on the earth's surface extending from north to south along the 90° E meridian (hence its name NER) and between 30° S to 17° N longitudes which is the western part of SA trench in the Indian Ocean. This ridge has an elevation of 2–3 km above the seafloor with ~200 km width (Sclater and Fisher 1974). The NER separates the central Indian Basin on the west from the Wharton Basin on the east (Sclater and Fisher 1974) and intersects the Broken Ridge in the south while it is completely buried under Bengal sediments north of 10° N (Curray et al. 1982).

The NER was emplaced when IA plate moved northward passing over the Kerguelen hot spot during 82–38 Ma (Coffin et al. 2002; Duncan 1991; Tiwari et al. 2003). The analysis of the core samples collected from deep sea drilling shows that the ridge is nearly 30 Ma in the south and 80 Ma in the north. The chronology data and the magnetic pattern studies suggested that the NER was emplaced at the rate of

118 km/Myr (Pringle et al. 2008). This rate is double as compared to the accretion rate (48–58 km/Myr) of the adjacent ocean floor contributing to the lengthening of the NER. The NER is a complex zone of deformation within the central Indian plate, which has experienced high seismic activity in the past (Petroy and Wiens 1989). The northern portion (up to 10° S) of the NER is more seismically active because it has undergone NW–SE compression while the southern part of the NER is seismically less active.

2.3 Fore-Arc

The tectonic elements of Andaman fore-arc are an accretionary prism, outer-arc ridge, a series of fore-arc basins and north-south faults. The accretionary prism comprises of scrapped-off oceanic sediments with highly deformed Cretaceous ophiolites along with older metamorphic rocks, which were uplifted during the Oligocene Epoch. The western outer slope of the accretionary prism is steeper and filled with progressively younger Neogene sediments as compared to the eastern side. This structure is controlled by dip of subducting plate, physical properties of plate and sediments (pore pressure, thickness, accretion rate, etc.), basal boundary and convergence rate (Davis et al. 1983; Saffer and Bekins 2002; VonHuene and Scholl 1991; Lallemand et al. 1994; Clift and Vannucchi 2004). Another essential part of fore-arc is the outer-arc ridge (Coco, Andaman, Nicobar, and Sumatra Islands). The rock types of these islands are Cretaceous serpentinites, ophiolites with radiolarian chert, Cretaceous to Eocene cherty pelagic limestone and a thick section of Eo-Oligocene flysch overlain by Neogene shallow water sediments (Chatterjee 1967; Roy 1983). The last parts of fore-arc are basins and faults, which contain west basin, Invisible Bank (IB), and West Sewell ridge (WSR). West Basin, a part of Andaman fore-arc basin is south-sloping plain of sediment from the continental slope. The IB extends along with the Andaman-Nicobar Islands segments with ~300 km long and 5 km wide. The interpretation of dredging samples (Rodolfo 1969a, b) and drilling results (Roy and Chopra 1987) suggested that IB is a cuesta that formed by sliver fault (here it is WAF) with last 6 Ma (Curray 2005, Moeremans and Singh 2014). The western boundary of the fore-arc basin is marked by the Eastern Margin Fault (EMF), which is associated with subsidence containing the Pliocene to Pleistocene sediments (Moeremans and Singh 2014). The boundary line between the outer-arc ridge and fore-arc basin is Diligent Fault (DF). The EMF and DF show apparently normal faults at times in past and today, and there may be the possibility of a dextral strike-slip motion (Roy 1983; Curray et al. 1979). The north of Andaman Island, DF is connected to Kabaw Fault (KF) of Myanmar (Hla Maung 1987; Goli and Pandey 2014).

2.4 Volcanic-Arc

The volcanic-arc in the Sumatra–Andaman Sea lies between fore-arc and back-arc. The geochemical data analysis of mainland Sumatra suggested that volcanism was initiated during the pre-Tertiary period (Rock et al. 1982) and intensity went up rapidly during the Tertiary. In the Palaeocene period, this volcanic-arc was situated along the southern margin of Sunda microplate (Crow 2005) and the inner arc was located on North Sumatra back-arc basin.

In the late mid-Eocene period, volcanic rocks got distributed along the west coast of Sumatra (Crow 2005) and volcanic rocks in the Aceh area might have resulted from back-arc volcanism at the same time (Cameron et al. 1980). The Barren and the Narcondam Islands are the two andesite volcanic islands in the northern Andaman Sea (Bhattacharya et al. 1993). The Barren Island is still active and showed activity in March 1991, followed by more recent eruptions in 1994–1995, 2005–2006 and 2008–2009 (Rajendran et al. 2003; Sheth et al. 2010). However, no historically known activity was reported in the Narcondam Island (Pal et al. 2007).

2.5 Back-Arc

Andaman back-arc is Cenozoic pull-apart basin formed by dextral shear between the right-stepping Sumatra, West Andaman and Sagaing fault systems (Curray 1989). This back-arc extension in west-east direction is from the Malay Peninsula to the Andaman Nicobar Islands, which is ~650 km while its average length between north-south direction between Burma and Sumatra is ~1200 km. The back-arc Andaman extensional basin comprises of AR, SR, CAB, EB and some other smaller topographic features of unknown character that are placed east of the Andaman and Nicobar Islands (Curray 2005). The AR and the SR have an average depth of 1200 m and 2000 m respectively. The CAB is the deepest part of the Andaman Sea with a water depth of ~4400 m (Kamesh Raju et al. 2004), separated the AR and the SR. The spreading centre in CAB, the origin is related to leaky transform tectonics (Uyeda and Kanamori 1979), evolved due to the collision of NER and Andaman trench in middle or late Miocene (Eguchi et al. 1979). This collision exerted a drag in the back-arc leading to the opening of Andaman Sea starting with a rate of 3.72 cm/year in about 13 Ma (Curray et al. 1982). Kamesh Raju et al. (2004) postulated that present full rate spreading in the CAB is about 38 mm/year and it has opened up to 118 km in about the last 4 Ma. Consequently, the extrusion tectonics prompted extension and rifting along the plane joining the Sagaing and Sumatran fault systems (Curray et al. 1979; Swe 1972).

Kamesh Raju et al. (2004) attributed magnetic anomalies in the western and eastern parts of the CAB to the blanket of sediment. The Morley and Alvey (2015) have interpreted that central basin opened in an episodic manner and the alternative model proposed that the central trough is equivalent to magnetically active narrow troughs in the attenuated continent crust for transitional continental-oceanic such as those found in the northern Ethiopian Rift or the Suguta Valley in Kenya.

The MB in the north of the offshore is an extension of the North Sumatra Basin, which is formed by the rifting, transtension, thinning of continental crust; began during early Oligocene, and be made up of horsts and grabens structures. The seismic refraction studies indicate that a thin continental crust underlies the basin. Curray (2005) interpreted that the crust of East Andaman, AR, and SR are of the early Miocene age that formed an oceanic crust. However, with the addition of seismic data by oil industries, Morley and Alvey (2015) proposed it to be a continental crust. The Shan Plateau in Burma, Malay Peninsula and continental margin in the eastern Andaman Sea (Suess 1904) consists of the Palaeozoic and the Mesozoic rocks overlying continental crust (Acharya 1994). The Shan Plateau is assumed to be a fault known as the Shan Scarp Fault, and its western margin is interpreted to be an old suture (Acharya 1998; Mitchel 1989).

2.6 Seismicity

The Sumatra–Andaman subduction is seismically active regions and—has produced a destructive earthquake in the past (Engdahl et al. 2007). Kondul 1847 ($7.5 < M < 7.9$), Off Car Nicobar 1881 (Mw 7.9) and West of Middle Andaman 1941 (Mw 7.7) earthquakes are among the most significant earthquakes occurred in Andaman subduction zone (Bilham 2005). A large earthquake also occurred in the North Andaman on January 28, 1679, with a magnitude of 7.5 (Iyengar et al. 1999). The Car Nicobar (1881) earthquake produced a tsunami surge at Car Nicobar (Rogers 1883) with height of 0.25 m in Chennai on the east coast of India (Oritz and Bilham 2003). The source mechanism of this earthquake was analysed using the tsunami travel times (Oritz and Bilham 2003). Jhingran (1953) reported that the Middle Andaman earthquake (1941) has uplifted (by ~1.5 m) along the western margin and subsided along the eastern margin of middle Andaman regions. South Andaman and Sumatra subduction zone experienced five earthquakes with more than 8.3 magnitudes (Mw) (Briggs et al. 2006) in the last ten years. The megathrust earthquake of December 2004 (Mw 9.3) is one the largest ever recorded by seismic instruments) and produced a devastating tsunami across the Indian Ocean. This earthquake started at a depth of 30–40 km with a ruptured over a length of 1300 km and width of 150 km from the Simeulue Island to the northern Andaman Island (Ammon et al. 2005; Subarya et al. 2006). After three months, on March 28, 2005, the second earthquake of Mw 8.7 occurred near Nias Island 150 km south, and it ruptured 350 km of the plate interface. Three years later, an Mw 8.5 earthquake struck on 12 September 2007 at 800 km further south near Bengkulu (Konca et al. 2007). On the 11 April 2012, two strike-slip earthquakes with magnitude Mw 8.6 (largest strike-slip event of the globe) and Mw 8.2 occurred just to the south of the 2004 earthquake on the subducting plate (Duputel et al. 2012). The 2004 and 2005 earthquakes inhibited failure north of the ASC, WAF and SFS and caused static Coulomb stress distribution change along the Sumatra–Andaman-Sagaing Fault system (Cattin et al. 2009). The maximum Coulomb stress varies in the northern part of the Great Sumatra Fault (GSF) (Cattin et al. 2009; McCloskey et al. 2005).

The aftershocks of 2004 earthquake dominated in the region from Sumatra to the Andaman Islands and a cluster of an aftershock occurred at the intersections of WAF and the seaward extension of GSF (Kundu et al. 2012). The aftershock activity is absent in the region north of about 15° N latitudes, a zone that has not generated many earthquakes in the past. Mishra et al. (2007) studied the focal mechanism of the aftershocks and suggested the presence of arc-normal compression (thrust faults) along the subduction front and extension (normal and strike-slip faults) in the back-arc region. Around thousands of shallow aftershock, events occurred in two months, and approximately 100 of them were of magnitude higher than five as reported by the National Earthquake Information Center (NEIC). Engdahl et al. (2007) suggested that this high aftershock seismicity associated with the back-arc transform fault indicates accompanying slip partitioning along the boundary and most of them were characterised by shallow depth right-lateral strike-slip, but also include a number of normal faulting mechanisms oriented perpendicular to the predominant strike-slip in the region along a linear trend. There is a transition zone of aftershock activity at ~5.5° N marked by Kennet and Cummins (2005), and it broadly corresponds to changes in the physical properties of the plate interface.

3 Lithospheric Structure and Its Geodynamic Implications

Wegener (1966) was the first person who postulated a rift origin of Andaman Sea. The first systematic geophysical survey (gravity, magnetic and seismic) was conducted in the Andaman Basin as part of the International Indian Ocean Expedition (Weeks et al. 1967) on board D.S. Coast Guard ship Pioneer. They interpreted that the Barisan range of northern Sumatra extends up to 10° N into the Andaman Sea. Rodolfo (1969a, b) suggested that the central Andaman Sea is 100–200 km wide trough marked by steep and elongated sea valleys and seamounts such as the Nicobar Deep, Barren–Narcondam volcanic islands, IB, AR and SR. A seismic velocity model is suggested for the Andaman region from passive seismic tomography (Kayal et al. 2004), A regional subsurface structure model of the subduction zone is proposed by utilizing all the available data from seismic profiles in the sea, satellite gravity, magnetic, and bathymetry data (Curray 2005).

The negative trench parallel gravity anomalies (TPGA; Fig. 2) are observed over Sumatra–Andaman region, whereas Java region shows positive TGPA (Grevemeyer and Tiwari 2006). Combining previous observations of TPGA over the other subduction zones by Song and Simons (2003) and modelling of the seismogenic zone, Grevemeyer and Tiwari (2006) have proposed that Sumatra–Andaman region is more vulnerable for large earthquakes compared to Java region (Gravemeyer and Tiwari 2006). Their study have shown that forearc structure plays an vital role in defining width of coupling zone and suggested that coupling zone is more (135 km) in Sumatra region as compared to Java region (35 km; Fig. 4). The GPS Studies (Stein and Okel 2005; Gahalaut et al. 2006) suggested that the probability of occurrence of the

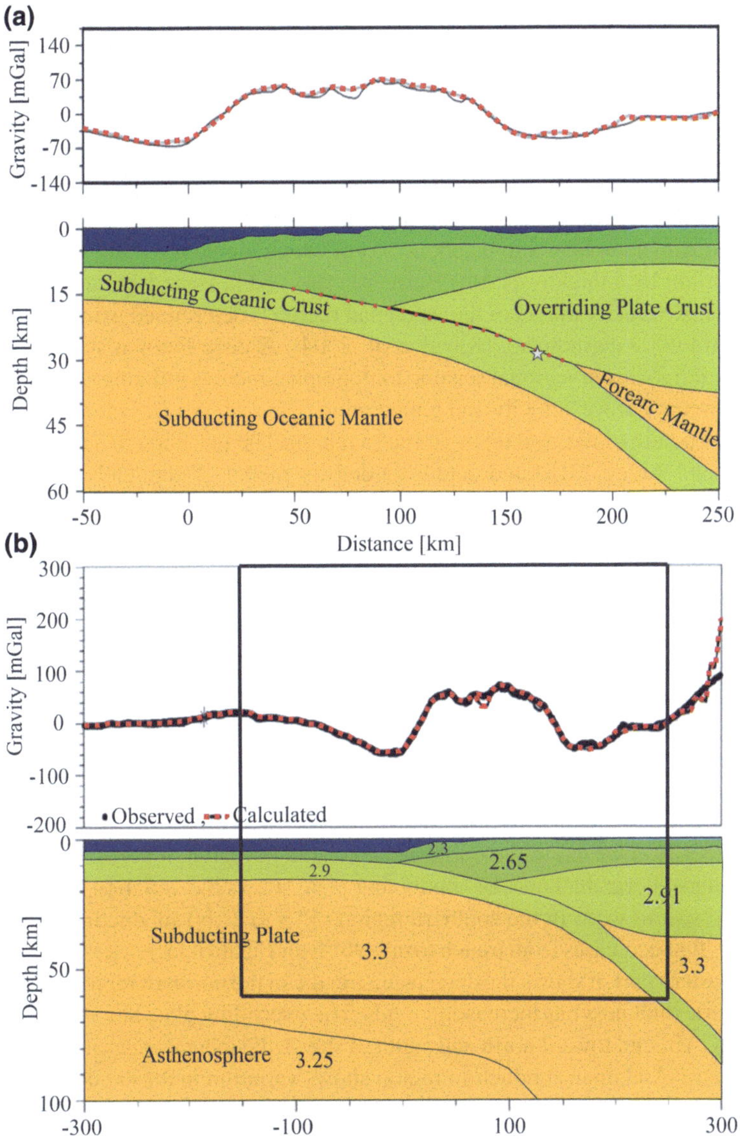

Fig. 4 A 2D density model (modified after Grevemeyer and Tiwari 2006) along a profile (P1) marked on the Fig. 2

mega earthquake between 14° N and 21° N is minimum as compared to the southern part of subduction.

Morley and Alvey (2015) presented the crustal thickness map (Fig. 5) of the region based on the inversion of satellite gravity using lithosphere gravity anomalies. The map shows significant lateral variations in thickness of crust both along and across the arc. The 2D density structure (Fig. 6) derived by Yadav and Tiwari (2019) from the combined gravity-geoid modelling across profile C1 (see location in Fig. 2) reveals significant mass heterogeneities within the upper plate across the Andaman arc. Further, the crustal thickness associated with the epicentral region of Coco earthquake (Mw ~ 7.5) of 2009 could be very well demarcated using the horizontal gravity gradient (N-S) component computed from sediment corrected Bouguer gravity anomaly by Rao et al (2011). Their study reveals maximum crustal thickness below the Andaman Island to be about 30 km which includes a thin (~9 km) Indian crust overlain by a thick (~21 km) Burmese crust and possess 5 km thick sedimentary strata. A velocity model of the Andaman region is determined using aftershocks of Sumatra 2004 earthquake. (Kayal et al. 2004). A deep thermal structure of the region is derived from regional seismic tomographic models utilizing empirical relations between S-wave velocity and temperature (Yu et al. 2017). Figure 7 shows the thermal structure of the Andaman–Sumatra region. The inversion of satellite gravity (Morley and Alvey 2015) and detail 3D density model (Yadav and Tiwari, 2018) showed that the thickness of crust under NER, AR and SR is more compared to the adjacent regions (Fig. 8a) whereas less crustal thickness is found under the CAB and MB (Fig. 8b).

The seismic reflection studies (Moeremans et al. 2014; McNeill and Henstock 2014) showed that the oceanic crust is highly disturbed due to numerous faults, associated with varying topographic relief features along the margin. In addition, the frontal zone is highly folded having a series of thrusts bounding the blocks of layer preserved sediments whose vergence varies from seaward in the south to landward in the north (Fig. 2). Interestingly, the weak reflection was observed over the accretionary prism indicating a high dissipation of seismic energy due to a high degree of faulting and compacting of sediments. The accretionary prism is broader near the 10° N however, in the northern and southern region, it becomes narrow.

The lithospheric thickness of subducting plate shows the variation from south to north and east to west. In the southern region (5° S to 2° N) subducting lithosphere thickness first decreases from trench to up ~96° E and again it increases up ~92° E and further eastern part, it shows the decreasing trends. In the northern region, subducting lithospheric thickness has increasing trends. The overriding plate has a low thickness near the CAB, and toward north and south, it shows the increasing trend. The slab of the Sumatra–Andaman subduction region shows variation in dips and depth of slab from south to north region. In the southern region, the slab extends to large depth and shows the folding nature, and toward north, slab depth decreases up to 14° N and between 14° N to ~16° N, the slab is absent. Further north, depth of the slab shows an increasing trend in the study area.

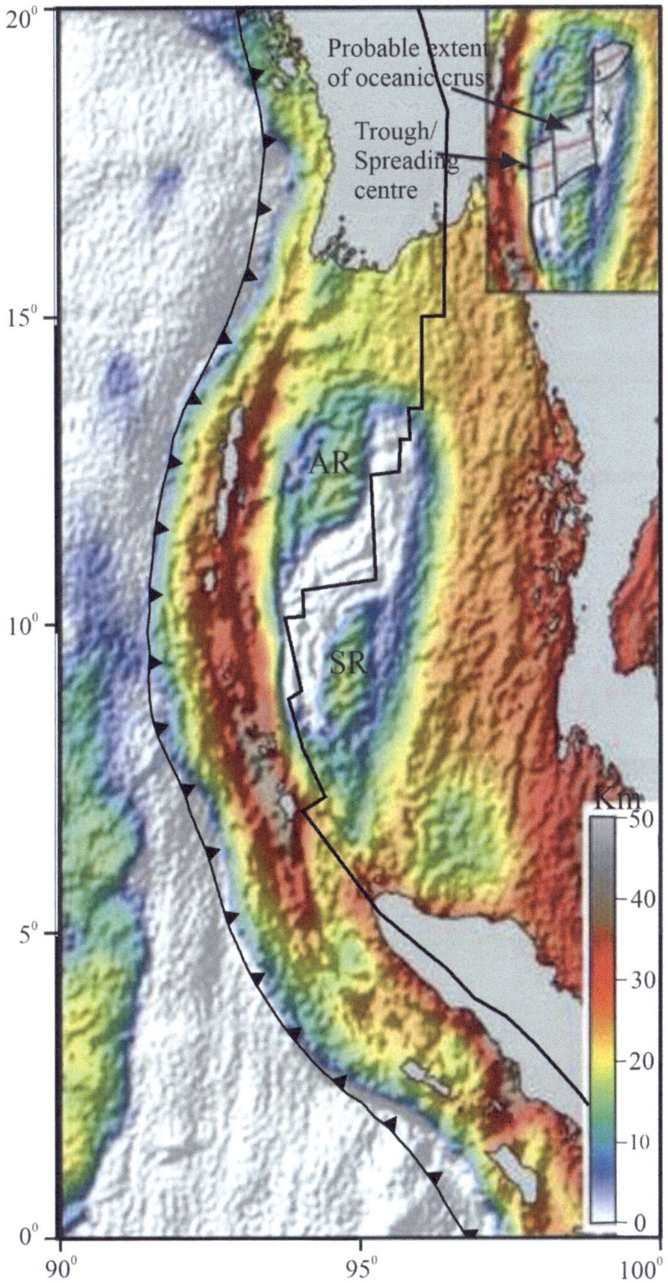

Fig. 5 Crustal thickness map derived from gravity inversion using a lithosphere thermal gravity anomaly (after Morley and Alvey 2015)

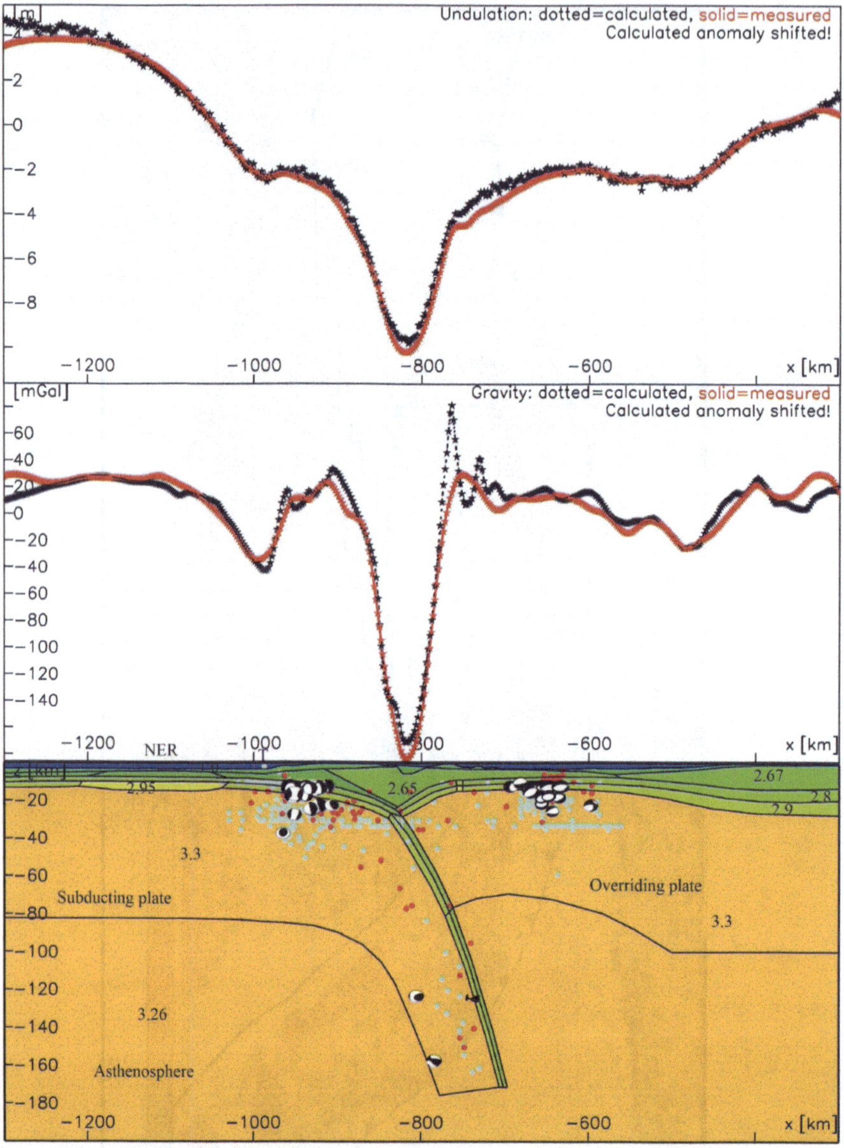

Fig. 6 A lithospheric density model cross-section (C1) along with observed and computed anomalies. The Moho depth under NER is more as compared to the surrounding regions (after Yadav and Tiwari 2019)

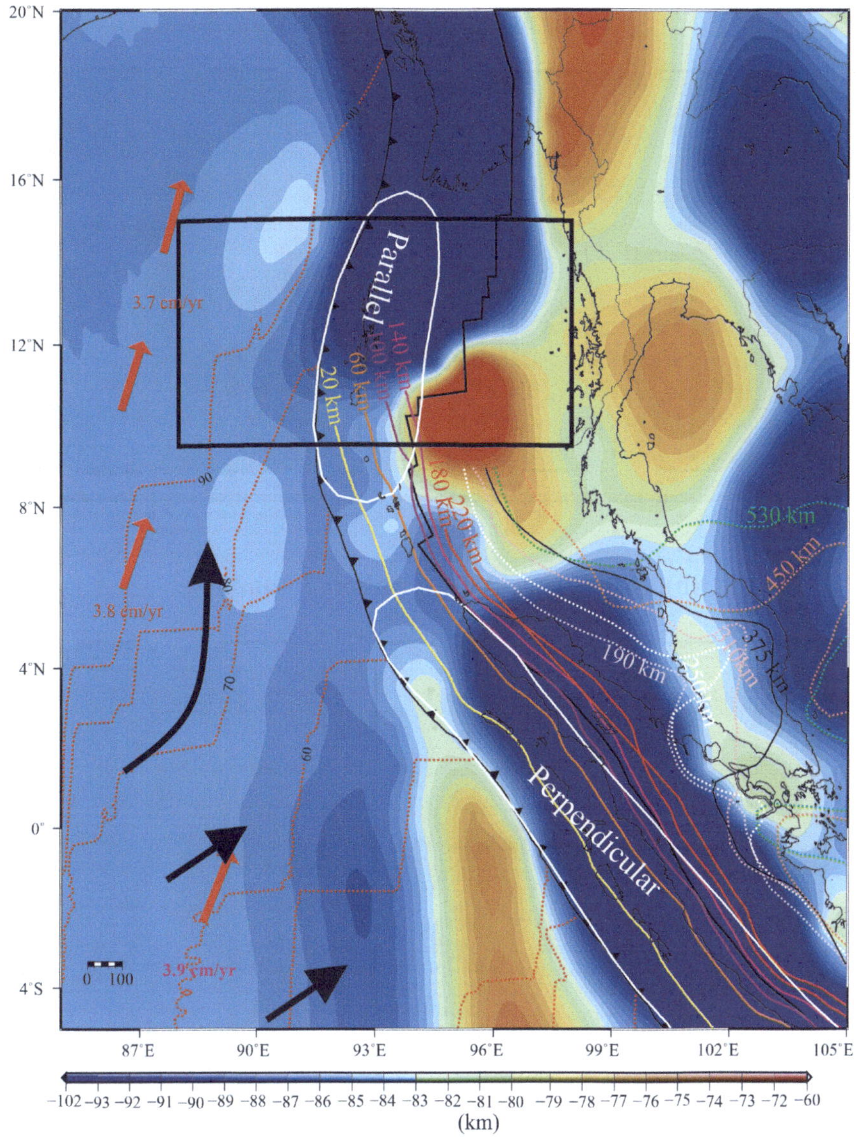

Fig. 7 Maps of thermal lithosphere thickness in the Sumatra-Andaman subduction region (Yu et al. 2017) with subducting slab contours superimposed (USGS and Pesicek et al. 2008). The white line bounded area is the subslab anisotropy: along the arc in the Andaman region and perpendicular in Sumatra region. The black arrows show the sub mantle flow directions (after Srijayanthi and Ravikumar 2019)

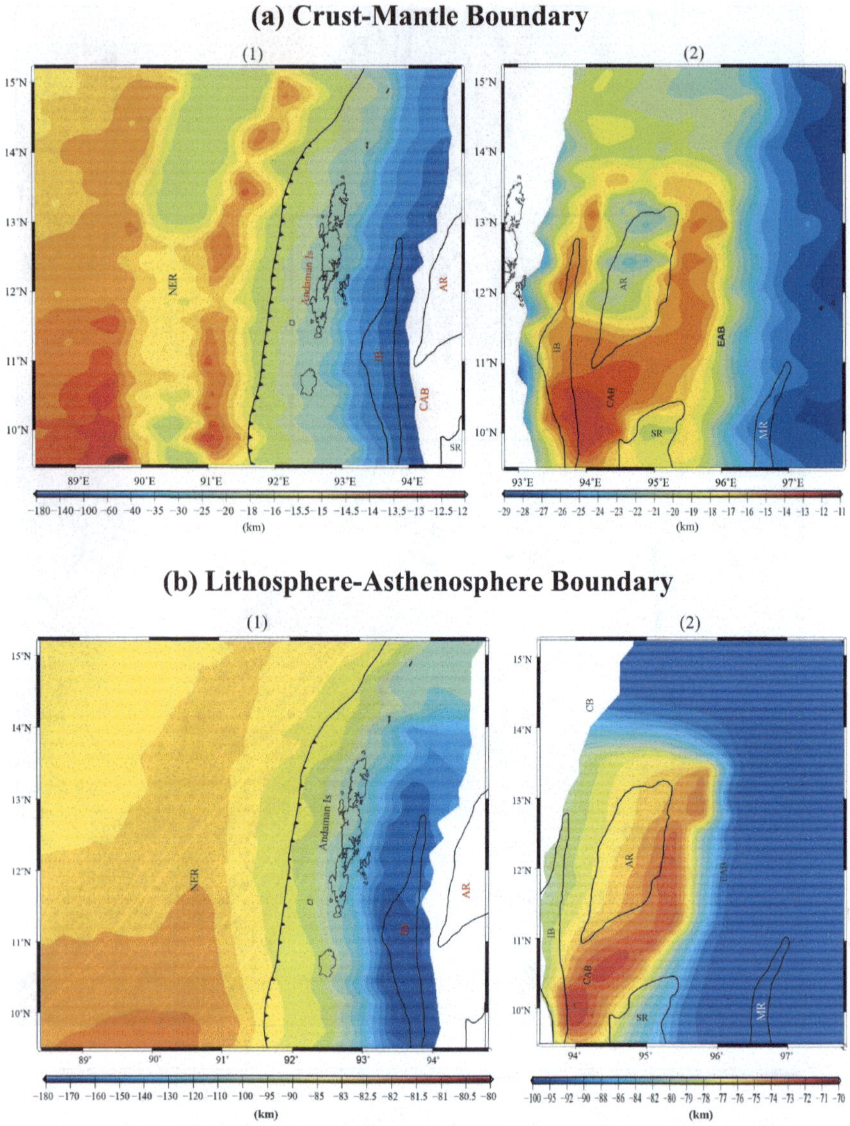

Fig. 8 These are the **a** crust-mantle boundary maps **b** lithosphere–asthenosphere boundary maps of subducting and overriding plate. These maps are computed from the joint modelling of free air gravity and geoid anomalies (after Yadav and Tiwari 2018)

Lay et al. (2005) suggested that the mechanical coupling on the thrust plane might be influenced by increasing age northward of the subducting plate. The young subducting lithosphere tends to result in interplate faults with shallow dips and broad contact areas and generates great mega earthquakes while the older lithosphere is not expected to experience any such events (Ruff and Kanamori 1980; Scholz and Campos 1995). The NNE–SSW trending linear features like NER situated adjacent to the Sumatra–Andaman trench are reflected by the positive free-air gravity anomaly (Chakraborty and Mukhopadhyay 2006). From the analysis of seismicity in the frontal arc of the Andaman-Sumatra subduction zone, Gahalaut et al. (2010) suggested that the number of earthquake in the north Andaman fore-arc (10–15° N) is about 28% less than that in South or Nicobar–Sumatra fore-arc because of the proximity of NER to the Andaman trench (Subrahmanyam et al. 2008; Gahalaut et al. 2010). The buoyant NER has an irregular topography that induces complexities in the subduction process because the pressure of the buoyant force increases the normal stress across the interface. The varying age of materials including NER entering the subduction zone may influence the physical properties along the slab. It is best illustrated by the ratio of shear wave speed and bulk sound speed (Kennett and Cummins 2005) at a depth of 75 km, which significantly reduces at the upper side of the plate at 7, 8.5 and 10.5° N and are associated with the localised features (Kennett and Cummins 2005).

Fitch (1972) reported that the NE movement of Indian plate was resolved or partitioned into two large components: strike-slip on the SF (5 cm/year) and high rate normal subduction along the Sunda-Andaman trench (4 cm/year). The structure of the Andaman arc is dominated by the east dipping nappies having a folding, however, intense folding is observed off Sumatra (Weeks et al. 1967; Curray et al. 1979). The NER collided with the Sumatra–Andaman trench in the middle or late Miocene (Eguchi et al. 1979) and transmitted its compressional stress into the back-arc area causing the opening of the Andaman Sea.

Several researchers have concluded that the nature of the regional seismicity is controlled by tectonic process. The structure of subduction is reflected in the seismicity pattern like thrust dominated subduction front, the strike-slip faulting along the WAF and the extensional processes with the ASC. The Geophysical studies (Guzman-Spezial and Ni 1996; Dasgupta et al. 2003; Khan and Chakraborty 2005) have attempted to map the contour of top surface of identified Benioff zone of Andaman representing the dipping lithosphere using either Hypocentral Trend Surface (HTS) analysis or closer grid basis of the shallow-depth Benioff zone hypocenters. Radhakrishana et al. (2008) have adopted the procedure for plotting the vertical sections and grid areas of Dasgupta et al. (2003) for configuration of Andaman Benioff Zone (ABZ) and calculated the slab gravity anomalies in the Andaman arc region. With the help of available earthquake location data sets, Dasgupta et al. (2003) constructed the 29 depth sections of the Burmese-Andaman arc and 28 of them falls in the study area. They proposed that the dip of ABZ is not uniform throughout; rather it varies from 43° to 53° in the North Andaman and 38°–50° in the south Andaman. Each of the vertical section reveals the distinct seismic character of the arc. The Seismological studies (Pesicek et al. 2008; Kennett and Cummins 2005; Shapiro et al. 2008) have also explained the variation of dip of the slab from south to north and found that the dip of slab varies from moderately (~25°–50°) below Burma and northern Sumatra to steeply dipping (60°–90°) below the Andaman and south. The P and S wave velocities perturbation models (Pesicek et al. 2010; Kennett and Cummins 2005) suggest variations of slab depth from south to north of the Sumatra–Andaman region. After 15° northwards, no slab appears and is characterized by a recent cessation of subduction process and transformation into a strike-slip zone (Le Dain et al. 1984; Rao and Kalpana 2005; Gahalaut and Gahalaut 2007). The S wave tomography study by Kennet and Cummins (2005) has presented east-west cross-sections of the area (2 to 16° N) with 2° cross-section intervals. These cross-sections at 2 and 4° N reveals the presence of a shallow dipping segment down to around 150 km with a steeper dip at greater depth. Between 6 and 8° N, dip of the slab steepens in deeper part and further north from 10° N, the slab becomes nearly vertical at the intermediate depth. Further, they interpreted that structure of subducted slab is the result of roll-back associated with the formation of the Andaman back-arc basin and spreading. Pesicek et al. (2008) deciphered P-wave seismic velocity structure below North Sumatra, which reveals that slab, is folded at depth exhibiting geometry similar to that of the volcanic-arc and trench at the surface (Fig. 7). Shapiro et al. (2008) calculated mantle temperature at 50 km depth from the study of surface-wave tomography of the upper mantle. The conversion from isotropic seismic velocities to anisotropic is explained with temperature variation estimated from the thermo-elastic properties of mantle materials obtained in the laboratory. They interpreted that during 2004 Sumatra–Andaman earthquake; rupture propagation was slow in the Andaman segment due to weak seismic coupling resulting from a combination of the less buoyant subducting plate and weak back-arc lithosphere. Kumar et al. (1996) suggested that P and T axis orientations in the southern part indicate an active subduction whereas the orientations in the northern part are in disagreement with the local trend exhibiting two distinct tectonic patterns. They also found that events with greater depth (depth > 90 km) in the Burmese arc are associated with thrust faulting and shallower depth (<90 km) are dominated by normal and strike-slip faulting. Recently, Kumar et al. (2016) examined Andaman subduction zone based on Receiver Function (RF) analysis and identified the evidence for tearing in the slab. They have suggested that the thickness of crust and lithosphere are deeper in the region between 7 and 8° N latitude (~40 and ~70 km) compared to either side where crustal and lithospheric thicknesses are 20 and 50 km respectively. This teared slab area coincides with the most significant horizontal coseismic offset of the 2004 Sumatra earthquake inferred from GPS studies (Gahalaut et al. 2006; Yadav et al. 2013). Seismic anisotropy directions are trench orthogonal in Sumatra and trench Parallel in Andaman, which are likely to be caused due to lithospheric structure, age of the lithosphere plate and direction of plate movement (Fig. 7; Srijayanthi and Ravikumar 2019).

4 Summary

The lithospheric structure in Sumatra–Andaman region is complex and varies along as well as across the trench. The lithospheric thickness of IAP increases from south to north and from west to east. The lithospheric thickness of overriding plate is minimum near the CAB and increases toward the north and south. The dip and depth of subducting slab also vary significantly from south to north region. In the southern region, subducting slab penetrates deeper compared to northern region. Subduction depth decreases up to 14° N and between 14 and ~16° N, the slab appears to be absent. The crustal model shows thick crust beneath the features like NER, AR and SR as compare to their surrounding regions. Near sea surface structures, such as thickness of accretionary prism, style of faulting in subducting as well as overriding plates also differ substantially along the subduction axis. The structural variations are probably caused due to variations in the age of the lithosphere, oblique convergence of the IAP beneath the BSP, NER-subduction interaction, and sedimentation rates etc.

Acknowledgements We thank to Dr Chuanhai Yu, who has provided the thermal lithospheric thickness data. We also thank CSIR-National Geophysical Research Institute for permission to publish this paper. The figures are created with GMT software (Wessel and Smith, 1995). Authors would like to thank M. Radhakrishna and Bijendra Singh for critical review and suggestions.

References

Acharya SK (1994) Accretion of Indo-Australian Gondwanic blocks along peri-Indian collision margins. In: Ninth International Gondwana Symposium, Hyderabad, India, Geological Society of America, pp 1029–1049

Acharyya SK (1998) Break-up of the greater Indo-Australian continent and accretion of blocks framing south and east Asia. J Geodyn 26(1):149–170

Ammon CJ, Ji C, Thio HK, Robinson D, Ni S, Hjorleifsdottir V, Ichinose G (2005) Rupture process of the 2004 Sumatra–Andaman earthquake. Science 308(5725):1133–1139

Bhattacharya A., Reddy CSS, Srivastav SK (1993) Remote sensing for active volcano monitoring in barren Island, India. In: Ninth thematic conference on geologic remote sensing, Pasadena, California, pp 993–1003

Bilham R (2005) A flying start, then a slow slip. Science 308(5725):1126–1127

Briggs RW, Sieh K, Meltzner AJ, Natawidjaja D, Galetzka J, Suwargadi B, Prayudi D (2006) Deformation and slip along the Sunda megathrust in the great 2005 Nias-Simeulue earthquake. Science 311(5769):1897–1901

Cameron NR, Clarke MCG, Aldiss DT, Aspden JA, Djunuddin A (1980). The geological evolution of northern Sumatra. In: 9th annual convention proceedings, pp 149–187

Cattin R, Chamot-Rooke N, Pubellier M, Rabaute A, Delescluse M, Vigny C, Fleitout L, Dubernet P (2009) Stress change and effective friction coefficient along the Sumatra–Andaman–Sagaing fault system after the 26 December 2004 (Mw = 9.2) and the 28 March 2005 (Mw = 8.7) earthquakes. Geochem Geophys Geosyst 10(3)

Chakraborty B, Mukhopadhyay R (2006). Imaging trench-line disruptions: swath mapping of the Andaman subduction zone. Curr Sci 1418–1421

Chatterjee PK (1967) Geology of the main islands of the Andaman area. In: Proceedings of the symposium on Upper Mantle Project, pp 348–362
Christensen UR, Hofmann AW (1994) Segregation of subducted oceanic crust in the convecting mantle. J Geophys Res Solid Earth 99(B10):19867–19884
Clift P, Vannucchi P (2004) Controls on tectonic accretion versus erosion in subduction zones: implications for the origin and recycling of the continental crust. Rev Geophys 42(2)
Coffin MF, Pringle MS, Duncan RA, Gladczenko TP, Storey M, Müller RD, Gahagan LA (2002) Kerguelen hotspot magma output since 130 Ma. J Petrol 43(7):1121–1137
Crow MJ (2005) Tertiary volcanicity. In: Sumatra geology, resources and tectonic evolution. The Geological Society, London, pp 98–119
Curray JR (1989) The Sunda Arc: a model for oblique plate convergence. Neth J Sea Res 24(2–3):131–140
Curray JR (2005) Tectonics and history of the Andaman Sea region. J Asian Earth Sci 25(1):187–232
Curray JR, Moore DG, Lawver LA, Emmel FJ, Raitt RW, Henry M, Kieckhefer R (1979) Tectonics of the Andaman Sea and Burma: convergent margins
Curray JR, Emmel FJ, Moore DG, Raitt RW (1982) Structure, tectonics, and geological history of the northeastern Indian Ocean. In: The ocean basins and margins. Springer, New York, pp 399–450
Curray JR, Emmel FJ, Moore DG (2002) The Bengal Fan: morphology, geometry, stratigraphy, history and processes. Mar Pet Geol 19(10):1191–1223
Dasgupta S, Mukhopadhyay M, Bhattacharya A, Jana TK (2003) The geometry of the Burmese–Andaman subducting lithosphere. J Seism 7(2):155–174
Davidson JP, Arculus R (2006) The significance of Phanerozoic arc magmatism in generating continental crust. Cambridge University Press, Cambridge
Davis D, Suppe J, Dahlen FA (1983) Mechanics of fold-and-thrust belts and accretionary wedges. J Geophys Res Solid Earth 88(B2):1153–1172
Duncan RA (1991) Age distribution of volcanism along aseismic ridges in the eastern Indian Ocean. Proc Ocean Drill Program Sci Results 121:507–517
Duputel Z, Kanamori H, Tsai VC, Rivera L, Meng L, Ampuero JP, Stock JM (2012) The 2012 Sumatra great earthquake sequence. Earth Planet Sci Lett 351:247–257
Eguchi T, Uyeda S, Maki T (1979) Seismotectonics and tectonic history of the Andaman Sea. Tectonophysics 57(1):35–51
Engdahl ER, Villaseñor A, DeShon HR, Thurber CH (2007) Teleseismic relocation and assessment of seismicity (1918–2005) in the region of the 2004 Mw 9.0 Sumatra–Andaman and 2005 Mw 8.6 Nias Island great earthquakes. Bull Seismol Soc Am 97(1A):S43–S61
Fitch TJ (1972) Plate convergence, transcurrent faults, and internal deformation adjacent to southeast Asia and the western Pacific. J Geophys Res 77(23):4432–4460
Forsyth D, Uyeda S (1975) On the relative importance of the driving forces of plate motion. Geophys J Int 43(1):163–200
Gahalaut VK, Gahalaut K (2007) Burma plate motion. J Geophys Res Solid Earth 112(B10)
Gahalaut VK, Nagarajan B, Catherine JK, Kumar S (2006) Constraints on 2004 Sumatra Andaman earthquake rupture from GPS measurements in Andaman–Nicobar Islands. Earth Planet Sci Lett 242(3):365–374
Gahalaut VK, Subrahmanyam C, Kundu B, Catherine JK, Ambikapathy A (2010) Slow rupture in Andaman during 2004 Sumatra–Andaman earthquake: a probable consequence of subduction of 90°E ridge. Geophys J Int 180(3):1181–1186
Goli A, Pandey DK (2014) Structural characteristics of the Andaman forearc inferred from interpretation of multichannel seismic reflection data. Acta Geologica Sinica (English Edition) 88(4):1145–1156
Grevemeyer I, Tiwari VM (2006) Overriding plate controls spatial distribution of megathrust earthquakes in the Sunda–Andaman subduction zone. Earth Planet Sci Lett 251(3):199–208
Guzman-Speziale M, Ni JF (1996) Seismicity and active tectonics of the western Sunda Arc. World Reg Geol 1(8):63–84

Hla Maung H (1987) Transcurrent movements in the Burma–Andaman Sea region. Geology 15(10):911–912
Iyengar RN, Sharma D, Siddiqui JM (1999) Earthquake history of India in medieval times. Indian J Hist Sci 34:181–238
Jarrard RD (1986) Relations among subduction parameters. Rev Geophys 24(2):217–284
Jhingran AG (1953) A note on an earthquake in the Andaman Islands (26th June 1941). Rec Geol Surv India 82(2)
Kamesh Raju K, Ramprasad T, Rao PS, Rao BR, Varghese J (2004) New insights into the tectonic evolution of the Andaman basin, northeast Indian Ocean. Earth Planet Sci Lett 221(1):145–162
Katili JA (1973) Geochronology of West Indonesia and its implication on plate tectonics. Tectonophysics 19(3):195–212
Katili JA (1975) Volcanism and plate tectonics in the Indonesian island arcs. Tectonophysics 26(3–4):165–188
Kayal JR, Gaonkar SG, Chakraborty GK, Singh OP (2004) Aftershocks and seismotectonic implications of the 13 September 2002 earthquake (Mw 6.5) in the Andaman Sea basin. Bull Seismol Soc Am 94(1):326–333
Kennett BLN, Cummins PR (2005) The relationship of the seismic source and subduction zone structure for the 2004 December 26 Sumatra–Andaman earthquake. Earth Planet Sci Lett 239(1):1–8
Khan PK, Chakraborty PP (2005) Two-phase opening of Andaman Sea: a new seismotectonic insight. Earth Planet Sci Lett 229(3):259–271
Klootwijk CT, Gee JS, Peirce JW, Smith GM (1992) Neogene evolution of the Himalayan-Tibetan region: constraints from ODP Site 758, northern Ninetyeast Ridge; bearing on climatic change. Palaeogeogr Palaeoclimatol Palaeoecol 95(1–2):95–110
Konca AO, Hjorleifsdottir V, Song TRA, Avouac JP, Helmberger DV, Ji C, Sieh K, Briggs R, Meltzner A (2007). Rupture kinematics of the 2005 Mw 8.6 Nias–Simeulue earthquake from the joint inversion of seismic and geodetic data. Bull Seismol Soc Am 97(1A):S307–S322
Kumar MR, Rao NP, Chalam SV (1996) A seismotectonic study of the Burma and Andaman arc regions using centroid moment tensor data. Tectonophysics 253:155–165
Kumar P, Srijayanthi G, Ravi Kumar M (2016) Seismic evidence for tearing in the subducting Indian slab beneath the Andaman arc. Geophys Res Lett 43(10):4899–4904
Kundu B, Legrand D, Gahalaut K, Gahalaut VK, Mahesh P, Kamesh Raju KA, Catherine JK, Ambikapthy A, Chadha RK (2012) The 2005 volcano-tectonic earthquake swarm in the Andaman Sea: triggered by the 2004 great Sumatra–Andaman earthquake. Tectonics 31(5)
Lallemand SE, Schnürle P, Malavieille J (1994) Coulomb theory applied to accretionary and nonaccretionary wedges: possible causes for tectonic erosion and/or frontal accretion. J Geophys Res Solid Earth 99(B6):12033–12055
Lay T, Kanamori H, Ammon CJ, Nettles M, Ward SN, Aster RC, Beck SL, Bilek SL, Brudzinski MR, Butler R, DeShon HR, Ekström G, Satake K, Sipkin S (2005) The great Sumatra–Andaman earthquake of 26 December 2004. Science 308(5725):1127–1133
Le Dain AY, Tapponnier P, Molnar P (1984) Active faulting and tectonics of Burma and surrounding regions. J Geophys Res Solid Earth 89(B1):453–472
McCaffrey R (1992) Oblique plate convergence, slip vectors, and forearc deformation. J Geophys Res Solid Earth 97(B6):8905–8915
McCaffrey R (2009) The tectonic framework of the Sumatran Subduction zone. Annu Rev Earth Planet Sci 37:345–366
McCloskey J, Nalbant SS, Steacy S (2005) Indonesian earthquake: earthquake risk from co seismic stress. Nature 434(7031):291
McNeill LC, Henstock TJ (2014) Forearc structure and morphology along the Sumatra-Andaman subduction zone. Tectonics 33(2):112–134
Mishra OP, Kayal JR, Chakrabortty GK, Singh OP, Ghosh D (2007) Aftershock investigation in the Andaman-Nicobar Islands of India and its seismotectonic implications. Bull Seismol Soc Am 97(1A):S71–S85

Mitchell AHG (1989) The Shan Plateau and western Burma: Mesozoic-Cenozoic plate boundaries and correlations with Tibet. Tectonic Evol Tethyan Region. Springer, Netherlands, pp 567–583

Morley CK, Alvey A (2015) Is spreading prolonged, episodic or incipient in the Andaman Sea? Evidence from deepwater sedimentation. J Asian Earth Sci 98:446–456

Ortiz M, Bilham R(2003) Source area and rupture parameters of the 31 December 1881 Mw = 7.9 Car Nicobar earthquake estimated from tsunamis recorded in the Bay of Bengal. J Geophys Res Solid Earth 108(B4)

Pal T, Mitra SK, Sengupta S, Katari A, Bandopadhyay PC, Bhattacharya AK (2007) Dacite–andesites of Narcondam volcano in the Andaman Sea—an imprint of magma mixing in the inner arc of the Andaman-Java subduction system. J Volcanol Geoth Res 168(1):93–113

Peltzer G, Tapponnier P (1988) Formation and evolution of strike-slip faults, rifts, and basins during the India-Asia collision: an experimental approach. J Geophys Res Solid Earth 93(B12):15085–15117

Pesicek JD, Thurber CH, Widiyantoro S, Engdahl ER, DeShon HR (2008) Complex slab subduction beneath northern Sumatra. Geophys Res Lett 35(20)

Pesicek JD, Thurber CH, Zhang H, DeShon HR, Engdahl ER, Widiyantoro S (2010) Teleseismic double-difference relocation of earthquakes along the Sumatra–Andaman subduction zone using a 3-D model. J Geophys Res Solid Earth 115(B10)

Petroy DE, Wiens DA (1989) Historical seismicity and implications for diffuse plate convergence in the northeast Indian Ocean. J Geophys Res Solid Earth 94(B9):12301–12319

Pringle M, Frey FA, Mervine EM, Sager WW (2008) New Ar/Ar ages from the Ninetyeast Ridge, Indian Ocean: beginning of a robust Indo-Atlantic hotspot reference frame. In: AGU Fall meeting abstracts

Radhakrishna M, Lasitha S, Mukhopadhyay M (2008) Seismicity, gravity anomalies and lithospheric structure of the Andaman arc, NE Indian Ocean. Tectonophysics 460(1):248–262

Rajendran CP, Earnest A, Rajendran K, Das RD, Kesavan S (2003) The 13 September 2002 North Andaman (Diglipur) earthquake: an analysis in the context of regional seismicity. Curr Sci 84(7):919–924

Rao NP, Kalpna (2005) Deformation of the subducted Indian lithospheric slab in the Burmese arc. Geophys Res Lett 32:L05301

Rao P, Rao N, Hazarika P, Tiwari V, Kumar R, Singh A (2011) Structure and tectonics of the Andaman subduction zone from modeling of seismological and gravity data. In: New Frontiers in Tectonic Research-General Problems, Sedimentary Basins and Island Arcs. InTech

Rock NM, Syah HH, Davis AE, Hutchison D, Styles MT, Lena R (1982) Permian to Recent volcanism in northern Sumatra, Indonesia: a preliminary study of its distribution, chemistry, and peculiarities. Bull Volc 45(2):127–152

Rodolfo KS (1969a) Bathymetry and marine geology of the Andaman Basin, and tectonic implications for Southeast Asia. Geol Soc Am Bull 80(7):1203–1230

Rodolfo KS (1969b) Sediments of the Andaman basin, northeastern Indian Ocean. Mar Geol 7(5):371–402

Rogers WH (1883) U.S. Patent No. 272,327. U.S. Patent and Trademark Office, Washington, DC

Roy TK (1983) Geology and hydrocarbon prospects of Andaman–Nicobar basin. Petroliferous basins of India petroleum. Asia J 37–50

Roy TK, Chopra NN (1987). Wrench faulting in Andaman forearc basin, India. In: Offshore technology conference

Ruff L, Kanamori H (1980) Seismicity and the subduction process. Phys Earth Planet Inter 23:240–252

Saffer DM, Bekins BA (2002) Hydrologic controls on the morphology and mechanics of accretionary wedges. Geology 30(3):271–274

Scholz CH, Campos J (1995) On the mechanism of seismic decoupling and back arc spreading at subduction zones. J Geophys Res Solid Earth 100(B11):22103–22115

Sclater JG, Fisher RL (1974) Evolution of the east: central Indian Ocean, with emphasis on the tectonic setting of the Ninetyeast Ridge. Geol Soc Am Bull 85(5):683–702

Shapiro NM, Ritzwoller MH, Engdahl ER (2008) Structural context of the great Sumatra–Andaman Islands earthquake. Geophys Res Lett 35(5)
Sheth HC, Ray JS, Bhutani R, Kumar A, Awasthi N (2010) The latest (2008–09) eruption of Barren Island volcano, and some thoughts on its hazards, logistics and geotourism aspects. Curr Sci 98(5):620–662
Sieh K, Natawidjaja D (2000) Neotectonics of the Sumatran fault, Indonesia. J Geophys Res Solid Earth 105(B12):28295–28326
Moeremans R, Singh, C, Satish (2014). Seismic evidence of continental margin influence on the Ninety East ridge in the Bay of Bengal. Geophys Res Lett 41:7143–7150
Socquet A, Vigny C, Chamot-Rooke N, Simons W, Rangin C, Ambrosius B (2006) India and Sunda plates motion and deformation along their boundary in Myanmar determined by GPS. J Geophys Res Solid Earth 111(B5)
Song TRA, Simons M (2003) Large trench-parallel gravity variations predict seismogenic behavior in subduction zones. Science 301(5633):630–633
Srijayanthi G, Ravikumar M (2019) Subslab anisotrophy in Andaman subductin zone controlled by slab dip? Phys Earth Planet Interors 286:21–28
Stein S, Okal EA (2005) Seismology: speed and size of the Sumatra earthquake. Nature 434(7033):581–582
Subarya C, Chlieh M, Prawirodirdjo L, Avouac JP, Bock Y, Sieh K, McCaffrey R (2006) Plate-boundary deformation associated with the great Sumatra–Andaman earthquake. Nature 440(7080):46–51
Subrahmanyam C, Gireesh R, Chand S, Raju KK, Rao DG (2008) Geophysical characteristics of the Ninetyeast Ridge-Andaman island arc/trench convergent zone. Earth Planet Sci Lett 266(1):29–45
Suess E (1904) The face of the Earth, vol 5. Clarendon Press, Oxford
Swe W (1972) Strike-slip faulting in Central Belt of Burma. In: Haile NS (ed) Regional conference on the geology of Southeast Asia. Geological Society of Malaysia, Kuala Lumpur
Tiwari VM, Diament M, Singh SC (2003) Analysis of satellite gravity and bathymetry data over Ninety-East Ridge: variation in the compensation mechanism and implication for emplacement process. J Geophys Res Solid Earth 108(B2)
Uyeda S, Kanamori H (1979) Back-arc opening and the mode of subduction. J Geophys Res Solid Earth 84(B3):1049–1061
Van Huene R, Scholl DW (1991) Observations at convergent margins concerning sediment subduction, subduction erosion, and the growth of continental crust. Rev Geophys 29(3):279–316
Weeks LA, Harbison RN, Peter G (1967) Island arc system in Andaman Sea. AAPG Bull 51(9):1803–1815
Wegener A (1966) The origin of continents and oceans. Dover Publications, New York, p 246
Yadav R, Tiwari VM (2018) Lithospheric density structure of Andaman Subduction Zone from joint modelling of gravity and geoid data. IJMG 47(5):931–936
Yadav R, Tiwari VM (2019) A three-dimensional density model of north Andaman subduction zone. J Geodyn 129:247–261
Yadav RK, Kundu B, Gahalaut K, Catherine J, Gahalaut VK, Ambikapthy A, Naidu MS (2013) Coseismic offsets due to the 11 April 2012 Indian Ocean earthquakes (Mw 8.6 and 8.2) derived from GPS measurements. Geophys Res Lett 40(13):3389–3393
Yu C, Shi X, Yang X, Zhao J, Chen M, Tang Q (2017) Deep thermal structure of Southeast Asia constrained by S-velocity data. Res, Mar Geophys Res. https://doi.org/10.1007/s11001-017-9311-x

Crustal Model for the Andaman Outer Arc: Constraints from Earthquake, Gravity and Receiver Function Data

K. Haripriya, M. Radhakrishna and Manoj Mukhopadhyay

Abstract Here we present an integrated model for the descending crust of the Indian Plate by reinterpreting the gravity anomalies and constraining the model by earthquake data and Receiver Function (REF) results. The Slab Residual Bouguer Gravity Anomaly (SRBGA) calculated for the Andaman arc region reaches its maximum amplitude towards the eastern flank of Andaman Outer Arc (AOA) below the Andaman Forearc. The integrated crustal model for the AOA presented here is based on the following data constraints: (a) Digital Elevation Model (DEM) for AOA, (b) IRIS seismicity data for top lithospheric earthquakes that allows to trace the overall pattern of seismically active crust underlying three E-W profiles crossing the plate margin, (c) the Moho-map for AOA prepared by making use of the Parker Oldenburg's Algorithm to invert the SRBGA, and (e) by undertaking integrated modeling for two E-W geophysical traverses, AA′ and BB′, taken across the AOA in areas of the North and South Andaman Islands at a gap of ~400 km distance. Gravity interpretation for the two profiles is constrained by results available from REF analysis on (i) Moho depth, (ii) their further derivation into crustal density by using Brocher 2005-empirical relationship and (iii) Poisson's ratio. These values are deduced from REF data monitored at eight broadband seismic stations distributed north-south on the AOA. These results demonstrate a systematic increase in average crustal density from the Middle to South Andaman; while, an increase in Poisson's ratio demarcates a locally thickened crust in the intervening region between these islands. Main results from the integrated model are: (i) Crustal model beneath the AOA approaching the subduction zone, (ii) A low density zone at shallow depth within the Andaman Sedimentary Arc (ASA), a partial support for which comes from a Low Velocity Zone (LVZ) detected by REF having Vs = 1.3–2.5 km/s, (iii) Deeper continuity of the Jarwa Thrust contained within ASA to crustal depths, and (iv) the refined Moho-margin against the top-mantle. Gross differences in crustal properties, namely; Vs-values, thickness, density and Poisson's ratio, are noticed between the Middle and South

K. Haripriya · M. Radhakrishna (✉)
Department of Earth Sciences, Indian Institute of Technology Bombay, Powai, Mumbai 400076, India
e-mail: mradhakrishna@iitb.ac.in

M. Mukhopadhyay
Department of Applied Physics, PNG University of Technology, Lae, Papua New Guinea

J. S. Ray and M. Radhakrishna (eds.), *The Andaman Islands and Adjoining Offshore: Geology, Tectonics and Palaeoclimate*, Society of Earth Scientists Series,
https://doi.org/10.1007/978-3-030-39843-9_8

Andaman Islands which are separated by a locally thickened crust by ~8 km. Depth-slices are presented based on 32 focal mechanism solution for crustal earthquakes occurring below the traverses AA′ and BB′, which reveal: the prevalence of trench-parallel normal faulting under western parts of AOA, while, thrust faulting is more conspicuous closer to the subduction zone as defined by the Jarwa Thrust and the Eastern Margin Fault.

Keywords Andaman outer arc · Seismotectonics · Gravity anomalies · Seismological data · Andaman-Nicobar islands

1 Introduction

The AOA presents nearly 1100 km long margin between the Indian and Burma plates where the major tectonic features are: Andaman Trench, AOA (comprised of the ASA), Nicobar Deep—Andaman Forearc, and Andaman Backarc with its spreading ridge (Rodolfo 1969; Curray et al. 1979) (Fig. 1). The Andaman-Nicobar Trench represents the northernmost portion of active plate subduction in NE Indian Ocean, further north the plate margin is diffused. The AOA formed in Oligo-Miocene times due to east-west compression of sediments derived from the Malayan Shelf to the east and the Irrawaddy Delta at the Burmese continental margin to the north (Rodolfo 1969). A compiled list from published literature on the stratigraphic succession of rocks constituting the ASA infers that the minimum sediment thickness expected here is of the order of 4 km (Table 1). The sediments together with their underlying crust are dragged down the subduction zone along the east margin of the AOA. We have discussed elsewhere that the deepest penetration depth of the subducting lithosphere is ~150 km below the Andaman Sea (Dasgupta and Mukhopadhyay 1993; Dasgupta et al. 2003).

Structures in the AOA are mostly dominated by east-dipping nappes showing gentler folding in northern areas as compared to tighter folding towards the south (Weeks et al. 1967). Structures found in the Cretaceous-Oligocene sediments are generally more deformed than those developed in younger strata (Eremenko and Sastri 1977). Lithology of the Andaman Islands comprises of dismembered ophiolite slices and Cretaceous and Tertiary sediments, and the forearc is dominated by siliciclastic turbidites (Andaman Flysch Group) and Pyroclastic flow (Archipelago Group) (Pal et al. 2003; Cochran 2010; Bandopadhyay and Carter 2017). Several north-striking faults and thrusts within AOA are known from surface mapping and also from offshore seismic surveys; the best known of them is the Jarwa thrust developed in the Andaman Islands. Marine seismic reflection data presented by Curray et al. (1979) and Moeremans and Singh (2015) mapped a set of north-south faults along the eastern edge of the AOA; most prominent of them are the Eastern Margin Fault and the West Andaman Fault (Fig. 1).

Crustal configuration and subsurface mass distribution below the AOA are poorly studied at present. Our objective in the present study is to focus on the structures

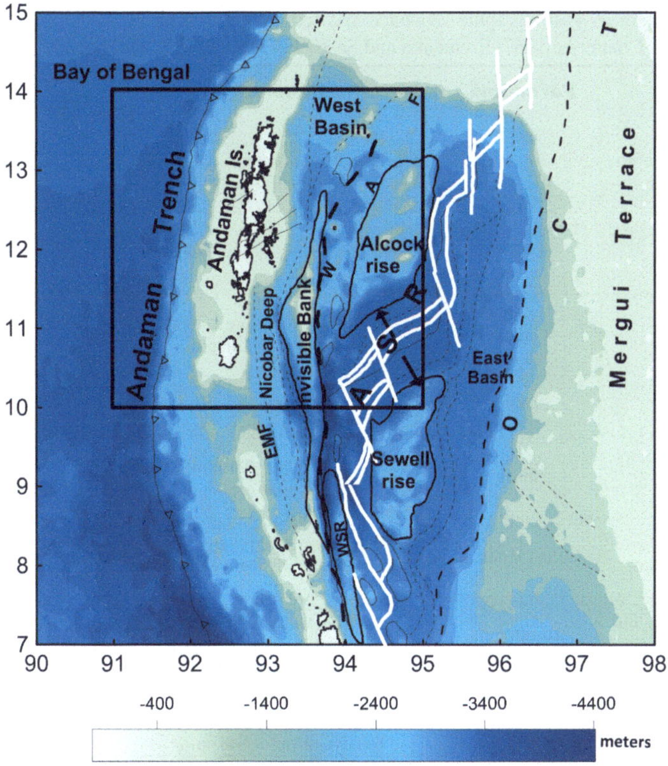

Fig. 1 Bathymetric map for the Andaman Outer Arc as the Indian Ocean plate subducts under the Burma plate in the Andaman Sea. EMF: Eastern Margin Fault. WAF: West Andaman Fault. WSR: West Sewell Ridge. ASR: Andaman Spreading Ridge. OCT: Ocean Continent transition

underlying the AOA and the configuration of the crust at the Andaman subduction zone. For this, we firstly obtain the 3-D gravity inverted Moho geometry below AOA and then carry out 2-D gravity forward modeling for two traverses across the Islands by utilizing constraints from receiver function analysis based on the data from eight broadband seismic stations on the ASA. This helps quantifying the subsurface mass distribution for the structural model as prepared by gravity interpretation for the two traverses crossing the Middle and South Andaman Islands. Focal mechanism solutions of crustal earthquakes that occurred in intimate association with the descending crust is interpreted in the context of the geometric configuration deduced by the gravity models. The REF results provide significant indications on crustal velocity and thickness values beneath AOA.

Table 1 Generalized stratigraphic sequence and sediment character of Andaman Island region (*Data source* Chatterjee 1967; Eremenko and Sastri 1977; Roy 1983; Mukhopadhyay 1988)

Age	Formation	Lithology	Thickness (Max.) in meter
Pleistocene to Recent		Beach and tidal deposits coral reef and raised beaches	
Unconformity			
Miocene to Pliocene	Long	Foraminiferal clay, thin bands of silt	60
	Guitar	Foraminiferal limestone, calcareous sandstone and siltstone	450
Unconformity			
	Round	Chalk, sandstone and siltstone	520
	Strait	Sandstone, grit, conglomerate, marl and siltstone	500
Unconformity			
Late Eocene to Oligocene	Port Blair	Thick to massive sandstone, shale and siltstone	750
Unconformity			
Late Cretaceous to Paleocene	Baratang	Shale, associated gray wackes and limestone	1370
Unconformity			
Cretaceous	Port Meadow	Radiolarian chert, jaspers, quartzite, limestone, marble and oceanic basement/ophiolite suite	>500

2 Digital Elevation Map in Relationship to Seismicity Below the Andaman Outer Arc

Seismicity is superposed on the DEM for AOA, prepared using Sandwell and Smith Satellite data (Fig. 2a). The map illustrates the regional topography surrounding the ASA and its correlation to plate margin seismicity. It shows higher topography of the AOA in respect of the trench to the west; the topography steeply decreases towards the Nicobar deep to its east.

The most conspicuous tectonic features appearing on the map are: the Andaman Trench, ASA, the Invisible Bank, and the Alcock Rise. Topographic low is observed along the trench region and high along the island. Further, the West Andaman Fault is visible as a continuous fault in N-W direction and Alcock Seamount as a topographic high forming the volcanic arc.

Seismicity map for the study area is prepared using IRIS data for the period 1978 through 2012. As it can be seen from the map that earthquakes of moderate to medium sizes originate both below the AOA, Andaman Forearc and the Backarc spreading region. Three profiles SS1–SS3 are selected for studying the seismicity-depth pattern, (location on Fig. 2a). The northernmost profile SS1 is trending in

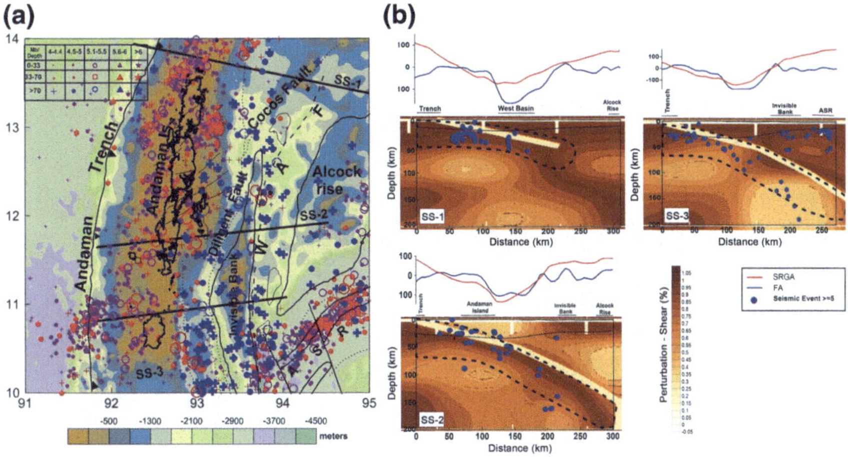

Fig. 2 **a** Seismotectonic map for the Andaman Outer Arc superposed on Digital Elevation Map. **b** Seismicity-depth sections, SS-1, SS-2 and SS-3, are generated by projecting the earthquakes within a narrow band of ±0.5° in latitude on either side of the profiles and comparison with the wave speed models of Kennett and Cummins profiles (2005)

WNW-ESE' direction crossing the Trench, West Basin, Cocos fault and WAF. The middle profile SS2 trending in WSW-ENE direction crossing the Trench, Andaman Island, Diligent fault, Invisible Bank and Alcock Rise. The southernmost profile SS3 trending in a similar direction as SS2 crosses the Diligent fault, Invisible Bank and the Andaman spreading centre on the far east end. Structure of the Benioff zone is discernible by seismicity pattern. Additional knowledge on the subducting slab is provided by the S-wave speed model prepared by Kennett and Cummins (2005) using inversion of S waves generated by aftershocks of the 2004 Andaman Sumatra Earthquake. East-west cross-sections of S wave speed model are displayed for different latitudes in the region. The subducting slab is demarcated by faster shear wave speed. We have generated Benioff zone geometry underlying these profiles by projecting the earthquakes within a narrow band of ±0.5° in latitude on to the profiles; where the configuration is constrained by the shear wave speed model proposed by Kennett and Cummins (2005) (Fig. 2b).

3 Integrated Crustal Model for the Andaman Outer Arc

Any gravity interpretation study for AOA crust is bound to remain speculative at the present stage since there is no gravity coverage available directly over the Andaman Islands. The gravity field for the Andaman arc, first described by Peter et al. (1966), illustrates a bipolar gravity signature of average amplitude of 180 mgal, higher gravity anomaly near the Invisible Bank in the north Andaman Sea and gravity low over the

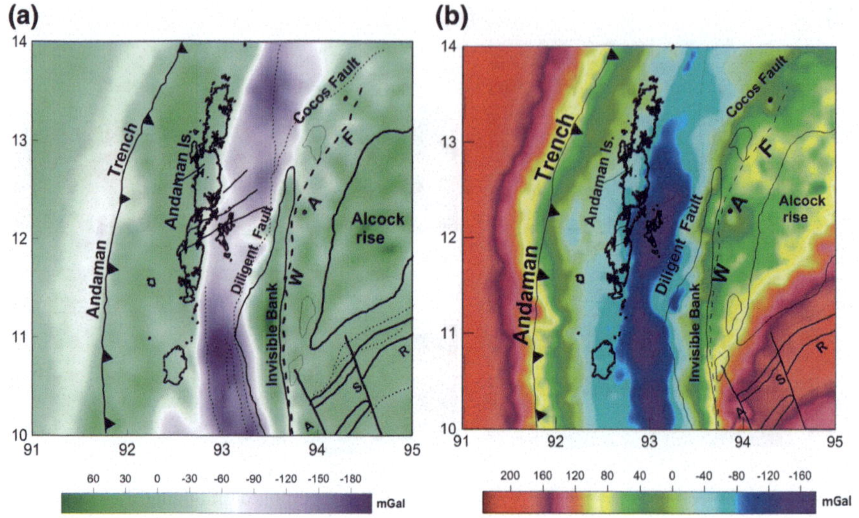

Fig. 3 **a** Tectonic features of the Andaman Outer Arc superposed on Free-Air anomalies as overlay. **b** Tectonic features of the Arc are superposed on Slab Residual Bouguer Gravity Anomaly map. Refer text

Nicobar Deep. We present a revised Free-air anomaly (FA) map (Fig. 3a) for the Andaman Arc based on the available gravity data. Elsewhere we have given an interpretation.

We have postulated elsewhere that large scale mass transfer occurs at the leading edge of the subducting Indian Plate based on a refined crustal model (Radhakrishna et al. 2008). Here we improve upon this model by considering a 3D configuration of subducting Indian plate where the method of Talwani and Ewing (1960) is utilized to calculate the 3-D subducting slab gravity below Andaman Benioff zone. The actual calculation was done by dividing the slab into several horizontal cross-sections and gravity effect of these sections are calculated and integrated over depth from top to deepest point. Using the computed value of slab gravity anomaly, a contour map of slab gravity has been prepared with smooth long wavelength high over the east of Little Andaman and reduces towards the trench region. The so-called 'SRBGA' (Fig. 3b) map is prepared by subtracting the computed slab gravity anomaly from Complete Bouguer Anomaly. We have studied elsewhere the overall gravity anomaly pattern in surrounding offshore areas of the AOA (Mukhopadhyay 1988; Mukhopadhyay and Krishna 1995). Notwithstanding such limitations, 2D-Gravity modeling is undertaken for the AOA by using the SRBGA for the Andaman Arc. Our initial study using SRBGA for the region mainly focused on the deeper lithosphere (Radhakrishna et al. 2008). Here our contention is to focus on the AOA: Accretionary Complex and crustal column underneath the exposed sedimentary ridge. The SRBGA shows a major regional low at the centre with second-order residual highs, anomaly variation is by almost 250 mgal (from −124 to 125 mgal) (Fig. 3b). The regional gravity low is attributed by the subducting lithosphere.

The 3D-Moho map (Fig. 4a) for the area is next prepared by taking help of the Parker Oldenburg's Algorithm (Gomez Ortiz and Agarwal 2005) by inverting SRBGA. The input parameters are the average Moho depth, assumed density for the crust and sub-crustal layers. Here the Moho-depth ranges from 15 to 25 km; of average thickness 20 km shows the best fit to the depth derived from the receiver function results (Gupta et al. 2016). The average density for crust and sub-crustal layers are adopted as 2.9 and 3.3 g/cm^3 respectively. We make use of the 3D-Moho-map for purposes of integrated 2D modelling underlying the two traverses, AA′ and BB′, taken across the Middle and South Andaman Islands (location on Fig. 4a). Two available geologic sections, PP′ and QQ′ (source: Roy 1983, 1992), which constitute part of Profiles AA′ and BB′, elucidate the sediment deposition pattern underlying the ASA. A comparison between the calculated Moho values and that obtained from receiver functions has been shown in Fig. 4b.

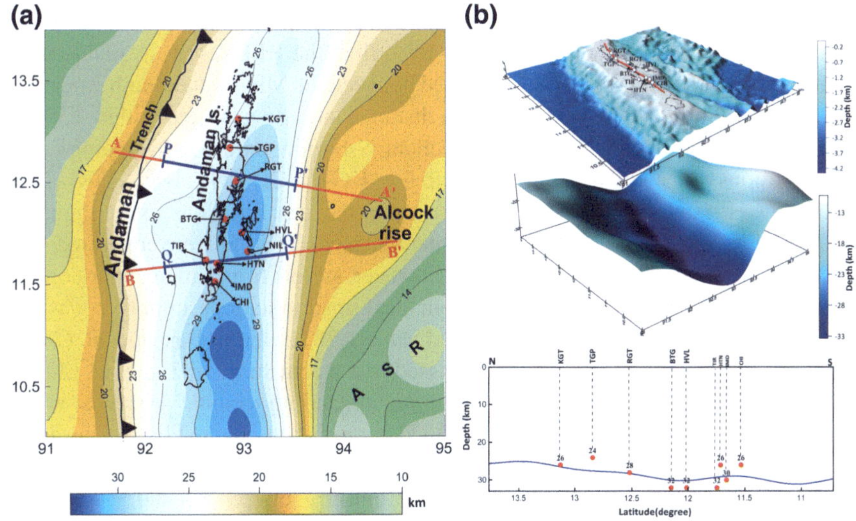

Fig. 4 **a** Moho map with tectonic features of the Andaman Outer Arc showing the traverse locations, AA′ and BB″. Profile segments PP′ and QQ′ are discussed in the text. **b** Top panel: Digital elevation model for the AOA, broadband seismic stations are identified, Middle panel: 3D- representation of Moho-depth map, Bottom panel: Comparison of Moho-depths derived from receiver function analysis corresponding to the seismic stations identified on the top panel (source: Gupta et al. 2016) and corresponding to the latitudinal section (line marked in red) in the top panel

4 Receiver Function Analysis for the Andaman Outer Arc—Inferred Values on Crustal Density, Poisson's Ratio and Thickness

REF results are of much significance in Integrated Modeling for subducting crust due to three primary reasons: REF itself is a function of seismic impedance which is a product of velocity and density, it can estimate the Poisson's ratio, and it can derive crustal thickness values. An incoming near-vertical wave generated by a teleseismic event is used for REF analysis as it enters a seismic station; hence the vertical component in a 3- component seismogram can be deconvolved from the E- and N-components rotated to tangential and radial for isolating the signal from the underlying crust underneath the station. REF results can resolve the crustal depths where distinct horizontal layering is suspected, particularly the Moho-horizon. For the AOA region, the results on active-source seismic surveys, published so far, are confined to offshore areas and that too, they do not quite penetrate to great depths into the crust. To overcome such limitations, we utilize the REF results available from eight broadband seismic stations distributed between the Middle and South Andaman Islands belonging to AOA over a 200 km distance latitude-wise and within a 30 km span in longitude, thus providing an excellent N-S view on the east-dipping crust. These stations, located from north to south are: KGT, TGP, RGT, BTG, TIR, HTN, IMD and CHI (source: Gupta et al. 2016) where receiver functions for earthquakes of magnitude 5.5 and above having epicenter distances 30°–95° were computed. The REF data analysis given by these authors, however, suffers from a major shortcoming that no reliable Vp value was available from local seismic stations, rather Vp was calculated from assumed Vp/Vs ratio of fixed value 1.78 similar to Bassin et al. (2000) and Rao et al. (2011). Moreover, as discussed in a recent study by Liu et al. (2017), that though Vp/Vs calculation is crucial in understanding tectonic processes which modify the crust, the lack of reliable Vp/Vs determinations is most likely caused by the strong reverberations on the REFs with the loose sedimentary layer. We calculate Vp using a new empirical regression relationship between Vp and Vs given by Brocher (2005). $Vp = 0.9409 + 2.0947 * Vs - 0.8206 * Vs^2 + 0.2683 * Vs^3 - 0.0251 * Vs^4$.

Brocher (2005) made a review on various nonlinear, multivalued and quantitative relations involving: Poisson's ratio (σ), Vp and Vs values, which are often employed to infer density for a wide range of lithology present in the crust. In order to infer the probable density for the AOA crustal layers, we make use of Vs values reported by Gupta et al. (2016), and evaluate the density ρ as well as the Poisson's ratio, by the revised 'Brocher-2005-relationship'. Poisson's ratio for AOA crust is estimated in the present case by using the Brocher (2005) relation: $0.5\ [(Vp/Vs)^2 - 2]/(Vp/Vs)^2 - 1]$. Density is calculated by the Nafe Drake Relation $\rho = 1.6612\ Vp - 0.4721\ Vp^2 + 0.0671\ Vp^3 - 0.0043\ Vp^4 + 0.000106\ Vp^5$ (Brocher 2005). These are listed in Table 2, and plotted on Figs. 5 and 6, showing the average density and Poisson's ratio for the upper and lower crustal rocks while their summary results are presented in Fig. 7 between the Middle and South Andaman Islands. Poisson's ratio is a useful

Table 2 Vs-values and crustal thickness obtained from the seismic stations located on Andaman Island

Station	Vs (km/s)	Crustal thickness	Vp (km/s)	Poisson's ratio	Density (g/cc)
KGT_crust	3.2	26	5.40	0.23	2.60
TGP_crust	3.1	24	5.22	0.23	2.57
RGT_crust	3.2	28	5.40	0.23	2.60
BTG_Upper crust	3.8	6	6.54	0.24	2.84
BTG_Lower crust	3.6	16	6.15	0.24	2.75
BTG_crust	3.25	34	5.49	0.23	2.62
TIR_Upper crust	3.5	12	5.96	0.24	2.71
TIR_Lower crust	3.9	12	6.74	0.25	2.89
TIR_crust	3.3	32	5.58	0.23	2.63
HTN_Upper crust	3.3	13	5.58	0.23	2.63
HTN_Lower crust	3.6	8	6.15	0.24	2.75
IMD_crust	3.35		5.67	0.23	2.65
CHI_crust	3.1	26	5.22	0.23	2.57

Showing the inferred Vp, Poisson's ratio and density in the remaining columns

elastic parameter that quantifies the shearing property of rocks. For common rock types, the Poisson's ratio varies from 0.2–0.35 and especially sensitive to the composition: increasing silica content lowers the Poisson's ratio and the mafic content increase it (Zandt and Ammon 1995). Anisotropy also can be a reason for low Poisson's ratio (Zandt et al. 1994). The average estimated Poisson's ratio for AOA crust is 0.23, a value that indicates crust with high silica content. For the upper crust (Fig. 5) there is a systematic increase in the value of the crustal thickness from the North to South Andaman, while, the average density and Poisson's ratio follow the opposite trend, high in the north which decreases gradually towards the south. Unlike upper crust, thickness for the lower crust is greater to the north but decreases towards the south (Fig. 6). In the case of average density and poisson's ratio they rather follow an east west trend, low in the east towards high in the west. For the whole crust (Fig. 7) crustal thickness is maximum in the middle Andaman and decreasing in either direction which is also evident on the Moho map (refer Fig. 4a). Notice that there is also a gradual increase towards the south for both the average crustal density as well as for Poisson's ratio.

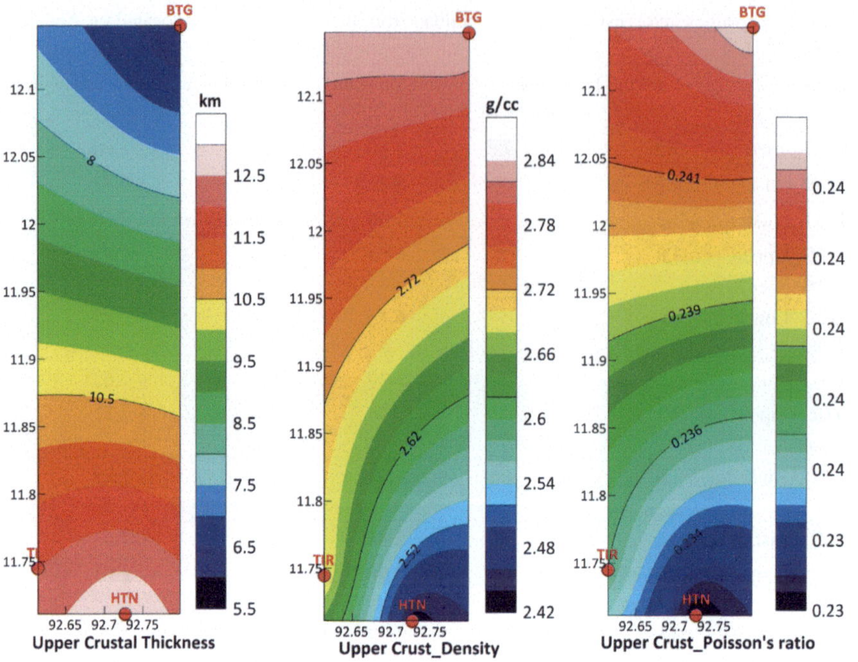

Fig. 5 Summary results on Receiver function analysis based on three seismic stations situated on the Andaman Islands: (i) Thickness of upper crust up to 14 km depth, (ii) Average density for the upper crust (in g/cm^3) and (iii) Poisson's ratio. X-axis, Longitude; Y-axis, Latitude. Refer text for station abbreviations

5 Integrated Crustal Model for the Andaman Outer Arc

Integrated crustal model is presented below for two profiles AA′ and BB′ that cross the AOA across the Middle Andaman and South Andaman Islands respectively, profile separation is by ~400 km. In forward modeling approach using Oasis Montaj software, the densities of various geological domains are fixed whereas the shapes of the bodies are changed in order to get a reasonable fit between the observed and calculated anomalies. This forward modeling procedure is based on the line integral method of Talwani et al. (1959).

Profiles AA′ and BB′ represent nearly 300 km long gravity profiles taken across the Middle Andaman and South Andaman Islands respectively, where the plate-margin tectonic features commonly transgressed are: the Andaman Trench, Accretionary Complex constituting the ASA on the down-going Indian Plate, interplate margin along the ASA in contact with the Nicobar Deep on the overriding plate to the east (Figs. 8 and 9). The SRBGA shows a major regional low at the centre and some residual highs; the anomaly variation along Profile AA′ is by 248 mgal (from −124 to 124 mgal), while, that for BB′ is by 226 mgal (from −148 to 77 mgal). Here, the

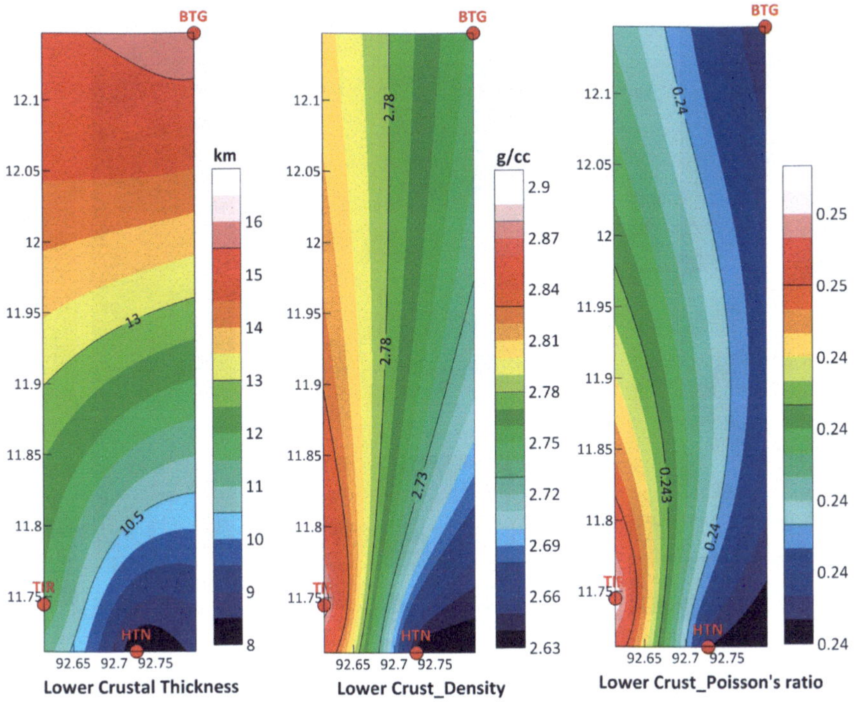

Fig. 6 Summary results on Receiver function analysis based on three seismic stations situated on the Andaman Islands: (i) Lower crust up to 10 km thick, (ii) Average density for the lower crust (in g/cm^3) and (iii) Poisson's ratio. X-axis, Longitude; Y-axis, Latitude. Refer text for station abbreviations

inherent assumption being that the gravity low is developed in consequence of mass deficiency created by the subducting lithosphere.

The topmost layer in both models is the water layer, of density 1.03 g/cm^3. Next underlying is the sediment layer of density 2.35 g/cm^3 (Table 3). Stratigraphic estimate on sediment thickness is ~4 km (Table 1), but gravity modeling infers a much greater thickness at the median part of AOA where, the sediments and ophiolites which mainly constitute the accretionary complex, attain a thickness almost double this value. The Vs-values obtained from REF data analysis for the seismic stations situated over ASA have detected a Low Velocity Zone at shallow depth, 2–6 km (Gupta et al. 2016) (in this context also refer the work on REF Data analysis for sediment strata and anisotropy in Ireland Basin, by Licciardi and Agostinetti 2017). This we interpret to be suggestive for 'Low Density Sediments', presumably fluid-saturated, whose, broad outline beneath the corresponding REF Seismic Stations is sketched in both the gravity-models. Fluids expelled from the descending crust at the subduction zone probably got trapped at shallower depth within the ASA. In support of this contention, we refer the readers to the Groundwater Investigation Field Reports prepared by the Central Ground Water Board (2013) for the water

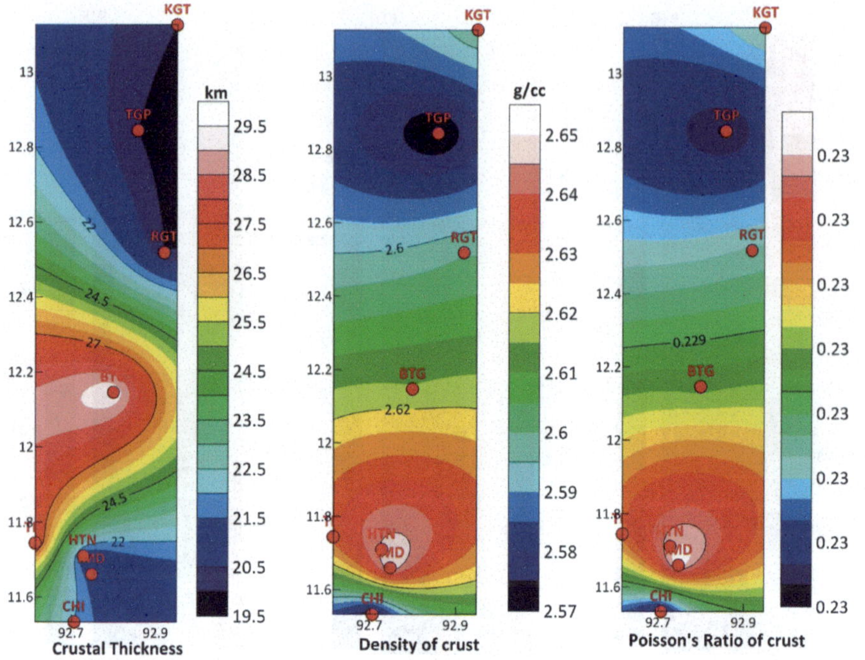

Fig. 7 Summary results on receiver function analysis based on eight seismic stations located on the Andaman Islands: (i) Crust up to Moho-margin, (ii) Average density for the entire crust (in g/cm^3) and (iii) Poisson's ratio. X-axis, Longitude; Y-axis, Latitude. Refer text for station abbreviations

springs mapped in the Andaman-Nicobar Islands. Since we have no direct access to the fluid-saturated sediments at depth, we have modeled this lesser density sediment column, of assumed density 2.3 g/cm^3, by including it within the sediment strata of overall density of 2.35 g/cm^3 for both Profiles AA′ and BB′.

In order to explain the relative gravity high along the east margin of the ASA, the gravity model requires the existence of a higher density mass of assumed density 3.0 g/cm^3, at ~4 km depth, where the Eastern Margin Fault (EMF) dissect the plate margin. EMF marks the western boundary of the Andaman forearc basin (Moeremans and Singh 2015). Inference on the nature of the mafic mass at the plate margin remains uncertain. It is worth noting here that ophiolite bodies are mapped along the east-margin of ASA, refer Roy (1983, 1992) and gravity interpretation reported by us (Mukhopadhyay 1988). Since the denser mass occupies an analogous tectonic location under both profiles, it may be taken as evidence for tectonic shearing between the descending crust against the ocean floor of the overriding plate. For simplicity, we have assumed a 2-layered oceanic crust, possessing an identical density of 2.7 and 2.9 g/cm^3, both in the descending and overthrust plates. The Moho-depths under both profiles are constrained by the REF-depths. Moho-depth control is available from three broadband seismic stations (KGT, RGT and BTG) on the Middle Andaman Island (Fig. 8). Similarly, three seismic stations (TIR, IMD and BTG)

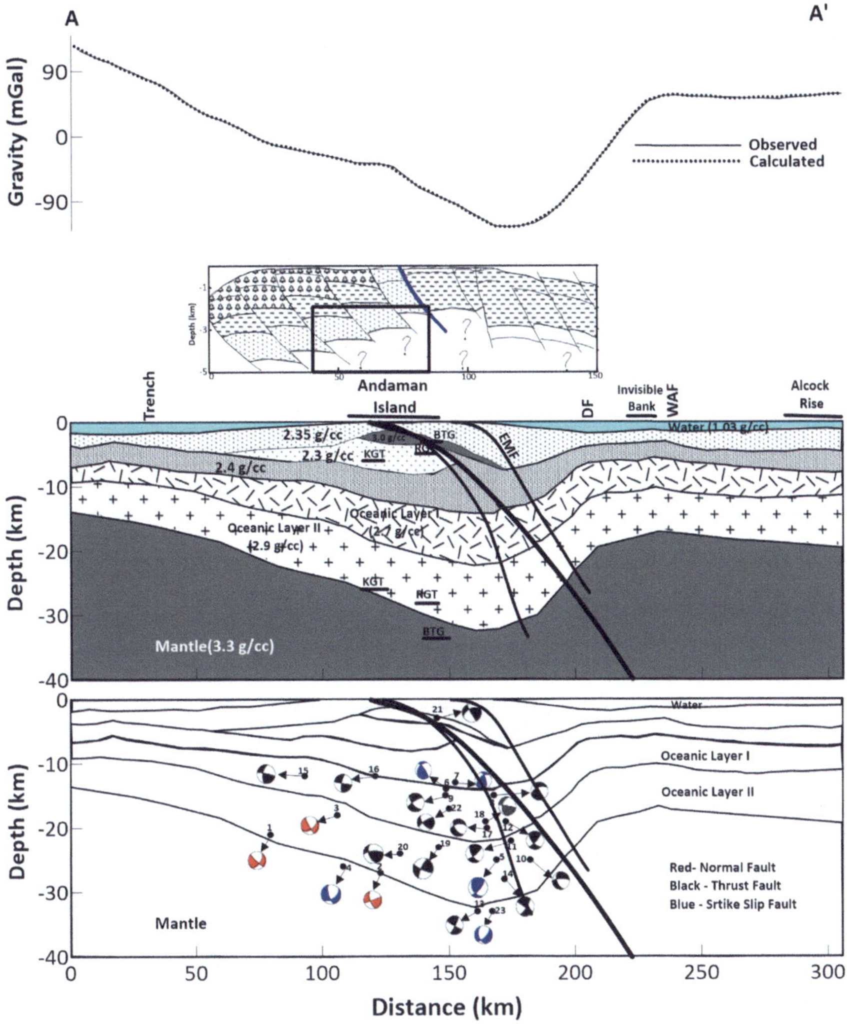

Fig. 8 Integrated Crustal Model illustrating the results of gravity interpretation underlying the Traverse AA′. Upper panel: Slab Residual Bouguer Gravity Anomaly for the Traverse AA′; location on Fig. 4a. Geological section is drawn after Roy 1992. The Jarwa thrust is marked in blue line in the geological section. Depth-slice for focal mechanism solutions for crustal earthquakes is plotted in the lower panel; refer text. Crustal thickness and sediment thickness values from the receiver functions (Gupta et al. 2016) are plotted by horizontal bars. DF: Diligent Fault, WAF: West Andaman Fault, EMF: Eastern Margin Fault

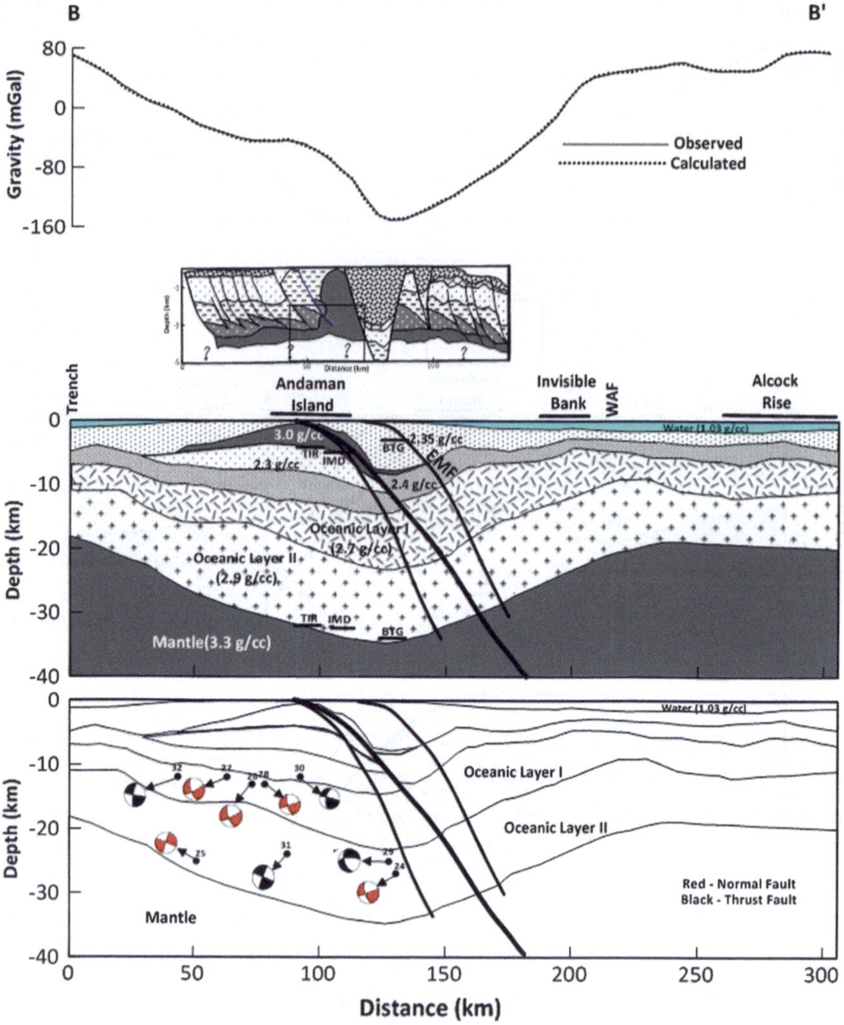

Fig. 9 Integrated Crustal Model illustrating the results of gravity interpretation underlying the Traverse BB′. Upper panel: Slab Residual Bouguer Gravity Anomaly for the Traverse BB′; location on Fig. 4a. The geological section is drawn after Roy 1983. The Jarwa thrust is marked in blue line in the geological section. Depth-slice for focal mechanism solutions for crustal earthquakes is plotted in the lower panel; refer text. Crustal thickness and sediment thickness values from the receiver functions (Gupta et al. 2016) are plotted by horizontal bars. WAF: West Andaman Fault, EMF: Eastern Margin Fault

Table 3 Density constraints used for the 2D gravity modeling

Layer	Density (g/cm^3)
Water	1.03
Sedimentary layer	2.35
Ophiolitic body	3.0
Low velocity zone	2.30
Compact sedimentary layer	2.4
Oceanic layer I	2.7
Oceanic layer II	2.9

provide constraints on Moho-depth for Profile BB′ taken across the South Andaman Island (Fig. 9).

Results for a total number of 32 focal mechanism solutions for earthquakes, with magnitude $\geq$5, originating beneath the AOA are presented next in order to gain an insight into the stress regime affecting the subducting crust. The solution source parameters are taken from our earlier study (Dasgupta et al. 2003) and the Global Centroid Moment Tensor Catalog (https://www.globalcmt.org) (Table 4). The focal mechanism solutions are schematically represented underlying the two Traverses, AA′ and BB′, in the form of depth-slices, where, the faulting pattern is differentiated in colour (Figs. 8 and 9). As it can be seen from the plots that the thrust fault mechanisms having east-dipping planes mostly define the interplate margin between the descending crust below the AOA as against the overriding crust below the Andaman Forearc. 15 such thrust faults are identified under the Traverse AA′ that closely refer to the Jarwa Thrust and the Eastern Margin fault further east, corresponding to crustal shocks. A total number of eight normal-faulting events (3 events for Traverse AA′ and 5 events for Traverse BB′) is seen on the depth-slices mostly designate the crustal areas which reflect Trench-parallel faulting due to the bending of the crust. A similar distribution of faulting and P, T and B-axes are known from other subduction zones (cf. Sandiford 2008).

6 Discussion

The Integrated Model proposed for subducting crust under the AOA is clearly an improvement on previous studies, since, we could use here constraints from the followings: DEM for AOA, revised seismicity data from IRIS, an updated SRBGA map, and derived values on crustal thickness, average crustal density, Poisson's ratio and 3D-Moho map as obtained from the available analysis on receiver function studies. This allowed us to prepare the refined crustal model for the AOA crust to reasonable details, whose, primary findings focus on the sediment configuration below the ASA and the geometry for the crustal layer representing top part of the subducting lithosphere to about 34 km depth.

Table 4 Focal mechanism source parameters for crustal earthquakes originating below the Andaman Outer Arc

Sl No.	Date (mm/dd/yyyy)	Time (hh:mm:ss)	Long	Lat	Mb	Ms	Depth (km)	Strike1	Dip1	Strike2	Dip2	Tpl	Taz	Bpl	Baz	Ppl	Paz	Nature of fault
1*	12/27/2004	00:49:32	92.37	13.05	6	5.5	21	359	42	229	60	10	297	27	32	62	189	Normal
2*	12/27/2004	14:46:49	92.72	12.39	5.8	5.3	27	191	19	344	73	28	81	8	346	61	242	Normal
3*	04/10/2007	13:56:56	92.59	13.13	5.5	5.4	18	238	49	360	60	6	117	34	23	56	216	Normal
4*	3/26/2003	04:22:35	92.67	12.78	5.8	5.3	26	245	51	348	74	15	111	47	5	39	214	Strike slip
5*	03/06/2004	10:21:35	93.19	13.05	5.2	5.2	25	214	52	319	72	41	184	46	339	13	82	Strike slip
6*	12/26/2004	11:05:06	92.84	13.49	6.3	6.3	14	29	56	132	72	38	355	50	155	10	257	Strike slip
7*	12/26/2004	14:48:44	92.86	13.52	5.8	5.7	13	137	56	38	78	33	352	53	201	14	91	Strike slip
8*	02/07/1978	12:30:48	93.19	12.98	5.5	5.3	15	129	43	2	61	60	321	28	164	10	69	Thrust
9*	02/07/1978	20:32:01	93.03	12.83	5.6	5.6	15	162	36	336	54	81	230	3	338	9	68	Thrust
10*	11/02/1981	21:10:32	92.96	11.77	5.7	5.5	25	109	21	331	74	59	259	13	147	28	50	Thrust
11*	9/13/2002	22:28:36	93.17	13.33	6.2	6.7	22	212	42	337	63	58	199	29	353	12	90	Thrust
12*	9/14/2002	19:58:41	93.15	13.33	5.7	5.6	19	206	50	336	53	62	184	28	359	2	90	Thrust
13*	04/10/2004	15:57:04	93.09	13.19	5	5.4	33	359	30	198	62	72	129	8	14	16	281	Thrust
14*	05/03/2004	17:41:33	93.19	13.18	5.3	5	28	1	34	207	59	72	151	12	20	13	287	Thrust
15*	12/30/2004	17:58:15	92.35	12.28	5.8	5.2	12	29	13	160	81	53	59	10	161	36	258	Thrust
16*	02/03/2006	20:34:15	92.17	11.83	5.6	5.9	12	356	10	175	80	55	85	0	175	35	265	Thrust
17*	6/18/2010	23:09:34	93.12	13.21	6.1	5.9	20	116	38	333	58	69	289	18	141	10	48	Thrust
18+	2/16/1974	01:51:11	92.32	11.47	5.2		19	10	10	140	84	50	42	9	142	38	238	Thrust
19+	9/30/1993	17:04:48	92.58	11.84	5.3	4.8	23	344	25	173	65	70	90	3	351	20	260	Thrust

(continued)

Table 4 (continued)

Sl No.	Date (mm/dd/yyyy)	Time (hh:mm:ss)	Long	Lat	Mb	Ms	Depth (km)	Strike1	Dip1	Strike2	Dip2	Tpl	Taz	Bpl	Baz	Ppl	Paz	Nature of fault
20+	11/02/1981	21:10:26	92.87	12.81	5.7	5.5	24	110	21	331	74	59	259	13	147	28	50	Thrust
21+	02/07/1978	12:30:40	93	12.81	5.5	5.3	3	129	43	2	61	60	321	28	164	10	69	Thrust
22+	02/07/1978	20:31:55	93.04	12.89	5.6	5.6	17	162	36	336	54	81	230	3	338	9	68	Thrust
23+	01/12/1968	04:17:43	93.12	13.27	5.5		33	322	78	222	52	17	88	50	338	35	190	Strike slip
24*	12/27/2004	14:46:49	92.72	12.39	5.8	5.3	27	191	19	344	73	28	81	8	346	61	242	Normal
25*	12/27/2004	19:13:22	92.28	11.72	5.5	5.1	25	22	20	188	71	26	282	4	190	64	91	Normal
26*	01/06/2005	11:55:48	91.99	11.01	5.4	5.3	13	211	26	348	70	23	92	17	354	61	232	Normal
27*	6/27/2008	13:07:15	91.95	11.09	5.7	5.8	12	216	28	344	72	24	91	21	351	58	225	Normal
28*	6/28/2008	15:29:14	92.1	11	5.5	5.3	13	206	19	351	75	29	90	10	354	59	247	Normal
29*	11/02/1981	21:10:32	92.96	11.77	5.7	5.5	25	109	21	331	74	59	259	13	147	28	50	Thrust
30*	12/30/2004	17:58:15	92.35	12.28	5.8	5.2	12	29	13	160	81	53	59	10	161	36	258	Thrust
31*	10/11/2005	03:38:01	92.02	10.89	5.5	5.1	24	1	22	169	68	66	71	4	171	23	263	Thrust
32*	02/03/2006	20:34:13	92.17	11.83	5.6	5.9	12	356	10	175	80	55	85	0	175	35	265	Thrust

Solutions are schematically represented as depth-slices on Figs. 8 and 9. The symbols * and + against the Sl. No. indicate the source of data. *—From Global Centroid Moment Tensor. +—From Dasgupta et al. 2003

The gravity model predicts almost 8 km thickness for the accretionary wedge under AOA represented by the ASA; this is nearly double the stratigraphic estimate for the sediments mapped on surface. Further, the gravity model predicts a Low Density Zone (LDZ) within the sediment column extending to 7 km depth, corresponding to a LVZ, underneath the ASA. Geological field mapping data on water-springs for the South and Middle Andaman Islands prove that median-parts of both the islands are dotted with plentiful of water springs (Central Ground Water Board 2013). Most likely, the water squeezed out of the subducting crust escaped and was trapped in the sediments at a shallower level that is held responsible for the LVZ and LDZ (Toya et al. 2017).

Passive seismology using distant earthquakes has played an increasingly important role in crustal structure investigation, the REF technique has extensive application in this regard since the 1980s (Prodehl et al. 2013). We have utilized the REF results from eight broadband seismic stations located on AOA to constrain the gravity interpretation presented here. Like any other subduction zone, the Moho in the present study area exists in two distinct zones: the subducting Moho below the AOA and the Moho below the overriding Burma plate. Seismic expression of both forms is supposedly linked to the landward-dipping LVZ that is noticed worldwide and which generally coincides with the Benioff-zone seismicity (Bostock 2013). The seismological traverses described in the foregoing for AOA (Fig. 2) are in excellent agreement with this observation. The oceanic igneous crust is highly deformed at the subduction zone, where it descends to depths more than 34 km. REF data on Moho-margin is utilized to constrain the Integrated Model. This is accomplished for both profiles, crossing the Middle and South Andaman Islands, having a separation by 400 km.

The 3D-Moho map illustrates the deformation experienced by the deeper crustal rocks due to the dominance of compressive stresses (Fig. 4a). Between the Middle and South Andaman, the crust is abnormally thick, by about 7–8 km, as compared to the areas to its immediate north and south. The down-bulged crust is nearly 30 km thick below the Middle Andaman Island, what thickens to more than 34 km below the South Andaman.

The structural model for the two profiles crossing the Middle and South Andaman Islands presents a view of the two most conspicuous faults defining the margin between the downgoing Indian Ocean and overriding Burma plates, namely, the Jarwa Thrust and the Eastern Margin Faults; the latter is studied in details by Moreamans and Singh (2015). Compressive stress acting along east-dipping planes are common to both from shallow to crustal depths, as shown by a number of focal mechanism solutions displayed on the gravity models. Sediment packets, as well as the oceanic igneous crust, are sliced by the two thrust faults at the plate margin. Consequently, not only the sediments are deformed, the oceanic igneous crust is, in part, dismembered and overthrust along the plate margin and taken to a shallower depth, only to be exposed in part along the east margin of ASA. We have used the empirical relationship given by Brocher (2005) to derive average crustal density and Poisson's ratio from the Vs- and Vp-values deduced from REF data analysis. The results reveal a higher density crust beneath the South Andaman Island, where, the Poisson's ratio is also found to be the highest.

7 Conclusions

Major conclusions arrived from the present study are:

(a) Three seismologic sections across the North, Middle and South Andaman Islands representing the exposed segments of the AOA demonstrate that the subducting crust up to about 38 km is seismically active throughout its entire spectrum, in so much so, that a number of earthquakes of magnitude 5.0 and above have originated from shallow to great crustal depths. Focal mechanism solution for several crustal earthquakes illustrates that the Accretionary Complex and the underlying crust experience normal or thrust faulting under NE-SW compressive stress. Thrust Fault mechanisms along east-dipping planes mostly corroborate in defining the interplate margin between the descending crust below the AOA as against the overriding crust below the Andaman Forearc. The sediments and oceanic crustal layers are intensely deformed to crustal depths between the Middle and South Andaman Islands.
(b) The Slab Residual SRBGA for AOA is prepared following the Talwani- Ewing approach. The 3D-Moho-map underlying AOA is generated by inversion of this SRBGA by making use of the Parker Oldenburg's Algorithm. Reasonable correspondence is noticed between the 3D-Moho map and the Moho-depths obtained from REF data analysis from eight broadband seismic stations distributed on AOA. Such credence allows us to utilize the 3D- Moho map as an important input for interpreting the Integrated Model between the downgoing and overriding plates in the study area.
(c) Next, we deduce the average density and Poisson's ratio for the subducting AOA crust and their variations over a distance of 200 km between the Middle and South Andaman Islands based on the available REF data, for which, the 'Brocher 2005-Empirical Relationships' are deployed. Average crustal density is found to be the highest beneath the South Andaman Island, while, the Poisson's ratio attains its highest value in the intervening region between the Middle and South Andaman, where, crust thickens abnormally by ~7 km.
(d) Integrated Model for AOA crust is presented underlying two gravity profiles for which we make use of the data constraints from seismologic profiles and the REF analysis on the crustal thickness and density. Both profiles support a model for subducting crust down to 34 km depth at the plate margin. The Eastern Margin Fault separating the AOA against the Andaman Forearc appears to be the most significant tectonic feature in this scenario. At shallower depth, a Low Velocity Zone, up to 6 km thick, detected initially by Vs-values, acts as a low density zone within the sediment package that forms the Accretionary Complex. It is conjectured that the origin of the low density sediments is due to higher

fluid content that was squeezed out of the subduction zone and trapped at a shallower depth.

(e) Focal mechanism solution for 32 crustal earthquakes, of magnitude 5.0 and above originating below the gravity profiles, helps further define the distribution of primarily the thrust or normal faulting down to ~34 km depth beneath AOA. From this exercise, it appears that the Andaman Outer Arc contains a packet-of-thrusts under its eastern margin closer to the subduction-zone. It may not be just the Jarwa Thrust or Eastern Margin Fault, rather, a series of parallel-thrusts that are commonly inclined eastward, thus slicing the descending crust. The Jarwa Thrust and the Eastern Margin Fault are by far the most prominent of these; their crustal penetration is traceable from the surface-exposure to crustal depths. Such Thrust-packet probably evolved in consequence to continuing subduction of the Indian plate under the influence of E-W compressive stress as the crust descends into the subduction zone into the Nicobar Deep.

Acknowledgements We sincerely acknowledge the University Grant Commission (UGC) for providing fellowship to Haripriya K. We were benefitted by valuable comments given by two anonymous reviewers.

References

Bandopadhyay PC, Carter A (eds) (2017) The Andaman–Nicobar accretionary ridge: geology, tectonics and hazards. Geol Soc London Memoir, vol 47, pp. 75–93

Bassin C, Laske G, Masters G (2000) The current limits of resolution for surface wave tomography in North America EOS. Trans Am Geophys Union 81:F897

Bostock MG (2013) The Moho in subduction zones. Tectnophys 609(8):547–557

Brocher TM (2005) Empirical relations between elastic wave speeds and density in the earth's crust. Bull Seism Soc Am 95(6):2081–2092

Central Ground Water Board, Eastern Region Kolkata (2013) Ground water information booklet Andaman & Nicobar Islands Sci Rep Series E 47 & 50

Chatterjee PK (1967) Geology of the main islands of the Andaman Sea. In: Proceedings symposium on upper mantle project, geophysical research board National Geophysical Research Institute, Hyderabad, India, pp 348–360

Cochran JR (2010) Morphology and tectonics of the Andaman Forearc northeastern Indian Ocean. Geophys J Int 182:631–651

Curray JR, Moore DG, Lawver LA, Emmel FJ, Raitt RW, Henry M, Kieckhefer R (1979) Tectonics of Andaman Sea and Burma. In: Watkins JS, Montadert L, Dickerson PW (eds) Geological and geophysical investigations of continental margins, vol 29. American Association Petroleum Geologists. Memoir, pp 189–198

Dasgupta S, Mukhopadhyay M (1993) Seismicity and plate deformation below the Andaman arc northeastern Indian Ocean. Tectonophys 225:529–542

Dasgupta S, Mukhopadhyay M, Bhattacharya A, Jana TK (2003) The geometry of the Burmese-Andaman subducting lithosphere. J Seismology 7:155–174

Eremenko NA, Sastri VV (1977) On the petroleum geology of Andaman Islands. Bull ONGC 14:1–13

Gomez Ortiz D, Agarwal BNP (2005) 3DINVER. M: a MATLAB program to invert the gravity anomaly over a 3D horizontal density interface by Parker–Oldenburg's algorithm. Comp Geosci 31:513–520

Gupta S, Borah K, Saha G (2016) Continental like crust beneath the Andaman Island through joint inversion of receiver function and surface wave from ambient seismic noise. Tectoniphys 687:129–138

Kennet BLN, Cummins PR (2005) The relationship of the seismic source and subduction zone structure for the 2004 December 26 Sumatra-Andaman earthquake. Earth Planet. Sci. Lett 239:1–8

Licciardi A, Agostinetti PA (2017) Sedimentary basin exploration with receiver functions: seismic structure and anisotropy of the Dublin Basin (Ireland). Geophysics 82(4)

Liu L, Gao SS, Liu KH, Mickus K (2017) Receiver function and gravity constraints on crustal structure and vertical movements of the Upper Mississippi Embayment and Ozark Uplift. J Geophys Res Solid Earth 122:4572–4583

Moeremans RE, Singh SC (2015) Fore-arc basin deformation in the Andaman-Nicobar segment of the Sumatra-Andaman subduction zone: Insight from high-resolution seismic reflection data. Tectonics 34:1736–1750

Mukhopadhyay M (1988) Gravity anomaly and deep structure of Andaman arc. Marine Geophys Res 9:197–210

Mukhopadhyay M, Krishna MBR (1995) Gravity anomalies and deep structure of Ninetyeast Ridge north of the equator, Eastern Indian Ocean—a hot spot trace model. Marine Geophys Res 17:201–216

Pal T, Chakraborty PP, Gupta TD, Singh CD (2003) Geodynamical evolution of the outer arc-forearc belt in the Andaman Islands, the central part of the Burma and Java Subduction complex. Geol Mag 140:289–307

Peter G, Weeks LA, Burns RE (1966) A reconnaissance geophysical survey in the Andaman Sea and across the Andaman Island Arc. J Geophys Res 2:495–509

Prodehl C, Kennett B, Artemieva IM, Thybo H (2013) 100 years of seismic research on the Moho Tectonophys 609(8):9–44

Radhakrishna M, Lasitha S, Mukhopadhyay M (2008) Seismicity, gravity anomalies and lithospheric structure of the Andaman arc, NE Indian Ocean. Tectonophys 460:248–262

Rao NP, Rao CN, Hazarika P, Tiwari VM, Kumar MR, Singh A (2011) Structure and tectonics of the Andaman subduction zone from modeling of seismological and gravity data (Chapter 11). New Frontiers in Tectonic Research General Problems, Sedimentary Basins and Island Arcs, pp 249–268

Rodolfo KS (1969) Bathymetry and marine geology of the Andaman basin, and tectonic implications for Southeast Asia. Geo Soc Am Bull 80:1203–1230

Roy TK (1983) Geology and hydrocarbon prospects of Andaman-Nicobar Basin. Pet Asia J 1:37–50

Roy SK (1992) Accretionary prism in Andaman Forearc. Geol Surv Ind Spl Pub 29:273–278

Sandiford M (2008) Seismic moment release during slab rupture beneath the Banda Sea. Geophys J Int 174:659–671

Talwani M, Ewing M (1960) Rapid computation of gravitational attraction of threedimensional bodies of arbitrary shape. Geophysics 25:203–225

Talwani M, Worzel JL, Landisman M (1959) Rapid gravity computations for two dimensional bodies with application to the Mendocino submarine fracture zone. J Geophys Res 64:49–59

Toya M, Kato A, Obara K, Takeda T, Yamaoka K (2017) Down-dip variations in a subducting low-velocity zone linked to episodic tremor and slip: a new constraint from ScAp waves. Scientific Reports 7 Article number: 2868

Weeks LA, Harbison RN, Peter G (1967) Island arc system in the Andaman sea. Am Assoc Pet Geol Bull 51:1803–1815

Zandt G, Ammon CJ (1995) Continental crust composition constrained by measurements of crustal Poisson's ratio. Nature 374:152–154

Zandt G, Velasco AA, Beck SL (1994) Composition and thickness of the southern Altiplano crust Bolivia. Geol 22:1003–1007

Geomagnetic Deep Sounding in Andaman Islands

P. B. V. Subba Rao, M. Radhakrishna, Santu Ghoshal, P. V. Vijaya Kumar and A. K. Singh

Abstract The Andaman-Nicobar Islands form a major part of the outer arc sedimentary complex that stretches nearly 1100 km along the Andaman-Sunda trench-arc system and separate the back-arc Andaman basin from the Bay of Bengal. Transient geomagnetic field variations recorded along 25 stations covering North-, Middle- and South- Andaman Islands during different phases have been analyzed. Hypothetical event analyses (HEA) denote anomalous Z/H along the eastern margin of Andaman Islands. This is associated with the contact zone between ophiolite bodies and the sedimentary basin containing terrigenous sediments, clay and fluids. An interpreted model along three different profiles brings out anomalous conductivity zones along Jawar thrust and eastern margin fault (EMF). Anomalous zone along Jawar thrust could be related to the formation of a fractured, cataclastic zone along a fault that can enhance the percolation of fluids. Also, these fluids are generated by the metamorphosis of the subducting crust and sedimentary rocks. Emplacement of obducted sediments and fluids through eastern margin fault has been reflected as an anomalous zone. The above said anomalies could be resolved after separating the island effect from the observed induction arrows.

Keywords Electromagnetic induction · Geomagnetic deep sounding · Conductivity anomalies · Andaman outer arc · Obducted sediments · Terrigenous sediments

P. B. V. Subba Rao (✉) · S. Ghoshal · P. V. Vijaya Kumar · A. K. Singh
Indian Institute of Geomagnetism, New Panvel, Navi Mumbai 410218, India
e-mail: srao@iigs.iigm.res.in

M. Radhakrishna
Department of Earth Sciences, Indian Institute of Technology, Powai, Mumbai 400076, India

Present Address:
S. Ghoshal
ONGC, Jorhat, Assam, India

J. S. Ray and M. Radhakrishna (eds.), *The Andaman Islands and Adjoining Offshore: Geology, Tectonics and Palaeoclimate*, Society of Earth Scientists Series,
https://doi.org/10.1007/978-3-030-39843-9_9

1 Introduction

Electromagnetic (EM) fields are widely used to study the geodynamic processes of the earth due to their ability to penetrate deep into the earth and resolve the electrical conductivity of complex geological structures (Gough 1989; Arora 2002). Electrical conductivity profile is obtained from magnetovariational (MV) or magnetotelluric (MT) methods. Both these methods are complimentary in the sense that MV has better lateral resolution and MT, a better vertical resolution. In MV, three components of the time varying earth's magnetic field are recorded by using an array of magnetometers. In MT, apart from recording three component magnetic field variations, simultaneous measurement of two orthogonal components of the horizontal electric field (telluric currents) is carried out. MV method is also known as geomagnetic deep sounding (GDS).

GDS uses natural geomagnetic transient variations to image the earth's interior in terms of lateral electrical conductivity contrasts. Data are collected by an array of magnetometers that are operated simultaneously in the study area to map electrical conductivity anomalies in terms of regional scale. Magnetometer array studies in Himalayan region was useful in mapping trans Himalaya conductor (THC) in the frontal Himalaya and contiguous Indian shield (Lilley et al. 1981; Arora et al. 1982; Reddy 1997; Arora 2002), Garhwal Lesser Himalaya conductivity anomaly (GLHCA, Reddy and Arora 1992, 1993), South Indian offshore conductivity anomaly (SIOCA, Arora and Subba Rao 2002), induction pattern observed at Andaman–Nicobar region have been explained in terms of high conducting Cretaceous–Tertiary sediments filling the Andaman–Nicobar deep along with partial melts/volatile fluids that are intruding into the eastern margin of fore-arc basin (Subba Rao 2008), etc.

After Sumatra earthquake of December 2004, a regional scale GDS (Subba Rao 2008) and pilot MT (Gokarn et al. 2006) studies were carried out. Followed by their success in providing new insight in constraining the tectonic model for the region, a fresh GDS campaign in the form of closely spaced magnetometers 25 stations have been covered in Andaman Islands during different phases in 2008–2011. The focus of the present article is to describe the frequency and polarization characteristics of the induction response observed in Andaman Islands and modeling results obtained along three different profiles (North, middle and south Andaman region).

With an aim to understand the mapped conductivity structures, the evolution of Andaman Islands has been discussed. Next the principles of GDS, data acquisition, processing and modeling steps of time varying magnetic fields collected from Andaman Islands are briefly outlined. The factors responsible for mapped electrical conductivity anomalies and its tectonic implications are discussed.

2 Geology and Tectonics of Andaman Region

Figure 1a. displays the major tectonic features of Andaman arc region (Curray 2005) superimposed on Bouguer anomaly map (after Radhakrishna et al. 2008; Haripriya et al. 2019). The major tectonic features of Andaman arc from east to west are

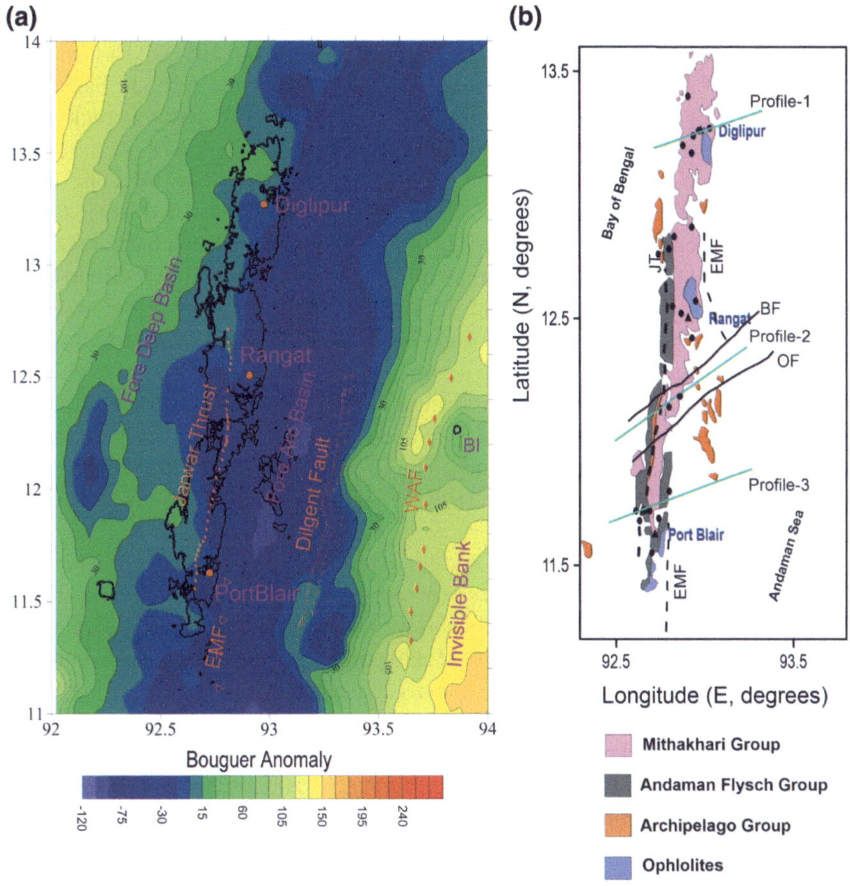

Fig. 1 **a** Regional tectonic features of Andaman Islands and surrounding region superimposed on gravity anomaly map (after Curray 2005; Radhakrishna et al. 2008; Haripriya et al. 2019). Steep negative gravity anomaly observed over forearc basin denotes that the basin consists of thick sedimentary basin. Positive anomalies over Invisible bank (IB) and Barren Island (BI) denote mafic and ultramafic intrusive bodies. Different tectonic features shown are EMF: Eastern Margin Fault and WAF: West Andaman Fault. **b** Geology map of Andaman Islands and the adjoining region (modified after Pandey et al. 1992: GSI 2001; Curray 2005) showing the locations of GDS sites along with the distribution of ophiolites and different sedimentary units. BF: Button fault, OF: Outram fault, EMF: Eastern margin Fault and JT: Jawar thrust. Interpreted 3 profiles are shown in Fig. 1b. Profile-1 stands for Diglipur profile, Profile-2 stands for Baratang profile and Profile-3 for PortBlair profile

back-arc Andaman Sea, West Andaman fault (WAF), Nicobar deep (fore arc basin), Andaman-Nicobar ridge (outer arc) and Andaman trench. This fore arc basin is bounded by EMF (eastern margin fault) on the west and on east by the Diligent faults (DF) (Cochran 2010; Singh et al. 2013; Moeremans and Singh 2015). According to Cochran (2010), DF divides forearc basin into two parts (a) shallow one towards west and (b) deep seated basin towards east. The EMF forms the boundary between the Andaman-Nicobar ridge (outer arc region) and a fore arc basin to the east and coincides with a steep gravity gradient forming the western edge of the deep gravity low over the foreland basin (Cochran 2010). West Andaman fault (WAF) forms the eastern limit of Nicobar deep. Invisible Bank (IB) within the fore arc basin is a cuesta formed by the WAF and high positive gravity anomaly over this region has been attributed to mafic/ultramafic intrusives (Roy and Chopra 1987).

The Andaman-Nicobar Islands have formed during the Oligo-Miocene times (Rodolfo 1969). The stratigraphy of Andaman Islands are classified as (i) Cretaceous to Palaeocene Ophiolite suite, (ii) Lower to Middle Eocene Mithakhari Group, (iii) Late Eocene to Oligocene Andaman turbidities (Andaman Flysch Group), and (iv) Mio-Pliocene Archipelago Group (Ray 1985; Ray et al. 1988; Pal et al. 2003; Bandopadhyay 2005; Bandopadhyay and Carter 2017). Two prominent N-S trending faults/thrusts have been observed over the Andaman Islands (Fig. 1b) (a) Jarawa thrust is the most significant morpho-tectonic feature trending N-S all along the main islands (Eremenko and Sastry 1977) and (b) Eastern Margin Fault (EMF) to the immediate east of Andaman Islands. The EMF acts as a boundary between the outer-arc ridge and the fore-arc basin (Roy 1983; Cochran 2010; Singh et al. 2013; Moeremans and Singh 2015).

Ophiolites occur along the eastern margin of the Andaman and Nicobar Islands in N–S to NE–SW trending bodies and are exposed as isolated patches in the west. These ophiolites are derived from the ocean floor as well as ultra- basic/volcanic/pelagic sediments along with older metamorphics (Mukhopadhyay et al. 2003). These ophiolite rocks have contact with underlying sediments and denote thrust-controlled emplacement in an accretionary environment (Pal et al. 2003; Ghosh et al. 2017) during subduction or westward propagated nappes thrust over a passive continental margin. Ophiolites mainly consist of serpentinite, dunite, harzburgite, pyroxenite, olivine gabbro, anorthositic gabbro and gabbroic anorthosite (Dhana Raju and Negi 1984; Jafri and Sheikh 2013).

Mud volcanoes are known from Baratang Island of the middle Andaman and Diglipur of the north Andaman (Banerjee 1975). One of the important north-south oriented thrust fault (Jarawa fault) lies close to the Baratang mud volcano (Chaudhuri et al. 2012; Mishra et al. 2007; Roy 1992) and is composed of smectite–illite– kaolinite–chlorite–plagioclase–quartz–calcite assemblage derived both from sediments and altered oceanic crust plus water released from dehydration of clay minerals (Ray et al. 2013).

3 Data Collection and Processing

3.1 Principle of Geomagnetic Deep Sounding

Geomagnetic Deep Sounding (GDS) uses natural geomagnetic transient variations in three orthogonal components, (geographic north (X), east (Y), and vertical (Z) components) to image the earth's interior in terms of lateral electrical conductivity contrasts. These time-varying magnetic fields have their origin in the current systems in the ionosphere and the distant magnetosphere which act as a natural source of electromagnetic (EM) field and induce secondary currents in the conductive layers of the earth. The depth penetration of induced currents is controlled by the skin-depth relationship and is expressed as

$$\delta = 503.0 * \sqrt{\rho T}$$

where δ is skin depth in meters for a given time period T in seconds and ρ is the resistivity of the medium in Ωm. Thus, depth of penetration depends on the unknown electrical conductivity distribution. If EM signal encounters a conductive material in subsurface, it will decay more quickly with depth than a resistive medium.

As the EM waves associated with these currents undergo attenuation and induce eddy currents, which in turn contribute to the magnetic field components measured at the surface. If earth is devoid of lateral variations in conductivity, the induced currents are flowing horizontally in a finite shell, the horizontal geomagnetic field variations (X and Y) are in phase with the inducing external part and out phase with vertical geomagnetic field variation (Z) (Gough and Ingham 1983; Arora 1997). Thus, the horizontal fields are doubled to that of the external fields and the vertical component is reduced outside the perfectly conducting half-space. The detection and interpretation of the anomalous parts of magnetovariational fields leads to the mapping of lateral conductivity anomalies. This is the basic objective of GDS studies. This technique is also known as magnetovariational method because of the physical parameters used being transient magnetic field variations (Rokityansky 1982).

3.2 Induction Arrows

The main goal of processing technique is to transform magnetovariational data into frequency-dependent response functions symbolizing estimate of normalized anomalous field. This is achieved by calculating the transfer functions that show the relationship between anomalous and normal field components.

In practice, time-varying magnetic fields observed at any recording site can be considered to be composed of normal and anomalous parts (Schmucker 1970). If X_n, Y_n & Z_n and X_a, Y_a & Z_a represent the components of the normal field and anomalous field respectively, then magnetic field components at any site (X_s, Y_s

and Z_s) can be separated into normal and anomalous parts (e.g. $X_s = X_{n+} X_a$). In such a case, at a given site and for a given frequency interstation transfer functions that relate the anomalous field components and the normal field components can be expressed as:

$$\begin{bmatrix} X_a \\ Y_a \\ Z_a \end{bmatrix} = \begin{bmatrix} T_{xx} & T_{xy} & T_{xz} \\ T_{yx} & T_{yy} & T_{yz} \\ T_{zx} & T_{zy} & T_{zz} \end{bmatrix} \begin{bmatrix} X_n \\ Y_n \\ Z_n \end{bmatrix} \tag{1}$$

The tensor (as shown in the above equation) relating the anomalous and normal fields contain the nature of the electrical conductivity distribution.

As noted earlier due to the presence of anomalous body, (a) the external and internal constituents of the normal magnetic field reinforce each other in horizontal components, such that $X_n \gg X_a$; $Y_n \gg Y_a$ and (b) but annul each other in case of a vertical magnetic field component such that the observed Z variations (Z_a = Z) signifies the signatures of induction in lateral conductivity inhomogeneities. The wave length of the external source fields is very large (several thousand kilometers) and permits plane wave approximation for EM source fields. Thus above equation reduces to

$$Z_a = Z_s = T_{zx}.X_n + T_{zy}.Y_n \tag{2}$$

The magnitude of Induction (real/quadrature) arrow is given by

$$S = \sqrt{(T_{zx})^2 + (T_{zy})^2} \tag{3}$$

and azimuth of the arrow

$$\phi = \tan^{-1}\left(\frac{T_{zy}}{T_{zx}}\right) \tag{4}$$

The azimuths of the arrow signify the direction of maximum correlation between upward and horizontal field components whereas the magnitude of the arrow represents the ratio of vertical field amplitude to the horizontal components with which it correlates. The induction arrow, their orientation, magnitude, spatial distribution and frequency dependence are important tool for outlining extent and depth of lateral inhomogeneity (Arora 1997). The direction of induction arrows is reversed so that it points towards region of high electrical conductivity. Hence, when these arrows are displayed at all the sites, gives information about the trend of the involved conductivity structures.

We have selected local night-time variations to ensure that the inducing field is uniform. The vertical field transfer functions are estimated by using 8–12 events (depending on data availability at each site). Using the above Eqs. (2) and (3), the induction arrows were computed for the period range of 8–128 min using robust technique (Egbert and Booker 1993) and are discussed below.

Salient features of real and quadrature induction arrows corresponding to periods 8, 16, 19, 26 and 43 min are shown in Fig. 2a, b and discussed below.

- Andaman Islands are surrounded by Seawater where bathymetry varies from 50–4000 m. The observed induction arrows have the influence of conducting Seawater and subsurface structure. In spite of deeper sea in the foredeep region as compared

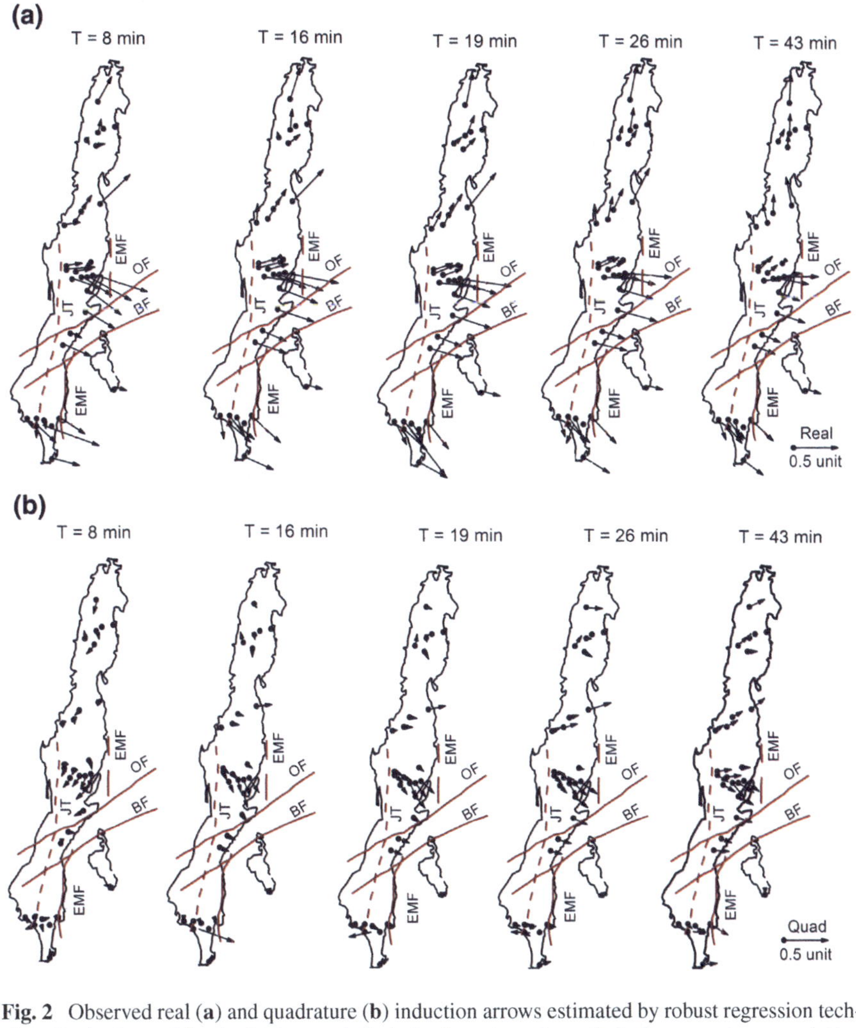

Fig. 2 Observed real (**a**) and quadrature (**b**) induction arrows estimated by robust regression technique for Andaman Islands. At short periods, induction arrows in north Andaman region are pointing towards Barren-Narcondum basin. As periodicity increases these arrows turn anticlock wise and point towards north of Andaman. Along Rangat and Port Blair profiles, the arrows constantly pointing in SE orientations at all periods towards the fore-arc basin

to forearc region, induction arrows are pointing towards forearc region suggesting anomalous conductivity anomaly beneath it. In north Andaman region, the real induction arrows (along Diglipur and Maybunder profiles) with their NNE-NE orientation point towards ophiolite rocks and sedimentary basin at 8 min periodicity. With increasing period, arrows rotate anti-clockwise and point towards deep Sea & sedimentary basins of outer arc region.

- In middle and south Andaman Islands (Rangat and Port Blair profiles), the arrows constantly have SSE to SE orientations at all periods and uniformly point towards deeper part of the fore-arc basin where maximum gravity low is observed. The observed induction arrows along Rangat profile are partially suppressed (especially at Panchavati) due to the contact zone between sedimentary basin and magmatic bodies (ophiolite exposures near Panchavati).
- Behavior of induction arrows in middle and south Andaman Islands suggest that the forearc region is more conducting due to the presence of highly conducting Cretaceous–Tertiary sediments filling the Andaman–Nicobar deep. Also volatile fluids derived from the subducting Indian plate that intrude into the fore-arc basin through the West Andaman Fault (WAF), DF and EMF form a source for enhanced conductivity.

3.3 Frequency Characteristics of Observed Induction Arrows

Amplitude of real and quadrature response for different periods are shown in Fig. 3a–c for Diglipur, Baratang and Port Blair profiles respectively. The real induction arrows acquire peak values in the period range of 16 min for Diglipur region and 19 min for south Andaman (Port Blair) region as shown in Figs. 3a, c. In this period range, the quadrature arrows have nearly vanishing amplitudes and the direction of quadrature arrows reverse sign above and below this period. This critical period (T_c) at which real arrows tend to be maxima and quadrature arrows flip their direction is characteristic of anomalous body. Using the relation given below, the critical period ($T_c = 19$ min) is used to estimate longitudinal conductance (product of conductivity (σ in S/m and Q cross-sectional area in m^2) of the anomalous body (Rokityansky 1982; Chen and Fung 1987).

$$G = \sigma * Q = 5 * 10^4 (T_c(s))^{1.2} = 2.32 * 10^8 \text{ Sm} \quad (5)$$

Using the relation given above, the estimated longitudinal conductance of the anomalous conductivity structure is about $2.32 * 10^8$ Sm and is attributed to forearc basin. A notable feature of the observed induction effects is that the characteristic period increases from north to south. On the assumption that the electrical conductivity of the sedimentary fill in the basin remains the same, the increase of the T_c from N-S would qualitatively imply that the cross sectional area (width and thickness) of the involved structure increases from N–S. In the present case, this can be

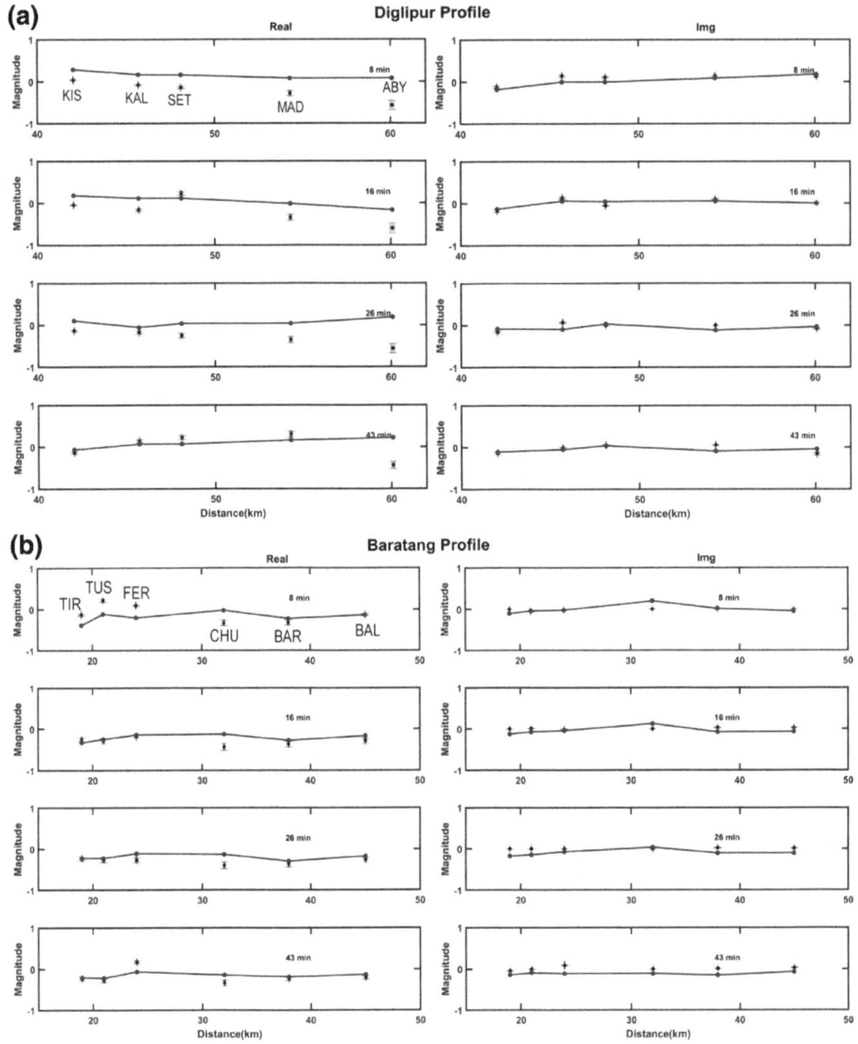

Fig. 3 **a** Amplitude of real and quadrature Z/H for Diglipur profile. Real induction arrows acquire peak values in the period range of 16 min for Diglipur region. Blue line denotes the shielding effect due to Sea water column. **b** Amplitude of real and quadrature Z/H for Baratang profile. **c** Amplitude of real and quadrature Z/H for Port Blair Profile. Real induction arrows acquire peak values at a periodicity 19 min for which quadrature arrows have nearly vanishing amplitude

explained by the widening and deepening of the Andaman-Nicobar basin filled with conducting sediments.

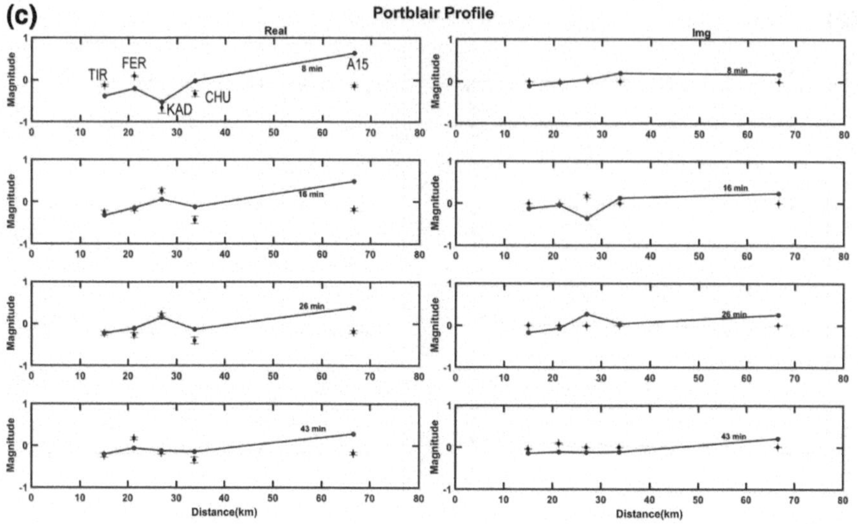

Fig. 3 (continued)

3.4 Hypothetical Event Technique

Projecting the induction arrows on to the line of the transect of GDS is carried out by the hypothetical event technique. In a hypothetical event analysis (HEA), vertical field at all sites is computed with respect external horizontal field of one unit with specified polarization (Bailey et al. 1974). Uniform horizontal field of one unit with an azimuth θ in clockwise direction with respect to the north will induce currents such that vertical field can be written as

$$Z_r = (T_{zx})_r.\cos\theta + \left(T_{zy}\right)_r.\sin\theta \quad (6)$$

$$Z_i = (T_{zx})_i.\cos\theta + \left(T_{zy}\right)_i.\sin\theta \quad (7)$$

Magnitude and phase of anomalous field is given by

$$Z_m = \sqrt{(Z_r^2 + Z_i^2)} \quad (8)$$

$$Z_\phi = \tan^{-1}\left(\frac{Z_i}{Z_r}\right) \quad (9)$$

The advantage of HEA is that for different polarization of the hypothetical event, the induction response of the conducting body can be studied to get the information about the strike and dimensionality of the body. The 2D structure would cause maximum anomaly at a particular polarization and would excite no response (give rise to minimum anomalous field) at the polarization opposite to the earlier.

Ingham et al. (1983) suggested a different presentation of transfer functions Z_r and Z_m in terms of Z/H pseudosetions. In order to know the variations of conductivity both with depth and along the profile (qualitatively), values of Z_m or Z_r from a number of sites along a traverse are contoured with distance along the horizontal axis and the square root of the period along the vertical axis. The additional advantage of this method is that the predicted field component in horizontal and vertical field can be used to formulate maps of anomalous field in three components. So the comparison between the observed and formulated maps gives the additional information than observed induction arrows (Arora 1987). These Z/H pseudo-sections do not provide true depths of the conductive structures, the dominance of the observed response at different periods can be understood by inferring the involved conductive structures.

Using the HEA, the induction response in vertical field was computed at all the stations for the selected period bands in the range of 8–128 min. Figures 4a–h gives a pseudo-section of Z_m for four different periods 8, 19, 26 and 43 min respectively. Different ophiolite bodies exposed on Andaman Islands are also plotted on the anomaly map and are shown as a grey shaded portion. As displayed in Fig. 4a–h, the polarization angle for which the induction response is maximum is for N12° W, N10° E, N18° E and N20° E for periods 8, 19, 26 and 43 min respectively. Examination of Z/H pseudo-sections along Andaman Islands brings out the strongest induction anomaly on the eastern part of the Andaman Islands (Fig. 4). These anomalies are observed at (A), (B), (C) and (D) at the contact zone between ophiolites and sedimentary basin. These anomalies are observed for lower periods and get diffused for higher periods suggesting the shallow nature of the anomalous body. Another interesting feature is the presence of high Z/H at Rangat in middle Andaman Islands, persists for all periods, suggesting a deep seated source. It has been suggested that this anomalous zone consists of Andaman flysch extending to a depth of about 17 km and is a contact zone between ophiolites and sedimentary basin as deduced from the MT studies (Gokarn et al. 2006).

4 Thin Sheet Modeling for Island Effect

Island effect arises due to the conductivity contrast between land and seawater of variable depth. The sea water acts as a low pass filter in transmitting electromagnetic waves. Thus, the surrounding sea water will tend to mask the effect of the induction arrows associated with a local anomalous conductor. If the Island effect is removed, one can expect to see the characteristics of any local subsurface anomalous conductor in the induction arrows. These Island effect is computed by using a thin sheet conductance model for the Andaman-Nicobar region by incorporating bathymetry of the region.

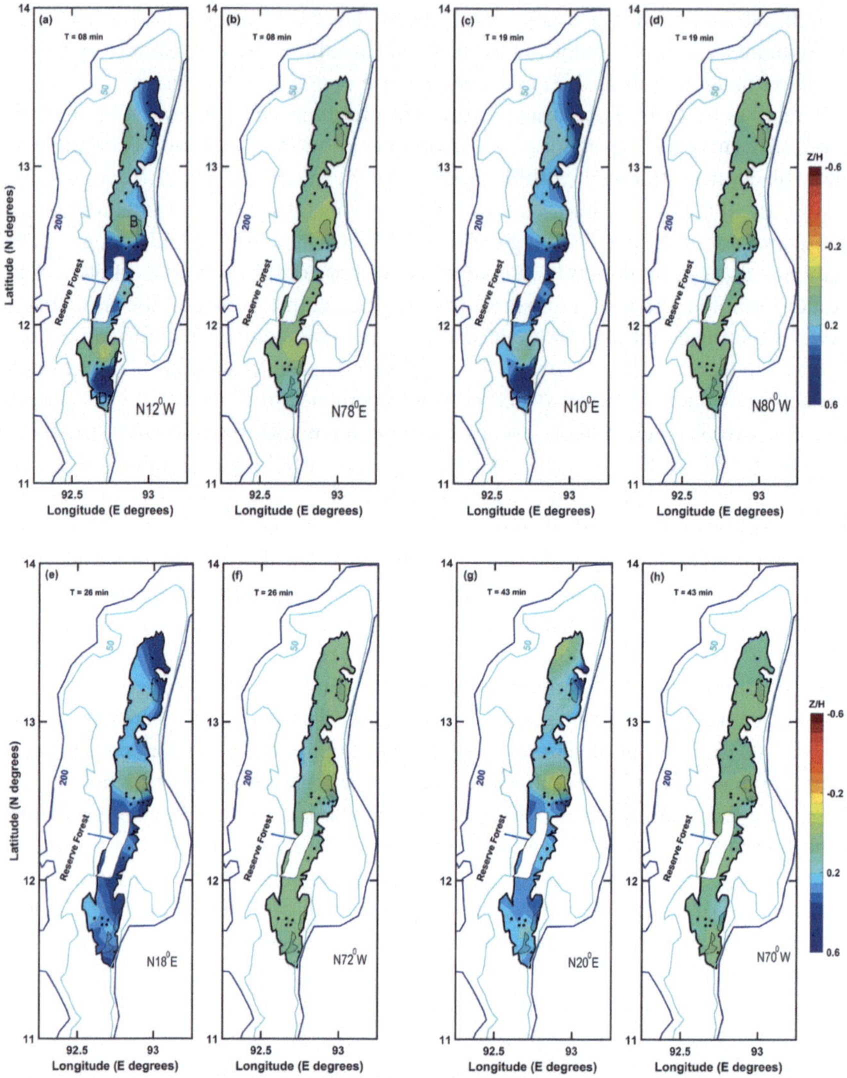

Fig. 4 **a–d** Examination of Z/H pseudo-sections of north, central and south Andaman Islands brings out strongest induction anomaly on the eastern parts of Andaman Islands at 8 and 19 min. Shaded portions A, B, C and D deonte ophiolite bodies exposed on Islands. These anomalies are observed between the contact zone of ophiolite bodies and sedimentary basin and disappear for opposite polarization. **e–h** Examination of Z/H pseudo-sections of north, central and south Andaman Islands brings out strongest induction anomaly on the eastern parts of Islands at 26 min periodicity and it diminishes for 43 min indicating that these anomalies are shallow in nature

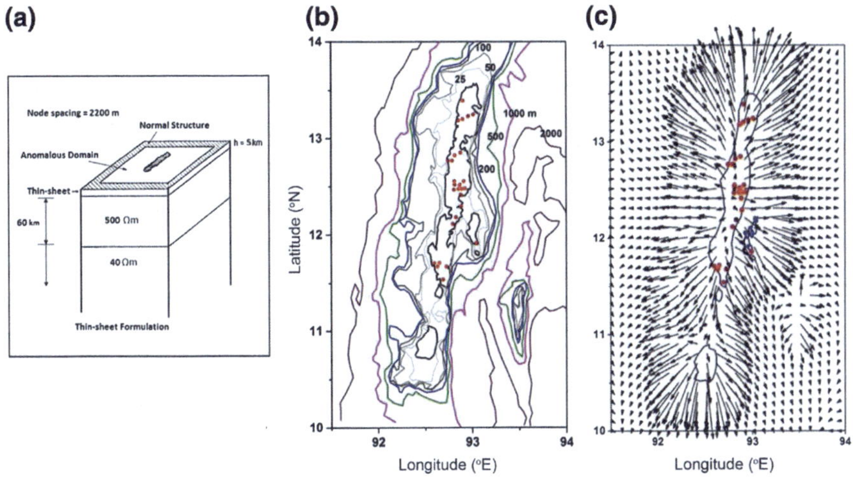

Fig. 5 **a** Shows thin-sheet formulation adopted for the present study and is underlain by different layers based on LMT studies carried in Andaman Islands. **b** Bathymetry of Andaman-Nicobar Islands and surrounding regions (after Amante and Eakins 2009). **c** Real induction arrows at a period of 26 min corresponding to thin sheet model depicting the lateral variations due to seawater of variable depth. Arrows are shown for alternative grid points

In the present study, 3D thin sheet formulation developed by Vasseur and Weidelt (1977) has been used to compute the Island effect. The thin sheet formulation considers that the conductivity anomalies are confined to a single layer at the surface of the Earth, which overlies a 1D layered structure. The choice of 1D model underlying the thin sheet is guided by LMT survey carried out in mid-Andaman region as shown in Fig. 5a. This model shows top layer consists of highly conducting sediments extending to a depth of about 5 km followed by 60 km thick crust having a resistivity of about 500 Ωm and considered to overlie a half space having a resistivity of about 40 Ωm.

For calculating the Island effect, the thickness of the water column has been adopted from the xy-topo maps (Fig. 5b) available from National Geophysical Data Center website (Amante and Eakins 2009). As available from literature, the resistivity of seawater is adopted to be 0.33 Ωm (Baba 2006). The above thin sheet was considered to be overlain by 1-D structure as discussed above. Since a region of normal structure surrounds the anomalous zone, the artefacts due to boundary are minimized by extending the grid to sufficiently large distances away from observational domain (Mareschal et al. 1987).

In the present case, the skin depth of the underlying layer for period 8 min is about 250 km and is much larger than the assumed thickness (5 km) of the anomalous surface layer. Similarly, the skin depth of the seawater is approximately 6.5 km, again about 3–4 times larger than the maximum depth of the seawater. Thus, the thin sheet conditions (Weaver 1982) are met everywhere. The grid spacing is 2.2 km and satisfies the condition that it should be less than one-third of the skin depth of the underlying layer. Thus, the thin sheet satisfies the conditions necessary to validate the thin sheet approximation as described in the beginning of this section.

For numerical computations, the area between 91.5°–94° E and 10°–14° N was divided into a grid of 125 by 200 cells with a node spacing of 2.2 km. Numerical solutions have been worked out for non-uniform thin sheet representing lateral variations arising from island and seawater of variable depths (Fig. 5b). Resulting induction arrows for alternative grid points are shown in Fig. 5c. As seen, all the calculated induction arrows are deflected from island and are pointing towards deep sea (2 km depth).

Residual induction arrows are obtained by subtracting vectorially the estimated island effect from the observed ones as suggested by Weaver and Agarwal (1991) and are shown in Fig. 6. Residual arrows acquire a peak value at 16 min along Diglipur profile (Fig. 6a) and 19 min along PortBlair profile (Fig. 6c). Low Z/H observed at 8 min along Baratang profile (Fig. 6b) suggests shallow nature of the conductivity anomaly. In order to identify these zones, 2D modeling of residual induction arrows have been carried out as discussed below.

5 2D Modeling

In order to test the induction response functions, the crustal model suggested by potential data has been considered (Haripriya et al. 2019, this issue) to construct the electrical conductivity model across the outer arc region. The crustal model, deduced from the joint analysis of the gravity and seismic data was adopted as a starting model. This model suggests depth of the subducting crust is about 32 km at the plate margin and low density sediments having a thickness of about 6 km. Resistivity values of seawater, basaltic crust and mantle were adopted from EMSLAB experiment across Juan de fuca subducting plate (EMSLAB 1990) and across New Zealand (Ingham et al. 2001). The EM induction response of the electrical conductivity model was obtained by using the numerical formulation of the Cerv et al. (1987). The resistivity of the sedimentary column was allowed to vary to explain the induction response. Since the effect of shielding from the seawater was eliminated from the observed data, the residual arrows were modeled by replacing the seawater with a material simulated by a resistivity of 10–20 Ωm (sedimentary layer). The resulting induction response of the models across Diglipur profile (profile-1), Baratang profile (profile-2) and PortBlair profile (profile-3) are shown in Fig. 7a–f. The resulting induction responses reproduce the most salient features of the residual observed induction response. The induction response along the profiles is dominated by the sedimentary layers. The resistivity values of 1 Ωm was assigned to top layer sediments that have been reflected as low density sediments/LVZ within the sedimentary basin in integrated gravity and seismic models (Haripriya et al. 2019) and is related to obducted sediments through faults and fissures. Both Jawar thrust (JT) and EMF (eastern margin fault) have been reflected as a high conductivity anomalies that is essential to reproduce the overall magnitude and spatial variation comparable with the residual response.

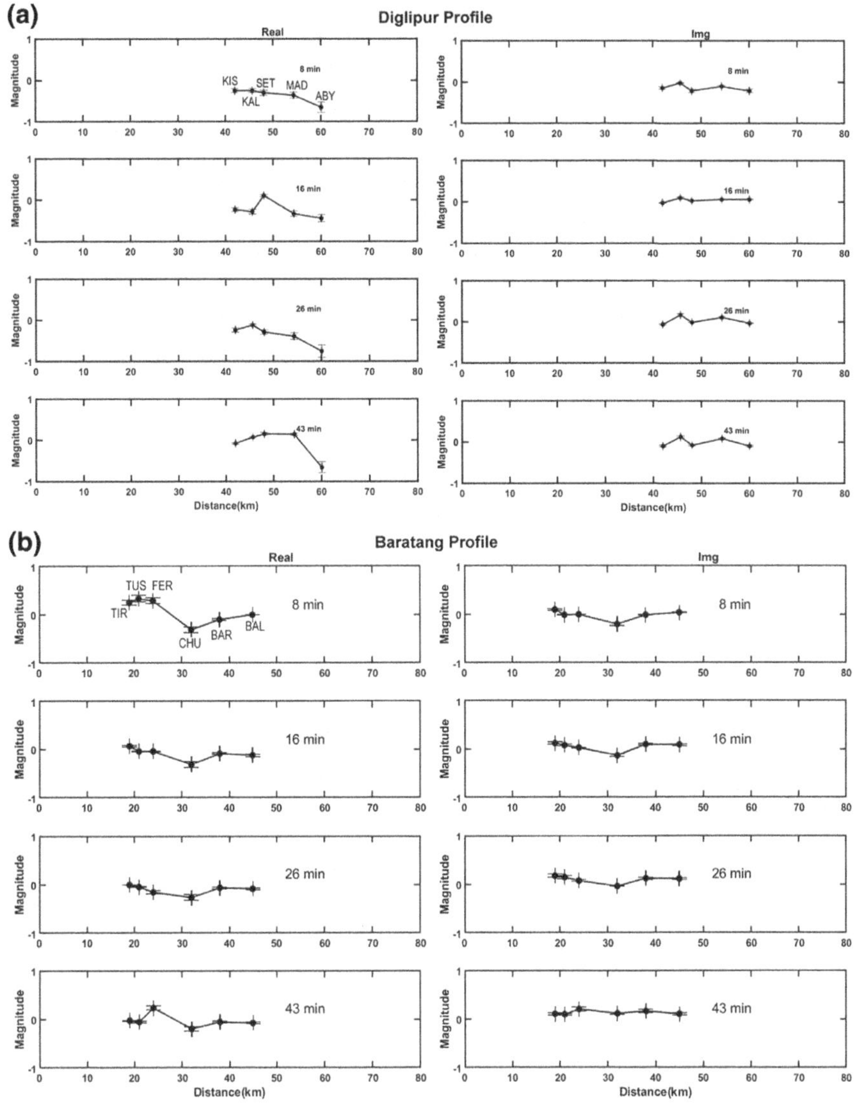

Fig. 6 **a** Residual induction arrows obtained for Diglipur profile. Real induction arrows acquire a peak value at 16 min denoting an anomalous zone between KIS and SET. **b** Residual induction arrows obtained for Baratang profile. **c** Residual induction arrows obtained for PortBlair profile. Real induction arrows acquire a peak value at 19 min denoting an anomalous zone between FER, KAD and CHU

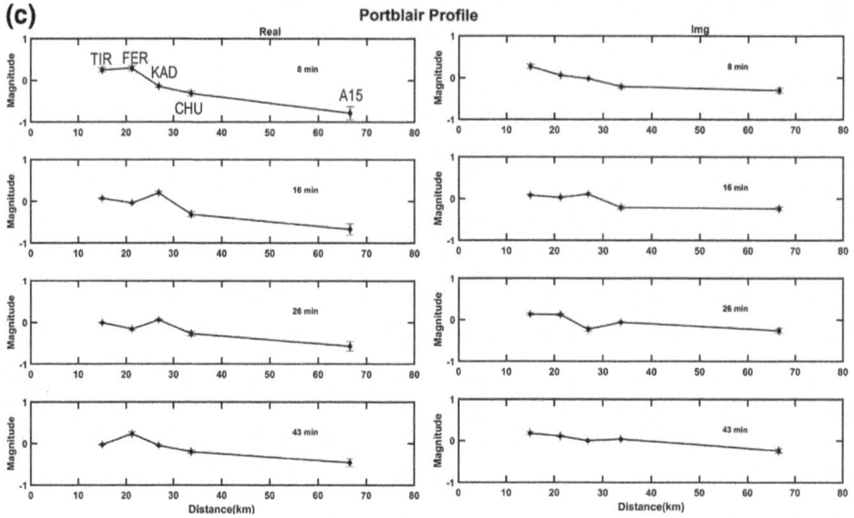

Fig. 6 (continued)

6 Results and Discussion

The observed induction pattern is dominated by high conducting sediments filling the Andaman fore arc basin and outer arc region. In middle Andaman region, the observed induction arrows along Rangat profile are partially suppressed (especially at Panchavati) due to the contact zone between sedimentary basin and magmatic bodies (ophiolite exposures near Panchavati). Ophiolite slices show thrust contact with underlying sediments and overlying sediments trending in N-S and NE-SW have been emplaced along the eastern margin fault (Pal et al. 2003; Acharya 2007; Allen et al. 2008; Ghosh et al. 2009; Pedersen et al. 2010; Pal 2011; Subba Rao et al. 2016; Morley and Searle 2017).

HEA brings out that the eastern margin of Andaman Islands is highly anomalous that could be related to emplacement of N-S trending ophiolites along the eastern margin fault of Andaman Islands and obducted sediments expelled from subducting plate.

EM anomalies (Fig. 7) in Andaman Islands may be caused by (a) fault zones between two different geological units (Jones 1987) as seen in accretionary complex, (b) the formation of a fractured, cataclastic zone along a fault can enhance the percolation of fluids and thereby enhance the conductivity anomaly (Unsworth 2010) is observed along the Jawar thrust and (iii) due to the presence of fluids generated by the metamorphosis of the subducting crust and obducted sediments through the eastern margin fault.

Modeling of GDS response brings out localized conductivity anomalies associated with Baratang mud volcano at a depth range of 2–4 km (Fig. 7b) and could be related to the presence of clayey terrigenous sediments. Enhanced conductivity may be

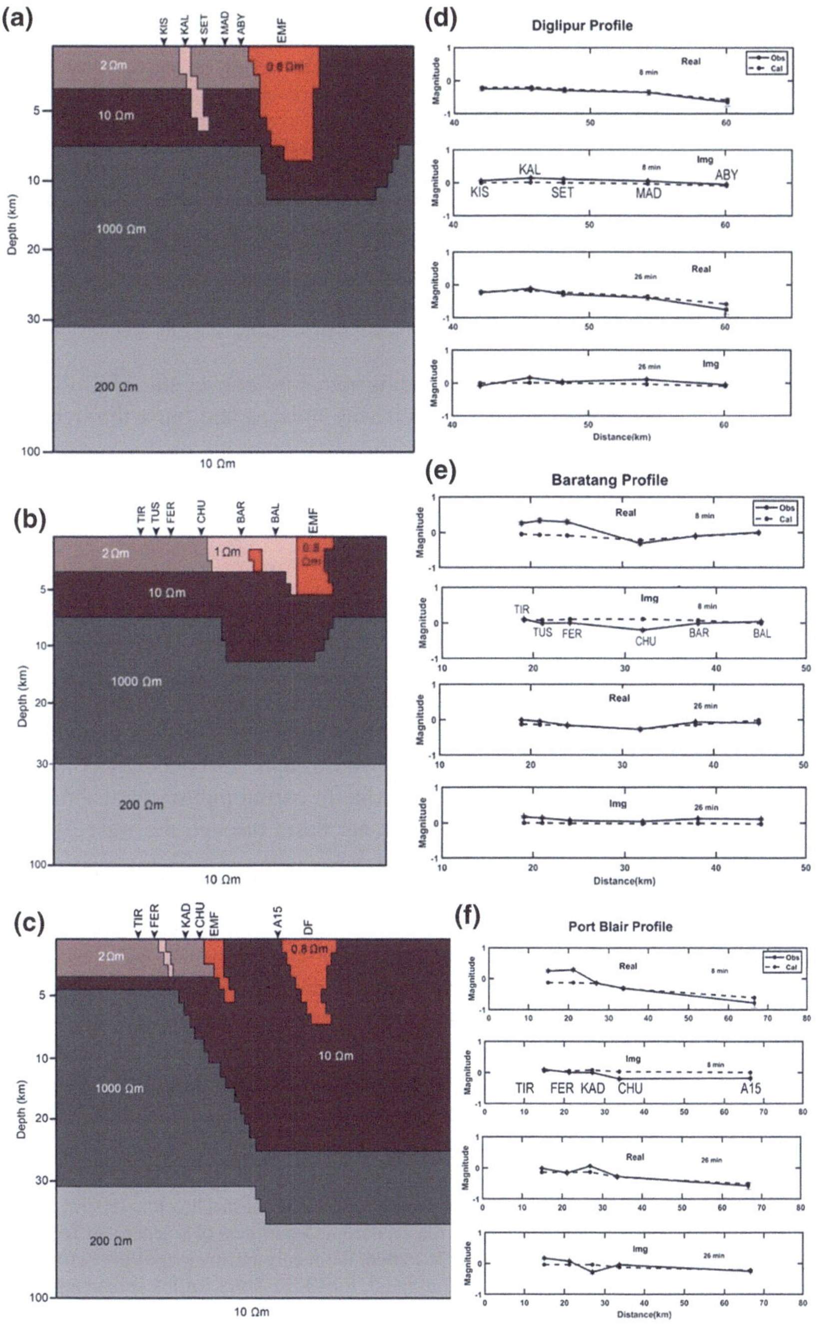
(a)
KIS KAL SET MAD ABY EMF
2 Ωm
0.8 Ωm
10 Ωm
1000 Ωm
200 Ωm
Depth (km)
10 Ωm
(b)
TIR TUS FER CHU BAR BAL EMF
2 Ωm
1 Ωm
0.8 Ωm
10 Ωm
1000 Ωm
200 Ωm
Depth (km)
10 Ωm
(c)
TIR FER KAD CHU EMF A15 DF
2 Ωm
0.8 Ωm
10 Ωm
1000 Ωm
200 Ωm
Depth (km)
10 Ωm
(d)
Diglipur Profile
Real
Img
8 min
26 min
Obs
Cal
KIS KAL SET MAD ABY
Magnitude
Distance(km)
(e)
Baratang Profile
Real
Img
8 min
26 min
Obs
Cal
TIR TUS FER CHU BAR BAL
Magnitude
Distance(km)
(f)
Port Blair Profile
Real
Img
8 min
26 min
Obs
Cal
TIR FER KAD CHU A15
Magnitude
Distance(km)

◂**Fig. 7** **a** Electrical crustal model across Diglipur profile, constrained by potential and seismic data. It brings out anomalous zone across EMF and Jawar thrust between KAL and SET that is related to emplacement of fluids & obducted sediments. **b** Electrical crustal model across Baratang profile, constrained by potential and seimic data. It brings out anomalous zone between BAR and BAL. This localized conductivity anomaly is located at a depth of about 2–4 km. It may represent a liquefaction zone associated with mud volcano that contains clayey terrigenous sediments along with fluids and is extending into forearc region. EMF also reflected as a anomalous zone. **c** Electrical crustal model across PortBlair profile brings out anomalous zones associated with Jawar thrust (JT, between FER and KAD), EMF and Diligent Fault (DF) zones. These anomalies zones may contain sediments along with fluids. **d–f** Comparison between observed and calculated GDS response at periods 8 and 26 min

attributed to the liquefaction zone extending into fore arc region. Mud-fluid mix eruption takes place when mud force their way upward and burst through to the surface.

7 Conclusions

From spatial distribution of these arrows, it is inferred that the sediments filling the fore arc basin (Andaman-Nicobar deep) are more conducting than the region of outer non-volcanic island arc. The high conductivity over forearc region (Andaman-Nicobar deep) can be explained in terms of (a) the Cretaceous–Tertiary sediments filling the Andaman–Nicobar sedimentary basin whose depth and conductivity increases towards south and (b) the Invisible bank is due to partial melts/volatiles that have intruded into the eastern margin of the fore-arc basin through the west Andaman fault (WAF).

Suppression of induction arrows between Rangat and Panchavati denotes the presence of anomalous conductivity zone that is reflected in Rangat profile as well as in MT studies carried out by Gokarn et al. (2006). It has been suggested that this suture consists of Andaman flysch extending to a depth of about 17 km.

EM modeling across three different profiles brings out anomalous zones across Jawar and eastern margin fault zones. These anomalies are caused by the presence of obducted sediments plus fluids that are expelled from the subducting sedimentary rocks at deeper levels. GDS response along Baratang profile brings out liquefaction zone (mud-fluid mix) at a depth of about 2–4 km extending into fore arc basin.

Acknowledgements The authors are thankful to Prof D S Ramesh Director, IIG for his kind encouragement and support. Finance support received from Department of Science and Technology (DST), New Delhi for the project no: **ESS/16/262/2006**. We deeply acknowledge Editors for inviting us to contribute to this issue. We would like to thank Prof B R Arora for his comments that has improved the earlier version of the manuscript.

References

Acharyya SK (2007) Collisional emplacement history of the Naga-Andaman ophiolites and the position of the eastern Indian suture. J Asian Earth Sci 29:229–242

Allen R, Carter A, Najman Y, Bandopadhyay PC, Chapman HJ, Bickle MJ, Garzanti E, Vezzoli G, Andò S, Foster GL, Gerring C (2008) New constraints on the sedimentation and uplift history of the Andaman-Nicobar accretionary prism, South Andaman Island. In: Draut AE, Clift PD, Scholl DW (eds) Formation and applications of the sedimentary record in arc collision zones: geological society of America special paper 436, pp 223–255. https://doi.org/10.1130/2008.2436(11)

Amante C, Eakins BW (2009) ETOPO1 arc-minute global relief model: procedures, data sources and analysis. In: NOAA technical memorandum NESDIS NGDC-24. National Geophysical Data Center, NOAA. https://doi.org/10.7289/V5C8276M

Arora BR (1987) Perturbation arrows and hypothetical event analysis in geomagnetic induction studies: experimental results from northwest India. Phys Earth Planet Inter 45:128–136

Arora BR (1997) Geomagnetic deep sounding. In: Arora BR, Sriniwas (eds) Natural source electromagnetic induction in the Earth. New Age International (P) Limited Publishers, New Delhi, pp 80–128

Arora BR (2002) Seismotectonics of the frontal Himalaya through the electrical conductivity imaging. In: Fujinawa Y, Yoshida A (eds) Seismotectonics in convergent plate boundary, pp 261–272

Arora BR, Lilley FEM, Sloane MN, Singh BP, Srivastava BJ, Prasad SN (1982) Geomagnetic induction and conductive structures in northwest India. Geophys J R Astron Soc 69:459–475

Arora BR, Subba Rao PBV (2002) Integrated modeling of EM response functions from Peninsular India and Bay of Bengal. Earth, Planets Space 54:637–654

Baba IK (2006) Electrical structure in marine tectonic settings. sp. Issue In: Weaver, JT, Arora BR (eds) 17th EM workshop on electromagnetic induction in the earth, vol 26. Surveys in Geophysics, pp 701–731

Bailey RC, Edwards RN, Garland GD, Kurtz R, Pitcher D (1974) Electrical conductivity studies over a tectonically active area in eastern Canada. J Geomagn Geoelectr 26:125–146

Bandopadhyay PC (2005) Discovery of abundant pyroclasts in the Namunagarh Grit, South Andaman: evidence for arc volcanism and active subduction during the Palaeogene in the Andaman area. J Asian Earth Sci 25:95–107

Bandopadhyay PC, Carter A (2017) Geological framework of the Andaman-Nicobar Islands. Geol Soc, Lond Memoirs 47:75–93

Banerjee A (1975) Mud volcanoes of the Andaman Islands. Indian J Earth Sci 2:11–17

Cerv V, Pek J, Praus O (1987) Numerical modeling of geoelectrical structures in Czechoslovakia. Phys Earth Planet Inter 45:170–178

Chaudhuri H, Ghose D, Bhandari RK, Sen P, Sinha B (2012) A geochemical approach to earthquake reconnaissance at the Baratang mud volcano, Andaman and Nicobar Islands. J Asian Earth Sci 46:52–60

Chen PF, Fung PCW (1987) Frequency response of the transfer functions of the current channelling between two oceans. J Geomagn Geoelectr 40:335–355

Cochran JR (2010) Morphology and tectonics of the Andaman Forearc, northeastern Indian Ocean. Geophys J Int 182:631–651

Curray JR (2005) Tectonics and history of the Andaman Sea region. J Asian Earth Sciences 25:187–232

Dhana Raju R, Negi BS (1984) Petrography and tectonic setting of the Andaman ophiolite suite with a note on its evolution. J Geol Soc India 25:655–665

Egbert GD, Booker JR (1993) Imaging crustal structure in southwestern Washington with small magnetometer arrays. J Geophys Res: Solid Earth 98:15967–15985

EMSLAB group (1990) The EMSLAB electromagnetic sounding experiment (14 papers). J Geophys Res 94:14093–14283

Eremenko NA, Sastri VV (1977) On the petroleum geology of Andaman Islands. Bull ONGC India 14:1–13

Ghosh B, Pal T, Bhattacharya A, Das D (2009) Petrogenetic implications of ophiolitic chromite from Rutland Island, Andaman—a boninitic parent in supra-subduction setting. Miner Petrol 96:59–70

Ghosh B, Morishita T, Ray J, Tamura A, Mizukami T, Soda Y, Ovung TN (2017) A new occurrence of titanian (hydro) andradite from the Nagaland ophiolite, India: implications for element mobility in hydrothermal environments. Chem Geol 457:47–60

Gokarn SG, Gupta G, Dutta S, Hazarika N (2006) Geoelectric structure in the Andaman Islands using magnetotelluric studies. Earth, Planets and Space 58:259–264

Gough DI (1989) Magnetometer array studies, earth structure and tectonic processes. Rev Geophys 27:141–157

Gough DI, Ingham MR (1983) Interpretation methods for magnetometer arrays. Rev Geophys 21:805–827

GSI (2001) Geological Map of Andaman-Nicobar region. Published by Geological Survey of India, Calcutta

Haripriya K, Radhakrishna M, Mukhopadhyay M (2019) Crustal model for the Andaman Outer Arc: constraints from Earthquake, Gravity and Receiver Function Data. this special issue

Ingham MR, Bingham DK, Gough DI (1983) A magnetovariational study of a geothermal anomaly. Geophys J R Astr Soc 72:507–618

Ingham MR, Whaler K, McKnight DD (2001) Magnetotelluric sounding of the Hikurangi Margin, New Zealand. Geophys J Int 144:343–355

Jafri SM, Sheikh JM (2013) Geochemistry of pillow basalts from Bompoka, Andaman-Nicobar Islands, Bay of Bengal, India. J Asian Earth Sci 64:27–37

Jones AG (1987) Are impact-generated lower-crustal faults observable? Earth Planet Sci Lett 85:248–252

Lilley FEM, Singh BP, Arora BR, Srivastava BJ, Prasad SN, Sloane MN (1981) A magnetometer array study in northwest India. Phys Earth Planet Inter 25:232–240

Mareschal M, Vasseur G, Srivastava BJ, Singh RN (1987) Induction models of southern India and effect of offshore geology. Phys Earth Planet Inter 45:137–148

Mishra OP, Kayal JR, Chakrabortty GK, Singh OP, Ghosh D (2007) Aftershock investigation in the Andaman-Nicobar Islands of India and its seismotectonic implications. Bull Seismol Soc Am 97:S71–S85

Moeremans RE, Singh SC (2015) Fore-arc basin deformation in the Andaman-Nicobar segment of the Sumatra-Andaman subduction zone: Insight from high-resolution seismic reflection data. Tectonics 34:1736–1750

Morley CK, Searle M (2017) Regional tectonics, structure and evolution of the Andaman—Nicobar Islands from ophiolite formation and obduction to collision and back-arc spreading. In: Bandopadhyay PC, Carter A (eds) The Andaman—Nicobar accretionary ridge: geology, tectonics and hazards, vol 47. Geological society, London, Memoirs, pp 51–74. https://doi.org/10.1144/M47.5

Mukhopadhyay B, Chakraborty PP, Paul S (2003) Facies clustering in turbidite successions: case study from Andaman Flysch Group, Andaman Islands, India. Gondwana Res 6:918–924

Pal T (2011) Petrology and geochemistry of the Andaman ophiolite: melt-rock interactions in a suprasubduction-zone setting. J Geol Soc Lond 168:1031–1045. https://doi.org/10.1144/0016-6492009-152

Pal T, Chakraborty PP, Gupta TD, Singh CD (2003) Geodynamic evolution of the outer-arc–forearc belt in the Andaman Islands, the central part of the Burma-Java subduction complex. Geol Mag 140:289–307

Pandey J, Agarwal PP, Dave A, Maithani A, Trivedi KB, Srivastava AK, Singh DN (1992) Geology of Andaman. Bull Oil Nat Gas Comm 29:19–103

Pedersen RB, Searle MP, Carter A, Bandophadhyay (2010) U-Pb zircon age of the Andaman ophiolite: implications for the beginning of subduction beneath the Andaman- Sumatra arc. J Geol Soc, Lond 167:1105–1112. https://doi.org/10.1144/0016-76492009-151
Radhakrishna M, Lasitha S, Mukhopadhyay M (2008) Seismicity, gravity anomalies and lithospheric structure of the Andaman arc, NE Indian Ocean. Tectonophys 460:248–262
Ray KK (1985) East coast volcanics: a new suite in the ophiolite of Andaman Islands. Rec Geol Surv India 116:83–87
Ray KK, Sengupta S, Van Den Hul HJ (1988) Chemical characters of volcanic rocks from Andaman ophiolite, India. J Geol Soc 145:393–400
Ray JS, Pande K, Awasthi N (2013) A minimum age for the active Barren Island volcano. Andaman Sea Curr Sci 104:934–939
Reddy CD (1992) A narrow conductive zone south of the Main Central Thrust of the Garhwal Himalaya. Memoir J Geol Soc India 24:121–131
Reddy CD (1997) An integral sign conductor in the frontal himalaya region and its tectonic interpretation. J Geomagn Geoelectr 49(11–12):1649–1658
Reddy CD, Arora BR (1993) Quantitative interpretation of geomagnetic induction response across the thrust zones of the Himalaya along the Ganga-Yamuna valley. J Geomagn Geoelectr 45:775–785
Rodolfo KS (1969) Bathymetry and marine geology of the Andaman Basin, and tectonic implications for Southeast Asia. Geol Soc Am Bull 80:1203–1230
Rokityansky II (1982) Geoelectromagnetic investigation of the earth's crust and mantle. Springer-Verlag, New York, 381pp
Roy TK (1983) Geology and hydrocarbon prospects of Andaman-Nicobar. Petroliferous Basins of India, Petroleum Asia Journal, Keshava Deva Malaviya Institute of Petroleum Exploration, pp 37–53
Roy SK (1992) Accretionary prism in Andaman forearc. Geol Surv India Spec Publ 29:273–278
Roy TK, Chopra NN (1987) Wrench faulting in Andaman forearc basin, India. In: Offshore Technology Conference. Offshore Technology Conference
Schmucker U (1970) Anomalies of geomagnetic variations in the southwestern United States. Bull Scripps Inst Oceanography 13:1–165
Singh SC, Moeremans R, McArdle J, Johansen K (2013) Seismic images of the sliver strike-slip fault and back thrust in the Andaman-Nicobar region. J Geophys Res: Solid Earth 118:5208–5224
Subba Rao PBV (2008) Regional conductance map of Andaman and Nicobar region. Gondwana Res 13:386–395
Subba Rao PBV, Radhakrishna M, Haripriya K, Rao BS, Chandrasekharam D (2016) Magnetic anomalies over the Andaman Islands and their geological significance. J Earth Syst Sci 125:359–368
Unsworth M (2010) Magnetotelluric studies of active continent-continent collisions. Surv Geophys 31:137–161. https://doi.org/10.1007/s10712-009-9086-y
Vasseur G, Weidelt P (1977) Bimodal electromagnetic induction in non-uniform thin sheets with an application to the northern Pyrenean induction anomaly. Geophys J Int 51:669–690
Weaver JT (1982) Regional induction in Scotland; an example of three-dimensional numerical modelling using the thin sheet approximation. Phys Earth Planet Inter 28:161–180
Weaver JT, Agarwal AK (1991) Is addition of induction vectors meaningful? Phys Earth Planet Inter 65:267–275

On the Trail of the Great 2004 Andaman-Sumatra Earthquake: Seismotectonics and Regional Tsunami History from the Andaman-Nicobar Segment

C. P. Rajendran and Kusala Rajendran

Abstract The 2004 earthquake on the Andaman-Sumatra trench provided an unprecedented opportunity to study the deformational mechanism in a subduction zone before, during and after a great plate boundary earthquake. It also gave an opportunity to map the coseismic changes during a great plate boundary-breaking earthquake as well as the effects of an unprecedented transoceanic tsunami. Multiple pieces of evidence obtained from the rupture zone and the distant sites of tsunami reach were used to identify older events. This paper reviews results of such studies that led to the identification of two previous events, dated at about 500 years and 1000 years B.P. In this paper, we also review the post-2004 intraplate seismicity in the Andaman and Nicobar segments and the associated mechanisms in terms of their state of stress before and after the great earthquake.

Keywords Andaman-Sumatra · Earthquakes · Tsunami · Indian Ocean

1 Introduction

The Sumatra–Andaman trench has witnessed several major earthquakes in the historical past, but the 2004 event (Mw 9.2) was unique by virtue of its unprecedented magnitude and the through-going rupture length of >1000 km (Fig. 1). It caused a mega-trans-oceanic tsunami, the first ever to be felt so widely along the shores of the Indian Ocean rim countries. It also triggered several large aftershocks along the plate boundary, and the outlying oceanic intraplate regions (Dewey et al. 2007; Engdahl et al. 2007; Rajendran et al. 2011b). Among the post-2004-seismic events, the 2012 pair of great earthquakes (Mw 8.6 and Mw 8.2) within the Wharton Basin was the largest among the intraplate events (Hill et al. 2015; Lay et al. 2016). As the first

C. P. Rajendran
Geodynamics Unit, Jawaharlal Nehru Centre for Advanced Scientific Research, Bengaluru 560064, India

K. Rajendran (✉)
Centre for Earth Sciences, Indian Institute of Sciences, Bengaluru 560012, India
e-mail: kusalaraj@gmail.com

J. S. Ray and M. Radhakrishna (eds.), *The Andaman Islands and Adjoining Offshore: Geology, Tectonics and Palaeoclimate*, Society of Earth Scientists Series,
https://doi.org/10.1007/978-3-030-39843-9_10

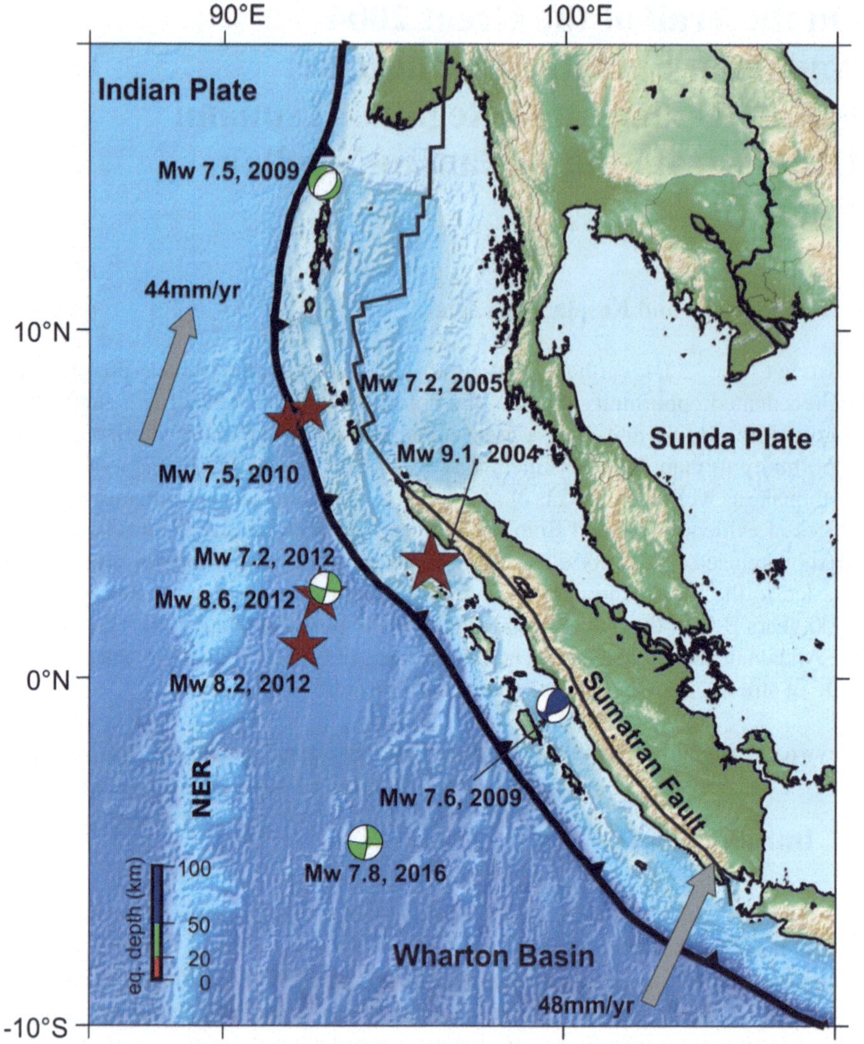

Fig. 1 The Andaman-Sumatra Subduction Zone and the Wharton basin showing the significant earthquakes discussed in this paper (after Parameswaran 2016). Grey arrow shows the plate convergence (DeMets et al. 2010)

great earthquake to have occurred in the era of digital broadband seismology, GPS geodesy and tsunami geology, the 2004 earthquake invoked much global interest and a number of papers have been published on various aspects of this earthquake. A broad review of all these papers is beyond the scope of the present review and here we only summarize some of the salient points on the tsunami history and the post-2004 intraplate earthquakes, mostly restricting to the Andaman and Nicobar segments of the subduction zone.

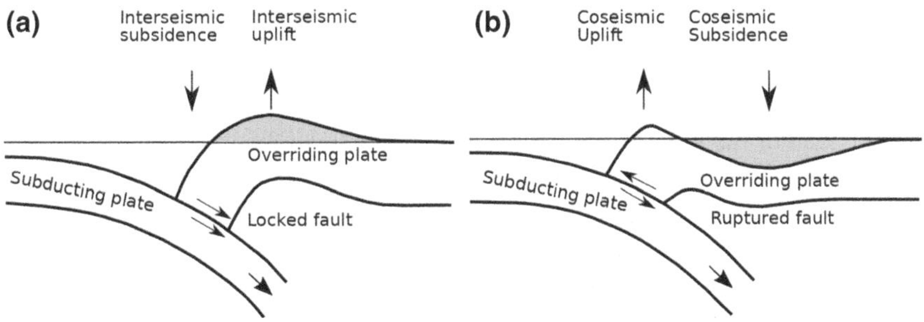

Fig. 2 Schematic diagrams showing the pattern of **a** inter-seismic and **b** coseismic deformation associated with a the deformation cycle of a subduction zone earthquake (after Plafker 1972)

Conventional wisdom informs us that a megathrust earthquake occurs at the end of a subduction cycle, with variability in the duration of the cycle and size of the earthquakes (e.g. Satake and Atwater 2007). Observations on the nature of deformation from global examples suggest that compressional earthquakes occur in the outer-rise before and tensional after the great event (e.g. Christensen and Ruff 1988). This style of faulting is considered as a manifestation of the state of stress of the subduction zone during the seismic cycle. The state of stress of the plate boundary before and after the megathrust is manifested also by uplift and subsidence on the overriding plate (Plafker 1972) (Fig. 2). The 2004 earthquake demonstrated these processes through the seismicity pattern as well as the associated land-level changes, which we discuss in this paper using examples from the Andaman and Nicobar segments.

The diverse characteristics of tsunami deposits within the rupture zone provided an opportunity to study their spatial distribution and for discriminating them from storm deposits, a challenge in the coastal regions that are affected by both these types of coastal surges (e.g. Morton et al. 2007; Switzer et al. 2012). Some of the 2004 deposits were also associated with land-level changes, and they served as proxies to develop the history of past tsunamigenic earthquakes. Intense aftershock activity followed the 2004 earthquake and their faulting mechanisms showed a mix of normal, strike-slip, and reverse faulting (e.g. Engdahl et al. 2007; Dewey et al. 2007). Post-earthquake relaxation is still continuing, as evident from the GPS-based geodetic models (Paul et al. 2012).

The 2004 earthquake sequence provides an excellent example to study the temporal and spatial pattern of seismicity before after a megathrust event. It also raised an important question about the timing of the previous tsunami, the true predecessor of the 2004 event. Evidence for two events were reported from India, Andaman Thailand and Sumatra dated as ~1000 yr and ~500 years before present (Rajendran et al. 2006, 2013; Jankaew et al. 2008; Monecke et al. 2008; Rajendran et al. 2013). This paper consists of two components, one, a review of source mechanisms of selected post 2004 earthquakes from the Andaman-Nicobar part of the subduction zone, and two, a summary some important observations on the chronology of past tsunamigenic earthquakes.

2 Regional Seismotectonics

The Andaman-Sumatra Subduction Zone defines the eastern, ocean-continent convergent zone of the India-Eurasia plate boundary where, the Indian Plate subducts beneath the Sunda Plate at ~48 mm/yr to ~44 mm/yr at an azimuth of ~30° (Fig. 1) (DeMets et al. 2010). The deformation of the subducting Indian Oceanic plate is believed to be influenced both by the plate-boundary stresses originating from the Himalaya collision zone and the slab pull due to subduction of the Australian plate along the Sunda megathrust (e.g. Delescluse et al. 2012). It is believed that the oblique subduction has led to the formation of a sliver and partitioning of the slip into trench parallel and perpendicular components (Fitch 1972; Ishii et al. 2013). To the east of the trench axis is the subducting Sunda plate, where deformation is partitioned along the northward continuation of the Sumatran Fault (SF) (Fig. 3). The dual slip components from the oblique subduction are portioned between the trench and the SF, leading to thrust events along the trench and right-lateral strike-slip events on the fault (McCaffrey 2009; Engdahl et al. 2007). The northern part of the Sumatra–Andaman subduction zone comprises the Nicobar and Andaman segments. The segment north of ~10° is named as the Andaman segment and the part between the Andaman segment and the source zone of the 2004 earthquake is regarded as the Nicobar segment (e.g. Dewey et al. 2007).

The motion along the Andaman-Nicobar part of the plate boundary is partitioned as a component of convergence across the subduction zone and as right-lateral strike-slip along the West Andaman Fault (WAF), an extension of the Sumatran Fault (Fig. 3), as discussed by Curray (2005). Normal and strike-slip faulting within the Andaman Sea occur due to the back arc spreading along the Andaman Spreading Ridge. The subducting plate off the Nicobar segment is marked by the N10° E trending, 90° E ridge, called the Ninety East Ridge (NER). This major regional linear oceanic feature has long been recognized as a potential kinematic plate boundary between India and Australia (Le Pichon 1968). The NER acts as the western boundary of the highly deforming Wharton Basin (Fig. 3), the triangular slice of the ocean plate west of the Nicobar and the southern Sumatran segments (Fig. 1). Considered as a part of the rigid India–Australia plate and displaying a wide band of left-lateral shearing, this region has experienced several earthquakes in the past and is believed to have accommodated at least 100 km of shearing during the last 7.5–8 Ma (e.g. DeMets et al. 1988). East of the trench axis, on the subducting plate, deformation is partitioned along the northward continuation of the Sumatra Fault (Fig. 3).

Seismic productivity varies along the strike of the trench, with no great earthquakes in the recent history, in the Andaman segment, except the 1941 (Mw 7.7) event (Jhingran 1953). The Car Nicobar Island is noted for the 1881 (Mw 7.9) earthquake, a thrust-faulting event at 15 km depth (Ortiz and Bilham 2003). Post 2004, the region experienced several large earthquakes, namely, the 10 August 2009 (Mw 7.5) Coco Island; 12 June 2010 (Mw 7.5) off Car Nicobar and the 2016 (Mw 7.8) Wharton Basin events (Lay et al. 2016 and references therein) among them. Post-2004, there was an increase in the level of activity also along the Andaman segment,

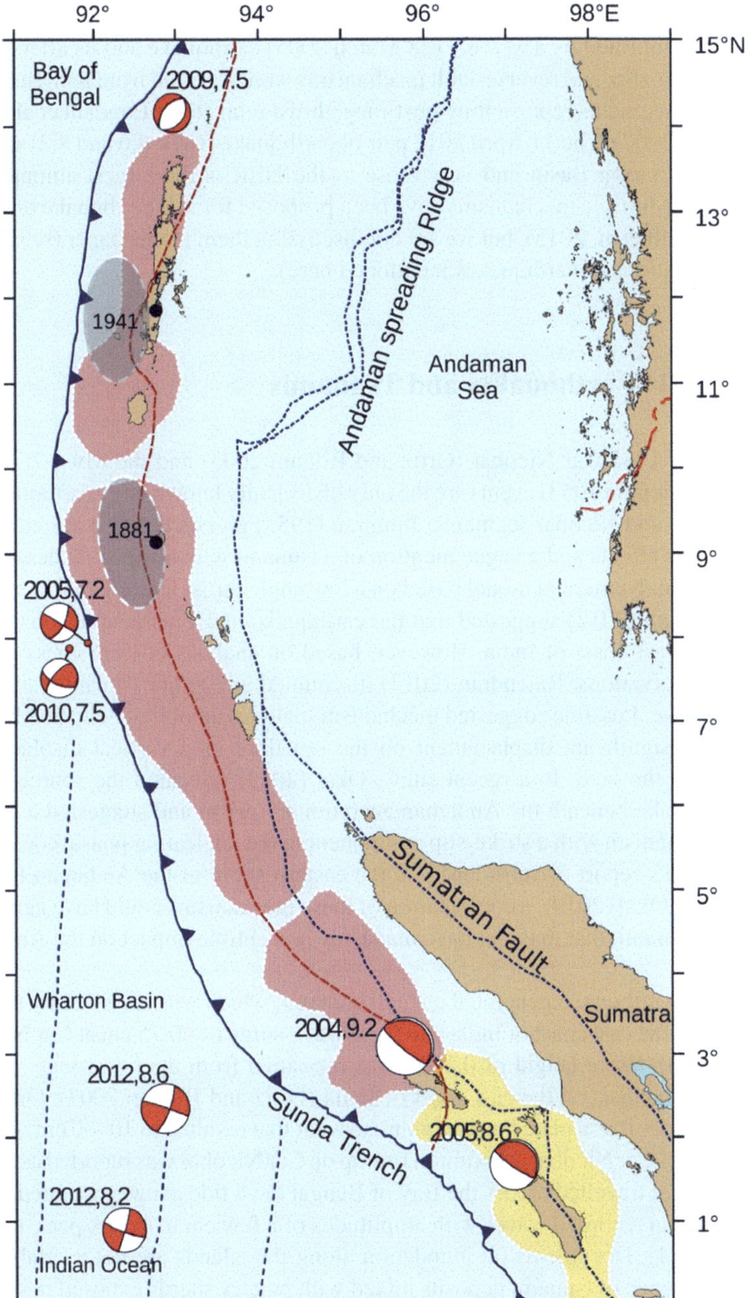

Fig. 3 Major tectonic features and ruptures of the recent (red, 2004; yellow, 2005) and historical earthquakes; shaded in grey are the estimated rupture zones of the 1881 and 1941 earthquakes. The GCMT solutions discussed in the text. Fossil transform faults are shown in broken lines (modified from Rajendran et al. 2013)

but no great earthquake occurred here. The southern Simeulue segment was the most active, as exemplified by a Mw 8.7 (28 March 2005) earthquake and its aftershocks. Normal, strike-slip, and reverse-fault mechanisms were reported from both Simeulue and Nicobar segments representing post-megathrust relaxation (Engdahl et al. 2007; Dewey et al. 2007). The 11 April 2012 pair of earthquakes (Mw 8.6 and 8.2) sourced within the Wharton Basin and very close to the NER is the largest among these earthquakes. Multiple mechanisms have been proposed for this earthquake (e.g. Yue et al. 2012; Hill et al. 2015), but we are not discussing them in this paper (see Figs. 1 and 3 for locations of earthquakes mentioned here).

3 Pre-2004 Earthquakes and Tsunamis

The Mw 7.9, 1881 Car Nicobar (Ortiz and Bilham 2003) and the Mw ~7.7, 1941 Andaman (Jhingran 1953) events are the only historically known large earthquakes in the Andaman and Nicobar segments. Jhingran (1953) gives a detailed account of its macroseismic effects and a vague mention of a tsunami with no specific description at any location. Numerical model based on a low angle thrust faulting mechanism by Srivastava et al. (2012) suggested that the earthquake had generated a tsunami that affected the east coast of India. However, based on analysis of eyewitness reports and field observations. Rajendran (2013) discounted any major tsunami caused by this earthquake. Possible suggested mechanism includes an oblique-slip mechanism without any significant displacement on the sea floor or a vertical displacement very close to the land. In a recent study, Okal (2019) relocated the source of the 1941 earthquake beneath the Andaman accretionary prism and suggested a normal faulting mechanism with a strike-slip component in the nucleating phase, concurring with Jhingran's report of subsidence on the eastern shore of the Andaman Islands. According to Okal (2019), a combination of these mechanisms could have generated a nominal tsunami, that may not have made any perceptible impact on the Andaman Coast.

The 1881 earthquake generated a small tsunami, which was recorded by the tide gauges along the east coast of India and there was a surge of ~0.75 cm at Car Nicobar (Rogers 1883). Wave height of 0.25 m was measured from the tide gauge stations at Madras (Chennai) on the east coast of India (Ortiz and Bilham 2003). Ortiz and Bilham consider it as a plate boundary thrust event that resulted in 10–60 cm of uplift of the island of Car Nicobar. Maximum run-up on Car Nicobar was reported as 75 cm, but as the wave travelled across the Bay of Bengal the 8 tide gauges operated by the Survey of India recorded waves with amplitudes of a few cm to 1.2 m, peak-to-peak (Oldham 1884). Documents on inundation along the islands are not available, but from the presence of tsunami deposits mixed with pottery sherds exposed in shallow trenches ~50 m inland of the Car Nicobar Island, Rajendran et al. (2013) inferred a minimum inundation of 50 m. The examples of 1941 and 1881 earthquakes suggest that faulting mechanisms and thus the tsunamigenic potential of large earthquakes on the Andaman and Nicobar Islands vary. Such diversities in the mechanisms are evident from the modern-day examples such as Mw 7.7, 2005 and 2010 earthquakes near Car Nicobar, discussed later in this paper.

4 Pre- and Post-2004 Deformation of the Outer Arc Prior to 2004

It has been documented that the complex processes that govern the behavior of subduction zones often lead to land level changes before, and during an earthquake (Plafker 1972; Atwater 1987). Convergence prior to great earthquakes causes bending and buckling of the crust, producing inland flexural bulge. These processes cause gradual uplift on the leading edge of the overriding plate and the mechanism of faulting is usually by thrusting. As the rupture releases stored elastic energy, the seaward edge of the continent springs back and moves upward, and the bulge collapses leading to subsidence of land (Fig. 2). The most compelling evidence for past tectonic movements along the plate margins comes from such ground deformation preserved in the geological records.

Rajendran et al. (2007) had presented evidence for the emergence of microatolls in the eastern margin of Andaman Island and Car Nicobar, observed during September 2003 and subsidence following the 2004 earthquake. The uplifted microatolls near Port Blair and Car Nicobar had submerged post-2004 due to the collapse of the bulge. Thus, the visual observations made 15 months prior to the 2004 suggesting that the microatolls show indications of emergence by 10 cm based on 2003 HSL and a previous HSL at Mundapahar near Port Blair. With the information that had been gathered from various sources, it is clear that Port Blair was emerging inter-seismically. This trend would have been preserved in the growth history of the contemporary microatolls before the abrupt co-seismic subsidence, in the vicinity of Port Blair. The pre-earthquake GPS model by Paul and Rajendran (2015) suggests that thrust fault along Sumatra and North Andaman had reached its peak in accumulating stress and was ready to rupture. Their model indicates that slow slip had started in the northern part sometime in 2003 and when the sudden release of energy took place on December 26, 2004 the slip propagated northward, and the thrust caught up the full rupture.

Variations in pre- and post-earthquake stress changes manifest also in the style of faulting, as evidenced by thrust faulting before and normal faulting earthquakes after the main megathrust (Spence 1987; Christensen and Ruff 1988). The Andaman segment had generated a few shocks prior to 2004, among them was the 13 September 2002 earthquake (Mw 6.5) near Diglipur, a thrust-faulting event at 21 km. The post-2004 normal-faulting event near Coco Island (10 August 2009, Mw 7.5) demonstrates the change to the post-2004 extensional stress (Fig. 4a, b) (Andrade and Rajendran 2011). Subduction zones also generate intra-slab tensional earthquakes at intermediate depths due to slab pull (Spence 1987). Further, intra-plate seismicity is known to originate due to coupling of compressional stresses in the over-riding plate and the

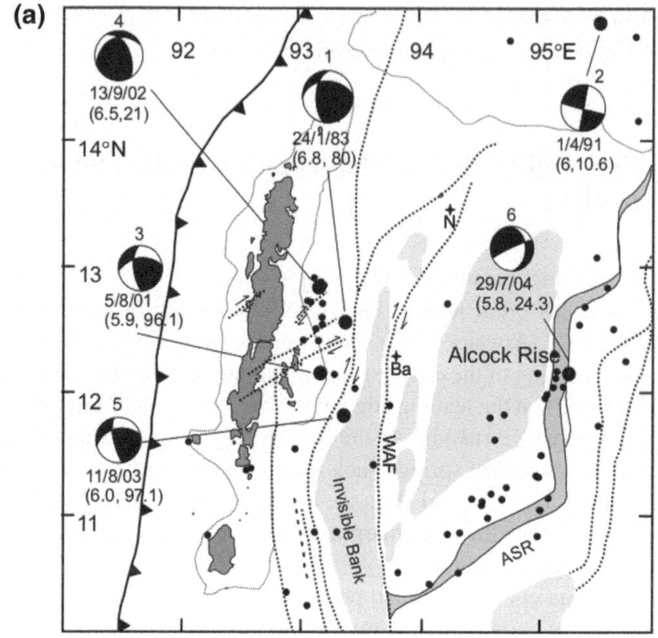
(a)
92
93
94
95°E
14°N
13
12
11
4
13/9/02
(6.5,21)
1
24/1/83
(6.8, 80)
2
1/4/91
(6,10.6)
3
5/8/01
(5.9, 96.1)
5
11/8/03
(6.0, 97.1)
6
29/7/04
(5.8, 24.3)
Alcock Rise
Ba
N
WAF
Invisible Bank
ASR

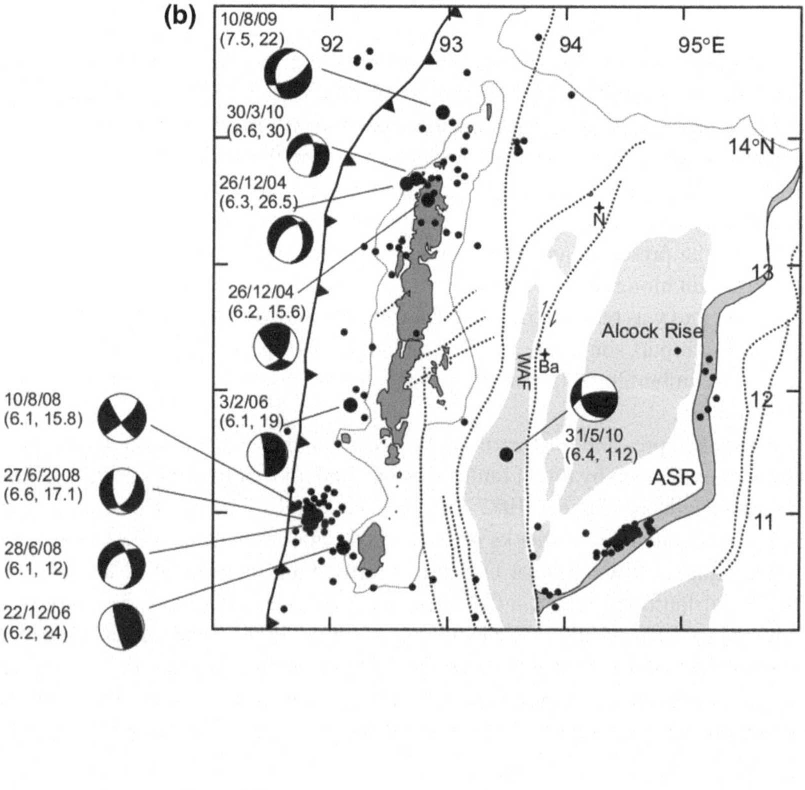
(b)
92
93
94
95°E
14°N
13
12
11
10/8/09
(7.5, 22)
30/3/10
(6.6, 30)
26/12/04
(6.3, 26.5)
26/12/04
(6.2, 15.6)
3/2/06
(6.1, 19)
10/8/08
(6.1, 15.8)
27/6/2008
(6.6, 17.1)
28/6/08
(6.1, 12)
22/12/06
(6.2, 24)
31/5/10
(6.4, 112)
Alcock Rise
Ba
N
WAF
ASR

◂**Fig. 4** **a** Epicentral locations of Mw ≥ 4.5, pre-2004 earthquakes with events Mw ≥ 5.9 identified (format dd-mm-yy), with their focal mechanisms **b** Epicentral locations of Mw ≥ 4.5, post-2004 earthquakes with events Mw ≥ 6.1. Numbers in the brackets below the date indicate moment magnitude and focal depth in km, respectively (data source: Global CMT catalog). Ba: Barren Island; N: Narcondam; ASR: Andaman Spreading Ridge; WAF: West Andaman Fault (after Andrade and Rajendran 2011)

tensional stresses within the subducting slab (Seno and Yoshida 2004). The accelerated seismicity within the Wharton Basin post-2004 may be attributed to transmission of plate boundary stresses to the interiors (Hill et al. 2015).

5 Coseismic Deformation

Land level changes as predicted by the model by Plafker (1972) were recorded from the emergence of microatolls prior to the megathrust (Fig. 5a). The same region registered coseismic subsidence as evidenced by the submergence of mangrove swamps (Fig. 5b). In the North Andaman, uplift of land was manifested mostly in the form of shoreline changes, and recession of watermarks showing the pre-earthquake survival levels of mussels and barnacles. For example, in Diglipur there was a rise in the ground level where mangrove swamps had been elevated by 50–60 cm above the pre-earthquake high-tide level. Several coral platforms were raised by more than 50 cm above the shoreline (Fig. 5c) (see Rajendran et al. 2007, 2013 for details). Although the northern Andaman generally registered uplift, Port Blair, on the eastern margin of South Andaman, subsided by 1 m, as recorded by the tide gauge. This was evident also from the land-level changes, such as the submergence of mangrove vegetation. In one such region we found evidence for a previously subsided line of trees (Fig. 5d). The southern islands of the Car Nicobar, and Great Nicobar registered more subsidence compared to the Andaman—about 3 m at Great Nicobar (Malik and Murty 2005) reported maximum subsidence (3 m) at Great Nicobar and ~1 m near Kamorta (Thakkar and Goyal 2006).

6 Post-2004 Wharton Basin Seismicity

The post-2004 seismicity suggests that a megathrust earthquake can lead to enhanced activity due to transfer of plate boundary stresses. The oceanic plate adjoining the trench south of 10° N has a history of strike-slip faulting (Stein and Okal 1978; Deplus et al. 1998). More striking among the post-2004 intraplate earthquakes are the great 2012, Mw 8.2 and M.8.6 events, discussed widely in the literature (Yue et al. 2012; Delescluse et al. 2012; Hill et al. 2015, for example). Deplus et al. (1998) had suggested reactivation of the near N-S oriented left-lateral faults as a

Fig. 5 **a** Emergence of microatolls by about 10 cm, near Port Blair as recorded in September 2003, 15 months before the 2004 earthquake; **b** coseismic subsidence near Port Blair, as evidenced by the submergence of mangrove swamps **c** uplift of coral platforms along the shore of Interview Islands **d** site near Port Blair where the subsidence of mangrove swamps exposed a previous subsidence event

potential mechanism for the earthquakes within the WB. The 12 June 2010 (Mw 7.5) earthquake on the subducting plate on the Nicobar segment is another post-2004 event mimicking the same mechanism (Rajendran et al. 2011a). There are many more examples of intraplate earthquakes within the Wharton Basin and the subducting slab (Abercrombie et al. 2003; Lay et al. 2016 and references therein). Here we discuss a few selected Mw > 7 earthquakes that bring out their potential relation with the structural fabric of the Wharton Basin (Fig. 1).

6.1 The 2005 and 2010 Car Nicobar Earthquakes

The 2005 Car Nicobar earthquake (Mw 7.2; depth 12 km) is the first significant post-2004 event in the Nicobar segment, followed by the 2010 event (Mw 7.5; depth

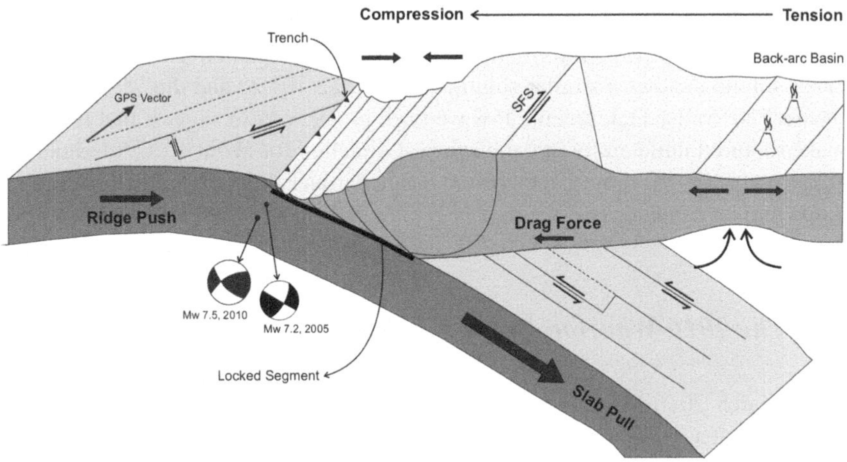

Fig. 6 A conceptual model of the oblique convergence of the Andaman subduction zone. The Global Positioning System (GPS) vector shows the motion of the Indian plate with respect to the Eurasian plate, with the subducting plate featuring a number of fracture zones. Slab pull leads to down-dip tensional behavior in the slab, whereas there exists a gradient in the maximum horizontal stress, from the back-arc basin (tension) to the locked interface (compression), of the overriding plate. The July 2005 Mw 7.2 and the June 2010 Mw 7.5 earthquakes are shown, after Mallick et al. 2017

33 km). The 2010 earthquake triggered a small tsunami that was recorded by some tide gauge stations and ocean bottom buoys (Rajendran et al. 2011a). Mallick et al. (2017) reported that both these events were related to the N-S fractures within the Wharton Basin and that both these events had oblique components in their left-lateral focal mechanisms. Aderhold and Abercrombie (2016) suggested a west-northwest–east-southeast (WNW–ESE) fault planes, but based on the variance analysis, Mallick et al. (2017) preferred the NNE–SSW plane. For the 2010 earthquake they had suggested two components of slip within the rectangular fault plane; one within the top 20 km where ~30% of the moment was released with mostly up-dip slip, and the other below 30 km depth where the motion was mostly strike-slip. The shallow oblique component was considered as responsible for the small tsunami. Oblique component was negligible for the 2005 slip model and the shallow strike-slip component was concentrated below the focus, possible reason for the lack of any tsunami. A schematic sketch of the mechanism operation within the slab is shown in Fig. 6.

6.2 *The 2012 Wharton Basin Earthquake*

The 10 January 2012 Mw 7.2 WB earthquake occurred three months before the great Mw 8.6 and Mw 8.2 earthquakes. The GCMT (Ekström et al. 2012) solution shows strike 103° and dip 81°; the finite-fault model from the U.S. Geological Survey National Earthquake Information Center (NEIC), based on a W-Phase moment tensor solution shows a similar solution with strike 101.2° and dip 75.1°. Fan and Shearer (2016) used teleseismic P-wave back projections to suggest that the earthquake ruptured unilaterally in the northwest direction for about 40 s. Aderhold and Abercrombie (2016) show that this event resulted in an EW-trending bilateral rupture, although they conclude that the near N-S fault plane cannot be ruled out.

6.3 The 2016 Wharton Basin Earthquake

The 2 March 2016 Mw 7.8 event (~36 km depth) is the largest to have occurred in the WB after the 2012 pair of great earthquakes off Sumatra. USGS reported its epicenter as 4.952° S, 94.330° E (24.0 km deep), sourced in a region with no historical seismicity. The location is offset from two prominent north-south trending fracture zones to the west and zones with subtle bathymetric expressions to the east with no clear bathymetric expression at its precise location. The USGS NEIC W phase inversion suggested two planes (strike 5°, dip 79°); (strike 274°; dip 84°). Global Centroid-Moment Tensor (GCMT) suggested slip on planes with similar geometry (strike 5°, dip 80°; strike 96°; dip 84°) indicating that the primary faulting geometry is strike-slip with north-south and east-west trending nodal planes (see Lay et al. 2016 and references). Lay et al. (2016) preferred north-south fault orientation, which is consistent with that of the nearby fracture zones, although the event does not locate directly on a prominent structure. Post-2012 seismic reflection profiles in the WB have exposed fault traces that go down to ~45 km, arguing for seismic sources within the oceanic lithosphere (Qin and Singh 2015).

7 Tsunami History: Records from the Andaman-Nicobar Islands

A question that was frequently asked following the 2004 tsunami was about the past history of such events. Reliable historical data for the last 500 years do not point to any coastal flooding leading to major loss of life on the east coast of India. The 2 April 1762 earthquake ($M \geq 7.6$) occurred off Myanmar (Chhibber 1934; Cummins 2007) and the tsunami that followed has possibly reached the north and south Andaman as evidenced by the sedimentary records near Port Blair (Malik et al. 2011). The east coast of India has experienced two previous transoceanic tsunamis in 1881 and 1883 (Dominey-Howes et al. 2007). Evidence of older tsunamis at Phra Thong (Thailand), Meulaboh (northern Sumatra, Indonesia), and Sri Lanka also suggest that the penultimate tsunami that affected these shores occurred during a time window of 780–1400 CE (Jankaew et al. 2008; Monecke et al. 2008; Jackson et al. 2014).

The Andaman and Nicobar Islands covers about two thirds of the length of the 2004 rupture and it has also preserved multiple proxies for the uplift and subsidence of land, along with a variety of tsunami deposits. Based on multiple lines of evidence from the Andaman and Nicobar Islands and the east coast of India, Rajendran et al. (2013) have estimated the timing and size of the previous tsunami-genic earthquakes. Here we discuss some results from the Andaman and Nicobar Islands.

The synchronous subsidence and the deposition of tsunami sands and debris along many parts of the south and north Andaman Islands provide a situation quite analogous to what has been observed in other regions with the history of large tsunamis (e.g. Atwater 1987; Cisternas et al. 2005). While coasts adjoining the trench may preserve synchronous evidence for coseismic subsidence together with tsunami deposits, only the latter would be present along the distant shores. Thus, following the 2004 earthquake, sites closer to the rupture zone presented evidence for subsidence and tsunami debris and sands, where the transoceanic sites were noted for the tsunami deposits, mostly in the form of inland sand sheets (Rajendran et al. 2013 and references therein).

7.1 Tsunami Chronology

Rajendran et al. (2013) used multiple proxies to develop the tsunami chronology and some of their results are discussed here. At Hut Bay, coseismically elevated by 30 cm, the 2004 tsunami traveled ~1 km through inland creeks and deposited sand, coral, and organic debris (Fig. 7a). The 50 cm-thick deposit of loose coral fragments mixed with assorted shells along the inland creek (~500 m away from the surf zone) inland suggested a previous tsunami deposition (Fig. 7b). The abrupt contact of the debris with the underlying fine-grained calcareous sand, devoid of any shells, was used to identify it as a high-energy wave deposition. Gastropods from the sample

Fig. 7 **a** A view of a creek at Hut Bay located on the first terrace above MSL, with its location marked on the map (inset). **b** A close view of the section at site HB-1 along the creek showing a debris layer underlain by calcareous sand (after Rajendran et al. 2013)

collected from the lower level of debris yielded an AMS date of 1480–1660 CE, which was considered as the maximum age for deposition.

At Rangachanga, located ~8 km south of Port Blair, destruction of trees following a previous subsidence event was found along the 10 m-wide creek known as Burmanallah, where remains of a fossil root zone were exposed during low tide. Roots projecting to the creek were exposed on both sides along the creek. Samples of in situ tree trunks from sites across the creek yielded ages ranging from 770–880 to 770–1040 CE (see Fig. 5d). At several locations close to Rangachanga, the 2004 tsunami had deposited sand sheets. Evidence from previous tsunamis is expected to be in the form of sand-peat couplets that mimics the 2004 deposition. Rajendran et al. (2013) have presented trench sections that exposed a veneer of fine sand deposited over organic-rich palaeosol ~50 cm below the present surface. A 20-cm-thick peat layer sandwiched between two calcareous sand beds yielded a date of. 1380–1520 CE. The site of this exposure is ~2 km south of the subsided mangrove swamp and is 50 m away from the current shore line. Such palaeosol-sand-association at similar depth was widely observed even at locations as far as 500 m inland of the surf zone.

Rajendran et al. (2013) have also discussed in detail how a storm-driven deposition cannot be a realistic mechanism for the inland sand deposition at these localities. These also point out the remarkable correspondence of the age of this deposit with that reported from Phra Thong and Meulaboh. Thus, the chronology obtained from the near and distant tsunami sand deposits suggests a window between 1290 and 1450 CE for the previous tsunami event that affected all these coastal regions. Rajendran (2019) suggests that this 1290–1450 CE derived from geological evidence must have occurred in 1343 CE—a conclusion based on the analyses of the historical data from the southwestern coast of India.

8 Discussion and Conclusions

The 2004 earthquake generated several near-field coseismic features including uplift and subsidence of land within the rupture zone. The closure of the interseismic cycle was signified by nominal pre-seismic uplift of land at various locations along the trench, but they were not recognized as precursors of an impending rupture (Rajendran et al. 2007). Analysis of the 2004 sequence highlighted how the subduction zone displayed signatures observed elsewhere-thrust and normal faulting events in the outer rise regions respectively before and after the megathrust. Analysis of the focal mechanisms of the pre- and post-earthquakes in the north Andaman by Andrade and Rajendran (2011) was useful to validate this idea and the underlying processes discussed by previous workers (e.g. Spence 1987; Christensen and Ruff 1988).

The post-2004 phase saw an unprecedented activation of the subducting oceanic plate, in particular the Wharton Basin. The largest of the events here were the 2012 great twin earthquakes off Sumatra, which has been discussed by several papers and we have not discussed them here. An important observation is that the post-2004 phase was dominated by activation of the N-S oriented fault systems within the

Wharton Basin and the subducting plate. Although the causative faults for some of the intraplate events in the Wharton Basin are not evident from the current level of understanding, the role of the N-S fabric of fractures in triggering earthquakes have been suggested by many studies (Lay et al. 2016; Mallick et al. 2017).

The coseismic deformation and the tsunami deposition gave an unprecedented opportunity to explore large areas affected by the earthquake and tsunami for past events. While the regions adjoining the rupture zone provide evidence for coseismic deformation as well as tsunami deposition, distant shores also offered evidence for the predecessor of the 2004 event. Among the previous tsunamis, the one between 8th and 11th centuries CE seems relatively better represented in the stratigraphy at various sites along the rupture zones as well as other distant coasts of mainland India and Sri Lanka (Rajendran et al. 2011b; Jackson et al. 2014). Evidence for one or more than tsunami during late medieval period (13th to 16th Century CE) has been reported from Thailand and Sumatra (Jankaew et al. 2008; Monecke et al. 2008, Meltzner et al. 2010). Evidence for the younger event obtained from parts of the Andaman and Nicobar part of the subduction zone, are less dramatic compared to what has been observed in Thailand.

Fifteen years after the great tsunami we have learned much about the subduction zone and what a great earthquake and tsunami can do. We understand that all great earthquakes do not generate tsunami, as observed in the Andaman-Nicobar part of the subduction zone. Their tsunamigenic power seems to depend on the nature of shallow slip- oblique slip with a shallow thrust component might have more potential compared to one with a strike-slip motion at the same depth. Research on tsunami geology has brought out many interesting data and these issues would remain topics of research for future.

Acknowledgements We thank Jyotiranjan Ray for the invitation to contribute this article. We are indebted to many of our collaborators including Anil Earnest, John Paul, S. Srinivasalu and Vanessa Andrade who have taken part in the work in the Andaman and Nicobar Islands at various stages. We also thank two anonymous reviewers for their suggestions.

References

Aderhold K, Abercrombie RE (2016) Seismotectonics of a diffuse plate boundary: observations off the Sumatra-Andaman trench. J Geophys Res 121:3462–3478. https://doi.org/10.1002/2015JB012721

Abercrombie RE, Antolik M, Ekström G (2003) The June 2000 M_W 7.9 earthquakes south of Sumatra: deformation in the India–Australia Plate. J Geophys Res 108(B1), 2018. https://doi.org/10.1029/2001jb000674

Andrade V, Rajendran K (2011) Intraplate response to the great 2004 Sumatra-Andaman earthquake: a study from the Andaman segment. Bull Seismol Soc Am 101:506–514. https://doi.org/10.1785/0120100155

Atwater BF (1987) Evidence of great Holocene earthquakes along the outer coast of Washington State. Science 236:942–944. https://doi.org/10.1126/science.236.4804.942

Christensen DH, Ruff LJ (1988) Seismic coupling and outer rise earthquakes. J Geophys Res 93:13421–13444. https://doi.org/10.1029/JB093iB11p13421

Chhibber HL (1934) The geology of Burma. MacMillan and Co. Limited, London, 538p, pp 47–70

Cisternas M, Atwater BF, Torrejón F, Sawai Y, Machuca G, Lagos M, Eipert A, Youlton A, Salgado C, Kamataki T, Shishikura M, Rajendran CP, Malik JK, Rizal Y, Husni M (2005) Predecessors of the giant 1960 Chile earthquake. Nature 437:404–407. https://doi.org/10.1038/nature03943

Curray JR (2005) Tectonics and history of the Andaman Sea region. J Asian Earth Sci 25:187–232. https://doi.org/10.1016/j.jseaes.2004.09.001

Cummins PR (2007) The potential for giant tsunamigenic earthquakes in the northern Bay of Bengal. Nature 449:75–78. https://doi.org/10.1038/nature06088

Delescluse M, Chamot-Rooke N, Cattin R, Fleitout L, Trubienko O, Vigny C (2012) April 2012 intra-oceanic seismicity off Sumatra boosted by the Banda-Aceh megathrust. Nature 490:240–244. https://doi.org/10.1038/nature11520

DeMets C, Gordon RG, Argus DF (1988) Intraplate deformation and closure of the Australia-Antarctica-Africa plate circuit. J Geophys Res 93:877–888. https://doi.org/10.1029/JB093iB10p11877

DeMets C, Gordon RG, Argus DF (2010) Geologically current plate motions. Geophys J Int 181:1–80. https://doi.org/10.1111/j.1365-246X.2009.04491.x

Deplus C, Diament M, Hébert H, Bertrand G, Dominguez S, Dubois J, Malod J, Patriat P, Pontoise B, Sibilla JJ (1998) Direct evidence of active deformation in the eastern Indian oceanic plate. Geology 26:131–134

Dewey JW, Choy G, Presgrave B, Sipkin S, Tarr AC, Benz HP, Earle Wald D (2007) Seismicity associated with the Sumatra-Andaman Islands earthquake of 26 December 2004. Bull Seismol Soc Am 97(1A):S25–S42

Dominey-Howes D, Cummins P, Burbidge D (2007) Historic records of teletsunami in the Indian Ocean and insights from numerical modeling. Nat Hazards 42:1–17. https://doi.org/10.1007/s11069-006-9042-9

Ekström G, Nettles M, Dziewonski AM (2012) The global CMT project 2004–2010: centroid-moment tensors for 13,017 earthquakes. Phys Earth Planet Inter 200–201:1–9. https://doi.org/10.1016/j.pepi.2012.04.002

Engdahl ER, Villasenor A, DeShon HR, Thurber CH (2007) Teleseismic relocation and assessment of seismicity (1918–2005) in the region of the 2004 Mw 9.0 Sumatra-Andaman and 2005 Mw 8.6 Nias Island great earthquakes. Bull Seismol Soc Am 97:43–61

Fan W, Shearer PM (2016) Fault interactions and triggering during the 10 January 2012 Mw 7.2 Sumatra earthquake. Geophys Res Lett 43:1934–1942. https://doi.org/10.1002/2016GL067785

Fitch TJ (1972) Plate convergence, transcurrent faults, and internal deformation adjacent to Southeast Asia and the Western Pacific. J Geophys Res 77:4432–4460

Hill EM, Yue H, Barbot S, Lay T, Tapponnier P, Hermawan I, Hubbard J, Banerjee P, Feng L, Natawidjaja D, Sieh K (2015) The 2012 Mw 8.6 Wharton basin sequence: a cascade of great earthquakes generated by near-orthogonal, young, oceanic mantle faults. J Geophys Res 120:3723–3747. https://doi.org/10.1002/2014JB011703

Ishii M, Kiser E, Geist EL (2013) Mw 8.6 Sumatran earthquake of 11 April 2012: rare seaward expression of oblique subduction. Geology 41:319–322. https://doi.org/10.1130/G33783.1

Jackson KL, Eberli GP, Amelung F, McFadden MA, Moore AL, Rankey EC, Jayasena HAH (2014) Holocene Indian Ocean tsunami history in Sri Lanka. Geology 42:859–862. https://doi.org/10.1130/G35796.1

Jankaew K, Atwater BF, Sawai Y, Choowong M, Charoentitirat T, Martin ME, Prendergast A (2008) Medieval forewarning of the 2004 Indian Ocean tsunami in Thailand. Nature 455:1228–1231. https://doi.org/10.1038/nature07373

Jhingran AG (1953) A note on the earthquake in the Andaman Islands (26 June 1941). Rec Geol Surv India 82(2):300–307

Lay TL, Ye CJ, Ammon A, Dunham Koper KD (2016) The 2 March 2016 Wharton Basin Mw 7.8 earthquake: high stress drop north-south strike-slip rupture in the diffuse oceanic deformation

zone between the Indian and Australian Plates. Geophys Res Lett 43:7937–7945. https://doi.org/10.1002/2016GL069931
Le Pichon X (1968) Sea-floor spreading and continental drift. J Geophys Res 73:3661–3697. https://doi.org/10.1029/JB073i012p03661
Mallick R, Parameswaran M, Rajendran K (2017) The 2005 and 2010 earthquakes on the Sumatra-Andaman trench: evidence for Post-2004 megathrust intraplate rejuvenation. Bull Seism Soc Am 107:1569–1581
Malik JN, Murty CVR (2005) Landscape changes in Andaman and Nicobar Islands (India) due to Mw 9.3 tsunamigenic Sumatra earthquake of 26 December 2004. Curr Sci 88:1384–1386
Malik JN, Shishikura M, Echigo T, Ikeda Y, Satake K, Kayanne H, Sawai Y, Murty CVR, Dikshit O (2011) Geologic evidence for two pre-2004 earthquakes during recent centuries near Port Blair, South Andaman Island, India. Geology 39:559–562. https://doi.org/10.1130/G31707.1
McCaffrey R (2009) The tectonic framework of the Sumatran subduction zone. An Rev Earth Planet Sci 37:345–366. https://doi.org/10.1146/annurev.earth.031208.100212
Meltzner AJ, Sieh K, Chiang H.-W, Shen C.-C, Suwargadi BW, Natawidjaja DH, Philibosian BE, Briggs RW, Galetzka J (2010) Coral evidence for earthquake recurrence and an A.D. 1390–1455 cluster at the south end of the 2004 Aceh–Andaman rupture. J Geophys Res 115, B10402. https://doi.org/10.1029/2010JB007499
Monecke K, Finger W, Klarer D, Kongko W, McAdoo BG, Moore AL, Sudrajat SU (2008) A 1,000-year sediment record of tsunami recurrence in Northeast Sumatra. Nature 455:1232–1234. https://doi.org/10.1038/nature07374
Morton RA, Gelfenbaum G, Jaffe BE (2007) Physical criteria for distinguishing sandy Tsunami and storm deposits using modern examples. Sediment Geol 200:184–207. https://doi.org/10.1016/j.sedgeo.2007.01.003
Okal E (2019) The large Andaman Islands earthquake of 26 June 1941: why no significant Tsunami? Pure Appl Geophys. https://doi.org/10.1007/s00024-018-2082-8
Oldham RD (1884) Note on the earthquake of 31 December 1881. Rec Geol Surv India 17(2):47–53
Ortiz M, Bilham R (2003). Source area and rupture parameters of the 31 December 1881 Mw = 7.9 Car Nicobar earthquake estimated from tsunamis recorded in the Bay of Bengal. J Geophys Res 108(B4):2215. https://doi.org/10.1029/2002jb001941
Parameswaran R (2016) Geodynamics of the India-Eurasia plate boundary from earthquake source models: examples from the Makran, central Himalaya, Burma, and Andaman Sumatra segments. Doctoral Thesis, Indian Inst. Science, Bangalore 246p
Paul J, Rajendran CP, Lowry AR, Andrade V, Rajendran K (2012) Andaman post-seismic deformation observations: still slipping after all these years? Bull Seism Soc Am 102:343–351. https://doi.org/10.1785/0120110074
Paul J, Rajendran CP (2015) Short-term pre-2004 seismic subsidence near South Andaman: Is this a precursor slow slip prior to a megathrust earthquake? Phys Earth Planet Int. https://doi.org/10.1016/j.pepi.2015.08.006
Plafker G (1972) Alaskan Earthquake of 1964 and Chilean Earthquake of 1960: implications for arc tectonics. J Geophys Res 77:901–925
Qin Y, Singh SC (2015) Seismic evidence of a two-layer lithospheric deformation in the Indian Ocean. Nat Comm 6:8298. https://doi.org/10.1038/ncomms9298
Rajendran CP (2013). Was the 1941 Andaman earthquake tsunamigenic? Comments on "Inundation studies for Nagapattinam region on the East coast of India due to tsunamigenic earthquakes from the Andaman region" by Srivastava et al. Nat Haz 65:981–994
Rajendran CP (2019) Historical accounts of sea-disturbances from south India and their bearing on the penultimate predecessor of the 2004 tsunami. Seism Res Lett 90(2A):774–783. https://doi.org/10.1785/0220180355
Rajendran CP, Rajendran K, Andrade V, Srinivasalu S (2013) Ages and relative sizes of pre-2004 Tsunamis in the Bay of Bengal inferred from geologic evidence in the Andaman and Nicobar Islands. J Geophys Res 118:1–18. https://doi.org/10.1002/jgrb.50122

Rajendran K, Andrade V, Rajendran CP (2011a) The June 2010 Nicobar earthquake: fault reactivation on the subducting oceanic plate. Bull Seismol Soc Am 101:2568–2577. https://doi.org/10.1785/0120110002

Rajendran CP, Rajendran K, Srinivasalu S, Andrade V, Aravali P, Sanwal J (2011b) Geoarcheological evidence of a Chola period tsunami from an ancient port at Kaveripattinam on the South-Eastern of India. Geoarchaeology 26:867–887

Rajendran CP, Rajendran K, Earnest A, Anu R, Machado T, Freymueller J (2007) The style of crustal deformation and seismic history associated with the 2004 Indian Ocean earthquake: a perspective from the Andaman-Nicobar Islands. Bull Seismol Soc Am 97:S174–S191. https://doi.org/10.1785/0120050630

Rajendran, CP, Rajendran K, Machado T, Satyamurthy T, Aravazhi P, Jaiswal M (2006). Evidence of ancient sea surges at the Mamallapuram coast of India and implications for previous Indian Ocean tsunami events. Cur Sci 91:1242–1247

Rogers RE (1883) Memorandum on the earthquake of the 31st December 1881 and the great sea-waves resulting therefrom, as shown on the diagrams of the tidal observatories in the Bay of Bengal, In: General report on the operations of the Survey of India 1881–1882. Govt. Print Office, Calcutta, pp 70–73

Satake K, Atwater BF (2007) Long-term perspectives on giant Earthquakes and Tsunamis at subduction zones. Ann Rev Earth Planet Sci 35:349–374

Spence W (1987) Slab pull and the seismotectonics of subducting lithosphere. Rev Geophys 25:55–69

Seno T, Yoshida M (2004) Where and why do large shallow intra-slab earthquakes occur? Phys Earth Planet Inter 141:183–206

Srivastava K, Krishna Kumar R, Swapna M, Swaroopa Rani V, Dimri VP (2012) Inundation studies for Nagapattinam region on the east coast of India due to Tsunamigenic earthquakes from the Andaman region. Nat Hazards 63:211–221

Stein S, Okal E (1978) Seismicity and tectonics of the Ninetyeast Ridge area: evidence for internal deformation of the Indian plate. J Geophys Res 83:2233–2245

Switzer AD, Srinivasalu S, Thangadurai N, Ram Mohan V (2012) Bedding structures in Indian Tsunami deposits that provide clues to the dynamics of Tsunami inundation. Geol Soc Londn Spec Pub 361(1):61–77

Thakkar MG, Goyal B (2006) Historic submergence and Tsunami destruction of Nancowrie, Kamorta, Katchall and Trinket Islands of Nicobar: consequences of 26 December 2004 Sumatra-Andaman earthquake. Cur Sci 90:989–994

Yue H, Lay T, Koper KD (2012) *En enchelon* and orthogonal fault ruptures of the 11 April 2012 great intraplate earthquakes. Nature 490:245–249. https://doi.org/10.1038/nature11492

A Glimpse of Crustal Deformation Through Earthquake Supercycle in the Andaman Region Using GPS Measurements

Harendra Dadhich, M. S. Naidu, Vineet K. Gahalaut, J. K. Catherine, Amit K. Bansal and V. Rajeshwar Rao

Abstract Surface measurements of crustal deformation and their modelling provide insights into the subsurface earthquake processes at depth. In this article we review the available geodetic measurements of crustal deformation which spans the interseismic, coseismic and postseismic phases of earthquake cycle corresponding to the 2004 Sumatra Andaman earthquake. The GPS measurements of interseismic period (during the preceding 10 year period from the 2004 Sumatra Andaman earthquake) show evidence of strain accumulation in the Andaman region. Coseismic offsets due to the 2004 earthquake are overwhelming which reach up to whopping 7 m and imply that coseismic slip at the plate boundary interface reach up to 15–20 m. Following the earthquake, the region experienced large postseismic deformation which is continuing even now, particularly in the Nicobar and Sumatra region. Total postseismic deformation until the end of 2016 corresponds to an earthquake of Mw 8.5. Aftershocks corresponds to less than 15% of this estimate, thus majority of the deformation is aseismic.

Keywords Subduction zone 2004 · Sumatra andaman earthquake · Crustal deformation · GPS measurements

1 Introduction

The oblique convergence in the Andaman Sumatra region between the India-Australia and Sunda plates results in slip partitioning, wherein predominantly trench normal motion is accommodated in the frontal arc region and trench parallel motion is accommodated in the back arc region, comprising the Sumatra Fault System, a major strike slip fault, and the Andaman Sea Ridge Transform fault region in the Andaman Sea (Fitch 1972; McCaffrey 1992). The Sumatra Fault System in the Sumatra region, joins with the West Andaman Fault, east of Great Nicobar and the

H. Dadhich
National Center for Seismology, New Delhi, India

M. S. Naidu · V. K. Gahalaut (✉) · J. K. Catherine · A. K. Bansal · V. Rajeshwar Rao
CSIR-National Geophysical Research Institute, Hyderabad, India
e-mail: vkgahalaut@yahoo.com

J. S. Ray and M. Radhakrishna (eds.), *The Andaman Islands and Adjoining Offshore: Geology, Tectonics and Palaeoclimate*, Society of Earth Scientists Series,
https://doi.org/10.1007/978-3-030-39843-9_11

latter extends northward up to east of North Andaman island (Curray 2005; Kamesh Raju et al. 2007). Recent investigations suggest that strain is fully partitioned in the Sumatra region whereas in the Andaman Nicobar region it is partial (Gahalaut and Gahalaut 2007). Thus oblique subduction occurs in the Andaman region. This region has long been identified as a region which has potential of generating large magnitude earthquakes. Indeed, the region produced 2 major earthquakes in 1881 and 1941 in Car-Nicobar and Andaman regions respectively. However, until the 2004 Sumatra Andaman earthquake, nobody expected that it can produce an earthquake as large as Mw 9.2. The occurrence of this and the 2011 Tohoku earthquake triggered the possibility of occurrence of superquakes or giant earthquakes in other parts of the convergent plate margins. It also gave rise to the realisation that such superquakes may have extremely large recurrence interval and hence may not be available in the historical records (Goldfinger et al. 2013). In a region it is not necessary that the entire accumulated strain be released in a major or great earthquake, a significant amount of strain may remain stored in the supercycle to be released in a superquake. Similar is the case in the Andaman Nicobar region. The 1881 Car Nicobar earthquake (Mw 7.9) and 1941 Andaman earthquake (Mw 7.7) might have released slip of up to ~3 m (Ortiz and Bilham 2003; Rajendran 2013). But the 2004 Sumatra Andaman earthquake that occurred after 123 and 63 years of the 1881 and 1941 events, released slip of more than 5 m in the two regions, clearly implying that the previous earthquakes did not fully release the strain in the two regions and the remnant and the accumulated strain of the intervening period was finally released by a superquake in 2004.

To study the earthquake supercycles, it is important to reliably and quantitatively extend the earthquake history through palaeoseismological studies and also to quantify the deformation through the cycle. In this article we review the crustal deformation prior, during and after the 2004 earthquake based on GPS measurements.

2 Interseismic GPS Measurements, Prior to the 2004 Earthquake

GPS measurements in the Sumatra region began in 1991 (Prawirodirdjo et al. 1997; Bock et al. 2003). These measurements provided evidence of plate boundary interface locking in the forearc region in the Nias, Batu, and Enggano islands, west of the Sumatra coast (McCaffrey 2002). However, these measurements did not extend into the source zone of the 2004 earthquake, even in the Sumatra region. Paul et al. (2001) provided results of GPS measurements at a site CARI, located at Port Blair, Andaman Island. They indicated that the measurements at this site provide evidence of strain accumulation; however, the extent of locking was unknown. Catherine et al. (2014) reported additional data acquired by the Survey of India. SOI established five campaign mode sites in March 1997 in the Andaman–Nicobar region. During the March 2004 survey, at least three of them (EAST, GOVI, and PORT in the

Andaman Islands) were reoccupied. With reference to the Indian plate, all sites show motion predominantly toward the west, slightly oblique to the trench (Fig. 1). The site motion varies from 7 mm/year at EAST, which is closest to the trench, to 12 mm/year at GOVI, which is farthest from the trench. The large error bars at GOVI and EAST are due to the smaller duration of data during 1995 campaign. The long-term convergence rate in the Andaman forearc, estimated from a plate circuit model is 30 mm/year (Gahalaut and Gahalaut 2007). In the Andaman region the partitioning is partial and the forearc motion is quite oblique to the trench, which is also evident from the azimuths of the coseismic offsets due to the 2004 earthquake (Gahalaut et al. 2006). The oblique motion of 30 mm/year gives a long term, trench-normal, and trench-parallel motion of about 23 and 19 mm/year, respectively. Catherine et al. (2014) used the deep slip model, in which the shallow part of the plate interface, the main thrust zone, is assumed to be locked, whereas the further down-dip part slips aseismically. In the Andaman region, the coseismic rupture extended almost up to the island belt and its down-dip edge was located at about 30 km depth (e.g., Shearer and Burgmann 2010). The main thrust zone has a gentle dip of about 10°–15°, whereas the deeper section slipping aseismically has a dip of about 20°, which becomes steeper with increasing depth (Engdahl et al. 2007). Catherine et al. (2014) assumed that the deep slip during the interseismic period occurred below 30 km depth on the plate interface with a dip of 20°, whereas the shallower part, the main thrust zone, was locked (Savage 1983). They found that the simulated curves corresponding to the trench normal motion fit the observations reasonably well whereas the fit is moderate for the trench-parallel motion. A better fit can be achieved by decreasing the trench-parallel motion from 19 to 15 mm/year. However, this will reduce the long-term motion of the forearc from the earlier estimated rate of 30–27.5 mm/year. Within the uncertainty involved in the previous estimate and in the observations of interseismic rates at the GPS sites, such a decrease in the estimated long-term rate is nominal.

The above analysis implies that, indeed, the main thrust zone was locked, and the region was accumulating strain. Thus, the GPS measurements at these sites provide evidence for strain accumulation in the region.

3 Coseismic Deformation Due to the 2004 Sumatra Andaman Earthquake

The Mw 9.2, 26 December 2004 Sumatra-Andaman earthquake ruptured about 1400 km of the Sunda-Andaman subduction boundary between the India-Australia and Sunda plates (Ammon et al. 2005; Lay et al. 2005; Stein and Okal 2005; Banerjee et al. 2007; Subarya et al. 2006; Chlieh et al. 2007). The 2004 earthquake not only provided the most extensive high quality broadband seismic waveform data set ever recorded for a great earthquake (Ammon et al. 2005), for the first time it also

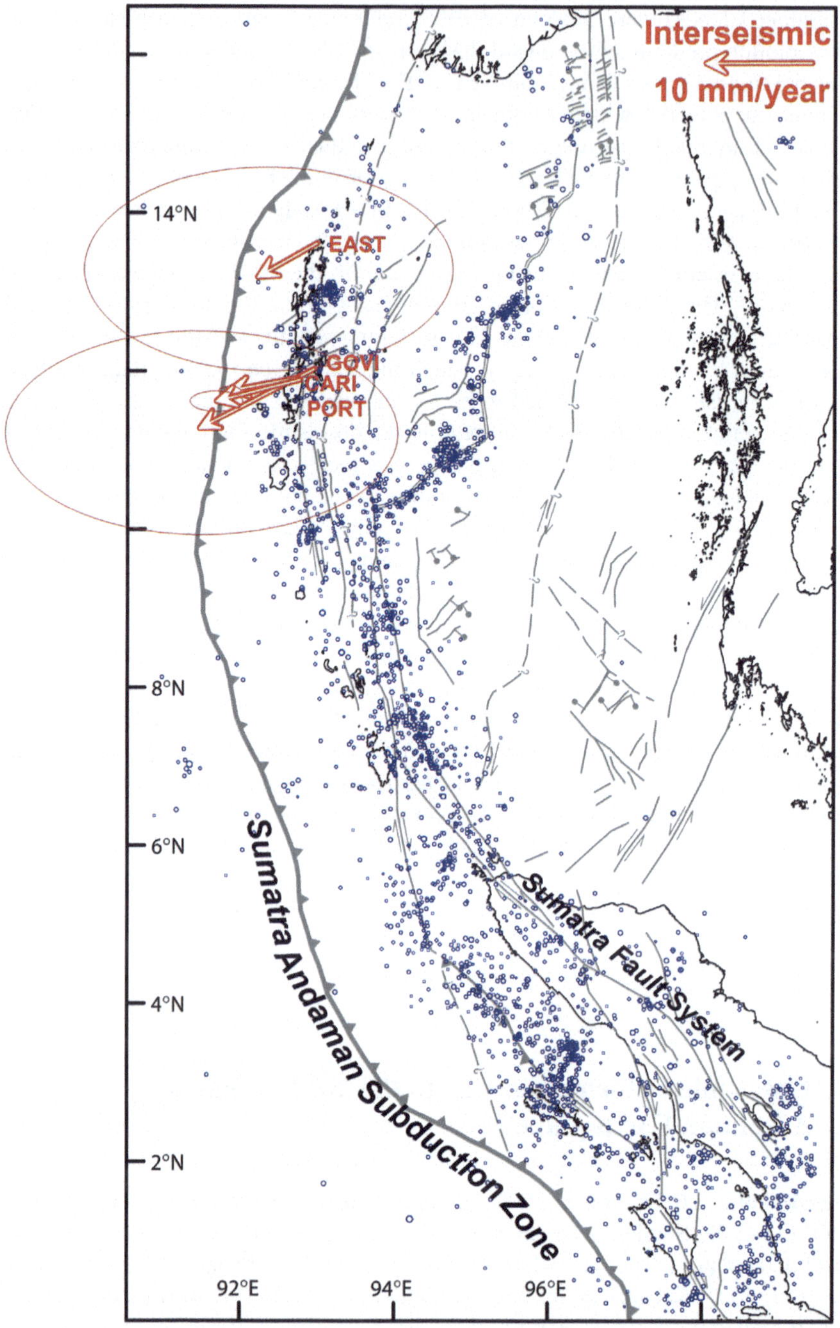

Fig. 1 Interseismic velocity in Indian reference frame at CARI, PORT, GOVI and EAST (Catherine et al. 2014). Seismicity prior to the 2004 Sumatra Andaman earthquake is shown by blue circles

provided the most extensive geodetic data for a great earthquake. In the Andaman-Nicobar region, the earthquake caused coseismic horizontal displacement of 1.6 (Havelock Island) to 6.6 m (Car Nicobar) in a predominantly WSW to southwest direction (Fig. 2) (Gahalaut et al. 2006; Banerjee et al. 2007). Coseismic subsidence of 1–2 m occurred on the east coast of the Andaman-Nicobar region, except in North and Little Andaman, where coseismic uplift of the order of 0.5–1.0 m was reported (Gahalaut et al. 2006). Comparable coseismic movements occurred in the Simeulue island and in the northern Sumatra region (Subarya et al. 2006).

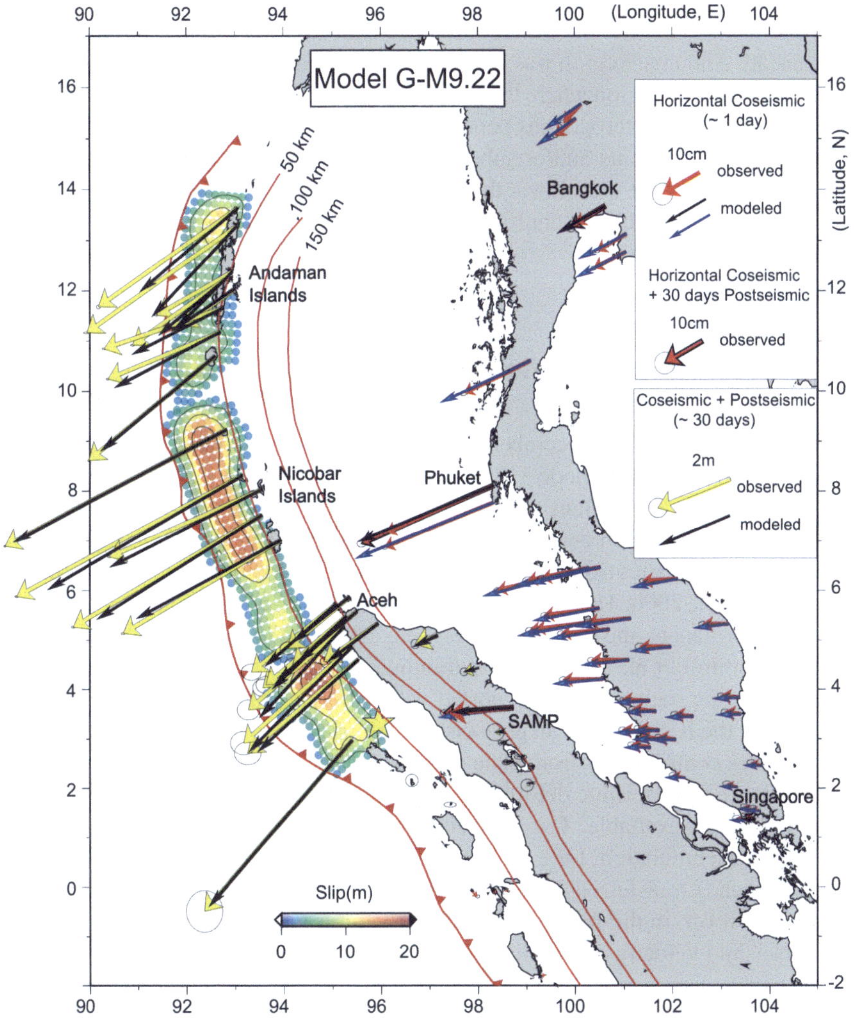

Fig. 2 Coseismic offsets at GPS sites in the Andaman Nicobar islands (Gahalaut et al. 2006) and in surrounding regions. Estimated coseismic slip distribution is also shown. Figure after Chlieh et al. (2007)

Coseismic vertical displacements in Andaman Nicobar region derived from the GPS observations are consistent with field observations and analyses of satellite imagery in the region (Meltzner et al. 2004). Even in the farfield region, e.g., at the IGS permanent GPS site IISC in south India that is located more than 2000 km from the earthquake epicenter, coseismic displacement of up to 1 cm was estimated (Banerjee et al. 2005; Catherine et al. 2005).

Gahalaut et al. (2006), Banerjee et al. (2007) and Chlieh et al. (2007) used these observations to estimate the coseismic slip on the causative rupture on the main thrust zone of the Andaman Sumatra subduction zone. They estimated a rupture length of ~1400 km and width varied from 100 to 180 km. Maximum slip exceeding 20 m occurred in two patches, off Banda Aceh and beneath Nicobar island (Fig. 2). The slip beneath Andaman region was relatively low (<10 m) but was relatively higher in the north Andaman region where the rupture terminated. Slip distribution and rupture appears to be more heterogenous beneath Andaman island which could also be due to presence of sediments and/or subducting 90° E ridge (Gahalaut et al. 2010). In this region the rupture was slower than in south in the Sumatra and Nicobar region. Apparently, this part did not contribute to tsunami generation because of the slower rupture (Singh et al. 2006).

4 Postseismic Deformation of the 2004 Earthquake

Immediately after the 26 December 2004 Sumatra-Andaman earthquake, we established several campaign mode GPS sites in the Andaman-Nicobar region during January-February 2005. Many of these sites were later converted to permanent continuous GPS sites. We processed daily observations files from these sites along with the IGS permanent sites using GAMIT software to obtain daily independent solutions (King and Bock 2000; Herring et al. 2010a, b). The daily coordinates and velocities of all permanent and campaign sites were estimated in the ITRF2008 reference frame (Altamimi et al. 2011) by constraining IGS reference station positions and velocities in the region to their reported values in that frame with standard errors provided by the IGS. Other than postseismic deformation, the estimated site velocity also includes contributions from plate motion, and the renewed strain accumulation. To estimate the postseismic deformation we need to subtract the interseismic velocity at that site, if available. The interseismic velocity at a site consists contribution from strain accumulation, long term convergence in the fore arc and the translatory motion which is transferred to the backarc. All these three estimates can vary along the arc. However, in the Andaman Nicobar region, such estimates (or the variation of interseismic velocity along the arc) are not available. In such circumstances we subtract interseismic velocity of PORT or CARI from the site velocity estimates along the arc, assuming that the interseismic velocity is uniform along the arc and it is same before and after the earthquake. Thus the residual velocity in the time series is assumed to be purely due to postseismic deformation which will become zero once the phase of postseismic deformation is over. The time series are shown in Fig. 3. The

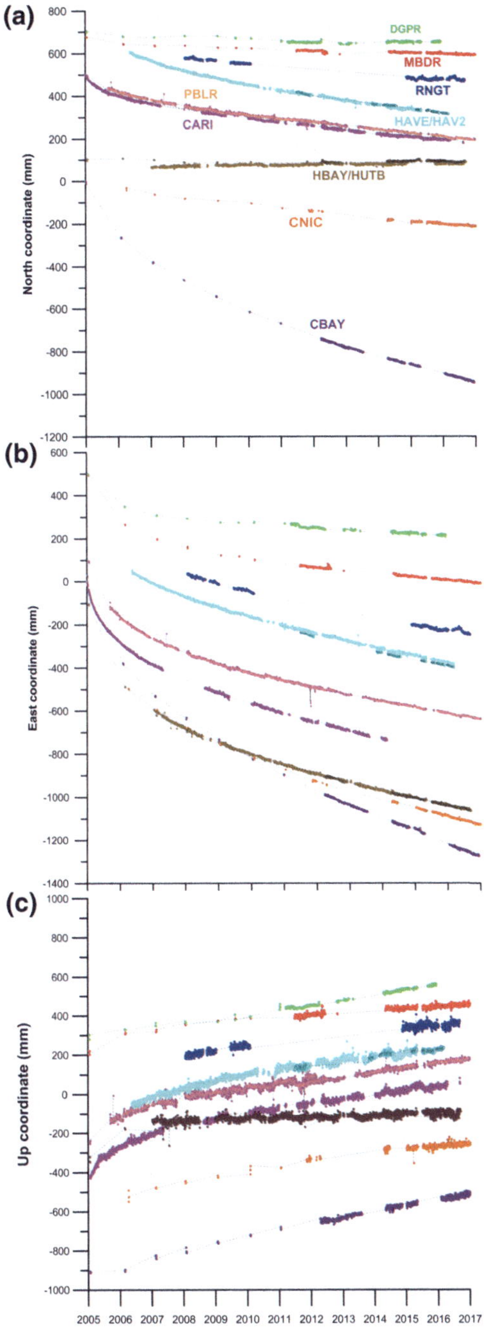
(a)
North coordinate (mm)
DGPR
MBDR
RNGT
PBLR
HAVE/HAV2
CARI
HBAY/HUTB
CNIC
CBAY
(b)
East coordinate (mm)
(c)
Up coordinate (mm)
2005 2006 2007 2008 2009 2010 2011 2012 2013 2014 2015 2016 2017

◂**Fig. 3** Postseismic temporal variations of N, E and Up coordinates of sites in the Andaman Nicobar region with respect to interseismic motion of CARI/PORT. We have also included a few sites (RNGT, HAVE, CARI, HBAY) installed by Paul et al. (2012) and available at http://geodesy.unr.edu/NGLStationPages/gpsnetmap/GPSNetMap.html. Note no variation at DGPR, the northern most site in Andaman Nicobar region since 2011, implying that there is insignificant postseismic deformation at this site now. Also note the large variation in postseismic deformation at CBAY, the southern most site in the Andaman Nicobar region. All sites show uplift. To avoid congestion, we have not plotted the error bars

velocity estimates are also shown in Fig. 4. It may be seen that immediately after the 2004 earthquake the region experienced very rapid deformation which is decreasing with time. It appears that postseismic deformation at sites in north Andaman has reduced significantly and it would have now become zero. Similar conclusion was suggested by Paul et al. (2012). However, sites in the Nicobar are still experiencing large deformation, even after more than a decade since the 2004 earthquake.

Gahalaut et al. (2008a, b) also analysed the postseismic deformation and the seismicity (aftershocks) of the region. Since both are diffusive process, they follow some similar characteristics. The afterslip, responsible for postseismic deformation, was mostly aseismic and did not contribute to the aftershocks. The aftershocks and postseismic displacements appear to follow a similar relationship, although with different decay times. The temporal dependence of the two differs only by a term linear in time. They concluded that the postseismic deformation of the first two years corresponds to a total moment release of Mw 8.5 and aftershocks during this period only contribute 13% of this moment release. Thus majority of the postseismic motion occurs aseismically.

Several investigators analysed the influence of 2004 Sumatra Andaman earthquake on the nearby region in terms of change in earthquake frequency using the concept of Coulomb stress change. Cattin et al (2009) identified regions of increase in stress and noted the increase in seismicity in the fore and back arc. Sevilgen et al. (2012) noted that the seismicity stopped on the transform faults but increased in the rift region, consistent with the analysis of Diehl et al. (2013). Wiseman and Bürgmann (2012) proposed that the occurrence of 2004 Sumatra Andaman earthquake led to increase in stress in the outer rise and northern Wharton basin where a couple of major earthquakes and the Wharton basin great earthquakes in 2012 occurred. Kundu et al. (2012) proposed that the 2004 earthquake actually triggered the volcanic swarm in the back arc, in a manner similar to the Barren volcanic eruption after the earthquake.

5 Simulation of Postseismic Deformation Due to 2004 Sumatra Earthquake

Gahalaut et al. (2008a) modelled postseismic deformation as the afterslip and suggested that the afterslip in the Andaman region is located further downdip of the

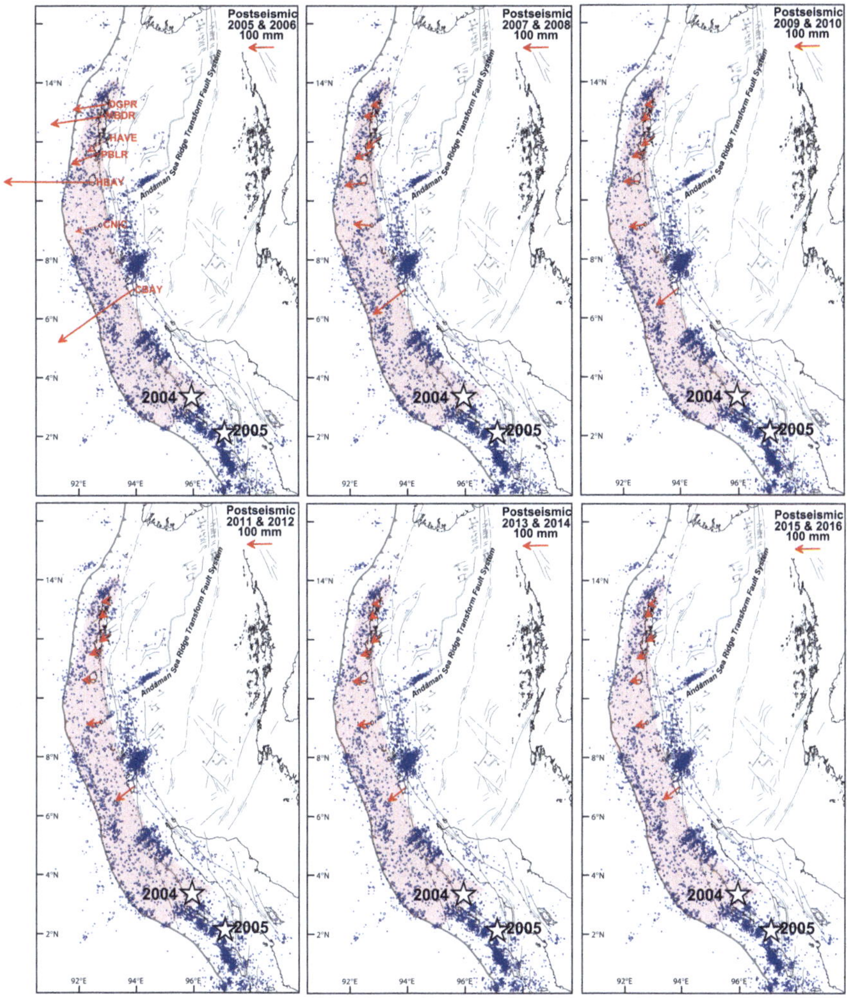

Fig. 4 Postseismic deformation rate (site velocity with respect to interseismic velocity of CARI/PORT during year 2005 & 2006, 2007 & 2008, 2009 & 2010, 2011 & 2012, 2013 & 2014 and 2015 & 2016) at GPS sites in Andaman Nicobar region. Velocity at CNIC and HAVE in 2005 is not known as the sites were not established in that year, hence they are shown as with light red colour in the first panel. Sites PBLR and CARI are very near to each other (within 2 km) and show similar motion, hence velocity of CARI is not shown here. Site RNGT has lots of data gaps and hence its velocity is not shown here. Two years aftershocks after the 2004 Sumatra Andaman earthquake are shown by blue dots in each panel. Red shading in each panel shows rupture area of the 2004 earthquake

coseismic rupture whereas in the Nicobar region, it partly overlaps with the coseismic rupture. Similar results were obtained by Paul et al. (2007). Wiseman et al. (2015) used a more complex and heterogeneous model of the subduction zone to simulate the postseismic deformation and suggested that addition of a rigid slab and viscosity contrast across the subducting oceanic plate improves fitting.

Following Paul et al. (2007) we tried to fit the exponential decay in the time series at each site using a function of the form

$$x(t) = x_o + Vt - A_o\left(1 - \exp\left(\frac{T_o - t}{\tau}\right)\right) \quad (1)$$

in which V is constant (interseismic) velocity, T_o is time of the Great Sumatra/Andaman earthquake, τ is characteristic timescale of the decay, and A_o is transient displacement in the limit as $t \rightarrow \infty$. We found that τ ranges from 2 to 14 years. But the most important result is the τ is different for the initial and later period, which in a way points to transient rheology. We show here the transient variation at PBLR. The required value of τ for the N, E and Up components for the period 2005–2010 varies between 1.7 and 2.6 while for later period it varies between 5.3 and 8.7 (Fig. 5). Similar behavior is seen at other sites also. We plan to take up a detailed modelling of the postseismic deformation soon as longer time series are now available from seven sites in the Andaman Nicobar region.

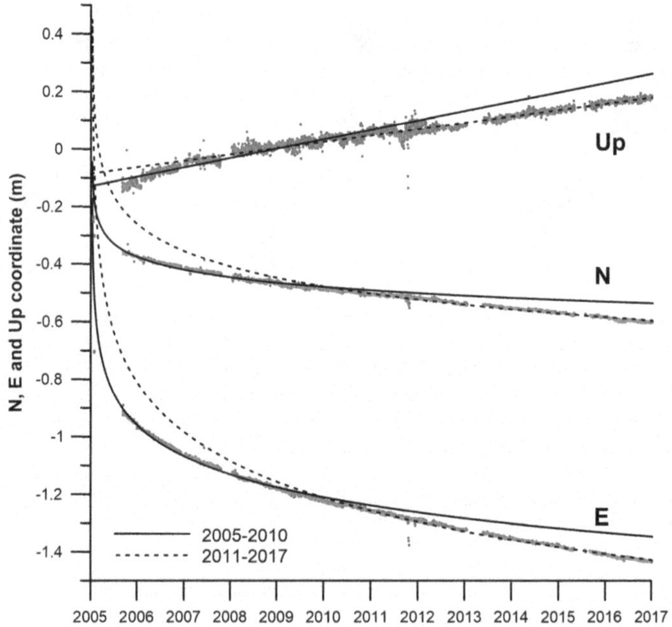

Fig. 5 Variation of N, E and Up components at PBLR and fitting of Eq. 1. Different values of τ and A_o are needed for the initial and later periods

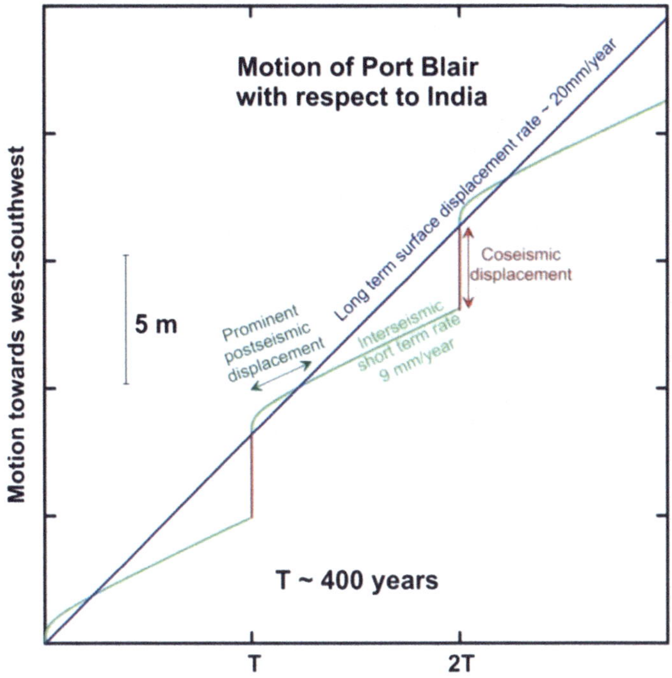

Fig. 6 Motion of Port Blair with respect to Indian plate through coseismic, postseismic and interseismic phases. In the interseismic period the site moves at a velocity of 9 mm/year towards WSW. The postseismic deformation is consistent with that observed at Port Blair. The coseismic displacement in WSW direction [corresponding to 3.07 m at Port Blair, Gahalaut et al. (2006)]is shown by steps. The long term surface movement of 20 mm/year at Port Blair is consistent with Gahalaut and Gahalaut (2007). The available estimates of interseismic, coseismic, and postseismic deformation provides a return period of ~400 years for the 2004 Sumatra Andaman type earthquake. Figure after Gahalaut et al. (2008a)

Gahalaut et al. (2008a) also attempted to simulate the recurrence interval of similar giant earthquakes in the Andaman region using the derived estimates of interseismic, coseismic and postseismic deformation (Fig. 6). They arrived at an estimate of ~400 years, which is surprisingly consistent with the palaeoseismological observations from the Andaman region (Malik et al. 2015).

6 Concluding Remarks

It is important to analyse the deformation cycle of earthquakes because the analysis of it provides constraints on the earthquake mechanism. We have reviewed the available information on the interseismic, coseismic and post seismic deformation in the Andaman Nicobar region due to the 2004 Sumatra Andaman earthquake. The ongoing postseismic deformation has been updated with new data, acquired through recent GPS measurements. Following are our conclusions.

The interseismic deformation, although very limited, provide evidence for strain accumulation but do not show any evidence of possible occurrence of such a large earthquake in the region. Coseismic deformation are spectacular in the sense that they provided evidence that the island belt shifted by 3–6 m towards southwest due to the earthquake which correspond to 10–20 m of coseismic slip on the subduction interface. The postseismic deformation due to such a large earthquake may continue for several decades and so is the case here. Even after 12 years of the earthquake, the region shows large postseismic deformation which may continue for several decades, at least in the Nicobar and southern Sumatra region.

Acknowledgements We acknowledge support from several colleagues contributed towards data acquisition and establishment of permanent GPS stations in the Andaman Nicobar region. The work was financially supported by the Ministry of Earth Sciences. We thank two anonymous reviewers for their constructive comments. The CSIR-National Geophysical Research Institute (CSIR-NGRI) reference number of the article is NGRI/Lib/2019/Pub-76.

References

Altamimi Z, Collilieux X, Métivier L (2011) ITRF2008: an improved solution of the international terrestrial reference frame. J Geodesy. https://doi.org/10.1007/s00190-011-0444-4

Ammon C et al (2005) Rupture process of the 2004 Sumatra-Andaman earthquake. Science 308:1133–1139

Banerjee P, Pollitz F, Bürgmann R (2005) Size and duration of the great 2004 Sumatra-Andaman earthquake from far-field static offsets. Science 308:1769–1772

Banerjee P, Pollitz F, Nagarajan B, Bürgmann R (2007) Coseismic slip distributions of the 26 December 2006 Sumatra-Andaman and 28 March 2005 Nias earthquakes from GPS static offsets. Bull Seismol Soc Am 97:S86–S102

Bock Y, Prawirodirdjo L, Genrich JF, Stevens CW, McCaffrey R, Subarya C, Puntodewo SSO, Calais E (2003) Crustal motion in Indonesia from global positioning system measurements. J Geophys Res 108(B8):2367. https://doi.org/10.1029/2001JB000324

Catherine JK, Gahalaut VK, Sahu VK (2005) Constraints on rupture of the December 26, 2004, Sumatra earthquake from far-field GPS observations. Earth Planet Sci Lett 237:673–679

Catherine JK, Gahalaut VK, Srinivas N, Kumar S, Nagarajan B (2014) Evidence of strain accumulation in the Andaman region for the giant 2004 Sumatra Andaman Earthquake. Bull Seismol Soc Am 104:587–591. https://doi.org/10.1785/0120130141

Cattin R, Chamot-Rooke N, Pubellier M, Rabaute A, Delescluse M, Vigny C, Fleitout L, Dubernet P (2009) Stress change and effective friction coefficient along the Sumatra-Andaman-Sagaing fault system after the 26 December 2004 (Mw = 9.2) and the 28 March 2005 (Mw = 8.7) earthquakes. Geochem Geophys Geosyst 10:Q03011. https://doi.org/10.1029/2008gc002167

Chlieh M et al (2007) Coseismic slip and afterslip of the great (Mw 9.15) Sumatra-Andaman earthquake of 2004. Bull Seismol Soc Am 97:S152–S173

Curray JR (2005) Tectonics and history of the Andaman sea region. J Asian Earth Sci 25:187–228

Diehl T, Waldhauser F, Cochran JR, Kamesh Raju KA, Seeber L, Schaff D, Engdahl ER (2013) Back-arc extension in the Andaman Sea: tectonic and magmatic processes imaged by high-precision teleseismic double-difference earthquake relocation. J Geophys Res 118:2206–2224. https://doi.org/10.1002/jgrb.50192

Engdahl E, Villasenor A, DeShon H, Thurber C (2007) Telesismic relocation and assessment of seismicity (1918–2005) in the region of the 2004 Mw 9.0 Sumatra-Andaman and 2005 Mw 8.6 Nias Island great earthquakes. Bull Seismol Soc Am 97:S43–S61. https://doi.org/10.1785/0120050614

Fitch TJ (1972) Plate convergence, transcurrent faults and internal deformation adjacent to Southeast Asia and the western Pacific. J Geophys Res 77:4432–4460

Gahalaut VK, Gahalaut K (2007) Burma plate motion. J Geophys Res 112:B10402. https://doi.org/10.1029/2007JB004928

Gahalaut VK, Nagarajan B, Catherine JK, Kumar S (2006) Constraints on 2004 Sumatra-Andaman earthquake rupture from GPS measurements in Andaman-Nicobar Islands. Earth Planet Sci Lett 242:365–374

Gahalaut VK, Jade S, Catherine JK, Gireesh R, Ananda MB, Dileep Kumar P, Narsaiah M, Jafri SSH, Ambikapathy A, Bansal A, Chadha RK, Gupta DC, Nagarajan B, Kumar S (2008a) GPS measurements of postseismic deformation in the Andaman-Nicobar region following the giant 2004 Sumatra-Andaman earthquake. J Geophys Res 113:B08401. https://doi.org/10.1029/2007JB005511

Gahalaut VK, Catherine JK, Jade S, Gireesh R, Gupta DC, Narsaiah M, Ambikapathy A, Bansal A, Chadha RK (2008b) No evidence of unusually large postseismic deformation in Andaman region immediately after 2004 Sumatra-Andaman earthquake. Geophys Res Lett 35:L10307. https://doi.org/10.1029/2008GL033704

Gahalaut VK, Subrahmanyam C, Kundu B, Catherine JK, Ambikapathy A (2010) Slow rupture in Andaman during 2004 Sumatra-Andaman earthquake: a probable consequence of subduction of 90° E ridge. Geophys J Int 180:1181–1186. https://doi.org/10.1111/j.1365-246X.2009.04449.x

Goldfinger C, Ikeda Y, Yeats RS, Ren J (2013) Superquakes and supercycles. Seismol Res Lett 84. https://doi.org/10.1785/0220110135

Herring TA, King RW, McClusky SC (2010a) Documentation of the GAMIT GPS Analysis Software release 10.4. Department of Earth and Planetary Sciences Massachusetts Institute of Technology Cambridge Massachusetts

Herring TA, King RW, McClusky SC (2010b) GLOBK, Global Kalman filter VLBI and GPS analysis program version 10.4. Department of Earth and Planetary Sciences Massachusetts Institute of Technology Cambridge Massachusetts

Kamesh Raju KA, Murty GPS, Amarnath D, Kumar MLM (2007) The west Andaman fault and its influence on the aftershock pattern of the recent megathrust earthquakes in the Andaman-Sumatra region. Geophys Res Lett 34:L03305. https://doi.org/10.1029/2006GL028730

King R, Bock Y (2000) Documentation for the GAMIT GPS analysis software Rel. 9.94. Mass Inst of Technol Cambridge Massachusetts

Kundu B, Legrand D, Gahalaut K, Gahalaut VK, Mahesh P, Kamesh Raju KA, Catherine JK, Ambikapthy A, Chadha RK (2012) The 2005 volcano-tectonic earthquake swarm in the Andaman Sea: triggered by the 2004 great Sumatra-Andaman earthquake. Tectonics 31:TC5009 https://doi.org/10.1029/2012tc003138

Lay T, Kanamori H, Ammon CJ, Nettles M, Ward SN, Aster RC et al (2005) The great Sumatra-Andaman earthquake of December 26, 2004. Science 308:1127–1133

Malik JN, Banerjee C, Khan A, Johnson FC, Shishikura M, Satake K, Singhvi AK (2015) Stratigraphic evidence for earthquakes and tsunamis on the west coast of South Andaman Island, India during the past 1000 years. Tectonophysics 661:49–65

McCaffrey R (1992) Oblique plate convergence, slip vectors, and forearc deformation. J Geophys Res 97:8905–8915

McCaffrey R, Stein S, Freymueller JT (2002) Crustal block rotations and plate coupling. Plate Boundary Zones Geodyn Ser 30:101–122

Meltzner A et al (2004) Uplift and subsidence associated with the Great Aceh-Andaman earthquake of 2004. J Geophys Res 111. https://doi.org/10.1029/2005JB003891

Ortiz M, and R Bilham (2003) Source area and rupture parameters of the 31 December 1881 Mw 7.9 Car Nicobar earthquake estimated from tsunamis recorded in the Bay of Bengal. J Geophys Res 108:2215 https://doi.org/10.1029/2002jb001941

Paul J, Burgmann R, Gaur VK, Bilham R, Larson KM, Ananda MB, Jade S, Mukul M, Anupama TS, Satyal G, Kumar D (2001) The motion and active deformation of India. Geophys Res Lett 28:647–651

Paul J, Lowry AR, Bilham R, Sen S, Smalley R Jr (2007) Postseismic deformation of the Andaman Islands following the 26 December, 2004 Great Sumatra-Andaman earthquake. Geophys Res Lett 34:L19309. https://doi.org/10.1029/2007GL031024

Paul J, Rajendran CP, Lowry AR, Andrade V, Rajendran K (2012) Andaman postseismic deformation Observations: still slipping after all these years? Bull Seismol Soc Am 102:343–351. https://doi.org/10.1785/0120110074

Prawirodirdjo L et al (1997) Geodetic observations of interseismic strain segmentation at the Sumatra subduction zone. Geophys Res Lett 24:2601–2604. https://doi.org/10.1029/97GL52691

Rajendran CP (2013) Was the 1941 Andaman earthquake tsunamigenic? Comments on "Inundation studies for Nagapattinam region on the east coast of India due to tsunamigenic earthquakes from the Andaman region" by Srivastava et al. 2012. Nat Hazards 65:981–984. https://doi.org/10.1007/s11069-012-0403-2

Savage JC (1983) A dislocation model of strain accumulation and release at a subduction zone. J Geophys Res 88:4984–4996. https://doi.org/10.1029/JB088iB06p04984

Sevilgen V, Stein RS, Pollitz FF (2012) Stress imparted by the great 2004 Sumatra earthquake shut down transforms and activated rifts up to 400 km away in the Andaman Sea. PNAS 109(38):15152–15156

Shearer P, Burgmann R (2010) Lessons learned from the 2004 Sumatra-Andaman megathrust rupture. Annu Rev Earth Planet Sci 38:103–131

Singh SK, Ortiz M, Gupta HK, Ramadass DGA (2006) Slow slip below Port Blair, Andaman, during the great Sumatra-Andaman earthquake of 26 December 2004. Geophys Res Lett 33:L03313. https://doi.org/10.1029/2005GL025025

Stein S, Okal EA (2005) Speed and size of the Sumatra earthquake. Nature 434:581–582

Subarya C et al (2006) Plate-boundary deformation associated with the great Sumatra-Andaman earthquake. Nature 440:46–51

Wiseman K, Bürgmann R (2012) Stress triggering of the great Indian Ocean strike-slip earthquakes in a diffuse plate boundary zone. Geophys Res Lett 39:L22304. https://doi.org/10.1029/2012GL053954

Wiseman K, Bürgmann R, Freed AM, Banerjee Paramesh (2015) Viscoelastic relaxation in a heterogeneous Earth following the 2004 Sumatra-Andaman earthquake. Earth Planet Sci Lett 431:308–317

Tectonics of the Andaman Backarc Basin—Present Understanding and Some Outstanding Questions

K. A. Kamesh Raju, K. K. Aswini and V. Yatheesh

Abstract The Andaman Backarc Basin is characterized by active seafloor spreading spurred by the right stepping dextral shear imposed on the trans-basinal Sagaing-Andaman-Nicobar- Sumatra fault system and presents a complex tectonic history owing to some unique geophysical characteristics such as the absence of magnetic anomalies and thickly sedimented spreading center. Recent advances such as high-resolution seafloor mapping, extensive deep penetrating multichannel seismic reflection studies, deep drilling by JOIDES Resolution, GPS investigations and passive Ocean Bottom Seismometer experiment have greatly contributed to our understanding of this complex backarc basin. However, there are many outstanding questions and uncertainties with respect to the evolutionary sequence, emplacement of crustal types and their ages. This brief review is an attempt to bring into focus the present status of our understanding of the tectonics of the Andaman Backarc Basin, to highlight the outstanding questions and to suggest possible ways to address them.

Keywords Andaman Sea · Back-arc basin · Seafloor mapping · Tectonics · Sumatra fault · Magnetic anomalies

1 Introduction

The Andaman Sea in the northeast Indian Ocean is an active marginal basin consisting of trench-arc-backarc elements. The extension in the Andaman Sea is primarily driven by oblique subduction of the Indian-Australian plate beneath the Southeast Asian/Sunda plate. The oblique subduction of the Indian plate in Andaman-Nicobar subduction zone induces complexity into the backarc basin and stimulates seafloor spreading over few segments of the trans-basinal fault system joining the Sagaing Fault in Myanmar and the Great Sumatra Fault of Sumatra Island (Fig. 1). Regional marine geophysical investigations conducted in the Andaman Sea have defined the

K. A. Kamesh Raju (✉) · K. K. Aswini · V. Yatheesh
CSIR-National Institute of Oceanography, Dona Paula, Goa 403004, India
e-mail: kamesh@ncaor.gov.in

K. A. Kamesh Raju
National Center for Polar and Ocean Research (NCPOR), Vasco-da-Gama, Goa 403804, India

J. S. Ray and M. Radhakrishna (eds.), *The Andaman Islands and Adjoining Offshore: Geology, Tectonics and Palaeoclimate*, Society of Earth Scientists Series,
https://doi.org/10.1007/978-3-030-39843-9_12

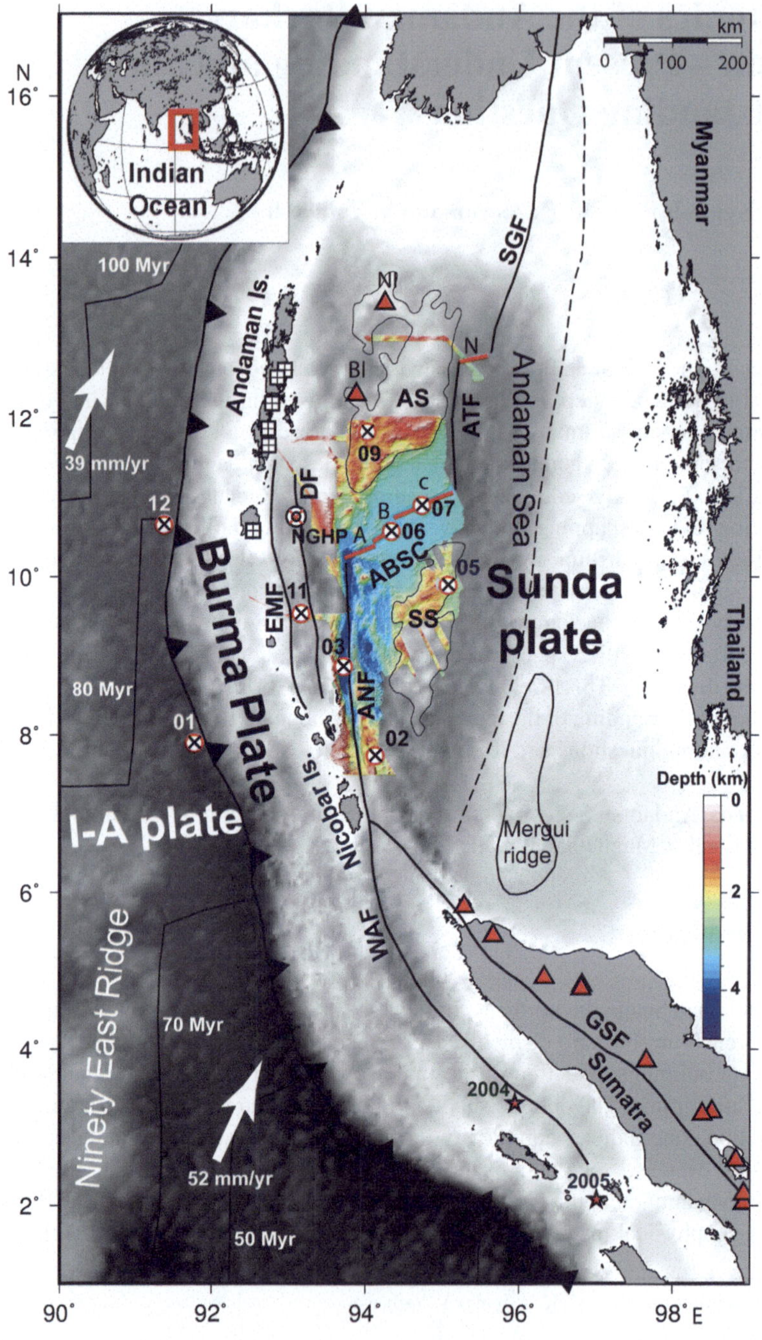
Indian Ocean
km
0 100 200
N
16°
14°
12°
10°
8°
6°
4°
2°
90°
92°
94°
96°
98° E
100 Myr
80 Myr
70 Myr
50 Myr
39 mm/yr
52 mm/yr
Andaman Is.
Nicobar Is.
Burma Plate
I-A plate
Ninety East Ridge
Sunda plate
Andaman Sea
Myanmar
Thailand
Sumatra
Mergui ridge
SGF
ATF
DF
EMF
ANF
WAF
GSF
ABSC
AS
SS
NI
BI
NGHP
N
A
B
C
01
02
03
05
06
07
09
11
12
2004
2005
Depth (km)
0
2
4

◂**Fig. 1** General tectonic framework of the Andaman Sea encompassing the Andaman Backarc Basin presented on a gray shaded bathymetric image (Smith and Sandwell 1997). Age contours and the plate motion directions (white arrows) are from Subarya et al. (2006). The major tectonic elements and the backarc spreading center and the part of multibeam bathymetry coverage are depicted (Kamesh Raju et al. 2004; Curray 2005; Singha et al. 2019). Black squares with cross denote broad band seismic stations from the ISLANDS network (Srijayanthi et al. 2012). Red circle with black cross represents the Ocean Bottom Seismometer deployments (Dewangan et al. 2018). Black double circle with red cross represents NGHP deep drilling site (Kumar et al. 2014). Great Sumatra Fault (GSF), West Andaman Fault (WAF), Eastern Margin Fault (EMF), Diligent Fault (DF), Andaman-Nicobar Fault (ANF), Andaman Backarc Spreading Centre (ABSC), Andaman Transform Fault (ATF), Sagaing Fault (SGF), Sewell Seamount (SS) and Alcock Seamount (AS) are marked. The dashed black line represents the ocean-continent boundary proposed by Curray (2005). Red triangles represent volcanoes. Red stars represent major earthquake events of 2004 and 2005

general tectonic setting (Rudolfo 1969; Curray et al. 1979) and the crustal structure (Mukhopadhyay 1988).The recent studies (Kamesh Raju et al. 2004, 2012a, b; Curray 2005; Gahalaut and Gahalaut 2007; Gahalaut et al. 2008; Radhakrishna et al. 2008; Cochran 2010; Diehl et al. 2013; Singh et al. 2005, 2011, 2013; Moeremans and Singh 2015; Singh and Moeremans 2017) provided greater insights into the morphotectonics, crustal architecture and geodynamic aspects of the Andaman Backarc Basin.

The newly acquired high-resolution multibeam swath bathymetry data together with gravity, magnetic and multichannel seismic reflection data described the crustal structure of important features of the Andaman Backarc Basin (Cochran 2010; Singh et al. 2013; Moeremans and Singh 2015; Singh and Moeremans 2017). The earthquake seismology information and its rigorous analysis yielded finer details about the spatio-temporal dynamics of active faults in the Andaman Sea, the ongoing orientation of the backarc spreading center (Diehl et al. 2013), the nature of the crust (Singha et al. 2019) and the configuration of the subducting slab (Dasgupta et al. 2003; Rao and Kalpana 2005; Engdahl et al. 2007; Yadav and Tiwari 2018; Singha et al. 2019). The crustal architecture within the backarc basin has been studied extensively (Singh et al. 2005; Singh et al. 2013; Moeremans and Singh 2015; Singh and Moeremans 2017). Jourdain et al. (2016) has provided subsurface image of the anomalous crustal structure associated with the sedimented spreading center (segment C) that explains the absence of magnetic anomalies (Kamesh Raju et al. 2004) over this spreading centre segment. GPS investigations have refined the plate motion vectors and provided constraints on geodynamics of the region (Subarya et al. 2006; Gahalaut and Gahalaut 2007). The recent ISLAND network initiative of establishment of seismological observatories over the Andaman Islands (Srijayanti et al. 2012) and the first passive Ocean Bottom Seismometer (OBS) experiment in the Andaman Sea (Dewangan et al. 2018) have provided refined subsurface imaging avenues (Singha et al. 2019). There has been considerable enhancement in our knowledge on the structure and tectonics of the Andaman Sea during the recent years due to the new initiatives that made the availability of high-resolution multibeam bathymetry, deep penetrating multichannel seismic reflection, GPS and OBS data sets (Fig. 1) and their interpretation. However, there are several unresolved issues and uncertainties in our understanding of the Andaman Backarc Basin. While there is general agreement on the backarc spreading center and the seafloor spreading, there are debates about the spreading activity as to whether it is prolonged, episodic or intermittent (Morley and Alvey 2015). The origin of the Alcock and Sewell rises is not conclusively established. The nature of the crust in the Andaman Backarc Basin has been inferred to vary from oceanic crust to hyper-thinned continental crust to extended continental crust, the boundaries of these crustal types and provinces have not been delineated in detail. An oceanic crust with a missing layer 2 and instead flooded with sill complexes has been inferred beneath the sedimented spreading center, that awaits confirmation by ground truth data. In this brief review, we attempt to provide an overview of the status of our understanding of the Andaman Backarc Basin, highlight some of the outstanding questions, and suggest possible ways forward to resolve them.

2 Tectonic Setting

The Andaman Backarc Basin in the northeast Indian Ocean is encompassed by the subduction zone in the west that extends from the Java Sumatra in the south and extends through Indo-Burma ranges in the north; Malay Peninsula and its continental shelf forms the eastern boundary; Irrawaddy-Martaban shelf represents the northern boundary; in the south it is bounded by the Sumatra Island and the Mergui Basin. The general physiographic features of the Andaman Basin were described by Rudolfo (1969). The convergence between the Indian and Southeast Asian plates in the Andaman Sea is characterized by varying degree of obliquity and rate of subduction (Sieh and Natawidjaja 2000; Subarya et al. 2006). Oblique subduction, the initiation of Andaman backarc spreading (Curray et al. 1979; Kamesh Raju et al. 2004; Curray 2005), and the volcanic arc consisting of Barren-Narcondam islands are the important tectonic elements in the Andaman Sea. Furthermore, the major fault systems connecting the Sagaing Fault in Myanmar to the Sumatra Fault (e.g., Sieh and Natawidjaja 2000; Vigny et al. 2003; Kamesh Raju et al. 2004, 2007) and the joining of Sumatra Fault (SF) with the West Andaman Fault (WAF) influence the geodynamics, and modulate the stress regime in this area. The motion along the major fault-systems (viz. Sagaing Fault, Andaman-Nicobar Fault and the Sumatra Fault) is predominantly strike-slip. The principal stress directions in general are consistent with the coseismic displacement of the two recent major earthquake events of December 2004 and March 2005 and the direction of convergence derived from the GPS measurements in the inter-seismic period (Gahalaut and Gahalaut 2007).

The varied velocity and direction of the subducting Indian plate and the increasing obliquity from south to north has influenced the geodynamics of the Andaman Sea. The subducting plate velocities vary from 5.7 cm/yr around 8° S off Sumatra to 3.7 cm/yr at around 16° N off north of Andaman Islands (Bock et al. 2003). Further, the age of the subducting Indian plate increases northwards, from ~50 Ma near the Simeulue Island off Sumatra to 80–120 Ma at the latitude of Andaman Islands (Subarya et al. 2006).

Barren Island is the only known active volcano in the northern Andaman Sea (Sheth et al. 2009). With respect to the known volcanoes in the region, there is a distinct gap in active volcanoes from 06° N to 12° N. Subduction of the base of the Ninetyeast Ridge was suggested as one of the possible reasons for this gap in volcanic activity (Subrahmanyam et al. 2008). Partial subduction of the Ninetyeast Ridge below the Andaman Trench was first suggested by Mukhopadhyay and Krishna (1995) based on the analysis of arc-trench gravity field. It is interesting to note the occurrence of several earthquake swarms at frequent intervals in different parts of the Andaman Sea (Kamesh Raju et al. 2012a, b; Diehl et al. 2013). These swarms have occurred along the backarc spreading center (during 1983–1984, 1993–1994, 2006, 2009 and 2012), near the Nicobar Island (during 1982, 1986, 1998, 2005, 2014 and 2015), along the offshore extension of the Sumatra Fault (in 2001) and near Simeulue Island (in 2002). All the earthquake swarms documented so far appear as minor events compared to the January 2005 swarm located off Nicobar Island. This was termed as the most energetic earthquake swarm that has ever been recorded globally (Lay et al. 2005). About 120 major ($M_W > 5.0$) and more than 570 minor ($1.0 < M_W < 5.0$) earthquakes were associated with the Nicobar earthquake swarm (http://earthquake.usgs.gov/regional/neic/). The highest rate of occurrence of earthquake swarm is located around Nicobar Island. This swarm has been considered as an aftershock effect of the tsunamigenic Great Sumatra-Andaman earthquake of 26 December, 2004 of magnitude M_W 9.3 that ruptured a 1300 km long stretch of the megathrust zone, the longest rupture zone of any recorded earthquake.

A cratered seamount has been identified based on high-resolution seafloor maps nearly at the center of the Nicobar swarm (Kamesh Raju et al. 2012a). Based on the identification of a chain of seamounts and the seabed samples collected from the crater, it is suggested that a combination of stress changes, and the tectonics have contributed to the occurrence of largest ever swarm in this region. An independent study (Kundu et al. 2012) also suggests that the 2005 Andaman Sea earthquake swarm was triggered by the 2004 great Sumatra-Andaman earthquake (Mw 9.3) and the delay in triggering is due to the poroelastic response of the medium to the elastic stress change caused by the megathrust earthquake.

3 Morphotectonics

Modern seafloor mapping methods have greatly contributed in detailing the seafloor features and their tectonic implications. The topographic configuration of the backarc spreading center (Kamesh Raju et al. 2004) provided distinctive evidence for the existence of an active spreading center. Several new seamounts have been mapped, among which the SM1 and SM2 seamounts are the major features. Besides precisely delineating the major Andaman-Nicobar Fault System (northern section of the West Andaman Fault), the high resolution mapping has revealed existence of several parallel faults and ridges.

The mapping efforts in the off Nicobar region provided evidence of active submarine volcanic arc in the Andaman Sea that links the Barren- Narcondam islands in the north to the Sumatran volcanoes in the south (Kamesh Raju et al. 2012a, b).The identification of a chain of seamounts extending towards the Sumatra Fault region in the south (Fig. 2) supports the existence of an active volcanic arc in the Andaman Sea. Deep drilling in the Andaman Sea with the D/V JOIDES Resolution under the National Gas Hydrate Program (NGHP-01) revealed extensive past volcanism. At site NGHP-01-17 (Fig. 1), a uniform 691.6 m thick sediment sequence of predominantly nanno fossil ooze deposited at an estimated sedimentation rate of ~5.6 cm/ky contains a remarkable record of the volcanic activity reflected in 382 horizons of pyroclastic materials, including layers and patches of white, gray, and black ash; white pumice fragments; and dispersed black ash (Kumar et al. 2014).

High resolution mapping off Sumatra region during 2005 and 2006, revealed the folding at the front of the accretionary wedge occurring primarily along the seaward dipping thrusts and demonstrated how the lower plate structure impacts on the deformation of the upper plate (Graindorge et al. 2008).

4 Seismicity

The oldest recorded earthquake event in the Andaman Sea dates back to 1847 (Oldham 1883). Several earthquakes have originated in the Andaman Sea region as a consequence of active convergent margin tectonics characterizing it as the highest hazard zone. The Seismotectonic Atlas of India and its Environs (Dasgupta et al. 2000) shows that the entire belt of the Andaman and Nicobar Islands fall under the highest hazard class (class V).

The 2004 tsunamigenic M_w 9.3 event that propagated NNW and ruptured over a 1300 km long stretch in about 8–10 min (Ammon et al. 2005) and the subsequent major events have changed the stress and seismicity pattern in the Andaman Sea. These events have also brought into focus the differences between the Andaman sector and the Sumatra–Java sector subduction zones. The magnitudes of displacements varied along-strike from uniform dip-slip in the southern (Sumatra) sector to dip-slip and strike-slip components in the Andaman and Nicobar sectors due to the increased obliqueness of rupture in the north (Carter and Bandopadhyay 2017). The variations also include a steeper dipping slab and a thicker pile of oceanic sediment cover in the Andaman sector compared to the Sumatra sector. The younger oceanic slab in the Sumatra sector (Muller et al. 1997) in the south is buoyant and dips less steeply compared to the slab in the Andaman sector and further north. This has resulted in stronger coupling to the overlying plate in the Sumatra sector than beneath the Andaman region. Kumar et al. (2016) suggested tearing of the subducting Indian plate at the Andaman trench, using the shear-wave receiver function technique and proposed that such slab discontinuities could be the locales for the intense seismicity.

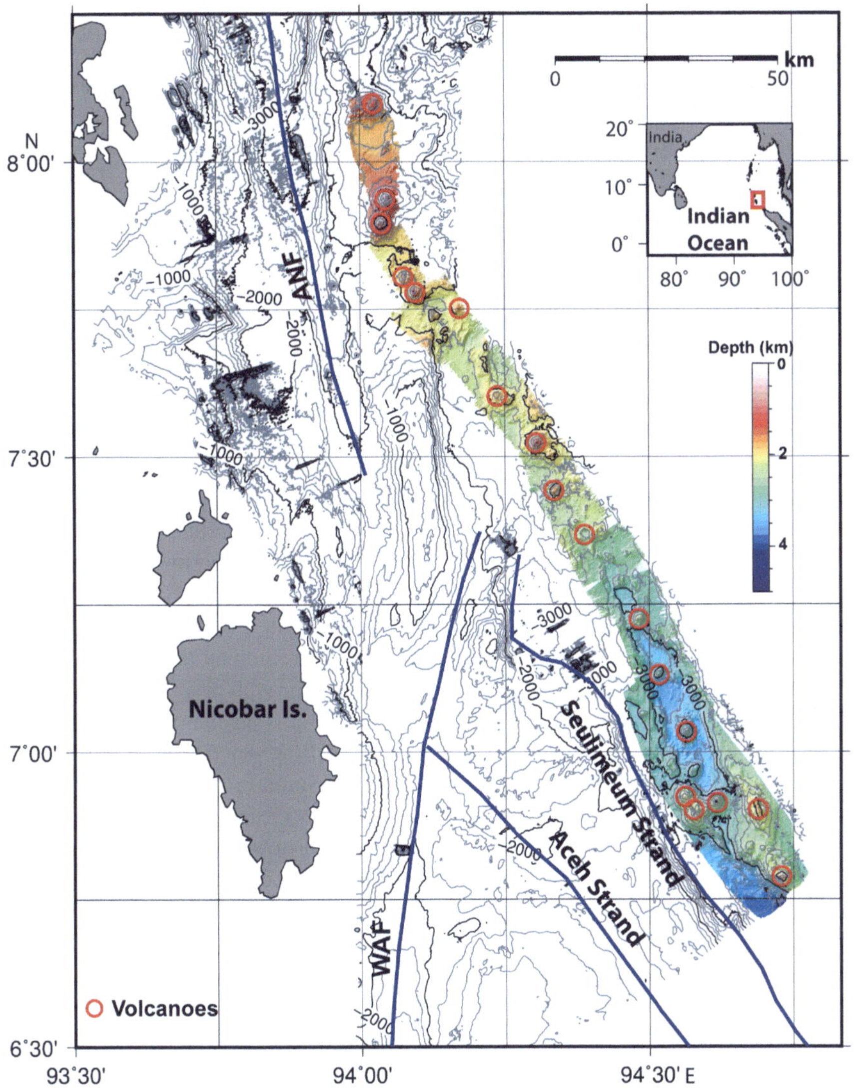

Fig. 2 Multibeam bathymetry from the off Nicobar region, contour interval 200 m. Colour coded bathymetric image highlights the identified volcanoes

The recent passive OBS experiment (Dewangan et al. 2018) has provided insights into the seismicity pattern. The study of Singha et al. (2019) enabled to refinement of the subducting slab geometry in the Nicobar-Andaman sector, the dip of the subducting Indian plate has varied from 26° in the off Nicobar region to 53° in the Andaman sector.

The other interesting feature that is apparent is the indentation of Ninetyeast Ridge on the trench at about 7° N that seems to modulate the curvature of the trench. A combination of strike-slip and normal thrust mechanisms for earthquakes in the convergent zone suggests partial subduction and partial shearing of the Ninetyeast Ridge near the Andaman arc–trench system (Subrahmanyam et al. 2008). Singha et al. (2019) noticed a seismic gap in the subducting Indian plate between 7.5° N and 9.5° N at depths greater than 100 km, which may be due to partial subduction of the Ninetyeast Ridge.

Distinctly different rupture patterns of December 2004 and March 2005 megathrust earthquakes have been observed in the Andaman-Sumatra region suggesting a strong influence of tectonic and structural elements. It has been suggested that the lithospheric scale boundary of the WAF and other tectonic elements modulate the occurrence of large earthquakes and their rupture pattern in the Andaman Sea (Singh et al. 2005; Kamesh Raju et al. 2007).

One more distinct seismicity event that followed the great mega-thrust event was the off Nicobar swarm that occurred during 27–30 January 2005. The off Nicobar swarm with more than 150 events with 5 or greater magnitude has been rated as the most energetic swarm ever observed globally. This swarm activity is suggested to be a part of the overall interplate motion partitioning (Lay et al. 2005). The 2004 earthquake greatly reduced the normal stress in the region of WAF from the planes or unclamped them. The unclamping of the region by the 2004 Sumatra-Andaman earthquake led to the initiation of the swarm through predominantly strike-slip faulting (Kamesh Raju et al. 2012a). Earthquake focal mechanisms suggest that the swarm was of volcano-tectonic origin and fluid flow played an important role in its occurrence and it was triggered by the 2004 Sumatra-Andaman earthquake through poroelastic relaxation of the coseismic stresses (Kundu et al. 2012). The 26 December 2004 Sumatra earthquake enabled study of how seismicity far from the rupture is promoted and inhibited by the main shock. These studies have shown that the 2004 Sumatra-Andaman earthquake altered seismicity in the Andaman backarc rift-transform system. Over a 300-km-long large transform section of the backarc, earthquakes of $M \geq 4.5$ stopped for five years, and the rate of transform events dropped by two thirds, while the rate of rift events increased eight fold (Sevilgen et al. 2012).

The Andaman Backarc Basin is also characterized by number of earthquake swarms (Fig. 3). These swarms have occurred along the backarc spreading center, near the Nicobar Island, along the offshore extension of the Sumatra fault and near Simeulue Island. Temporal and spatial patterns of seismicity in the Andaman Backarc Basin spreading centre are consistent with normal faulting. Regional seismicity is not only confined to the subduction zone; spreading in the Andaman Backarc Basin produced earthquake clusters linked to the episodes of magmatic dyke injection over cycles of c. 25 years (Diehl et al. 2013; Carter and Bandopadhyay 2017).

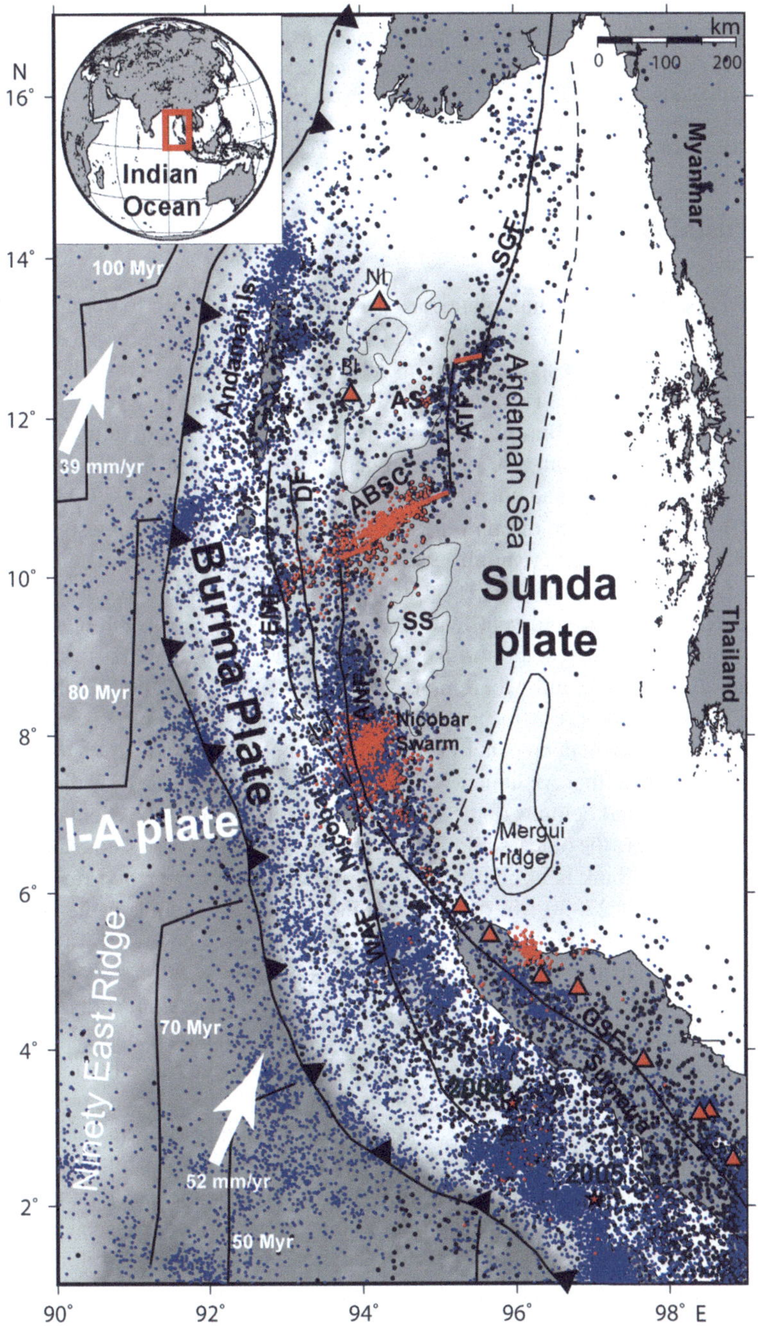
Indian
Ocean
km
0
100
200
N
16°
14°
12°
10°
8°
6°
4°
2°
90°
92°
94°
96°
98° E
100 Myr
80 Myr
70 Myr
50 Myr
39 mm/yr
52 mm/yr
Andaman Is
NI
BI
AS
ATF
SGF
ABSC
DF
EMF
SS
Andaman Sea
Sunda
plate
Myanmar
Thailand
Burma Plate
I-A plate
Nicobar Is
AMF
Nicobar
Swarm
Mergui
ridge
WAF
Ninety East Ridge
GSF
Sumatra
2004
2005

◂**Fig. 3** Seismicity of the region for the period 1965–2019, International Seismological Centre (ISC-Event Catalogue) data base (http://www.isc.ac.uk/iscbulletin/search/catalogue/). Black dots represent the earthquake epicenters of pre-tsunami event (1965 to 26th Dec 2004), Blue dots represent the earthquake epicenters of post-tsunami period. The red dots represent the documented earthquake swarms. The other notations are same as in Fig. 1

5 Crustal Architecture

Age of the subducting Indian plate varied from 50 Ma in south to about 80 Ma in the Andaman sector and to about 120 Ma towards further north (Subarya et al. 2006; Jacob et al. 2014). Lithospheric sections spaced at one degree across the 3500 km long Burmese-Andaman arc system based on seismicity data inferred variations in the depth and dip of the Benioff zone in correspondence to the curvature of the fold thrust belt (Dasgupta et al. 2003). The trench in the Andaman Subduction Zone is filled with thick Bengal and Nicobar fan sediments (Curray et al. 1979; Moeremans et al. 2014). The presence of ophiolites on the Andaman Islands suggests that these could be as old as 100 Ma (Rudolfo 1969; Pedersen et al. 2010; Sarma et al. 2010).

The East Margin Fault (EMF), Diligent Fault (DF), forearc basin and Invisible Bank (IB) are the prominent features in the Andaman-Nicobar forearc system. EMF defines the eastern margin of forearc and bounds the forearc basin in the west (Cochran 2010; Singh and Moeremans 2017). The EMF, and the DF are thought to have formed since pre-Miocene or maybe the middle Eocene (Moeremans and Singh 2015). Based on deep penetrating multichannel seismic reflection data, Singh et al. (2013) suggest that Andaman-Nicobar Forearc Basin is floored by the Malayan continental crust and belongs to the same regime as the floor of the Mentawai Basin further south, where the oldest sediments are up to 85 Ma old. These studies also imply that the Invisible Bank is of continental origin, and might have rifted from Malaya Peninsula at about 23–30 Ma during the initiation of spreading in the Andaman Sea. Forward gravity modeling carried out by Goli and Pandey (2014) supports the contention that the Andaman-Nicobar Forearc Basin is floored by continental crust. Morley and Alvey (2015) carried out the gravity inversion to determine the crustal thickness, their results suggest that the Central Andaman Basin is underlain by oceanic crust, while the adjacent regions of the Alcock and Sewell seamounts and the Eastern Andaman Basin are classified as extended continental crust.

The Andaman Backarc spreading centre is segmented into three distinct segments (Kamesh Raju et al. 2004; Jourdain et al. 2016): a western segment (consisting of segments A and B of Kamesh Raju et al. 2004); the central segment (segment C); and a northern segment (segment N). The northern segment seems to be a nascent spreading centre connected to the Sagaing Fault in the north and to the central segment in the south with a 180 km long transform fault. Both the northern and central segments are covered with thick sediments, whereas the western segment is devoid of sediments. Kamesh Raju et al. (2004) postulated that the rifting has started at the northeastern end and propagated southwesterly.

Based on bathymetric, single channel seismic and magnetic data, Kamesh Raju et al. (2004) suggested that the Andaman Backarc spreading centre (ABSC) initiated at c. 4.5 Ma, separating the Alcock and Sewell rises. However, Morley and Alvey (2015) recently re-examined the published geophysical data over the eastern part of the Central Andaman Basin (Kamesh Raju et al. 2004; Curray 2005) and suggested that the observed sedimentary geometries are incompatible with continual seafloor spreading since 4 Ma. Instead, these authors suggest that the Central Andaman Basin has formed by episodic seafloor spreading, mostly from the middle to late Miocene, followed by a recent re-activation. The conclusions of Morley and Alvey (2015) are debatable and remain inconclusive due to insufficient evidence in the form of new data or analysis (Curray 2015).

6 Absence of Magnetic Anomalies

Linear marine magnetic anomalies are the diagnostic signature of seafloor spreading process in the ocean basins and provide unequivocal evidence for the Vine-Matthews-Morley hypothesis. Study of marine magnetic anomalies provided understanding of the evolution of the ocean basins in different parts of the world oceans. However, some parts of the ocean basins have exhibited low amplitude incoherent magnetic anomalies; this is attributed to periods of magnetic quiet zone. There are yet, few other examples, where very young crust also failed to generate significant magnetic anomalies. Such unique situation arises in the Andaman Sea over the sedimented spreading center. The deep seismic imaging of the spreading center (Singh et al. 2010) provided compelling evidence for the presence of sill-sedimented structure and the absence of oceanic layer 2A beneath this spreading center. Here we consider some of the causes that might give rise to nonmagnetic young oceanic crust and suggest that melt sills altered by the interaction with sediments are responsible for the generation of young nonmagnetic oceanic crust resulting in the absence of magnetic anomalies.

Young oceanic crust apparently produced by the spreading centers that are not accompanied by the magnetic anomalies include, Gulf of California-Guyamas Basin (Larsen et al. 1972; Bischoff and Henyey 1974), Paul Revere Ridge and Winona Basin (Davis and Riddihough 1982), Middle valley of the Juan de Fuca Ridge (Davis and Lister 1977), Escanaba Trough—Southern Gorda Ridge (Raff and Mason 1961), Andaman backarc spreading center (Curray et al. 1979; Kamesh Raju et al. 2004).

In addition to the absence of magnetic anomalies, each of the above areas are overlain by thick sediments of several hundred meters ranging up to 1 km. It has been hypothesized that there is causal relation between the sediment cover and the absence/subdued magnetic anomalies. In the Yamato Basin of the Japan Sea, the absence of magnetic anomalies is attributed to the formation of sill-sediment complexes due to the emplacement of mafic magma into the unconsolidated sediments below the seafloor (Fukuma and Shinjoe Hamano 1998, ODP 794D). It was suggested that the mafic magma was emplaced into the unconsolidated sediments just below the ocean floor and formed sill-sediment complexes, instead of erupting on the seafloor and creating basalt lava flows as in the case of normal oceanic crust (Tamaki et al. 1990). Levi and Riddihough (1986) proposed that the magnetic anomalies can be suppressed due to the pervasive hydrothermal reactions underneath thick blanket of sediments.

Sediment filled rift at the northern-end of the Juan de Fuca ridge is characterized by a pronounced magnetic low in place of an expected positive anomaly, this was explained by the presence of non-magnetic crust. Based on the measured magnetization of the rocks at this location the anomalous magnetic feature is suggested to be the result of extensive hydrothermal alteration that has taken place under the hydrological seal of the sediments in the valley (Currie and Davis 1994, ODP139).

With respect to the absence of magnetic anomalies, the sedimented spreading center in the Andaman Sea is similar to the Yamato Basin of the Japan Sea. Kamesh Raju et al. (2004) suggested that early sediment burial is the cause of subdued magnetic anomalies resulting in low magnetization at greater depth. Hydrothermal alteration could be one more contributing factor. Magnetic remanence of the basalt was found to be subdued when samples of fresh basalts are subjected to temperature (350 °C) and pH (3.0) conditions of active hydrothermal environment in an experiment conducted by Ogishima and Kinoshita (1992).

Jourdain et al. (2016) have clearly imaged the sill-sediment complexes and provided evidence for the complete absence of the Layer 2A beneath the spreading center segment C. This interpretation was aided by the high-quality seismic reflection data (Fig. 4). This is probably the first ever documentation of the creation of nonmagnetic oceanic crust.

7 Discussion

From the observations and results, it is clear that there are unanswered questions and inconclusive inferences about the nature and type of the crust in the Andaman Sea and its evolution. The important questions can be classified as related to spreading center, nature of the crust and evolution history.

8 Backarc Spreading

It is generally accepted that the Central Andaman Basin (CAB) has evolved due to the seafloor spreading activity since last 4 Ma (Kamesh Raju et al. 2004; Curray 2005; Singh et al. 2013; Jourdain et al. 2016). However, Morley and Alvey (2015) opined that the observed sedimentary geometries are incompatible with continual seafloor spreading since 4 Ma. Their gravity analysis does indeed confirm that the CAB is underlain by oceanic crust. Magnetic anomalies in the CAB do not show any evidence that spreading has been episodic and no other evidence has been presented

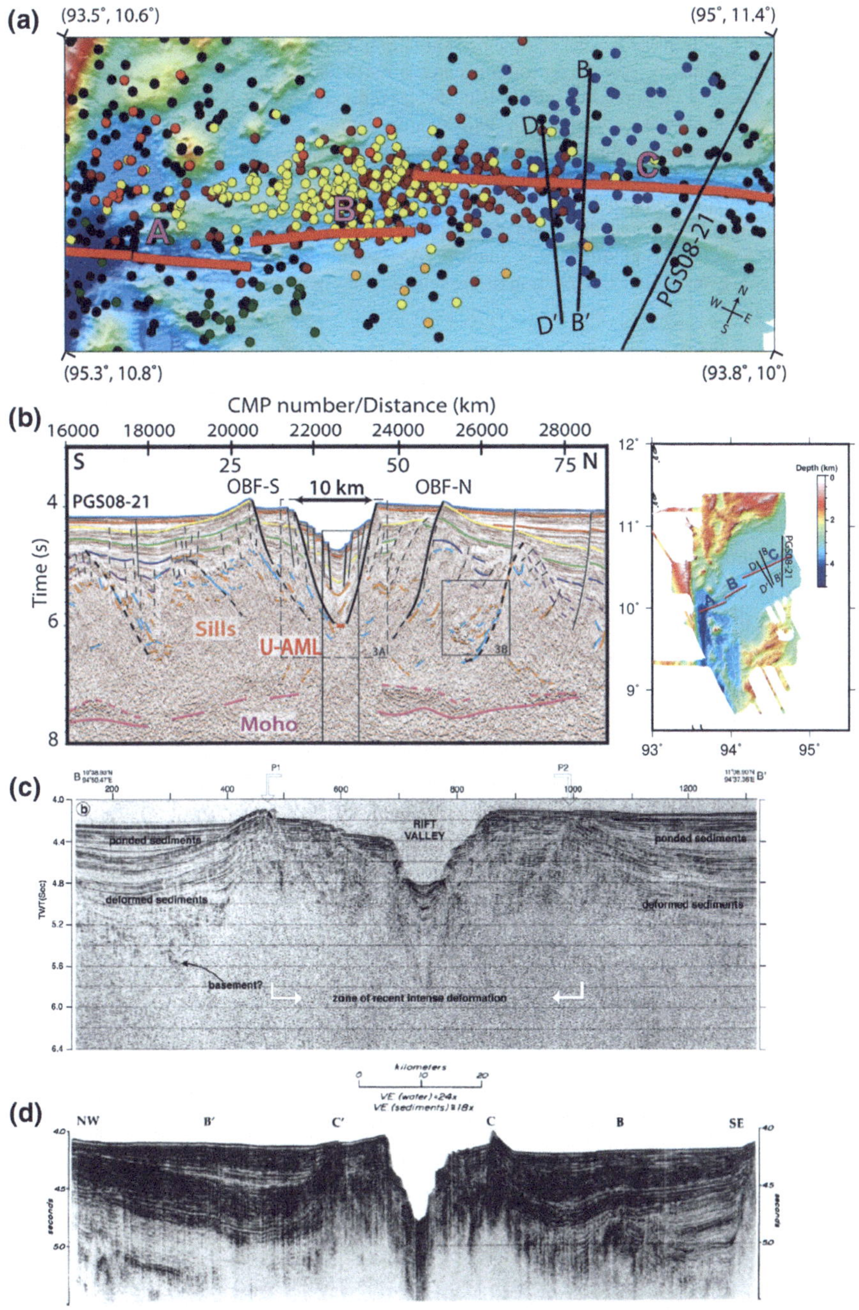
(a)
(93.5°, 10.6°)
(95°, 11.4°)
(95.3°, 10.8°)
(93.8°, 10°)
A
B
C
D
D'
B
B'
PGS08-21
N
S
W
E
(b)
CMP number/Distance (km)
16000
18000
20000
22000
24000
26000
28000
S
N
25
50
75
PGS08-21
OBF-S
10 km
OBF-N
Time (s)
4
6
8
Sills
U-AML
Moho
3A
3B
Depth (km)
12°
11°
10°
9°
93°
94°
95°
(c)
P1
P2
200
400
600
800
1000
1200
RIFT VALLEY
ponded sediments
ponded sediments
deformed sediments
deformed sediments
TWT(Sec)
basement?
zone of recent intense deformation
kilometers
0
10
20
VE (water)≈24x
VE (sediments)≈18x
(d)
NW
B'
C'
C
B
SE
seconds
seconds

◀**Fig. 4** Seismic sections across the sedimented spreading segment of the backarc spreading center along with the seismicity over the spreading center. **a** Earthquake events plotted on the multibeam bathymetry image of the spreading center. Blue circles represent events of 1984 swarm, dark green 1993 swarm, orange1994 swarm, brown 2006 swarm, yellow 2009 swarm and red 2012 swarm. **b** Deep penetrating multichannel seismic reflection section acquired with 8 km long streamer and an 81.93 L volume air guns as source (Jourdain et al. 2016), **c** single channel seismic reflection data acquired with 3 L airgun source (Kamesh Raju et al. 2004), **d** Single channel seismic reflection data (Curray et al. 1979). Note the Moho reflections in this section, OBF-N—Outer Bounding Fault North; OBF-S—Outer Bounding Fault South. U-AML—upper axial melt lens; L-AML—lower axial melt lens. Inset depicts the location of the seismic lines over the multibeam bathymetry map

showing that spreading could also be episodic. If spreading is episodic, one would expect the magnetic anomalies to show periods of spreading versus non-spreading. But two groups have independently studied the magnetics and neither interpreted any periods of non-spreading (Curray et al. 1979; Kamesh Raju et al. 2004; Curray 2005). Curray (2015) points out that the principal episodic character is the deposition and not the spreading activity. Episodic deposition can be caused by sea level fluctuations, that have occurred during glacial versus non-glacial times in the Pleistocene and late Pliocene (Curray et al. 1979; Curray 2005).

Further, the other unique character of the spreading center is the absence of magnetic anomalies over a segment of the spreading center that is overlaid by thick blanket of sediments. This anomalous character was documented by Curray et al. (1979) and Kamesh Raju et al. (2004) and they offered explanation in terms of rapid sedimentation over evolving oceanic crust (Curray et al. 1979) or alteration of magnetization due to rapid sedimentation and burial of the crust at greater depth (Kamesh Raju et al. 2004). The absence of magnetic anomalies observed in the Yamato Basin of the Japan Sea is attributed to the formation of sill-sediment complexes, thereby altering the formation of normal oceanic crust (Fukuma and Shinjoe Hamano 1998, ODP 794D). In the Andaman Backarc Basin, Jourdain et al. (2016) provided evidence on the basis of seismic reflection data for the formation of sill complexes within the sediments resulting in the absence of oceanic crustal layer 2A. This was proposed as the reason for the absence of magnetic anomalies. The exact nature of these sill complexes that are imaged by the deep penetrating multichannel seismic reflection data still remain unknown.

We need additional data or experiments to conclusively establish, whether the spreading is episodic or deposition is episodic and also to find out the exact nature of the sediment sill complexes. An IODP drilling into the sediment filled spreading center, may resolve these issues conclusively.

9 Volcanic Arc

Barren Island is the only known active volcano in the northern Andaman Sea (Sheth et al. 2009). In terms of the known volcanoes in the region, there is a distinct gap in active volcanoes from 06° N to 12° N. Subduction of the base of the Ninetyeast Ridge was suggested as one of the possible reasons for this gap in volcanic activity (Subrahmanyam et al. 2008).

Recent multibeam bathymetry investigations have mapped volcanic chains that constitute the volcanic arc (Kamesh Raju et al. 2012a, b; Tripathi et al. 2018). The recovery of fresh manganese globules from the cratered seamount off Nicobar Island (Kamesh Raju et al. 2012a), the recovery of pumice from the seamount tops (Kamesh Raju et al. 2004) adjacent to western tip of the spreading center, and several seamounts with well-defined craters (Tripathi et al. 2018) indicate recent active volcanism associated with the volcanic arc. The mapping of the new seamount chains (Fig. 2) established the link between the Barren and Narcondam sub-areal volcanoes in the north to the volcanoes of Sumatra in the south. Extensive volcanism in the geological past is documented in a 691.6 m deep drilling core recovered by D/V JOIDES Resolution (Kumar et al. 2014).

With the advent of the water column imaging along with multibeam swath bathymetry, it is now easier to locate active submarine volcanism and capture eruptions related to submarine activity. The Barren Island volcanic activity has been monitored on the Island, however, the near vicinity offshore region remains devoid of continuous monitoring and sampling.

10 Types of Crust in the Andaman Backarc Basin

Wide ranging inferences have been made regarding the nature of the crust leading to a complex mosaic of crustal types in the Andaman Sea (Fig. 5). These crustal types range from continental to extended continental, arc crust and the oceanic crust. There are knowledge gaps in terms of the emplacement process and the extent of these crustal types.

Most planets have basaltic crusts similar to Earth's oceanic crust, but the continental areas with thick buoyant silicic crust is a unique characteristic of Earth (Taylor and McLennan 1985; Rudnick and Gao 2003; Korenaga et al. 2002). It is believed that the continental crust is formed at the subduction zones (Taylor and McLennan 1985; Kelemen and Behn 2016). However, it is unclear why the subduction of dominantly basaltic oceanic crust would result in the formation of andesitic continental crust (Gazel et al. 2015). Since the continental crust is believed to have formed at the subduction zone processes, the understanding of the crustal types within the subduction factory encompassing the trench-arc-backarc system and their transformation process is critical to establish the evolutionary stages of the continental crust.

Continental crust is characterized by a lower-velocity upper crust (seismic P-wave velocity $V_P \approx 5$–6 km s^{-1}) and a higher-velocity lower crust ($V_P \approx 7$–7.5 km s^{-1}), separated from the underlying mantle ($V_P \approx 8$–8.5 km s^{-1}) by the sharp Moho (Christensen and Mooney 1995). With the exclusion of active orogenic belts or rifts, the continental Moho occurs at a relatively constant depth of 41 ± 6 km (Christensen and Mooney 1995). The Moho in a typical continental crust is more sharply defined

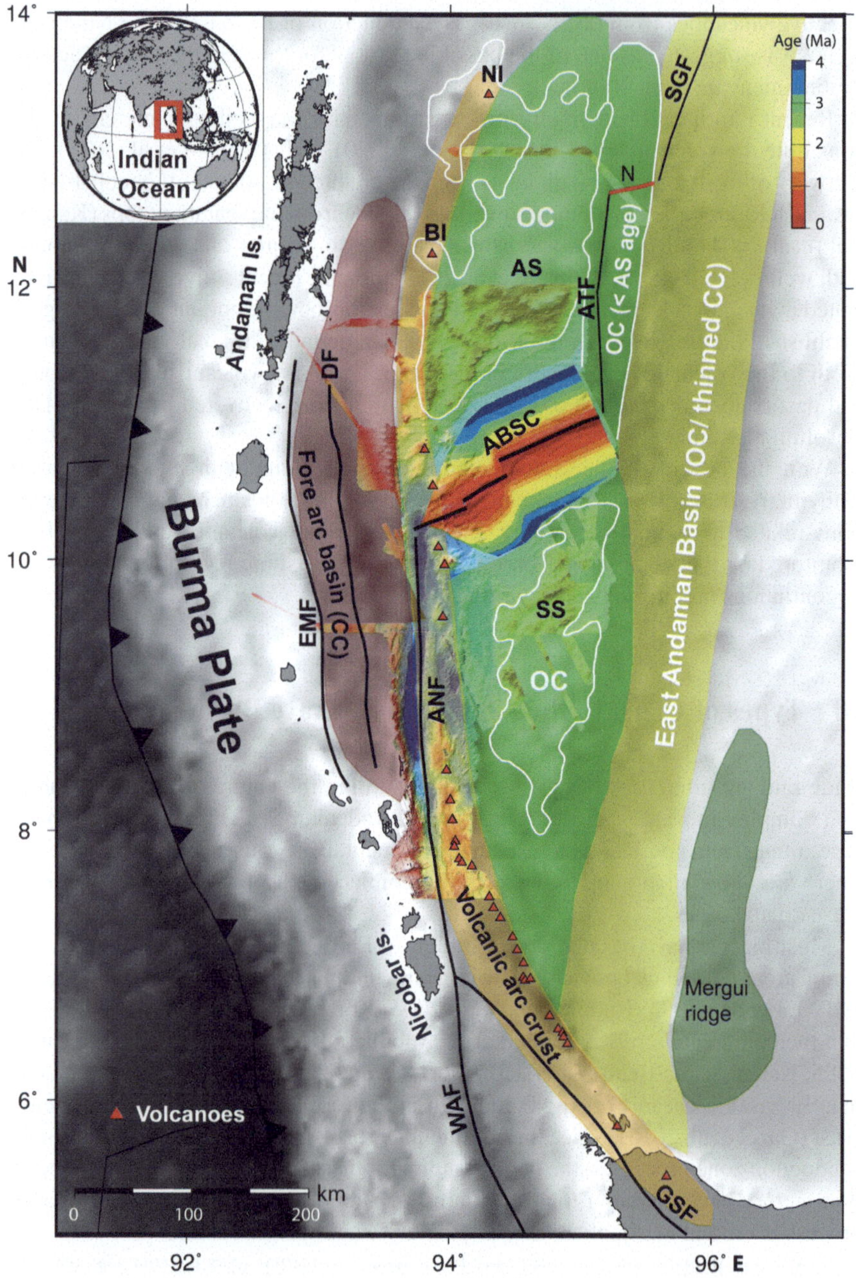

Fig. 5 Summary tectonic map with color shades representing crustal types that differ by age and nature. Major structural elements are marked. Age of the Central Andaman Basin (CAB) is derived from the magnetic anomaly identifications of Kamesh Raju et al. (2004). The age contours are extrapolated over the segment C, where magnetic anomalies could not be identified. CC denotes the continental crust (brown shade) and OC represents the oceanic crust (green shade). NI and BI represent Narcondam and Barren Island volcanoes respectively. Other notations are same as in Fig. 1

compared to the arc crust that is characterized by a weak discontinuity with less velocity contrast (increase in V_P from about 6.8 to 7.2 km s^{-1}) (Shillington et al. 2004; Takahasi et al. 2008; Tatsumi et al. 2008). This could be due to contact between mafic lower crust and unusually hot upper mantle or an intra-crustal contact between mafic and ultramafic cumulates (Takahashi et al. 2008). These studies suggest that there are distinct differences in continental and oceanic/arc Moho and significant reworking of the melts should take place in the transformation of the arc crust to the continental crust to convert the transitional lower-crust–mantle interface in arcs into the sharply defined crust–mantle discontinuity of continental regions (Jagoutz and Behn 2013; Jagoutz and Schmidt 2013).

With this background, let us look at the crustal types that have been inferred in the Andaman Sea. Singh et al. (2013) based on multichannel seismic reflection, gravity and earthquake seismological data suggested that the Andaman–Nicobar Forearc Basin is likely floored by continental crust rifted from the Malayan Peninsula, and the Invisible Bank could be a part of this continental block that rifted from the Malayan Peninsula 23–30 Ma ago. The eastern margin of the Invisible Bank is a deep rift basin which hosts the Andaman–Nicobar strike-slip sliver fault. There is general agreement on the existence of young oceanic crust in the Central Andaman Basin (Curray et al. 1979; Kamesh Raju et al. 2004; Curray 2005; Morley and Alvey 2015; Jourdain et al. 2016). There are differing views on the nature of crust beneath Alcock and Sewell rises that plays a critical role in the overall evolution of the Andaman Sea. While Curray (2015), supports the backarc/oceanic crust based on the dating of available sparse rock samples, Morley and Alvey (2015) favour thinned continental crust based on the analysis of gravity inversion. Gravity inversion may not be a good criterion to determine the type of crust solely based on crustal thickness estimates. Curray (2015) points out that the Ninetyeast Ridge also yielded a similar gravity inversion signature in the maps of Morley and Alvey (2015), highlighting the inconclusive nature of gravity inversion technique. As on date, no refraction or other measurements are available to decipher the nature of the crust with reasonable certainty.

Similar uncertainty also prevails regarding the nature and age of the crust in the East Andaman Basin. Curray (2005) interpreted it as oceanic crust of Middle to Late Miocene and Early Pliocene age that was compatible with the scenario of opening of East Andaman Basin. Morley and Alvey (2015) suggest that it is floored by thinned continental crust. There is severe lack of data in this region to derive meaningful inferences.

The derived velocity model of Singha et al. (2019) suggests the presence of Moho at a depth of 30 km beneath the Andaman and Nicobar Islands. The minimum 1D model shows a sharp jump in velocity from 6.7 to 8.1 km s^{-1} at about 30 km, which is interpreted as Moho beneath the accretionary prism. The velocity of oceanic crust (layer 3A and 3B) ranges from 6.8 to 7.3 km s^{-1} (Spudich and Orcutt 1980). Singha et al. (2019) interpret the lower velocity of the crust (6.7 km s^{-1}) to be of continental origin. The presence of continental crust beneath the accretionary prism and the forearc basin is also inferred from the analysis of seismic reflection and gravity data (Singh et al. 2013).

With respect to the Andaman Islands, models show the presence of Moho at 25–35 km. Radhakrishna et al. (2008) favored oceanic crust for the Andaman–Nicobar Outer Arc Ridge, based on modeling of gravity data. Rao et al. (2011) interpreted the velocity model in terms of 16 km thick overriding Burma plate underlain by a 9 km subducting Indian plate. Based on velocity structure computed from joint inversion of the teleseismic receiver function and Rayleigh wave group velocity analysis on 10 broadband seismographs on land, it is suggested that the crustal thickness beneath the Andaman Island varies from ~24 km in the north to ~32 km in the south. The crustal structure beneath the Andaman Islands is akin to continental crust and the subducting Indian plate may lie down below this overriding plate (Gupta et al. 2016). Singha et al. (2019) support the presence of continental crust beneath the Andaman and Nicobar Islands and the postulation of Curray (2005) that the Burmese microplate might have been separated from the Southeast Asian plate along the sliver fault system.

11 Concluding Remarks

There has been considerable enhancement in our knowledge during the recent years due to the new initiatives that made the availability of high-resolution multibeam bathymetry, deep penetrating multichannel seismic reflection, GPS and OBS data sets and their interpretation. This brief review brought out some of the unresolved issues and uncertainties in our understanding of the Andaman Backarc Basin. They are listed below with possible solutions:

With respect to the morphotectonics, the high-resolution mapping efforts have greatly improved our understanding of the Andaman Backarc Basin. The most crucial being the remarkable evidence about the existence of the backarc spreading center, the identification of series of volcanoes defining the active volcanic arc and the delineation of the Andaman-Nicobar Fault system.

The major difference between the Andaman–Nicobar and the Sumatra sectors of the subduction zone arises due to the presence of the backarc spreading center. Understanding the dynamics of the Andaman Sea spreading centre becomes critical to define the evolution of the Andaman Backarc Basin. Sedimented segment of the Andaman backarc spreading center provides a unique tectonic setting that lead to the creation of non-magnetic oceanic crust. The nature of the oceanic crust underneath the segment C and the composition of the sill complexes are of special interest. Further, the age and nature of the crust beneath the Alcock and Sewell seamounts is critical to define the recent evolutionary history of the Andaman Backarc Basin.

The age of the East Andaman Basin is the aspect that needs to be addressed to establish the initial opening of the Andaman Sea with more certainty. Close grid high resolution geophysical investigations aimed at identifying the nature of the crust are required.

Barren Island volcanism has been in focus for some time. What is needed may be a modern permanent observatory designed to cover the island and also the offshore region. Deployment of OBS clusters in the offshore region around the Barren Island would provide crucial information regarding the temporal variation of tectonic and magmatic activity associated with the Barren Island volcano.

Close grid and long-term monitoring of the Nicobar swarm is one more aspect that can provide insights into this hyperactive region. Several seamounts with cratered summit have been identified in this region indicating recent volcanic activity. These volcanoes are to be monitored with the deployment of clusters of OBS receivers. Water column chemistry by sampling, and water column imaging by multibeam bathymetry has to be carried out regularly to capture the occurrence of submarine volcanic eruptions.

Four decades have elapsed after the major geological and geophysical investigations initiative undertaken by the Scripps Institute of Oceanography. The 2004 tsunamigenic earthquake caught the attention of several international scientific groups and the spur in the investigations have provided considerable knowledge about the Andaman Backarc Basin. The region now awaits an IODP campaign to resolve some of the outstanding problems.

Acknowledgements We thank Prof. Sunil Kumar Singh, Director, CSIR-National Institute of Oceanography (CSIR-NIO), Goa and Dr. M. Ravichandran, Director, National Center for Polar and Ocean Research (NCPOR), Goa for their encouragement and support. We thank Prof. M. Radhakrishna, IIT, Mumbai for the invitation to write this review, which in fact, has actuated us to take-up of this work. All figures were drafted with the GMT software (Wessel and Smith 1995). This is a CSIR-NIO and NCPOR contribution. NCPOR contribution number B-3/2019-20.

References

Ammon CJ, Ji C, Thio H-k, Robinson D, Ni S, Hjorleifsdottir V, Kanamori H, Lay T, Das S, Helmberger D, Ichinose G, Polet J, Wald D (2005) Rupture process of the 2004 Sumatra-Andaman earthquake. Science 1133:1133–1140

Bischoff JL, Henyey TL (1974) Tectonic elements of the central part of the Gulf of California. Gol Soc Am Bull 85:1893–1904

Bock Y, Prawirodirdjo L, Genrich JF, Stevens CW, McCaffrey R, Subarya C, Puntodewo SSO, Calais E (2003) Crustal motion in Indonesia from global positioning system measurements. J Geophys Res 108:2367. https://doi.org/10.1029/2001JB000324

Carter A, Bandopadhyay PC (2017) Seismicity of the Andaman-Nicobar Islands and Andaman Sea. Geol Soc Lond Mem 47:205–213

Christensen NI, Mooney WD (1995) Seismic velocity structure and composition of the continental crust; a global view. J Geophys Res 100:9761–9788

Cochran JR (2010) Morphology and tectonics of the Andaman Forearc, northeastern Indian Ocean. Geophys J Int 182:631–651

Curray JR (2005) Tectonics and history of Andaman sea region. J Asian Earth Sci 25:187–232

Curray JR, Moore DG, Lawver LA, Emmel FJ, Raitt RW, Henry M, Kieckhefer R (1979) Tectonics of the Andaman Sea and Burma. Am Assoc Petrol Geol Mem 29:189–198

Curray JR (2015) Discussion: "Is spreading prolonged, episodic or incipient in the Andaman Sea? Evidence from deepwater sedimentation." by CK Morley and A Alvey. J Asian Earth Sci 111:113–119

Currie RG, Davis EE (1994) Low crustal magnetization of the middle valley sedimented rift inferred from sea-surface magnetic anomalies. In: Mottl MJ, Davis EE, Fisher AT, Stack JF (eds) Proceedings of the Ocean drilling program. Scientific results, vol 139, pp 19–27

Dasgupta S et al (2000) Seismotectonic Atlas of India and its environs (eds Narula, P L, Acharyya, S K & Banerjee J) 87pp. Geological Survey of India, Calcutta, India

Dasgupta S, Mukhopadhyay M, Bhattacharya A, Jana TK (2003) The geometry of the Burmese-Andaman subducting lithosphere. J Seismol 7:155–174

Davis EE, Lister CRB (1977) Tectonic structures on the Juan de Fuca Ridge. GSA Bull 88:346–363

Davis EE, Riddihough RP (1982) The Winona basin: structure and tectonics. Can J Earth Sci 19:767–788

Dewangan P, Reddy R, Kamesh Raju KA, Singha P, Aswini KK, Yatheesh V, Samudrala K, Shuhail M (2018) Nature of the ambient noise, site response, and orientation of ocean-bottom seismometers (OBSs): scientific results of a passive seismic experiment in the andaman sea. Bull Seismol Soc Am 108:248–259

Diehl T, Waldhauser F, Cochran JR, Kamesh Raju KA, Seeber L, Schaff D, Engdahl E (2013) Back-arc extension in the Andaman Sea: tectonic and magmatic processes imaged by high-precision teleseismic double-difference earthquake relocation. J Geophys Res 118:2206–2224. https://doi.org/10.1002/jgrb.50192

Engdahl ER, Villaseñor A, DeShon HR, Thurber CH (2007) Teleseismic relocation and assessment of seismicity (1918–2005) in the region of the 2004 Mw 9.0 Sumatra-Andaman and 2005 Mw 8.6 Nias Island great earthquakes. Bull Seismol Soc Am 97:43–61

Fukuma KH, Shinjoe Hamano Y (1998) Origin of the absence of magnetic lineations in the Yamato basin of the Japan Sea: magnetic properties of mafic rocks from Ocean drilling program hole 794D. J Geophys Res 103:17791–17805

Gahalaut VK, Gahalaut K (2007) Burma plate motion. J Geophys Res 112:B10402. https://doi.org/10.1029/2007JB004928

Gahalaut VK, Jade S, Catherine JK, Gireesh R, Ananda MB, Kumar P, Narsaiah M, Jafri SSH, Ambikapathy A, Bansal A, Chadha RK, Gupta DC, Nagarajan B, Kumar S (2008) GPS measurements of postseismic deformation in the Andaman-Nicobar region following the giant 2004 Sumatra-Andaman earthquake. J Geophys Res 113:B08401

Gazel E, Hayes JL, Hoernle K, Kelemen P, Everson E, Holbrook WS, Hauff F, van den Bogaard P, Vance EA, Chu S, Calvert AJ, Carr MJ, Yogodzinski GM (2015) Continental crust generated in oceanic arcs. Nat Geosci 8:321

Goli A, Pandey DK (2014) Structural characteristics of the Andaman Forearc inferred from interpretation of multichannel seismic reflection data. Acta Geol Sinica 88:1145–1156

Graindorge D, Klingelhoefer F, Sibuet JC, McNeill L, Henstock TJ, Dean S, Gutscher MA, Dessa JX, Permana H, Singh SC, Leau H, White N, Carton H, Malod JA, Rangin C, Aryawan KG, Chaubey AK, Chauhan A, Galih DR, Greenroyd CJ, Laesanpura A, Prihantono J, Royle G, Shankar U (2008) Impact of lower plate structure on upper plate deformation at the NW Sumatran convergent margin from seafloor morphology. Earth Planet Sci Lett 275:201–210

Gupta S, Borah K, Saha G (2016) Continental like crust beneath the Andaman Island through joint inversion of receiver function and surface wave from ambient seismic noise. Tectonophysics 687:129–138

Jacob J, Dyment J, Yatheesh V (2014) Revisiting the structure, age and evolution of the Wharton basin to better understand subduction under Indonesia. J Geophys Res 119:169–190

Jagoutz O, Behn MD (2013) Foundering of lower island-arc crust as an explanation for the origin of the continental Moho. Nature 131:504

Jagoutz O, Schmidt MW (2013) The composition of the foundered complement to the continental crust and a re-evaluation of fluxes in arcs. Earth Planet Sci Lett 371:177–190

Jourdain A, Singh SC, Escartin J, Klinger Y, Kamesh Raju KA, McArdle J (2016) Crustal accretion at a sedimented spreading center in the Andaman Sea. Geology 44:351–354

Kamesh Raju KA, Ramprasad T, Rao PS, Rao BR, Varghese J (2004) New insights into the tectonic evolution of the Andaman basin, northeast Indian Ocean. Earth Planet Sci Lett 221:145–162

Kamesh Raju KA, Murty GPS, Amarnath D, Mohan Kumar ML (2007) The West Andaman fault and its influence on the aftershock pattern of the recent megathrust earthquakes in the Andaman-Sumatra region. Geophys Res Lett 34:L03305. https://doi.org/10.1029/2006GLO28730

Kamesh Raju KA, Ray Durbar, Mudholkar A, Murty GPS, Gahalaut VK, Samudrala K, Paropkari AL, Ratheesh R, Surya Prakash L (2012a) Tectonic and volcanic implications of a cratered seamount off Nicobar Island, Andaman Sea. J Asian Earth Sci 56:42–53. https://doi.org/10.1016/j.jseaes.2012.04.018

Kamesh Raju KA, Mudholkar A, Murty G, Yatheesh V, Singh SC, Kiranmai S, Moeremans R (2012b) Multibeam Mapping of the West Andaman Fault, NW Sumatra fault, Andaman Volcanic Arc and their tectonic and magmatic implications. Abstract T41C-2608 presented at 2012 Fall Meeting AGU San Francisco Calif, 3–7 Dec

Kelemen PB, Behn MD (2016) Formation of lower continental crust by relamination of buoyant arc lavas and plutons. Nat Geosci 9:197–205

Korenaga J, Kelemen PB, Holbrook SW (2002) Methods for resolving the origin of large igneous provinces from crustal seismology. J Geophys Res 107(B9):2178

Kumar P, Collett TS, Boswell R, Cochran JR, Lalle M, Mazumdar A, Ramana MV, Ramprasad T, Riedel M, Saini K, Sathe AV, Vishwanath K, Yadav US, NGHP Expedition 01 Scientific Party (2014) Geologic implications of gas hydrates in the offshore of India: Krishna-Godavari Basin, Mahanadi Basin, Andaman Sea, Kerala-Konkan Basin. Mar Pet Geol 58:29–98

Kumar P, Srijayanthi G, Ravi Kumar M (2016) Seismic evidence for tearing in the subducting Indian slab beneath the Andaman arc. Geophys Res Lett 43:4899–4904

Kundu B, Legrand D, Gahalaut K, Gahalaut VK, Mahesh P, Kamesh Raju KA, Catherine JK, Ambikapthy A, Chadha RK (2012) The 2005 volcano-tectonic earthquake swarm in the Andaman Sea: triggered by the 2004 great Sumatra-Andaman earthquake. Tectonics 31:1–11

Larson PA, Mudie JD, Larson RL (1972) Magnetic anomalies and fracture zone trends in the Gulf of California. Geol Soc Am Bull 83:3361–3368

Lay T, Kanamori H, Ammon CJ, Nettles M, Ward SN, Aster RC, Beck SL, Bilek SL, Brudzinski MR, Butler R, DeShon HR, Ekstrom G, Satake K, Sipkin S (2005) The great Sumatra-Andaman earthquake of 26 December 2004. Science 308:1127–1132

Levi S, Riddihough RP (1986) Why are marine magnetic anomalies suppressed over sedimented spreading centers? Geology 14:651–654

Moeremans R, Singh SC, Mukti M, McArdle J, Johansen K (2014) Seismic images of structural variations along the deformation front of the Andaman-Sumatra subduction zone: implications for rupture propagation and tsunamigenesis. Earth Planet Sci Lett 386:75–85

Moeremans RE, Singh SC (2015) Fore-arc basin deformation in the Andaman-Nicobar segment of the Sumatra-Andaman subduction zone: insight from high-resolution seismic reflection data. Tectonics 34:1736–1750

Morley CK, Alvey A (2015) Is spreading prolonged, episodic or incipient in the Andaman Sea? Evidence from deepwater sedimentation. J Asian Earth Sci 98:446–456

Mukhopadhyay M (1988) Gravity anomalies and deep structure of the Andaman Arc. Mar Geophys Res 9(197):210

Mukhopadhyay M, Krishna MBR (1995) Gravity anomalies and deep structure of the Ninetyeast Ridge north of the equator, eastern Indian Ocean—a hot spot trace model. Mar Geophys Res 17:201–216

Muller RD, Roest WR, Royer JY, Gahagan LM, Sclater JG (1997) Digital isochrons of the world's ocean floor. J Geophys Res 102(B2):3211–3214

Ogishima T, Kinoshita H (1992) Alteration of ferromagnetic components of oceanic basalt under pressurized hydrothermal liquids in laboratory. J Geomagn Geoelectr 44:309–315

Oldham T (1883) A catalogue of Indian earthquakes from the earliest time to the end of AD 1869. Mem Geol Sur India 19:53

Pedersen RB, Searle MP, Carter A, Bandopadhyay PC (2010) U-Pb zircon age of the Andaman ophiolite: implications for the beginning of subduction beneath the Andaman-Sumatra arc. J Geol Soc Lond 167:1105–1112

Radhakrishna M, Lasitha S, Mukhopadhyay M (2008) Seismicity, gravity anomalies and lithospheric structure of the Andaman arc, NE Indian Ocean. Tectonophysics 460:248–262
Raff AD, Mason RG (1961) Magnetic survey off the west coast of North America, 40°N. latitude to 52°N. latitude. Geol Soc Am Bull 72:1267–1270
Rao NP, Kalpana (2005) Deformation of the subducted Indian lithospheric slab in the Burmese arc. Geophys Res Lett 32:L05301. https://doi.org/10.1029/2004GL022034
Rao NP, Rao CN, Hazarika P, Tiwari VM, Kumar MR, Singh A (2011) Structure and tectonics of the Andaman subduction zone from modeling of seismological and gravity data. In: New frontiers in tectonic research—general problems, sedimentary basins and Island Arcs. InTech Publications Janeza Trdine 9 51000 Rijeka Croatia 249–268
Rudnick RL, Gao S (2003) The composition of the continental crust. Treatise Geochem 3:1–64
Rudolfo KS (1969) Bathymetry and marine geology of the Andaman basin and tectonic implications for SE Asia. Geol Soc Am Bull 80:1203–1230
Sarma DS, Jafri SH, Fletcher IR, McNaughton NJ (2010) Constraints on the tectonic setting of the Andaman ophiolites, Bay of Bengal, India, from SHRIMP U-Pb zircon geochronology of plagiogranite. J Geol 118:691–697
Sevilgen V, Stein RS, Pollitz FF (2012) Stress imparted by the great 2004 Sumatra earthquake shut down transforms and activated rifts up to 400 km away in the Andaman Sea. Proc Natl Acad Sci USA 109:15152–15156
Sheth HC, Ray JS, Bhutani R, Kumar A, Smitha RS (2009) Volcanology and eruptive styles of Barren Island: an active mafic stratovolcano in the Andaman Sea, NE Indian Ocean. Bull Volc 71:1021–1039
Shillington DJ, Van Avendonk HJA, Holbrook WS, Kelemen PB, Hornbach MJ (2004) Composition and structure of the central Aleutian Island arc from arc-parallel wide-angle seismic data. Geochem Geophys Geosyst 5:Q10006
Sieh K, Natawidjaja D (2000) Neotectonics of the Sumatra fault, Indonesia. J Geophys Res 105(B12):28295–28326
Singh SC, Sumatra Aftershocks Team (2005) Sumatra earthquake research indicates why rupture propagated northward, Eos 86
Singh SC, Carton H, Chauhan AS, Androvandi S, Davaille A, Dyment J, Cannat M, Hananto ND (2011) Extremely thin crust in the Indian Ocean possibly resulting from Plume-Ridge interaction. Geophy J Int 184:29–42
Singh SC, McArdle J, Johansen K (2010) Seismic images of the axial melt lens, Moho and deep penetrating faults at the sedimented Andaman Sea Spreading Centre. Abstract V12A-05, presented at 2010 Fall Meeting AGU San Francisco Calif, 13–17 Dec
Singh SC, Moeremans R, McArdle J, Johansen K (2013) Seismic images of the sliver strike-slip fault and back thrust in the Andaman-Nicobar region. J Geophys Res 118:5208–5224
Singh SC, Moeremans R (2017) Chapter 13 Anatomy of the Andaman–Nicobar subduction system from seismic reflection data. Geol Soc Lon Memoirs 47:193–204
Singha P, Dewangan P, Kamesh Raju KA, Aswini KK, Ramakrushana Reddy T (2019) Geometry of the subducting Indian Plate and local seismicity in the Andaman Region from the passive OBS experiment. Bull Seismol Soc Am 109:797–811
Smith WHF, Sandwell DT (1997) Global sea floor topography from satellite altimetry and ship depth soundings. Science 277:1956–1962
Spudich P, Orcutt J (1980) A new look at the seismic velocity structure of the Oceanic-Crust. Rev Geophys 18:627–645
Srijayanthi G, Kumar MR, Sirisha T, Sushini K, Srihari Prasad G, Solomon Raju P, Singh A, Rao NP (2012) The ISLANDS network in the Andaman-Nicobar subduction zone. Seismol Res Lett 83:686–696
Subarya C, Chlieh M, Prawirodirdjo L, Avouac JP, Bock Y, Sieh K, Meltzner AJ, Natawidjaja DH, McCaffrey R (2006) Plate-boundary deformation associated with the great Sumatra-Andaman earthquake. Nature 440:46–51

Subrahmanyam C, Gireesh R, Chand Shyam, Kamesh Raju KA, Gopala Rao D (2008) Geophysical characteristics of the Ninetyeast Ridge-Andaman island arc/trench convergent zone. Earth Planet Sci Lett 266:29–45

Takahashi N, Kodaira S, Tatsumi Y, Kaneda Y, Suyehiro K (2008) Structure and growth of the Izu-Bonin-Mariana arc crust: 1. Seismic constraint on crust and mantle structure of the Mariana arc-back-arc system. J Geophys Res 113:B01104

Tamaki K, Pisciotto K, Allen J et al (1990) Background, objectives, and principal results, ODP Leg 127, Japan Sea Proc. ODP Init Rep 127:5–33

Taylor SR, McLennan SM (1985) The continental crust: its composition and evolution. Blackwell Scientific Publication Carlton, 312p

Tatsumi Y, Shukuno H, Tani K, Takahashi N, Kodaira S, Kogiso T (2008) Structure and growth of the Izu-Bonin-Mariana arc crust: 2. Role of crust-mantle transformation and the transparent Moho in arc crust evolution. J Geophys Res 113:B02203

Tripathi S, Nagendran G, Karthikeyan M, Tripathy K, Varghese S, Raghav SSR (2018) Morphology of submarine volcanic seamounts from inner volcanic arc of Andaman Sea. Indian J Geosci 71(3):451–470

Vigny C, Socquet A, Rangin C, Chamot-Rooke N, Pubellier M, Bouin M, Bertrand G, Becker M (2003) Present-day crustal deformation around Sagaing fault, Myanmar. J Geophys Res 108(B11):2533. https://doi.org/10.1029/2002JB001999

Wessel P, Smith WHF (1995) New version of the generic mapping tools released. EOS Trans AGU 76:329

Yadav R, Tiwari VM (2018) Lithospheric density structure of Andaman subduction zone from joint modelling of gravity and geoid data. Indian J Mar Sci 47:931–936

Miocene to Pleistocene Palaeoceanography of the Andaman Region: Evolution of the Indian Monsoon on a Warmer-Than-Present Earth

Wolfgang Kuhnt, Ann E. Holbourn, Janika Jöhnck and Julia Lübbers

Abstract The Andaman Sea is ideally situated to assess the sensitivity of the Indian Monsoon to insolation forcing and to changes in climate boundary conditions such as global ice volume and greenhouse gas concentrations on a warmer-than-present Earth. Sediment archives from this region record the monsoonal discharge of water and sediment from mountainous areas at the southern slope of the Eastern Himalayas, the western slopes of the Arakan Mountains and Indo-Burman Ranges and from the catchment of the Irrawaddy, Sittang and Salween Rivers. New sediment cores recovered during International Ocean Discovery Program (IODP) Expedition 353 (iMonsoon) provide the first complete millennial-scale resolution record of the Indian Monsoon over the late Neogene, when major changes in the Indian and East-Asian monsoonal subsystems occurred. Initial results from sediment natural gamma ray core-logging and foraminiferal stable isotope analyses indicate that a long-term increase in physical weathering and erosion coincided with a prolonged late Miocene cooling trend between ~7 and 5.5 Ma. Monsoonal erosion remained intense during the subsequent warming episode between 5.5 and 5.3 Ma, probably associated with a northward shift of the monsoonal rain belt. Ocean mixed layer oxygen isotope records indicate freshening of the Andaman Sea during late Miocene warm isotope stages as during Pleistocene interglacials. Knowledge of this past evolution of freshwater and sediment discharge from one of the core areas of Indian Monsoon precipitation is essential to better predict trends and consequences of monsoonal spatial variability with future climate change.

Keywords International Ocean Discovery Program · Miocene · Palaeoceanography · Indian monsoon · Andaman Sea

---

W. Kuhnt (✉) · A. E. Holbourn · J. Jöhnck · J. Lübbers
Institute of Geosciences, Christian-Albrechts-University, Kiel, Germany
e-mail: Wolfgang.Kuhnt@ifg.uni-kiel.de

J. S. Ray and M. Radhakrishna (eds.), *The Andaman Islands and Adjoining Offshore: Geology, Tectonics and Palaeoclimate*, Society of Earth Scientists Series,
https://doi.org/10.1007/978-3-030-39843-9_13

1 Introduction

Despite half a century of intense research, the physical forcing processes that drive the long-term variability of monsoonal precipitation within the Earth's strongest hydrological regime are still a matter of vigorous debate. These controversies largely stem from the fact that the Indo-Asian Monsoon consists of a series of regional monsoonal subsystems (Arabian Sea Monsoon, Indian Monsoon, East Asian Monsoon, Australian Monsoon), which exhibit differing intensity, temporal variability and sensitivity to driving mechanisms. The extent to which these subsystems are driven by an underlying single, large-scale, physical process has recently been the focus of discussions related to the evolving concept of a "global monsoon" (Wang and Ding 2008; An et al. 2015; Wang et al. 2017). These differing views largely stem from seasonal and regional differences in the manifestation of the monsoon (e.g., heavy rain over parts of the Indian subcontinent and the Bay of Bengal versus strong winds over the western Arabian Sea and parts of the open Indian Ocean) and from different assessments of the relative importance of external (orbital insolation) versus internal (boundary conditions such as ice volume, atmospheric greenhouse gases and topography) forcing factors for regional monsoon subsystems (e.g., Ruddiman 2006; Clemens and Prell 2007). Monsoonal records extending over orbital to tectonic timescales, in particular over warmer intervals of Earth's climate history characterized by fundamentally different internal boundary conditions, are crucial to evaluate the role of internal forcing factors in controlling the evolution of the monsoon.

To date, only a handful of records, generally at relatively low-resolution, retrace the deep time evolution of the Indian Monsoon. Micropalaeontological data from the Arabian Sea and from the central East Asian continent initially suggested that the intensity of monsoonal winds increased substantially at ~8–7 Ma (Kroon et al. 1991; Prell et al. 1992; Prell and Kutzbach 1997; An et al. 2001). The coeval emergence and expansion of arid-adapted C4 floras in South and Central Asia, indicative of reduced precipitation or decreased atmospheric CO_2 concentrations (e.g., Quade and Cerling 1995; Cerling et al. 1997; Huang et al. 2007), were also linked to an intensification of the dry boreal winter monsoon over the Asian continent. These changes were associated with concurrent changes in the erosional regime and tectonic evolution of the Himalayas and Tibetan Plateau during the late Miocene-Pliocene (Molnar et al. 1993, 2010; Molnar 2005). However, in contrast to the widely-held view of a late Miocene intensification of the Indo-Asian Monsoon, a much earlier onset of the monsoon has also been proposed, based on changes in Central Asian climate, weathering regime and related sediment fluxes (Guo et al. 2002; Sun and Wang 2005; Clift et al. 2008).

The mainstream view of a direct linkage between the uplift and expansion of the Tibetan Plateau, the intensification of the Indo-Asian Monsoon, global cooling and atmospheric pCO_2 decline has been challenged in recent years by new observations and refined palaeoclimate models (see discussion in Qiu 2013). For example, Rodriguez et al. (2014) suggested that the ~8–7 Ma intensification of monsoonal winds in the Arabian Sea, inferred on the basis of increased abundances of *Globigerina bulloides*, was an artifact of enhanced preservation related to uplift of the Owen Ridge and that the importance of the uplift of the Tibetan plateau was substantially overestimated in early climate models (Boos and Kuang 2010; Molnar et al. 2010; Liu and Yin 2011; Molnar and Rajagopalan 2012). In particular, the "classic" argument, that the rise of the Tibetan plateau near 10 Ma led to strengthening of the Indo-Asian Monsoon (Harrison et al. 1992; Molnar et al. 1993; Prell et al. 1992; Prell and Kutzbach 1997) has ben challenged by more recent studies, associating a higher eastern Tibet with decreasing rainfall over NW India between 10 and 7 Ma (Molnar and Rajagopalan 2012). A fundamental shortcoming in the debate on the late Miocene-Pliocene onset or intensification of the monsoon versus longer-term perspectives of variability is the lack of records from the core area of the Indian Monsoon in the Bay of Bengal. Well-dated proxy records of precipitation, salinity, terrigenous runoff and productivity from the Bay of Bengal and Andaman Sea are essential to monitor the intensification of the Indian Monsoon in its core region, where increased rainfall, and not strengthened wind, characterizes the monsoon.

IODP Expedition 353 (iMonsoon) targeted the monsoonal precipitation signal in its core geographic region of influence, including the western margin of the Andaman Sea and investigated long-term monsoonal circulation changes through the drilling of high-resolution sedimentary archives. The Bay of Bengal and Andaman Sea presently receive one of the largest freshwater and sediment discharges in the world from three major river systems that are primarily influenced by the Indian Monsoon. This region is, therefore, ideally located to monitor past monsoonal trends in relation to changes in insolation forcing and in climate boundary conditions such as orography, global ice volume and greenhouse gas concentrations. Two sites (U1447, 10° 47.40′ N/93° 00′ E, 1391 m water depth and U1448, 10° 38.03′ N/93° 00′ E, 1098 m water depth) in the vicinity of the Andaman Islands recovered extended upper Miocne to Pliocene successions, which provide an outstanding opportunity to assess the sensitivity of the Indian Monsoon to internal and external forcings on a warmer-than-present Earth. Here, (1) we integrate new isotope data from Sites U1447 and U1448 with proxy runoff data to track the evolution of the Indian Monsoon over the late Miocene to earliest Pliocene, an interval marked by important reversals in cooling and warming trends, and (2) we compare monsoonal trends on a Pleistocene bipolar glaciated world over the last million years to monsoonal variability on a warmer Earth in the late Miocene.

2 Neogene Sedimentary Archives of the Andaman Region: A Key to Reconstruct the Indian Palaeomonsoon

The Andaman Island arc system results from the oblique subduction of the Indo-Australian plate beneath the Eurasian plate. Subsequent stretching and rifting of the overriding plate since the early Miocene led to formation of an active spreading center in the deepest portion of the Andaman Sea during the late Miocene-early Pliocene (Raju et al. 2004; Curray 2005; Singh et al. 2013; Morley 2016). An accretionary wedge complex formed in the west of the spreading center, consisting of a series of shallower basins bordered by back thrust faults within the accretionary wedge. Garzanti et al. (2013) argued for distinctly different source areas for the Andaman flysch on the western side of the Andaman Islands. These authors suggested that it was associated with the Bengal Fan and consisted of re-deposited material from the Oligocene Irrawaddy deep sea fan (Allen et al. 2008) that was offscraped in the rear of the accretionary prism. In contrast, the terrigenous sediment input to the eastern margin of the Andaman Islands originates directly from river discharge from the Indo-Burman mountain ranges and the Irrawaddy-Salween river system and did not change since the early Miocene.

IODP Expedition 353 Sites U1447 and U1448 targeted the Nicobar-Andaman Basin between the Diligent and Eastern Margin faults, which provided a unique opportunity to recover undisturbed sedimentary successions in intermediate water depths deposited well above the calcite lysocline. Site U1447 is located at 10° 47.40′ N/93° 00′ E in 1391 m water depth, ~45 km east of Little Andaman Island, within a small basin on the eastern flank of a rise separating north-south oriented basins associated with the Eastern Margin and Diligent fault zones. Seismic profiles indicate that ~740 m of sediments overlay the accretionary wedge complex at this location (Fig. 1). The main drilling objective at this site was to recover Miocene to Holocene sediments in order to reconstruct changes in sea surface salinity and runoff associated

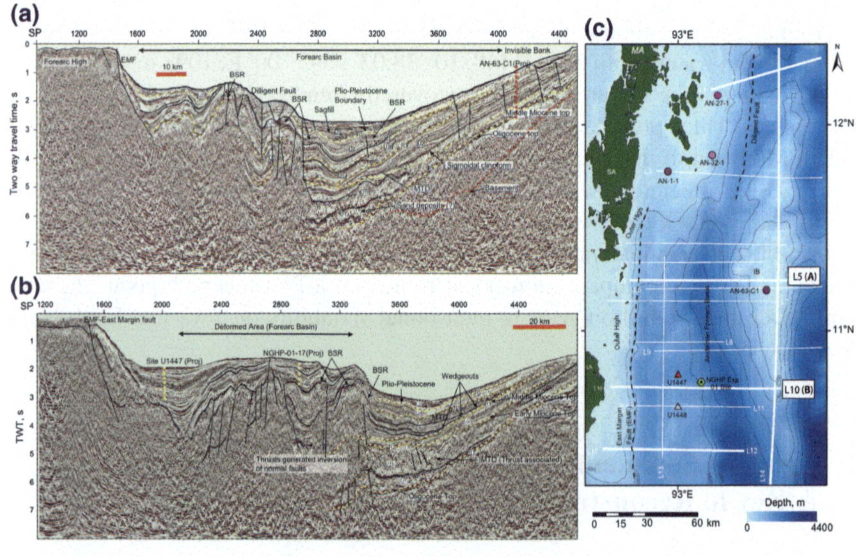

Fig. 1 Seismic stratigraphy of the eastern margin of the Andaman Ridge (from Pandey et al. 2017), under Creative Commons Attribution 4.0 International License (http://creativecommons.org/licenses/by/4.0/). E-W seismic line L5 at ~11° 45′N (**a**), E-W seismic line L10 at ~10° 45′N (**b**), Position of seismic lines on the satellite bathymetry map after Sandwell and Smith (2009) (**c**)

with local summer monsoon rainfall within the catchments of the Irrawaddy and Salween Rivers at tectonic to suborbital timescales. Site U1448 is located at 10° 38.03′ N/93° 00′ E in 1098 m water depth, ~44 km east of Little Andaman Island on a rise to the west of the deeper Site U1447. Seismic profiles indicate that ~420 m of sediments overlay the accretionary wedge complex at this location (Fig. 1). Site U1448 was selected due to its elevated position, protecting it from the influence of turbidites, which affect sedimentation through the Pliocene and lower Pleistocene at Site U1447.

Today, the terrigenous input to this part of the Andaman Sea stems dominantly from the Irrawaddy and Salween Rivers (Colin et al. 1999; Ali et al. 2015). Recent provenance studies using clay minerals and Sr- and Nd-isotopes indicate that this main sediment source has not fundamentally changed since the late Miocene (France-Lanord et al. 1993; Cawthern et al. 2014; Phillips et al. 2014). The variability in the amount and composition of this terrigenous discharge is linked to monsoon intensity in the source area through erosion, chemical weathering and transport at millennial, orbital and tectonic timescales. Calibration studies of core top sediments have shown that different elemental ratios can reliably trace different weathering regimes and that ratios of mobile versus immobile elements track the intensity of chemical weathering and/or erosion in the source areas on orbital and suborbital timescales (Liu et al. 2008; Wan et al. 2007, 2009). The Andaman Sea, thus, offers a unique opportunity to monitor changes in monsoon strength, weathering rates and transport of particulate materials to the ocean, in particular over the critical late Miocene to Pliocene interval, when major re-organizations of monsoonal patterns are documented in other monsoonal regions (e.g., Clift and Plumb 2008; Holbourn et al. 2018).

3 Modern (Late Pleistocene to Recent) Oceanographic and Climatic Setting

3.1 The Andaman Area and SE Asian Margin of Burma: A Core Area of Indian Monsoon Precipitation

The Andaman Sea is influenced by three major river systems, which give rise to one of the largest freshwater and sediment discharges in the world. The Ganges-Brahmaputra River system is ranked fourth among the world's rivers as a source of freshwater to the oceans (1120 km^3/yr) and supplies more sediment to the ocean than any other river in the world (1060 Mt/yr) (Milliman and Farnsworth 2011). The Irrawaddy and Salween Rivers together discharge up to 640 km^3/yr of freshwater and deliver up to 540 Mt/yr sediment (Figs. 2 and 3) (Milliman and Farnsworth 2011). The discharge of water and sediment to the Andaman Sea mainly originates from mountainous areas at the southern slope of the Eastern Himalayas (northern part of

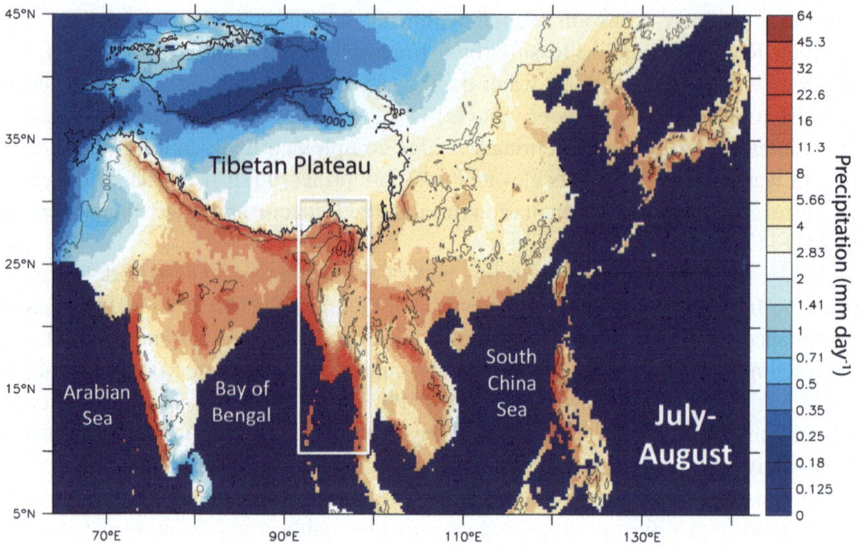

Fig. 2 Indian Summer Monsoon precipitation along the northeast margin of the Indian Ocean (July–August mean precipitation from APHRODITE (mm/day 1951–2007 modified after Day et al. 2015)). The white rectangle indicates the drainage area for runoff to the Andaman Sea

the Irrawaddy catchment), the western slopes of the Arakan Mountains and Indo-Burman Ranges and from the catchment of the Salween and Sittang Rivers, where orography creates one of the most important precipitation domains of the Indian Monsoon (Fig. 2) (Day et al. 2015).

During the last glacial cycle, sediment discharge from freshly eroded sources increased during intervals of Northern Hemisphere cooling such as the Younger Dryas (YD), Last Glacial Maximum (LGM) and Heinrich stadials of Marine Isotope Stage (MIS) 3 (Awasthi et al. 2014). Awasthi et al. (2014) suggested that this change in sediment provenance was linked to a southward shift of the main summer monsoon precipitation belt from the Himalayas during Northern Hemisphere cooling, without any major reduction in monsoon intensity. An alternate hypothesis is that enhanced physical erosion induced by a cooler and dryer climate and intensified glacial erosion in the highland of the Irrawaddy River may have produced larger volumes of unaltered sediment (Colin et al. 2006). The lower glacial sea level additionally increased the transport efficiency of rivers in the lower part of their catchment, thus contributing to increased discharge of freshly eroded unaltered sediment from the Himalayas into the Andaman Sea.

Over the last 280 kyr, clay mineralogy and chemical weathering proxy records in the deep central Andaman Sea were characterized by prominent 23 kyr periodicity (Colin et al. 1999; Cao et al. 2015), which is in phase with the Northern Hemisphere insolation driven climate in the lower Indo-Burman ranges and Irrawaddy River plain. Strong chemical weathering in the Irrawaddy plain must have dominated the terrigenous discharge during periods of intensified monsoon precipitation

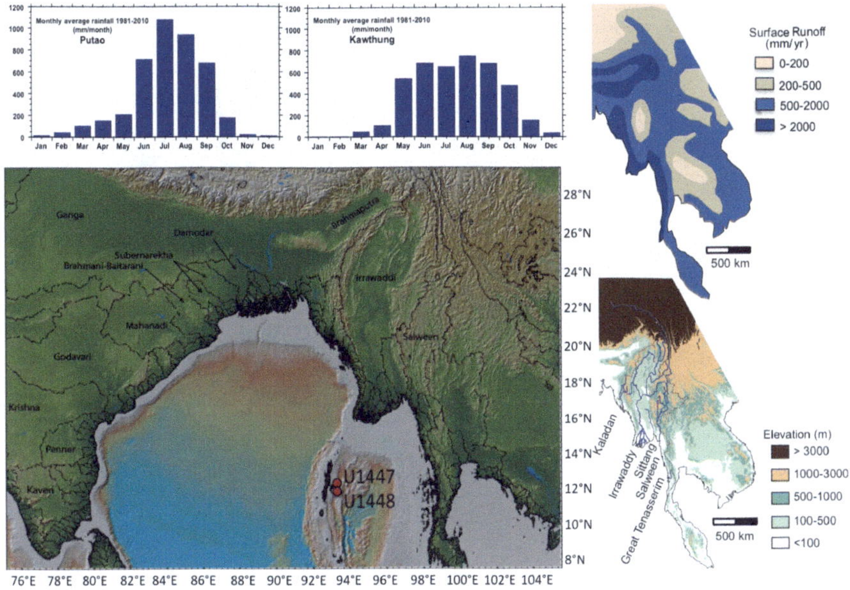

Fig. 3 Seasonal precipitation and runoff patterns of the Irrawaddy and Salween Rivers and location of IODP Sites U1447 and U1448 in relation to the main drainage basins on the Asian continent. Today, sediment and freshwater runoff at Sites U1447 and U1448 originates mainly from the Irrawaddy and Salween drainage basins on the Asian continent. Average monthly rainfall at Putao in Northern Myanmar and Kawthung at the southern end of Myanmar for the years 1981–2010 are from Aung et al. (2017). The extent, elevation and annual surface runoff of drainage basins are redrawn and simplified after Milliman and Farnsworth (2011)

in the lower Burman Ranges relative to input from the high Burman ranges, where physical erosion dominates. The northern Andaman Sea is characterized by unusually high suspended sediment concentrations due to high and seasonal sediment discharge from the Irrawaddy, Sittang and Salween Rivers and intense resuspension of sediments by strong tidal currents. This suspended sediment load is dispersed into the deep Andaman Sea as a mid-water nepheloid layer, which acts as the main source of sediment supply to the deep Andaman Sea (Ramaswamy et al. 2004). Sediment dispersed from the Bengal shelf and western and southern margins of Myanmar by intensified westerly surface currents during the winter season provides a supplementary source of terrigenous input from the Indo-Burman ranges into the Andaman Sea (Fig. 4). This input was already active during the LGM and increased in intensity during the last deglaciation (Awasthi et al. 2014). This direct coupling between monsoonal erosion and transport from a major hotspot of summer monsoonal precipitation in the Himalayan and Indo-Burman mountain ranges with its discharge into the sediment sink makes the Andaman Sea a prime area for monitoring the variability of the Indian Monsoon on suborbital, orbital and tectonic timescales.

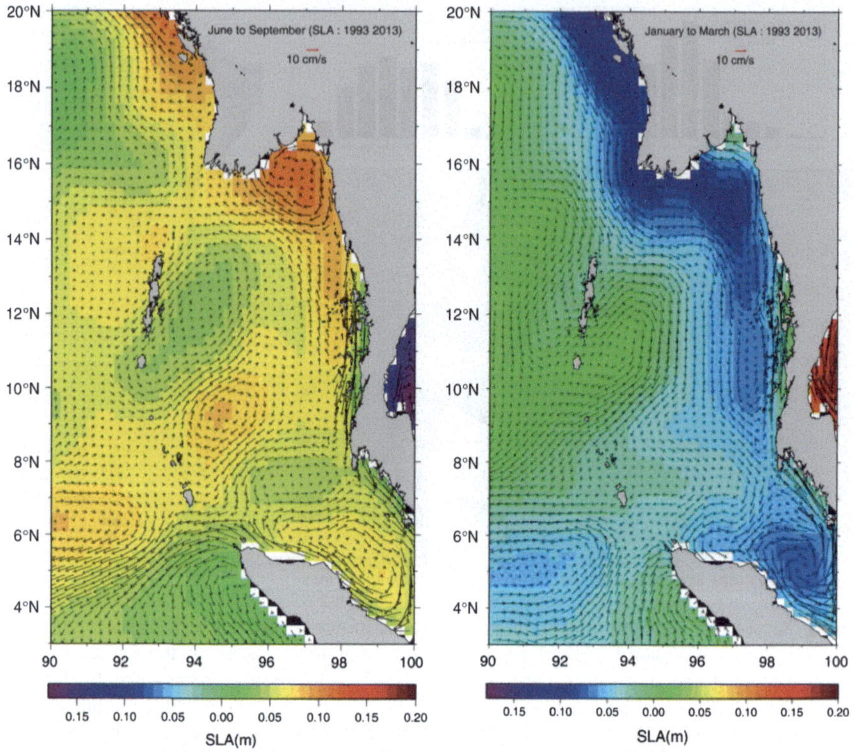

Fig. 4 Surface circulation in the Andaman Sea during summer (left) and winter (right) obtained by deriving geostrophic currents based on monthly mean climatology of sea level anomaly during the period 1993–2013 from BOBLME 2015 (Brewer et al. 2015). Note the important inflow of cool, low-salinity waters from the Bay of Bengal through the northern passage into the Andaman Sea and the outflow from the Andaman Sea through the middle passage into the southern Bay of Bengal during winter. In contrast, during the boreal summer months dominates the outflow of Andaman Sea water through the southern Great Channel into the equatorial Indian Ocean

3.2 Circulation and Water Mass Properties in the Andaman Sea

The Andaman Sea is connected by a broad passage at 6° N (Great Channel) to the open equatorial Indian Ocean and by two main passages at 10° N (Ten Degree Channel) and 15° N (Preparis Channel) to the Bay of Bengal (Fig. 4). Circulation through these passages affects the dynamic interaction between near-surface water masses of the eastern equatorial Indian Ocean and the Bay of Bengal (Chatterjee et al. 2017). Whereas the mean coastal circulation within the Andaman Sea and around the Andaman and Nicobar Islands is strongly influenced by remote atmospheric forcing from the equator, the seasonally changing current directions and intensities (Figs. 4 and 5) are to a large extent driven by local winds (Chatterjee et al. 2017).

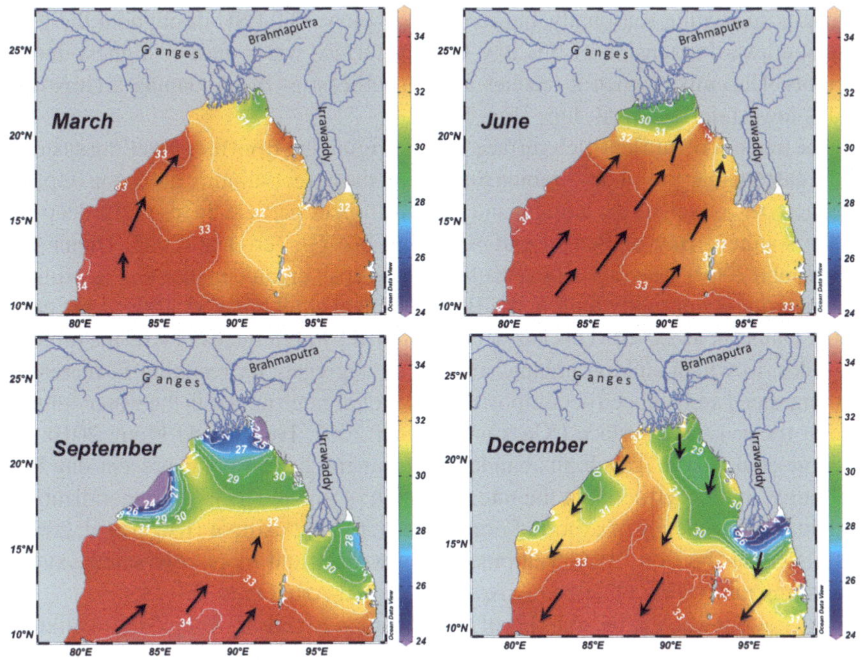

Fig. 5 Bay of Bengal and Andaman Sea salinity in September and December based on WOA 2013 data (Talley 2013) and simplified prevailing wind forcing based on Hastenrath and Lamb (1979) and Varkey et al. (1996)

Sea surface temperature and salinity in the Andaman Sea are largely controlled by the marked seasonality in monsoonal wind, precipitation and runoff induced by the switch between wet southwest (boreal summer) monsoon and dry northeast (boreal winter) monsoon (Fig. 5) (Varkey et al. 1996). Lowest salinities (~32 Practical Salinity Units (PSU)) in the Andaman Sea occur during peak monsoon runoff from smaller southern Myanmar rivers in September.

Surface salinity and ocean currents in the equatorial northeastern Indian Ocean also reflect the seasonal wind forcing by the southwest monsoon and the northeast monsoon (Schott et al. 2009). This monsoon driven circulation is characterized by a marked seasonal contrast between dominant freshwater export from the Bay of Bengal and northern Andaman Sea during boreal winter and eastward transfer of saline and nutrient-rich waters from the Arabian Sea during boreal summer (Fig. 5) (Gordon et al. 2016; Goswami et al. 2016). The influx of nutrient-rich, high-salinity water originating from a vigorous westerly current (Southwest Monsoon Current) that carries high-salinity waters eastward around the tip of India into the southern Bay of Bengal is reflected in the regional salinity and productivity patterns during July, August, and September (Fig. 5). Only the uppermost 100 m of the tropical eastern Indian Ocean in the Bay of Bengal and Andaman Sea consist of a low-salinity water mass derived from river runoff from India and Indochina with surface salinity strongly fluctuating seasonally, but remaining below 33.0 PSU throughout the year. The southward extension of this low-salinity surface water mass is highest during October–December, when it reaches as far southeast as the Andaman archipelago and is lowest during April–June (Fig. 5).

The transfer of nutrient-rich surface waters originating from the upwelling systems at the eastern margins of the Arabian Sea during the boreal summer monsoon impacts productivity as well as the amount and composition of particle fluxes to the deep sea in the southern Bay of Bengal and Andaman Sea (Unger et al. 2003; Unger and Jennerjahn 2009). The ratio of particulate deep-ocean organic matter flux to primary productivity is unusually high in the Bay of Bengal (Ittekkot et al. 1991; Lutz et al. 2002; Gauns et al. 2005). This has been explained by high fluxes of riverine mineral particles, to which sinking organic matter can aggregate to, resulting in increased sinking speed and reduced decomposition rates of organic matter in the water column and at the sediment surface (Klaas and Archer 2002; Iversen and Ploug 2010; Le Moigne et al. 2013). This high abundance of mineral ballast in the present-day Bay of Bengal influences not only the rate and depth of organic matter remineralization, and thus the efficiency of export of organic matter to the seafloor (Kwon et al. 2009), but also contributes to the development of a weaker and more contracted oxygen minimum zone (OMZ) than in the Arabian Sea (Al Azhar et al. 2017).

The bottom water at the margins of the Andaman Islands is occupied by a mixture of Indian Deep Water (IDW) originating from the equatorial Indian Ocean and local thermocline water. Two water masses occupy the thermocline of the tropical Indian Ocean in the Bay of Bengal and Andaman Sea, Indian Central Water (ICW), originating from the Southern Hemisphere Indian Ocean and Indonesian Throughflow Water (ITW), which is derived from North Pacific Intermediate Water modified by intense mixing during its passage through the Indonesian archipelago. The oxygen content of these intermediate water masses becomes strongly reduced during their transfer into the Bay of Bengal and Andaman Sea and the admixing of ITW results in some freshening of the ICW along its path into the Bay of Bengal. However, the low oxygen levels indicate an extremely low renewal rate for the thermocline waters of the Bay of Bengal (Tomczak and Godfrey 2005). Intermediate waters in the Andaman Sea are strongly influenced by the OMZ in the Bay of Bengal resulting in concentrations of dissolved oxygen below 50 μmol/kg and δ^{13}C values below zero (Fig. 6) (Talley 2013).

Analysis of Andaman Sea surface sediments show that planktic foraminifers are abundant and well preserved in water depth above ~1800 m (>100,000 individuals/gram) and decrease to <100 individuals/gram in water depth below 3000 m (Frerichs 1971), indicating that the foraminiferal lysocline hovers close to 3000 m. Pteropod preservation profiles in sediments off Middle Andaman Island indicate a sharp decline of pteropods between 1108 and 1147 m water depth, while planktonic foraminifers remain abundant and well-preserved (Bhattachjarjee 2005). These sediment surface distribution data also indicate a deepening of the aragonite compensation depth from the northern Andaman Sea to the south, suggesting that reduced carbonate production and carbonate ion concentration and the resulting shallowing

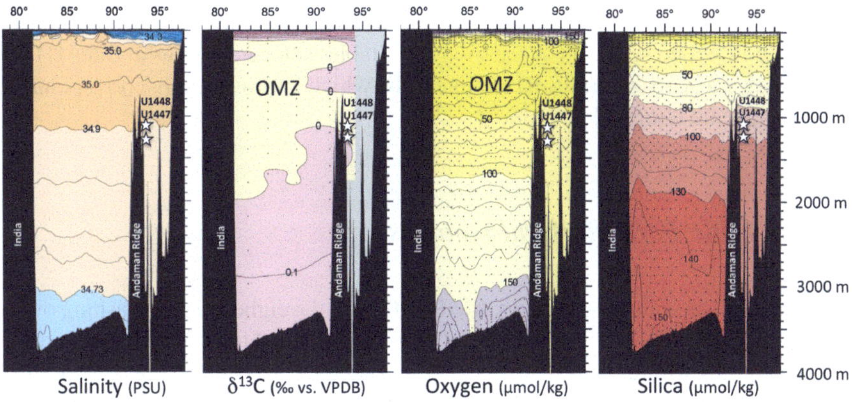

Fig. 6 Bay of Bengal and Andaman Sea salinity, $\delta^{13}C$, oxygen and silicate profiles along WOCE section I01 at 9° N (Talley 2013). Note the increasingly fresh, $\delta^{13}C$ and oxygen-enriched surface waters towards the Andaman Sea

of the aragonite compensation depth in the northernmost Andaman Sea are related to the freshwater discharge of the Irrawaddy River (Bhattachjarjee 2005).

4 Materials and Methods

4.1 Coring and Sampling Strategy

Three holes were drilled at Site U1447, ranging in depth from 24.4 to 738.0 m below seafloor (mbsf). In Hole A, 29 cores were retrieved with the Advanced Piston Corer (APC) down to 261.0 mbsf, with a total core recovery of 268.69 m. Below this interval, the half-length APC system recovered 15 cores down to 67.8 mbsf and 44 cores were cut with the Extended Core Barrel (XCB) system down to 409.2 mbsf. Holes U1447B and U1447C were only drilled with the APC system, down to 24.45 and 161.65 m.

At Site U1448, three holes were drilled down to 420.60 mbsf (U1448A), 358.22 mbsf (U1448B) and 34.71 mbsf (U1448C). Hole A was drilled to 204.32 mbsf with the APC system, then down to 343.60 mbsf with the half-length APC system and to a final depth of 420.60 mbsf with the XCB system. Hole B was drilled down to 177.71 mbsf with the APC system, then to 358.22 mbsf with the half-length APC system. A complete splice consisting of Hole A and B was constructed down to ~345 meters composite depth, whereas the deeper part of the U1448 splice consists of a single hole (U1448A) drilled with the XCB system. Core recovery in this lower part was extremely good (101%) with minimal gaps between cores and complete recovery of the stratigraphic succession. Postcruise sampling of the upper Miocene and lower Pliocene sediment successions for stable isotope analyses at Sites U1447 and U1448 was carried out in 10 cm intervals, which approximately corresponds to millennial-scale temporal resolution.

4.2 Benthic Foraminiferal Isotopes

Samples of 30 cc volume were oven dried at 40 °C and weighed before washing over a 63 μm sieve. Residues were oven dried at 40 °C on a sheet of filter paper, then weighed and sieved into different size fractions. We measured $\delta^{18}O$ and $\delta^{13}C$ in the epifaunal benthic foraminifers *Cibicidoides wuellerstorfi* and/or *Cibicidoides mundulus*. Well-preserved tests were broken into large fragments, cleaned in ultrapure ethanol in an ultrasonic bath, decanted and again dried at 40 °C. Stable carbon and oxygen isotopes were measured with a Finnigan MAT 253 mass spectrometer at the Leibniz Laboratory, University of Kiel. The instrument is coupled on-line to a Carbo-Kiel Device (Type IV) for automated CO_2 preparation from carbonate samples for isotopic analysis. Samples were reacted by individual acid addition (99% H_3PO_4 at 75 °C). On the basis of the performance of international and lab-internal standard carbonates, the precision is better than ±0.09‰. Paired measurements in Miocene samples from ODP Sites 1146 and 1237 previously indicated no significant offset in $\delta^{18}O$ and $\delta^{13}C$ between *C. wuellerstorfi* and *C. mundulus* (Holbourn et al. 2007, 2018). Results were calibrated using the NIST (National Institute of Standard and Technology, Gaithersburg, Maryland) carbonate isotope standard and NBS (National Bureau of Standard) 19 and NBS 20, and are reported on the Vienna PeeDee Belemnite (VPDB) scale.

5 Neogene Seismic Stratigraphy and Initial Drilling Results from the Andaman Forearc Basin

5.1 Upper Miocene to Pleistocene Sediment Sequence

Sediments recovered at Site U1447 consist of upper Miocene to Pleistocene hemipelagic clays, which include a significant biogenic component as well as thin, fine-grained distal turbidites, primarily composed of redeposited pelagic carbonate material (Clemens et al. 2016). Four distinct lithostratigraphic units were discriminated during Expedition 353. Unit I (0–126.00 m below sea floor (mbsf)) is composed of upper Pleistocene greenish gray clayey nannofossil oozes with foraminifers and of foraminifer-rich nannofossil oozes with clay and only rare occurrences of distal turbidites. Unit II (126.00–329.12 mbsf) consists of upper Pliocene to upper Pleistocene greenish gray foraminifer-rich nannofossil oozes with clay. This unit includes numerous light gray, foraminifer-rich, calcitic turbidites of varying thicknesses as well as bioclastic-rich layers with authigenic carbonate and foraminifers dominating in the sand or silt size fraction. Unit III (329.12–489.80 mbsf) is composed of upper Miocene to lower Pliocene greenish gray clayey nannofossil oozes to calcareous oozes with glauconite. Unit IV (489.80–740.46 mbsf) consists of upper Miocene greenish gray biosilica-rich clay with varying proportions of glauconite and nannofossils, reaching back to at least 9.83 Ma and with average sedimentation rates of 6 cm/kyr.

Hole U1448A recovered a continuous hemipelagic upper Miocene to Pleistocene succession, which contains all calcareous biostratigraphic markers and is not affected by gravity deposition (turbidites or fine-grained grain flows/debris flows) (Clemens et al. 2016). The upper ~180 m of this succession consist of dark greenish gray clay with varying proportions of nannofossils, foraminifers and biosiliceous microfossils such as sponge spicules and diatoms (Unit I of Pleistocene age). This upper unit is underlain by a ~160 m thick succession of greenish gray clay with nannofossils and foraminifers and extremely low biosilica content (Unit II of earliest Pleistocene to latest Miocene age). The base of this hemipleagic succession consists of ~40 m of dark to light greenish gray clay and nannofossil-rich clay with increased content of siliceous sponge spicules and bioturbation burrows and mottled patches (Unit III of late Miocene age). This lower unit shows distinct variability in lightness related to changes in carbonate content and does not include any turbidites or volcanic ash layers. Consistent sedimentation rates at Site U1448 (5–6 cm/ky) enable high-resolution palaeoclimatic reconstructions using calcareous and organic proxy indicators of palaeotemperature and salinity in a continuous succession reaching back to the late Miocene (~6.2 Ma).

5.2 *Middle to Late Miocene Unconformity*

Reflector R3 ("Mid-Miocene Top") is a critical marker in the regional sequence stratigraphic scheme of the western margin of the Andaman Sea (Pandey et al. 2017). This reflector forms the boundary between seismic sequence U3 (basin fill, late Miocene) and the sigmoidally prograding sequence U2 (middle Miocene), which includes mass transport deposits due to the regional uplift of basinal highs (Pandey et al. 2017). These authors correlated Reflector R3 with the major sea level regression at ~11 Ma in the sequence stratigraphic scheme of Haq et al. (1987). However, recent benthic foraminiferal isotope records indicate that a major expansion of the East Antarctic Ice Sheet and an associated sea-level fall of ~50 m occurred at ~13.8 Ma, followed by a further smaller expansion of the cryosphere at ~13.1 Ma (e.g., Holbourn et al. 2005, 2018). We therefore suggest that Reflector R3 represents a combination of regional uplift due to intensified compressive tectonic activity with a major eustatic sea-level fall between 13.9 and 13 Ma.

A major middle to late Miocene unconformity was revealed at 379.11 mbsf (Interval U1448-56X-5, 44–66 cm) during IODP Expedition 353 (Figs. 7 and 8). The

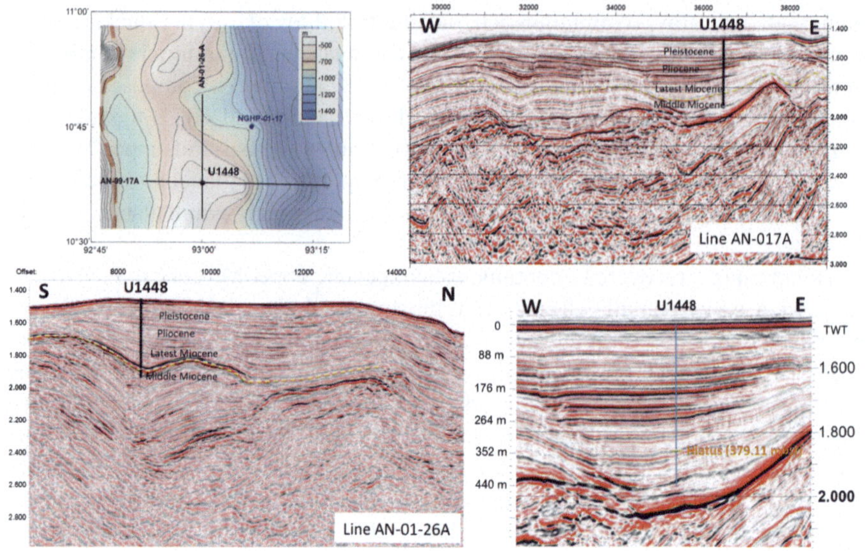

Fig. 7 Seismic expression of middle-late Miocene unconformity and hiatus at IODP Site U1448 from Clemens et al. (2016). Note the lateral N-S and E-W extension and the general northward dip of the unconformity, which may be reflected in the inclined unconformity plane in Hole U1448A

unconformity is expressed as an erosive sedimentary contact representing a major sedimentary hiatus that encompasses the time interval between >5.94 (first biostratigraphic datum above the contact, last occurrence of the calcareous nannoplankton marker *Reticulofenestra rotaria*) and <14.53 Ma (last biostratigraphic datum below the contact, last occurrence of the planktonic foraminifer *Praeorbulina sicana*) (Fig. 9) (Clemens et al. 2016). We suggest that this hiatus is the local expression of Reflector R3, indicating that tectonic changes in the late middle to early late Miocene resulted in regional deformation of the seafloor topography and to a local interruption of deep water sedimentation over periods of several million years at the eastern margin of the Andaman islands (Fig. 7).

Reflector R3 was not reached in Hole U1447A, which was terminated at 738 mbsf, close to the base of the sedimentary succession, dated as 9.83 Ma, on top of the accretionary wedge (Clemens et al. 2016). The earlier onset of pelagic drape sedimentation following the middle-late Miocene unconformity at this deeper site suggests that this unconformity may have formed on top of a northward inclined surface, which developed during a compressional phase in the late middle Miocene. However, sediments within the interval ~6.5 to 6.2 Ma, which corresponds to the onset of deep marine sedimentation at Site U1448 following the Reflector R3 unconformity were not recovered in Hole U1447A due to the loss of Core U1447A-70X. A lithological change, also suggested by a stepped increase in the potassium content of the sediment within this interval, may account for this unexpected loss.

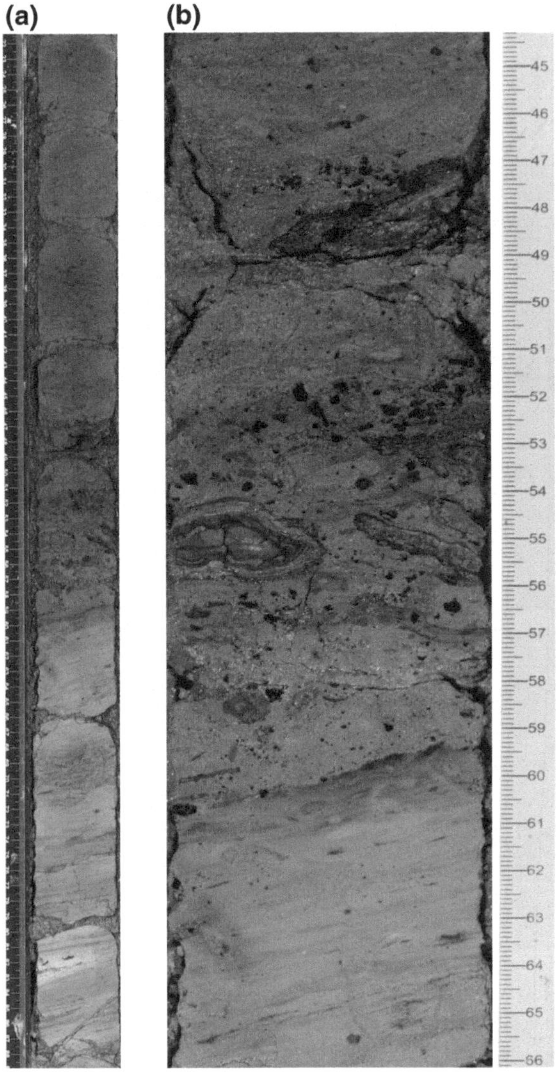

Fig. 8 Middle to late Miocene unconformity in Hole U1448A-56X-5, 44-66 cm (Clemens et al. 2016): sharp contact between nannofossil clay with dark green glauconite and pyrite nodules and underlying clayey biosiliceous ooze (**a**). Burrows in the upper few centimeters of the underlying unit contain nannofossil clay of the overlying unit, indicating the sedimentary nature of the contact (**b**). Nannofossils in the upper unit give a late Miocene age (NN11), whereas the lower unit is of middle Miocene age (NN4)

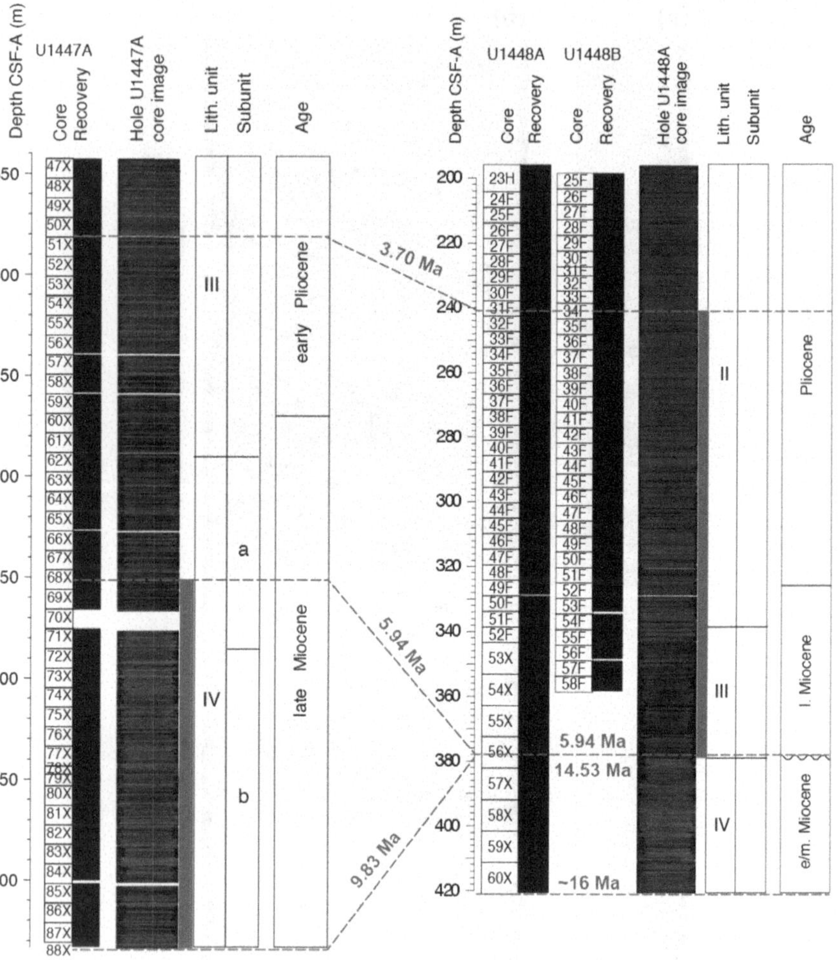

Fig. 9 Correlation of Miocene to lower Pliocene sedimentary successions of Sites U1447 and U1448. Gray bar indicates continuous spliced sequence spanning ~9.9 to 3.7 Ma. Age assignments are based on revised shipboard biostratigraphy of IODP Expedition 353 (Clemens et al. 2016, and Clara Bolton, pers. comm. 16/2/2016)

5.3 *Enigmatic Middle Miocene Sediment Sequence*

Middle Miocene sediments (seismic unit U2 of Pandey et al. 2017) recovered below the sedimentary hiatus at Site U1448 are composed of light greenish gray bioturbated biosiliceous ooze with varying proportions of clay and nannofossils. Calcium carbonate (between 20 and 30%) and TOC (between 0.2 and 0.4%) concentrations are lower than in the upper Miocene succession (Clemens et al. 2016). Benthic foraminifers are rare, poorly to moderately preserved, mainly consisting of taxa adapted to low oxygen conditions and enhanced levels of organic matter flux to the sea-floor. Sedimentologic evidence for gravity-driven mass transport deposition of large parts of this sequence includes irregular laminated bedding, microfaults and the presence of small allochthonous mottles and clasts (Clemens et al. 2016).

6 Neogene Evolution of the Indian Monsoon

6.1 Holocene Aridification Linked to Southward Shift of Inter-tropical Convergence Zone?

Most proxy records indicate a decrease in monsoonal precipitation from the Holocene Climate Optimum to the relatively cool pre-industrial late Holocene (Fleitmann et al. 2003; Gupta et al. 2003; Ponton et al. 2012). However, the intensity, spatial variability and causes of the Holocene decrease in wind and precipitation, leading to aridification of large parts of the Indian subcontinent, are still under debate. Southwest Indian records exhibit a long-term increase in summer monsoon rainfall, most probably driven by orographic precipitation in the western Ghats (Sarkar et al. 2000). In contrast, instrumental and proxy records of Holocene monsoonal precipitation over central India revealed increasing interannual variability of summer rainfall with prolonged periods of reduced monsoonal precipitation (monsoon breaks) (Gadgil 2003; Staubwasser and Weiss 2006; Sinha et al. 2011a, b). Speleothem-based precipitation reconstructions suggested that monsoon breaks may have lasted for decades or even centuries, leading to intense drought periods on the Indian subcontinent. Precipitation proxy data from speleothems in Oman show a more gradual decrease in precipitation during the Holocene associated with a southward migration of the Inter-Tropical Convergence Zone (ITCZ) (Fleitmann et al. 2003), which parallels the decrease in summer monsoon winds reconstructed offshore Oman (Gupta et al. 2003).

6.2 Pleistocene Variability on Suborbital and Orbital Timescales

A recent high-resolution proxy record of chemical alteration and provenance from the Andaman Sea indicates that the chemical weathering in the Myanmar watersheds intensified during the deglacial to mid-Holocene summer monsoon intensification (Miriyala et al. 2017). Increased silicate weathering started with the deglacial warming at ~17.7 ka and intensified at ~15.5 ka with Northern Hemisphere warming during the Bølling-Allerød. These findings are in broad agreement with deglacial salinity reconstructions, based on oxygen isotopes combined with temperature estimates from surface, thermocline and bottom dwelling foraminifers in the Andaman Sea along a latitudinal transect between 5 and 20° N (Gebregiorgis et al. 2016; Sijinkumar et al. 2016). These studies suggested that mid-Holocene (9–6 ka) mixed layer waters in the Andaman Sea were 3.8 PSU fresher than today, whereas they were essentially the same as today during the LGM, in contrast to the more saline water masses in the Bay of Bengal. Sijinkumar et al. (2016) explained the relative freshness of the Andaman Sea during the LGM through the lowered sea level, which reduced surface water interchange with the open Bay of Bengal and decreased the distance of the core locations to the river outflow. Surface- and thermocline-dwelling planktic foraminifers in the Andaman Sea and northern Bay of Bengal display increased $\delta^{18}O$ during the YD, indicating lower temperature and/or increased salinity linked to colder and dryer conditions in the hinterland due to weakened summer monsoon coupled with enhanced winter monsoon mixing of surface ocean water with deeper upper thermocline water. Interestingly, these suborbital changes in monsoonal intensity cannot be related to local insolation forcing, since Northern Hemisphere insolation consistently increased during the YD.

Orbital-scale mixed layer and thermocline temperatures and $\delta^{18}O$ records from the northern end of the Ninetyeast Ridge and Andaman Sea spanning the past 1 million years (Fig. 10) suggested that internal forcing mechanisms such as global ice volume and the temperature gradient between high and low latitudes exert major influence on the timing and spectral characteristics of monsoon precipitation proxies (Bolton et al.

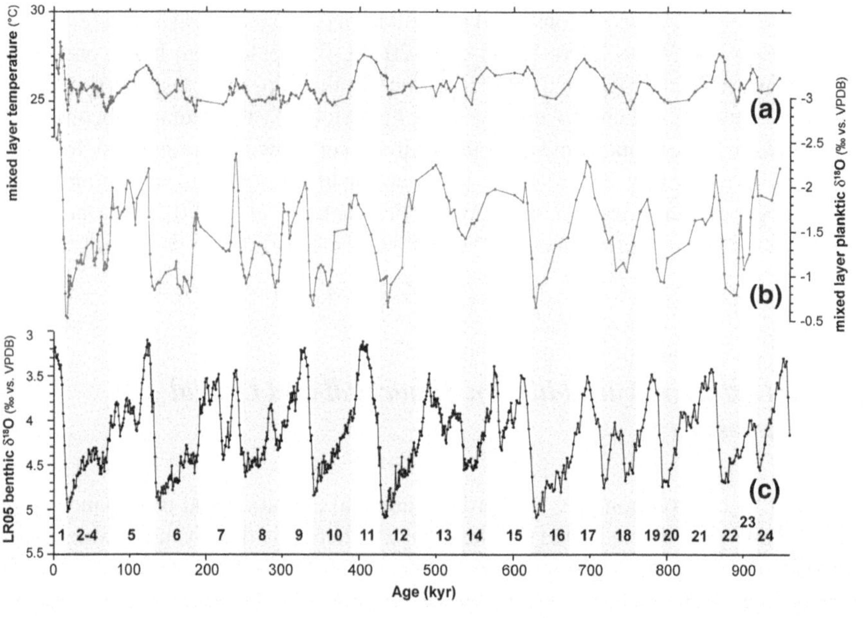

Fig. 10 Mg/Ca-derived mixed layer temperature at National Gas Hydrate Program (NGHP) Site 17 from Gebregiorgis et al. (2018) (**a**). Oxygen isotope ratios of the mixed-layer dwelling foraminifer *G. sacculifer* from Gebregiorgis et al. (2018) (**b**). Benthic oxygen isotope stack from Lisiecki and Raymo (2005) (**c**). Glacial stages are marked by gray shading

2013; Gebregiorgis et al. 2018). Spectral analyses of reconstructed seawater $\delta^{18}O$ time series indicated that upper water column stratification minima (corresponding to maxima in monsoonal wind strength) and Indian monsoon precipitation maxima lag Northern Hemisphere summer insolation maxima by ~9 kyr. This long time lag is at odds with a direct insolation forcing of the Indian summer monsoon in the Andaman Sea (Bolton et al. 2013; Gebregiorgis et al. 2018). However, precipitation maxima are in phase with obliquity minima, which coincide with intensified temperature gradients between the tropics and the Southern Hemisphere Indian Ocean, where the main moisture source for the Indian Monsoon is located (Ding et al. 2004; Clemens et al. 2008). This sensitivity of monsoonal precipitation in the Andaman area to Southern Hemisphere climate change suggests that the Indian Monsoon during the Pleistocene was mainly forced by internal processes, including changes in global ice volume and in the interhemispheric temperature gradient, rather than by direct insolation over the Asian landmass.

6.3 Late Miocene-Pliocene Evolution of the Indian Monsoon

The Miocene-Pliocene climate transition is characterized by a global cooling trend, which started at ~7.2 Ma and lasted until ~5.5 Ma (Herbert et al. 2016), culminating in two major high latitude cooling events (benthic $\delta^{18}O$ maxima TG14 and TG12) between 5.6 and 5.5 Ma (Holbourn et al. 2018). This cooling trend was followed by relatively rapid warming in the tropical northwestern Pacific Ocean between ~5.5 and ~5.3 Ma (Holbourn et al. 2018). These long-term climate fluctuations were associated with strengthening of the dry and cold East Asian winter monsoon between ~7.2 and ~5.5 Ma and intensification of the wet and warm summer monsoon between ~5.5 and ~5.3 Ma (Holbourn et al. 2018). The Andaman Sea Sites U1447 and U1448 recovered these intervals in unprecedented resolution (Fig. 9), allowing a first comparison of the response of the East Asian and Indian Monsoon subsystems to major episodes of global cooling and warming.

The lower part of Hole U1447A (Cores 68X to 88X, ~546.50–740.46 m drilled depth) provides insights into the evolution of the Indian Monsoon at a location directly influenced by the discharge of the Irrawaddy and Salween Rivers into the Andaman Sea (Fig. 11). Potassium (K) contents derived from spectral natural gamma ray (NGR) measurements in combination with shipboard biostratigraphy reveal a distinct change in sediment provenance and/or in climatic conditions and erosion in the sediment source area between ~7 and ~6 Ma. Spectral NGR logging allows to reliably estimate sedimentary contents of K from the characteristic gamma-ray energies of isotopes in the ^{40}K and ^{232}Th radioactive decay series by integrating counts over specific energy levels of the NGR spectrum (Dunlea et al. 2013; De Vleeschouwer et al. 2017). The elements Th and K are linked to the detrital clay fraction, in particular K commonly characterizes high illite content associated with dominant physical weathering and/or high erosion rates in the source area. Cross-plots of Th (ppm) versus K (%) are used in petrophysical log evaluation to discriminate between different clay types.

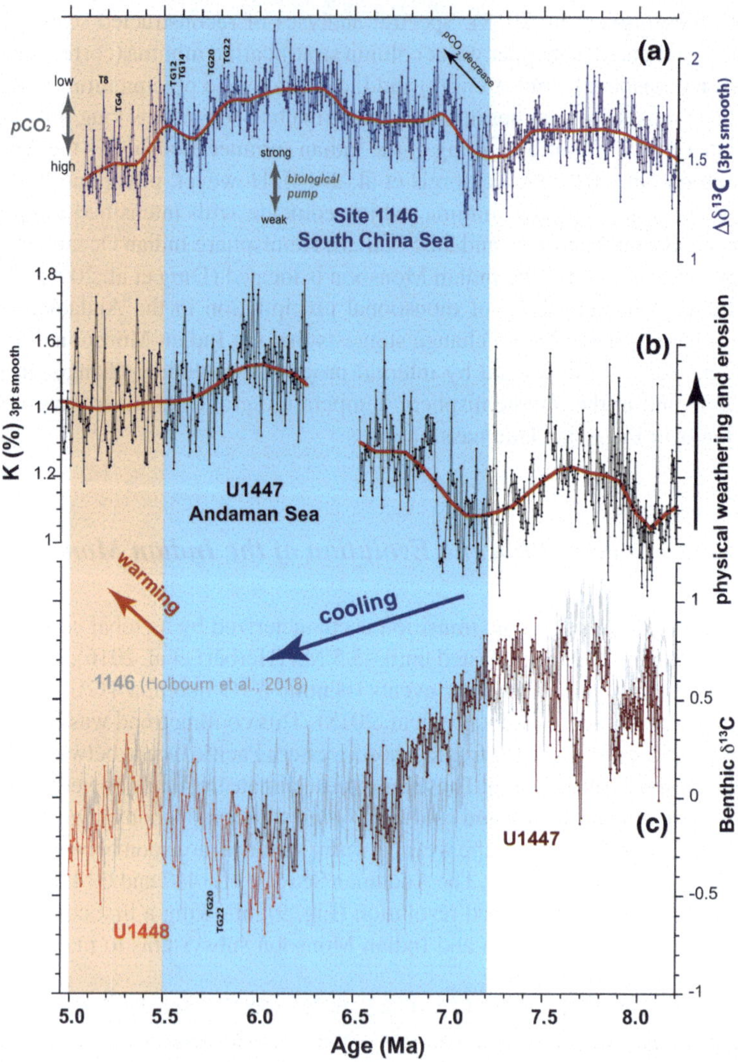

Fig. 11 Comparison of late Miocene (8.2–5 Ma) Indian and East Asian (South China Sea) Monsoon evolution. $\delta^{13}C$ gradient between mixed layer and epifaunal deep water benthic foraminifers at South China Sea ODP Site 1146 from Holbourn et al. (2018) (**a**), change in monsoonal terrigenous runoff (% K) derived from shipboard spectral gamma ray logging in Hole U1447A (**b**), and Sites U1447 and U1448 benthic foraminiferal $\delta^{13}C$ records from the Andaman Sea (**c**). Age models of Sites U1447 and U1448 are derived through correlation with ODP Sites 982 and 1146 orbitally-tuned benthic $\delta^{13}C$ records (Drury et al. 2018; Holbourn et al. 2018—gray background curve). Red smooth curves in (**a**) and (**b**) are fitted using the locally weighted least squared error (LOWESS) method

Generally, a Th/K ratio of >12 is suggested for kaolinitic clays and a Th/K ratio of >2 for illite (Schlumberger Crossplots for Porosity, Lithology and Saturation, CP19 Mineral Identification from NGS* Natural Gamma Ray Spectrometry Log).

In Hole U1447A, the change from K-depleted sediment discharge (indicating dominant chemical weathering in the source area) to K-enriched, illite-dominated discharge (from areas with prevalent physical weathering) coincides with the onset of global cooling at ~7.2 Ma (Fig. 11). The changes in the chemical composition of terrigenous runoff towards more K-rich sediments between ~7.1 and ~6.8 Ma and between ~6.5 and ~6.2 Ma may also relate to changes in erosional patterns in the sediment source area due to the onset of more seasonal monsoonal rainfall. Similar changes occur in the runoff of the Brahmaputra and Ganges Rivers towards the Bay of Bengal, where K-enriched, illite dominated terrigenous runoff is associated with increased erosion of physically weathered sediments during Pleistocene phases of strong monsoonal seasonality (Derry and France-Lanord 1997; Colin et al. 2006).

The long-term late Miocene cooling trend was also punctuated by transient Northern Hemisphere cooling events (TG isotope stages of Shackleton et al. 1995) between 6.5 and 5.5 Ma (Holbourn et al. 2018). These TG cold stages are strongly imprinted on the Andaman Sea mixed layer and intermediate water $\delta^{18}O$ (Fig. 12) and on the mixed layer Mg/Ca records. However, the amplitude of glacial-interglacial differences was substantially higher during the Pleistocene (~1.4‰ for MIS10 and MIS 11 compared to ~0.9‰ for TG21 and TG20), which likely reflects differences in Northern Hemisphere ice volume and hydrologic regimes over Asian mountain ranges during more extreme Pleistocene glacial conditions. The ~1‰ $\delta^{18}O$ difference between modern and late Miocene warm stages can only partly be explained by warmer mixed layer temperatures during the warmer Miocene, but also includes a component of increased monsoonal runoff and precipitation during periods of warmer climate.

7 Conclusions and Outlook

Climate models in combination with instrumental observations predict that over the next few centuries, global warming may significantly increase the spatial heterogeneity of monsoonal precipitation (Ashfaq et al. 2009; Loo et al. 2015). These model simulations suggest that enhanced greenhouse forcing results in suppression of summer precipitation and a delay in the onset of the Indian Monsoon over large (northern) parts of the Indian subcontinent and the Bay of Bengal, whereas the Andaman Sea and its Indo-Burman hinterland will receive enhanced precipitation (Ashfaq et al. 2009). The spliced sediment archive of Sites U1447 and U1448, recently drilled during IODP Expedition 353 (iMonsoon), will provide the first complete millennial-scale resolution mixed layer temperature and salinity records over the late Neogene in one of the core areas of the Indian Monsoon. Ongoing research on the sediment successions recovered by this expedition and other recent IODP Expeditions in the Indian Ocean and western Pacific marginal seas will improve understanding of the

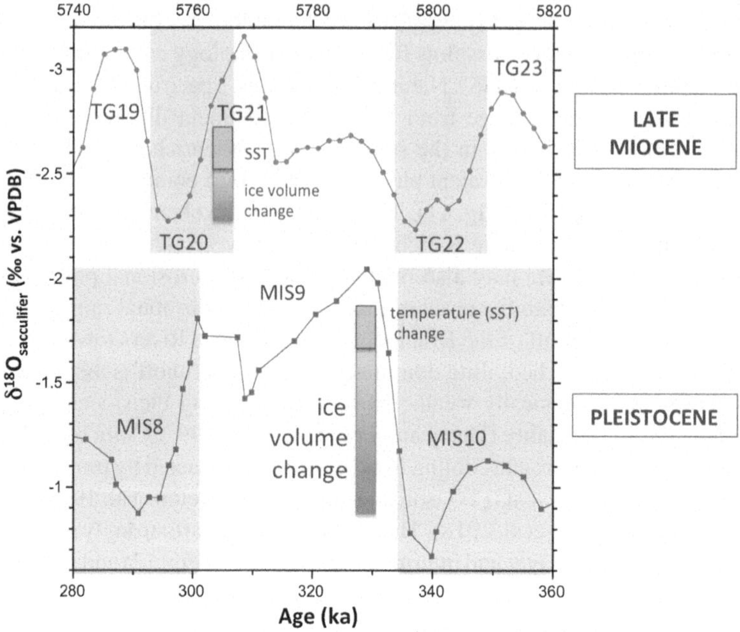

Fig. 12 Comparison of mixed layer (*Trilobatus sacculifer*) $\delta^{18}O$ across late Miocene TG Events 20 and 22 at Site U1448 (3 point moving average) and across Pleistocene MIS 9–13 at NGHP 17 (3 point moving average $\delta^{18}O$ data from Gebregiorgis et al. 2018). Change in Mg/Ca derived mixed layer temperature is based on Jöhnck et al. (unpublished data) for the late Miocene and on Gebregiorgis et al. (2018) for the Pleistocene. Ice volume change was estimated as 50 and 70% of benthic $\delta^{18}O$ variability for the late Miocene and Pleistocene, respectively. Glacial stages are shaded gray

primary forcing mechanisms on regional monsoonal subsystems and help constrain projections of future trends on a warmer Earth.

Acknowledgements This research used data and samples provided by the International Ocean Discovery Program (IODP). We are grateful to the IODP Expedition 353 shipboard party for all their efforts. Funding for this research was provided by the German Research Foundation (DFG) priority program (SPP) 527, grant KU649/35-1. We thank David De Vleeshouwer for making available potassium percentages from IODP Expedition 353 shipboard spectral natural gamma ray data and two anonymous reviewers for corrections and helpful suggestions to improve this manuscript.

References

Al Azhar M, Lachkar Z, Lévy M, Smith S (2017) Oxygen minimum zone contrasts between the Arabian Sea and the Bay of Bengal implied by differences in remineralization depth. Geophys Res Lett 44(21):11–106, 114. https://doi.org/10.1002/2017GL075157

Ali S, Hathorne EC, Frank M, Gebregiorgis D, Stattegger K, Stumpf R, Kutterolf S, Johnson JE, Giosan L (2015) South Asian monsoon history over the past 60 kyr recorded by radiogenic isotopes and clay mineral assemblages in the Andaman Sea. Geochem Geophys Geosyst 16(2):505–521. https://doi.org/10.1002/2014GC005586

Allen R, Carter A, Najman Y, Bandopadhya PC, Chapman HJ, Bickle J, Garzanti E, Vezzoli G, Andò S, Foster GL, and Gerring C (2008) New constraints on the sedimentation and uplift history of the Andaman-Nicobar accretionary prism, South Andaman Island. In: Draut A, Clift PD, Scholl DW (eds) Formation and applications of the sedimentary record in Arc Collision Zones, vol 436, pp 223–255. Geological Society of America Special Papers. https://doi.org/10.1130/2007.2436(11)

An ZS, Kutzbach JE, Prell WL, Porter SC (2001) Evolution of Asian monsoons and phased uplift of the Himalaya-Tibetan Plateau since Late Miocene times. Nature 411:62–66. https://doi.org/10.1038/35075035

An Z, Wu G, Li J, Sun Y, Liu Y, Zhou W, Cai Y, Duan A, Li L, Mao J, Cheng H, Shi Z, Tan L, Yan H, Ao H, Chang H, Feng J (2015) Global monsoon dynamics and climate change. Annu Rev Earth Planet Sci 43:29–77. https://doi.org/10.1146/annurev-earth-060313-054623

Ashfaq M, Shi Y, Tung W-W, Trapp RJ, Gao X, Pal JS, Diffenbaugh NS (2009) Suppression of south Asian summer monsoon precipitation in the 21st century. Geophys Res Lett 36:L01704. https://doi.org/10.1029/2008GL036500

Aung LL, Zin EE, Theingi P, Elvera N, Aung PP, Han TT, Oo Y, Skaland RG (2017) Myanmar climate report, MET report. Norwegian Meteorological Institute, Sept 2017, p 105

Awasthi N, Ray JS, Singh AK, Band ST, Rai VK (2014) Provenance of the Late Quaternary sediments in the Andaman Sea: implications for monsoon variability and ocean circulation. Geochem Geophys Geosyst 15(10):3890–3906. https://doi.org/10.1002/2014GC005462

Bhattacharjee D (2005) Pteropod preservation profiles in seabed sediments off Middle Andaman Island in Andaman Sea. Indian J Mar Sci 34(3):259–266. http://nopr.niscair.res.in/handle/123456789/1561

Bolton CT, Chang L, Clemens SC, Kodama K, Ikehara M, Medina-Elizalde M, Paterson GA, Roberts AP, Rohling EJ, Yamamoto Y, Zhao X (2013) A 500,000 year record of Indian summer monsoon dynamics recorded by eastern equatorial Indian Ocean upper water-column structure. Quat Sci Rev 77:167–180. https://doi.org/10.1016/j.quascirev.2013.07.031

Boos WR, Kuang Z (2010) Dominant control of the South Asian monsoon by orographic insulation versus plateau heating. Nature 463(7278):218–223. https://doi.org/10.1038/nature08707

Brewer D, Hayes D, Lyne V, Donovan A, Skewes T, Milton D, Murphy N (2015) An ecosystem characterisation of the Bay of Bengal. Bay of Bengal Large Marine Ecosystem Project, BOBLME-2015-Ecology-13

Cao P, Shi X, Li W, Liu S, Yao Z, Hu L, Khokiattiwong S, Kornkanitnan N (2015) Sedimentary responses to the Indian Summer Monsoon variations recorded in the southeastern Andaman Sea slope since 26 ka. J Asian Earth Sci 114:512–525. https://doi.org/10.1016/j.jseaes.2015.06.028

Cawthern T, Johnson JE, Giosan L, Flores JA, Rose K, Solomon E (2014) A late Miocene-early Pliocene biogenic silica crash in the Andaman Sea and Bay of Bengal. Mar Petroleum Geol 58:490–501. https://doi.org/10.1016/j.marpetgeo.2014.07.026

Cerling TE, Harris JM, MacFadden BJ, Leakey MG, Quade J, Eisenmann V, Ehleringer JR (1997) Global vegetation change through the Miocene/Pliocene boundary. Nature 389(6647):153–158. https://doi.org/10.1038/38229

Chatterjee A, Shankar D, McCreary JP, Vinayachandran PN, Mukherjee A (2017) Dynamics of Andaman Sea circulation and its role in connecting the equatorial Indian Ocean to the Bay of Bengal. J Geophys Res Oceans 122(4):3200–3218. https://doi.org/10.1002/2016JC012300

Clemens SC, Prell WL (2007) The timing of orbital scale Indian-monsoon changes. Quat Sci Rev 26(3–4):275–278. https://doi.org/10.1016/j.quascirev.2006.11.010

Clemens SC, Kuhnt W, LeVay LJ and the Expedition 353 Scientists (2016) Indian monsoon rainfall. In: Proceedings of the international Ocean discovery program 353, College Station, TX. http://dx.doi.org/10.14379/iodp.proc.353.108.2016

Clemens SC, Prell WL, Sun Y, Liu Z, Chen G (2008) Southern Hemisphere forcing of Pliocene δ18O and the evolution of Indo-Asian monsoons. Paleoceanography 23:PA4210. https://doi.org/10.1029/2008pa001638
Clift PD, Plumb RA (2008) The Asian Monsoon: causes, history and effects. Cambridge University Press, p 288. ISBN 978-0-521-84799-5
Clift PD, Hodges KV, Heslop D, Hannigan R, Van Long H, Calves G (2008) Correlation of Himalayan exhumation rates and Asian monsoon intensity. Nat Geosci 1(12):875–880. https://doi.org/10.1038/ngeo351
Colin C, Turpin L, Bertaux J, Desorairies A, Kissel C (1999) Erosional history of the Himalayan and Burman ranges during the last two glacial–interglacial cycles. Earth Planet Sci Lett 171(4):647–660. https://doi.org/10.1016/S0012-821X(99)00184-3
Colin C, Turpin L, Blamart D, Frank N, Kissel C, Duchamp S (2006) Evolution of weathering patterns in the Indo-Burman ranges over the last 280 kyr: effects of sediment provenance on 87Sr/86Sr ratios tracer. Geochem Geophys Geosyst 7(3):Q03007. https://doi.org/10.1029/2005GC000962
Curray JR (2005) Tectonics and history of the Andaman Sea region. J Asian Earth Sci 25:187–232. https://doi.org/10.1016/j.jseaes.2004.09.001
Day J, Fung I, Risi C (2015) Coupling of South and East Asian monsoon precipitation in July–August. J Clim 28(11):4330–4356. https://doi.org/10.1175/JCLI-D-14-00393.1
De Vleeschouwer D, Dunlea AG, Auer G, Anderson CH, Brumsack H, de Loach A, Gurnis M, Huh Y, Ishiwa T, Jang K, Kominz MA, März C, Schnetger B, Murra RW, Pälike H, and Expedition 356 Shipboard Scientists (2017) Quantifying K, U, and Th contents of marine sediments using shipboard natural gamma radiation spectra measured on DV JOIDES Resolution. Geochem Geophys Geosyst 18(3):1053–1064. https://doi.org/10.1002/2016GC006715
Derry LA, France-Lanord C (1997) Himalayan weathering and erosion fluxes: climate and tectonic controls. In: Ruddiman WF (eds) Tectonic uplift and climate change. Plenum Press, New York. https://doi.org/10.1007/978-1-4615-5935-1-12
Ding Y et al (2004) Overview of the South China Sea monsoon experiment. Adv Atmos Sci 21:343–360. https://doi.org/10.1007/BF02915563
Drury AJ, Westerhold T, Hodell D, Röhl U (2018) Reinforcing the North Atlantic backbone: revision and extension of the composite splice at ODP Site 982. Clim Past 14:321–338. https://doi.org/10.5194/cp-14-321-2018
Dunlea AG, Murray RW, Harris RN, Vasiliev MA, Evans H, Spivack AJ, D'Hondt S (2013) Assessment and use of NGR instrumentation on the JOIDES Resolution to quantify U, Th, and K concentrations in marine sediment. Sci Drill 15:57–63. https://doi.org/10.2204/iodp.sd.15.05.2013
Fleitmann D, Burns SJ, Neff U, Mangini A, Matter A (2003) Changing moisture sources over the last 330,000 years in Northern Oman from fluid-inclusion evidence in speleothems. Quat Res 60(2):223–232. https://doi.org/10.1016/S0033-5894(03)00086-3
France-Lanord C, Derry L, Michard A (1993) Evolution of the Himalayas since Miocene time—isotopic and sedimentological evidence from the Bengal Fan. In: Treloar PJ, Searle MP (eds) Himalayan tectonics, vol 74, pp 603–621. Geological Society Special Publications. https://doi.org/10.1144/GSL.SP.1993.074.01.40
Frerichs WE (1971) Planktonic foraminifera in the sediments of the Andaman Sea. J Foramin Res 1(1):1–14. https://doi.org/10.2113/gsjfr.1.1.1
Gadgil S (2003) The Indian monsoon and its variability. Annu Rev Earth Planet Sci 31(1):429–467. https://doi.org/10.1146/annurev.earth.31.100901.141251
Garzanti E, Limonta M, Resentini A, Bandopadhyay PC, Najman Y, Andò S, Vezzoli G (2013) Sediment recycling at convergent plate margins (Indo-Burman Ranges and Andaman-Nicobar Ridge). Earth Sci Rev 123:113–132. https://doi.org/10.1016/j.earscirev.2013.04.008
Gauns M, Madhupratap M, Ramaiah N, Jyothibabu R, Fernandes V, Paul JT, Prasanna Kumar S (2005) Comparative accounts of biological productivity characteristics and estimates of carbon fluxes in the Arabian Sea and the Bay of Bengal. Deep Sea Res Part 2 Top Stud Oceanogr 52(14–15):2003–2017. https://doi.org/10.1016/j.dsr2.2005.05.009

Gebregiorgis D, Hathorne EC, Sijinkumar AV, Nagender Nath B, Nürnberg D, Frank M (2016) South Asian summer monsoon variability during the last ~54 kyrs inferred from surface water salinity and river run off proxies. Quat Sci Rev 138:6–15. https://doi.org/10.1016/j.quascirev.2016.02.012

Gebregiorgis D, Hathorne EC, Giosan L, Clemens S, Nürnberg D, Frank M (2018) Southern Hemisphere forcing of South Asian monsoon precipitation over the past ~1 million years. Nat Commun 9(1):4702. https://doi.org/10.1038/s41467-018-07076-2

Gordon AL, Shroyer EL, Mahadevan A, Sengupta D, Freilich M (2016) Bay of Bengal: 2013 northeast monsoon upper-ocean circulation. Oceanography 29(2):82–91. https://doi.org/10.5670/oceanog.2016.41

Goswami BN, Rao SA, Sengupta D, Chakravorty S (2016) Monsoons to mixing in the Bay of Bengal: Multiscale air-sea interactions and monsoon predictability. Oceanography 29(2):18–27. https://doi.org/10.5670/oceanog.2016.35

Guo ZT, Ruddiman WF, Hao QZ, Wu HB, Qiao YS, Zhu RX, Peng SZ, Wei JJ, Yuan BY, Liu TS (2002) Onset of Asian desertification by 22 Myr ago inferred from loess deposits in China. Nature 416(6877):159–163. https://doi.org/10.1038/416159a

Gupta AK, Anderson DM, Overpeck JT (2003) Abrupt changes in the Asian southwest monsoon during the Holocene and their links to the North Atlantic Ocean. Nature 421(6921):354–357. https://doi.org/10.1038/nature01340

Haq BU, Hardenbol JAN, Vail PR (1987) Chronology of fluctuating sea levels since the Triassic. Science 235(4793):1156–1167. https://doi.org/10.1126/science.235.4793.1156

Harrison TM, Copeland P, Kidd WSF, Yin AN (1992) Raising Tibet. Science 255(5052):1663–1670. https://doi.org/10.1126/science.255.5052.1663

Hastenrath S, Lamb PJ (1979) Climatic Atlas of the Indian Ocean, Part I, Surface climate and atmospheric circulation. University of Wisconsin Press, Madison, p 116. ISBN 0299078140 9780299078140

Herbert TD, Lawrence KT, Tzanova A, Peterson LC, Caballero-Gill R, Kelly CS (2016) Late Miocene global cooling and the rise of modern ecosystems. Nat Geosci 9(11):843–847. https://doi.org/10.1038/ngeo2813

Holbourn AE, Kuhnt W, Schulz M, Erlenkeuser H (2005) Impacts of orbital forcing and atmospheric carbon dioxide on Miocene ice-sheet expansion. Nature 438(7067):483–487. https://doi.org/10.1038/nature04123

Holbourn AE, Kuhnt W, Schulz M, Flores J-A, Andersen N (2007) Orbitally-paced climate evolution during the middle Miocene "Monterey" carbon isotope excursion. Earth Planet Sci Lett 261(3–4):534–550. https://doi.org/10.1016/j.epsl.2007.07.026

Holbourn AE, Kuhnt W, Clemens SC, Kochhann KDG, Jöhnck J, Lübbers J, Andersen N (2018) Late Miocene climate cooling and intensification of southeast Asian winter monsoon. Nat Commun 9(1):1584. https://doi.org/10.1038/s41467-018-03950-1

Huang Y, Clemens SC, Liu W, Wang Y, Prell WL (2007) Large-scale hydrological change drove the late Miocene C4 plant expansion in the Himalayan foreland and Arabian Peninsula. Geology 35(6):531–534. https://doi.org/10.1130/G23666A.1

Ittekkot V, Nair RR, Honjo S, Ramaswamy V, Bartsch M, Manganini S, Desai BN (1991) Enhanced particle fluxes in Bay of Bengal induced by injection of fresh water. Nature 351(6325):385–387. https://doi.org/10.1038/351385a0

Iversen MH, Ploug H (2010) Ballast minerals and the sinking carbon flux in the ocean: Carbon-specific respiration rates and sinking velocity of marine snow aggregates. Biogeosciences 7(9):2613–2624. https://doi.org/10.5194/bg-7-2613-2010

Klaas C, Archer DE (2002) Association of sinking organic matter with various types of mineral ballast in the deep sea: implications for the rain ratio. Glob Biogeochem Cycles 16(4):1116. https://doi.org/10.1029/2001GB001765

Kroon D, Steens T, Troelstra SR (1991) Onset of monsoonal related upwelling in the western Arabian Se as revealed by planktonic foraminifers. In: Prell WL, Niitsuma N et al (eds) Proceedings of the

Ocean drilling program, scientific results, vol 117. College Station, TX (Ocean Drilling Program), pp 257–263. http://dx.doi.org/10.2973/odp.proc.sr.117.126.1991
Kwon EY, Primeau F, Sarmiento JL (2009) The impact of remineralization depth on the air–sea carbon balance. Nat Geosci 2(9):630–635. https://doi.org/10.1038/ngeo612
Le Moigne FAC, Gallinari M, Laurenceau E, De La Rocha CL (2013) Enhanced rates of particulate organic matter remineralization by microzooplankton are diminished by added ballast minerals. Biogeosciences 10(9):5755–5765. https://doi.org/10.5194/bg-10-5755-2013
Lisiecki LE, Raymo ME (2005) A Pliocene—Pleistocene stack of 57 globally distributed benthic δ18O records. Paleoceanography 20(1):PA1003. https://doi.org/10.1029/2004PA001071
Liu X, Yin Z (2011) Forms of the Tibetan Plateau uplift and regional differences of the Asian monsoon-arid environmental evolution—a modeling perspective. Earth Environ 2(3):401–416
Liu J, Wang B, Yang J (2008) Forced and internal modes of variability of the East Asian summer monsoon. Clim Past 4:225–233. www.clim-past.net/4/225/2008/
Loo YY, Billa L, Singh A (2015) Effect of climate change on seasonal monsoon in Asia and its impact on the variability of monsoon rainfall in Southeast Asia. Geosci Front 6:817–823. https://doi.org/10.1016/j.gsf.2014.02.009
Lutz M, Dunbar R, Caldeira K (2002) Regional variability in the vertical flux of particulate organic carbon in the ocean interior. Global Biogeochem Cycles 16(3):1037. https://doi.org/10.1029/2000GB001383
Milliman JD, Farnsworth KL (2011) River discharge to the Coastal Ocean. Cambridge University Press, p 384. ISBN 978-0-521-87987-3
Miriyala P, Sukumaran NP, Nath BN, Ramamurty PB, Sijinkumar AV, Vijayagopal B, Ramaswamy V, Tyson S (2017) Increased chemical weathering during the deglacial to mid-Holocene summer monsoon intensification. Sci Rep 7:44310. https://doi.org/10.1038/srep44310
Molnar P (2005) Mio–Pliocene growth of the Tibetan Plateau and evolution of East Asian climate. Palaeontol Electron 8(1):1–23. http://palaeo-electronica.org/2005_1/molnar2/molnar2.pdf
Molnar P, Rajagopalan B (2012) Late Miocene upward and outward growth of eastern Tibet and decreasing monsoon rainfall over the northwestern Indian subcontinent since ~10 Ma. Geophys Res Lett 39(9):L09702. https://doi.org/10.1029/2012GL051305
Molnar P, England P, Martinod J (1993) Mantle dynamics, uplift of the Tibetan Plateau, and the Indian Monsoon. Rev Geophys 31(4):357–396. https://doi.org/10.1029/93RG02030
Molnar P, Boos WR, Battisti DS (2010) Orographic controls on climate and paleoclimate of Asia: thermal and mechanical roles for the Tibetan Plateau. Annu Rev Earth Planet Sci 38(1):77–102. https://doi.org/10.1146/annurev-earth-040809-152456
Morley CK (2016) Cenozoic structural evolution of the Andaman Sea: evolution from an extensional to a sheared margin. In: Nemcok M, Rybar S, Sinha ST, Hermeston SA, Ledvenyiova L (eds) Transform margins: development, controls and petroleum systems, vol 431. Geological Society Special Publications, pp 39–61. http://doi.org/10.1144/SP431.1
Pandey DK, Anitha G, Prerna R, Pandey A (2017) Late Cenozoic seismic stratigraphy of the Andaman Forearc Basin, Indian Ocean. Pet Sci 14(4):648–661. https://doi.org/10.1007/s12182-017-0197-7
Phillips SC, Johnson JE, Underwood MB, Guo J, Giosan L, Rose K (2014) Long-timescale variation in bulk and clay mineral composition of Indian continental margin sediments in the Bay of Bengal, Arabian Sea, and Andaman Sea. Mar Petroleum Geol 58:117–138. https://doi.org/10.1016/j.marpetgeo.2014.06.018
Ponton C, Giosan L, Eglinton TI, Fuller DQ, Johnson JE, Kumar P, Collett TS (2012) Holocene aridification of India. Geophys Res Lett 39(3):L03704. https://doi.org/10.1029/2011GL050722
Prell WL, Kutzbach JE (1997) The Impact of Tibet-Himalayan elevation on the sensitivity of the monsoon climate system to changes in solar radiation. In: Ruddiman WJ (ed) Tectonic uplift and climate change. Plenum Press, New York, pp 171–201. https://doi.org/10.1007/978-1-4615-5935-1
Prell WL, Murray DW, Clemens SC, Anderson DM (1992) Evolution and variability of the Indian Ocean summer monsoon: evidence from the western Arabian Sea drilling program. In: Duncan

RA, Rea DK, Kidd RB, Rad U, Weissel JK (eds) Synthesis of results from scientific drilling in the Indian Ocean, vol 70. Geophysical Monograph, pp 447–469. http://dx.doi.org/10.1029/GM070p0447
Qiu J (2013) Monsoon Melee. Science 340(6139):1400–1401. https://doi.org/10.1126/science.340.6139.1400
Quade J, Cerling TE (1995) Expansion of C4 grasses in the late Miocene of northern Pakistan: evidence from stable isotopes in paleosols. Palaeogeogr Palaeoclimatol Palaeoecol 115(1–4):91–116. https://doi.org/10.1016/0031-0182(94)00108-K
Raju KAK, Ramprasad T, Rao PS, Ramalingeswara Rao BR, Varghese J (2004) New insights into the tectonic evolution of the Andaman basin, northeast Indian Ocean. Earth Planet Sci Lett 221:145–162. https://doi.org/10.1016/S0012-821X(04)00075-5
Ramaswamy V, Rao PS, Rao KH, Thwin S, Srinavasa Rao N, Raiker V (2004) Tidal influence on suspended sediment distribution and dispersal in the northern Andaman Sea and Gulf of Martaban. Mar Geol 208(1):33–42. https://doi.org/10.1016/j.margeo.2004.04.019
Rodriguez M, Chamot-Rooke N, Huchon P, Fournier M, Delescluse M (2014) The Owen Ridge uplift in the Arabian Sea: implications for the sedimentary record of Indian monsoon in late Miocene. Earth Planet Sci Lett 394:1–12. https://doi.org/10.1016/j.epsl.2014.03.011
Ruddiman WF (2006) What is the timing of orbital-scale monsoon changes? Quat Sci Rev 25(7):657–658. https://doi.org/10.1016/j.quascirev.2006.02.004
Sandwell DT, Smith WHF (2009) Global marine gravity from retracked Geosat and ERS 1 altimetry: ridge segmentation v. spreading rate. J Geophys Res 114:B01411. https://doi.org/10.1029/2008jb006008
Sarkar A, Ramesh R, Somayajulu BLK, Agnihotri R, Jull AJT, Burr GS (2000) High resolution Holocene monsoon record from the eastern Arabian Sea. Earth Planet Sci Lett 177(3–4):209–218. https://doi.org/10.1016/S0012-821X(00)00053-4
Schott FA, Xie S-P, McCreary JP Jr (2009) Indian Ocean circulation and climate variability. Rev Geophys 47:RG1002. http://dx.doi.org/10.1029/2007RG000245
Shackleton NJ, Hall MA, Pate D (1995) Pliocene stable isotope stratigraphy of Site 846. In: Pisias NG, Mayer LA, Janecek TR, Palmer-Julson A, van Andel TH (eds) Proceedings of ODP, scientific results, vol 138. College Station, TX (Ocean Drilling Program), pp 337–355. https://doi.org/10.2973/odp.proc.sr.138.117.1995
Sijinkumar A, Clemens S, Nath BN, Prell W, Benshila R, Lengaigne M (2016) $\delta^{18}O$ and salinity variability from the last glacial maximum to recent in the Bay of Bengal and Andaman Sea. Quat Sci Rev 135:79–91. https://doi.org/10.1016/j.quascirev.2016.01.022
Singh SC, Moeremans R, McArdle J, Johansen K (2013) Seismic images of the silver strike-slip fault and back thrust in the Andaman-Nicobar region. J Geophys Res 118:5208–5224. https://doi.org/10.1002/jgrb.50378
Sinha A, Berkelhammer M, Stott L, Mudelsee M, Cheng H, Biswas J (2011a) The leading mode of Indian summer monsoon precipitation variability during the last millennium. Geophys Res Lett 38:L15703. https://doi.org/10.1029/2011GL047713
Sinha A, Stott L, Berkelhammer M, Cheng H, Edwards RL, Buckley B, Aldenderfer M, Mudelsee M (2011b) A global context for megadroughts in monsoon Asia during the past millennium. Quat Sci Rev 30:47–62. https://doi.org/10.1016/j.quascirev.2010.10.005
Staubwasser M, Weiss H (2006) Holocene climate and cultural evolution in late prehistoric-early historic west Asia—Introduction. Quat Res 66(3):372–387. https://doi.org/10.1016/j.yqres.2006.09.001
Sun X, Wang P (2005) How old is the Asian monsoon system?—Palaeobotanical records from China. Palaeogeogr Palaeoclimatol Palaeoecol 222(3–4):181–222. https://doi.org/10.1016/j.palaeo.2005.03.005
Talley LD (2013) Hydrographic atlas of the world Ocean circulation experiment (WOCE), vol 4. Indian Ocean. ISBN: 0904175588

Tomczak M, Godfrey JS (2005) Regional Oceanography: an introduction, pdf version 1.1. Chapter 12, hydrology of the Indian Ocean, pp 199–214. http://www.incois.gov.in/Tutor/regoc/pdfversion.html

Unger D, Jennerjahn T (2009) Impact of regional Indian Ocean characteristics on the biogeochemical variability of settling particles. American Geophysical Union Geophysical Monograph Series, vol 185, pp 257-280. https://doi.org/10.1029/2008GM000703

Unger D, Ittekkot V, Schäfer P, Tiemann J, Reschke S (2003) Seasonality and interannual variability of particle fluxes to the deep Bay of Bengal: influence of riverine input and oceanographic processes. Deep Sea Res Part 2 Top Stud Oceanogr 50(5):897–923. https://doi.org/10.1016/S0967-0645(02)00612-4

Varkey MJ, Murty VSN, Suryanarayana A (1996) Physical oceanography of the Bay of Bengal and Andaman Sea. Oceanogr Mar Biol 34:1–70

Wan SM, Li AC, Clift PD, Stutt JB (2007) Development of the East Asian monsoon: mineralogical and sedimentologic records in the northern South China Sea since 20 Ma. Palaeogeogr Palaeoclimatol Palaeoecol 254(3–4):561–582. https://doi.org/10.1016/j.palaeo.2007.07.009

Wan S, Kürschner WM, Clift PD, Li A, Li T (2009) Extreme weathering/erosion during the Miocene Climatic Optimum: evidence from sediment record in the South China Sea. Geophys Res Lett 36:L19706. https://doi.org/10.1029/2009GL040279

Wang B, Ding QH (2008) Global monsoon: dominant mode of annual variation in the tropics. Dyn Atmos Oceans 44(3–4):165–183. https://doi.org/10.1016/j.dynatmoce.2007.05.002

Wang PX, Wang B, Cheng H, Fasullo J, Guo ZT, Kiefer T, Liu ZY (2017) The global monsoon across time scales: mechanisms and outstanding issues. Earth Sci Rev 174:84–121. https://doi.org/10.1016/j.earscirev.2017.07.006

Late Quaternary Chronostratigraphy, Carbonate Mass Accumulation Rates and Palaeoceanography of the Andaman Sea

A. V. Sijinkumar, B. Nagender Nath and Pavan Miriyala

Abstract The Andaman Sea, the least explored semi-enclosed marginal sea in the northern Indian Ocean, is one of the great repositories of palaeo-sedimentary archives for Quaternary climate and oceanographic studies due to its physiographic setting. This marginal sea has some unique features with respect to its response to the past climate change and the difference in the deep-water characteristics with the adjacent Bay of Bengal (BoB). Himalayan Rivers including those traversing Indo-Burmese mountain ranges have one of the world's highest physical and chemical erosion rates contributing enormous quantity of sediments to the BoB and the Andaman Sea which provides a continuous record of climate variability. Based on ten available accelerator mass spectroscopy (AMS) dated cores retrieved from deeper water depths (>1240 m), the estimated average sedimentation rate for the Andaman basin is ~13 cm/ka and thus the basin is a valuable sedimentary archive for Quaternary climate both in the Andaman Sea and the source areas in Myanmar watersheds. The basin is also characterized by high variability in spatial as well as temporal sedimentation rate. The highest sedimentation rate of ~860 cm/ka reported in the northern part of the Andaman Sea at the Ayeyarwady River mouth with a water depth of ~500 m, whereas the lower rate of sedimentation (~3.1 cm/ka) is witnessed on some rises and mounts in the Western Andaman Sea. These spatial and temporal variations are mainly attributed to the complex bottom topography with valleys and seamounts and the varying oceanography and climate. The basin contains a substantial amount of carbonate sediments which are valuable archives for palaeoceanographic data. Mass accumulation rate (*MARs*) of carbonate skeletons (Pteropods, planktonic and benthic foraminifera), $CaCO_3$ and total organic carbon in this basin was quantified and were found to be highly comparable with temporal variations in terrigenous input to the

A. V. Sijinkumar (✉)
Department of Geology, Central University of Kerala, Kasaragod, India
e-mail: sijingeo@gmail.com

B. Nagender Nath
CSIR-National Institute of Oceanography, Dona Paula, Goa, India

P. Miriyala
CSIR-National Geophysical Research Institute, Hyderabad, India
e-mail: pavanmiriyala@ngri.res.in

J. S. Ray and M. Radhakrishna (eds.), *The Andaman Islands and Adjoining Offshore: Geology, Tectonics and Palaeoclimate*, Society of Earth Scientists Series,
https://doi.org/10.1007/978-3-030-39843-9_14

basin suggesting that the carbonate productivity within the water column and erosion and weathering processes on Myanmar watersheds were in tandem reflecting the response to the variations in Indian Summer Monsoon (ISM). Very high (~3.0×10^5 no./cm^2/ka) *MARs* of foraminifera was seen during last glacial-Holocene transition, which coincides with relatively higher sedimentation rate and summer insolation at 30° N latitude. The *MARs* of skeletal carbonate (planktonic and benthic foraminifera) was high during early to mid-MIS 3 and characteristically low during late MIS 3 and late Holocene, possibly driven by the ISM variability.

Keywords Carbonate · Mass accumulation rate · Palaeoclimate · Indian Summer Monsoon · Late Quaternary

1 Introduction

Marine sediments have the potential to provide high-resolution records of palaeoclimatic and palaeoceanographic changes. Such records are essential for understanding the climatic variability on time scales, ranging from Holocene to glacial-interglacial periods. Between 6 and 11 billion metric tons of sediment accumulate in the ocean basins annually, and they provide an archive of climatic conditions near the ocean surface or on the adjacent continents (Bradley 1999). In the Andaman Basin, fairly good sedimentation has been taking place since Miocene with a significant contribution from biogenic sources (Rodolfo 1969). But a recent study in the Western Andaman Sea has shown a significant hiatus in sedimentation during the latest Miocene and Pliocene (Flores et al. 2014; Clemens et al. 2016). However, there are no reports of break-in sedimentation history during the Quaternary. The late Quaternary sediments of the Andaman Sea were used to infer the climatic history and the vegetation cover on the continents, changes in water column chemistry, biological activity, deep water circulation, North Atlantic teleconnections etc. (e.g., Naqvi et al. 1994; Colin et al. 1998, 1999; Alagarsamy et al. 2010; Sijinkumar et al. 2010, 2016b).

Interpreting sedimentary archives in the ocean floor requires accurate dating, which is of fundamental importance to palaeoclimatic and palaeoceanographic studies. To place the geological changes in a time perspective, one needs to know absolute dates of such events recorded in a sedimentary sequence. The major problem in the Quaternary studies is that each of the chronostratigraphic method used for dating the sediments has either time or material limitations. Without reliable estimates on the timing of events in the past, it is impossible to place the changes in a regional/global perspective. Accurate dating is also important to decipher the lead and lag of events recorded in different sequences and regions. The linear sedimentation rate is also useful in estimating mass accumulation rates (*MARs*) of the components at the seafloor. From the dated horizons, the ages for other horizons can be assigned by interpolation/extrapolation of calibrated dates, assuming constant sedimentation rates which would be useful for calculating sedimentation rate for a period of interest. The rate of sedimentation varies with water depth and distance of land to the basin. The accumulation of sediments on the seafloor is not evenly distributed and depends basically on the bottom topography and hydrographical conditions. The Andaman Sea topography is very irregular with seamounts and valleys (Rodolfo 1969). The late Quaternary sediments of the Andaman Sea are dated mainly based on AMS radiocarbon techniques (e.g., Rashid et al. 2007; Awasthi et al. 2010; Sijinkumar et al. 2011, 2015; Cao et al. 2015; Kumar et al. 2018; Sebastian et al. 2019), combined with oxygen isotope stratigraphy and biostratigraphy (Colin et al. 1998; Sijinkumar et al. 2010; Marzin et al. 2013; Flores et al. 2014; Ali et al. 2015; Clemens et al. 2016; Gebregiorgis et al. 2018).

The carbon (>95%) lost from the ocean reservoir to the sedimentary sink appears to be transferred as skeletal $CaCO_3$, produced in the surface waters and carries a productivity signal which can be preserved in the underlying pelagic carbonate sediments (Brummer and van Eijden 1992). In these skeletal carbonate sediments, foraminifera is the most dominant components. They are abundant, widespread, and have an excellent preservation potential. The shells of planktonic foraminifera are abundant in most oceanic sediments. The average rate of total carbonate content in the deep Andaman basin is nearly 14% and the rate of carbonate sedimentation is 2 cm/ka (Rodolfo 1969). The Sewell Rise, Alcock Rise and Mergui Ridge in the West Andaman Sea are covered by calcareous oozes whereas the East Basin is dominated by a siliceous ooze (Tripathi 2014). The average carbonate content represented by planktonic foraminifera, benthic foraminifera, and pteropods in sediments of the Middle Andaman and Barren island region is 58%, 7% and 11% respectively (Bhattacharjee 2005).

The accumulation and preservation of carbonate are controlled by dilution of terrigenous material, depth of water column and nature of bottom water mass. Based on the distribution and preservation of planktonic foraminifera in surface sediments of the northern Indian Ocean, foraminiferal lysocline has demarcated deepest (3800 m) in the equatorial region, rises abruptly to 3300 m in the Arabian Sea, and varies from 2600 to ~2000 m in the Bay of Bengal (BoB), with shallow lysocline occurring in the northern Bay (Cullen and Prell 1984). Well preserved foraminifera with high species diversity were present in the sedimentary record of 48 ka from a core (AAS 11) retrieved at a water depth of ~2900 m in the southwestern Andaman Sea which was not expected in the similar water depth in the BoB (Sijinkumar 2011). The better preservation of carbonate in the Andaman Sea was attributed to the presence of relatively warm bottom water than that of the BoB. The presence of pteropods in the Andaman sediments was reported by various researchers (Bhattacharjee 2005; Panchang et al. 2007; Sijinkumar et al. 2010, 2015; Tripathi 2014) and the position of aragonite compensation depth (ACD) is placed at ~1300 m. The ACD is shallow in the northern part of the Andaman Sea (~1150 m) and deepest in the southern part (1500 m) (Bhattacharjee 2005).

The recovery of sediment cores from the marginal Andaman Sea has provided an opportunity to construct high resolution records of climatic and oceanographic variability during the late Quaternary especially during last glacial to Holocene time span (Naqvi et al. 1994; Rashid et al. 2007; Alagarsamy et al. 2010; Sijinkumar et al.

2010, 2011, 2015, 2016a, b; Miriyala et al. 2017; Sebastian et al. 2019) and few studies on million years time scale (Flores et al. 2014; Clemens et al. 2016; Gebregiorgis et al. 2018). While the Arabian Sea is well studied for climate and oceanographic studies on different time scales, very little is known about the variability of *MARs* of carbonate deposition and their application in past monsoon in the BoB and even less about the Andaman Sea. Probably, this region would provide information about the linkages between South Asian and East Asian monsoon systems and one of the best places for studying the Indian Summer Monsoon (ISM) variability and resultant erosional history of the Himalayan-Burman ranges. In this study, we attempted to reconstruct late Quaternary spatial and temporal variability in the sediment accumulation by using available AMS dated cores and present detailed temporal record of *MAR* of $CaCO_3$, organic carbon, planktonic foraminifera, benthic foraminifera and pteropods for the last 54 ka.

2 Studies on Carbonate Accumulation in the Andaman Sea

There are fewer studies on the accumulation and preservation of carbonate in the Andaman Basin, but there were no records on *MARs* of carbonate (Bhattacharjee 2005; Tripathi 2014). The *MARs* of carbonate is directly related to productivity. *MARs* were widely used for quantifying palaeoproductivity induced by ISM variability. *MARs* of organic carbon content (Müller and Suess 1979; Sarnthein et al. 1988), benthic foraminifer accumulation rate (Herguera 1992) and the species composition of foraminiferal assemblages (Mix 1989; Loubere 1991) have been used to quantify palaeoproductivity rates. To our knowledge, there are no estimates of palaeoproductivity for the Andaman Sea. Reconstructions of productivity pattern are of great interest because of important links to nutrient status, current patterns, mixing of water mass, wind, the global carbon cycle, and biogeography. The important proxies which are used to assess palaeoproductivity are fertile species of planktic foraminifera, organic carbon, $CaCO_3$, and biogenic opal contents in marine sediments (Berger 1989; Prahl et al. 1989; Shimmield et al. 1994; Singh et al. 2011). Organic carbon is derived from land as organic detritus and from oceanic surface productivity. The concentration of organic matter in the sediments depends on the quantity of supply of organic matter and environment of deposition. There is a relationship between productivity in surface waters and organic carbon accumulation content in underlying sediments (Sarnthein et al. 1988). The relation between the organic carbon content of the sediments and the productivity in upper ocean has led to the use as a productivity proxy (e.g., Müller and Suess 1979; Lyle 1988; Pedersen and Calvert 1990). However, it was debated in the early nineties whether the organic carbon variation is due to productivity or preservation (Pedersen and Calvert 1990; Paropkari et al. 1993). Only a very small part of the organic matter produced in the upper ocean reaches the abyssal seafloor (Heath et al. 1977). Organic carbon undergoes microbial and chemical degradation during its transition through the water column. Sedimentary organic matter mainly carbon, nitrogen and phosphorus, calcium carbonate, and opal constitute the bulk of the biogenic fraction of the sediments. Palaeoproductivity studies are intended to document temporal and spatial changes in the production of these biogenic components with time and also to discern the causes and effects of the productivity and its link with the Earth's climate.

A first detailed study on sediments of the Andaman Sea was undertaken by Rodolfo (1969). Rodolfo (1969) has studied sources of sediments into the Andaman basin and inferred the contribution of the Ayeyarwady, Salween and Sittang Rivers from Myanmar. Rao (1983) has studied the clay mineralogy of the sediments around the Andaman Islands to decipher their provenance. Recent studies on the Myanmar shelf sediments (Ramaswamy et al. 2004; Rao et al. 2005) have shown that the sediment influx from these rivers is transported along the shelf by monsoon currents and carried to the deep sea-floor through submarine canyons. Studies were also carried out on the distribution of sediments and turbidity in the continental shelf of the Ayeyarwady by Rodolfo (1969), Ramaswamy et al. (2004) and Rao et al. (2005), Girishbai et al. (2017). Ramaswamy et al. (2008) have reported the nature of organic matter from stable carbon isotopes in the surface sediments from the Myanmar shelf and modelled the relative contributions of terrigenous and planktonic sources to the sedimentary organic matter. Their findings concluded that the outer shelf relict sands were low in total organic carbon compared to the inner shelf mud belt. The major source of organic matter in the inner shelf sediments is terrigenous and this terrigenous organic carbon percentage reduces gradually offshore reflecting the greater influence of freshwater and terrigenous sediment discharge. Evidence for hydrothermal activity in the Andaman Back Arc Basin was presented using multibeam bathymetric, magnetic, gravity and seismic methods (Rao et al. 1996). Apart from fluvial and biological sources (Rodolfo 1969; Roonwal et al. 1997), the Andaman basin was found to receive material from pelagic, aeolian, weathering of sea-floor rocks, and possible hydrothermal sources (Kurian et al. 2008). Hydrothermal signatures from geological material recovered (Rao et al. 1996) and hydrocarbons of hydrothermal origin are also reported in sediments from the Andaman Back-arc basin (Chernova et al. 2001; Venkatesan et al. 2003).

Among the carbonate deposits, foraminifera was used widely in the Andaman Sea as a proxy to understand carbonate accumulation and palaeoclimate (Frerichs 1968, 1970, 1971; Mazumdar and Sharma 1991; Sharma and Mazumdar 1993; Saidova 2008). Frerichs (1970) has studied the distribution and ecology of foraminifera in the sediments from the Andaman Sea. Majority of the earlier studies were carried out to understand taxonomy, taxonomy with ecology and surface distribution. Based on foraminifera, Frerichs (1968) demarcated Pleistocene-Holocene boundary and Wisconsin Glacial biostratigraphy. Frerichs (1971) have reconstructed palaeobathymetric map of the Andaman Sea by using Neogene foraminiferal assemblages along with new inferences on seafloor tectonism in the Andaman Sea. Srinivasan and Azmi (1979) have correlated late Cenozoic marine sections in the Andaman-Nicobar Islands with Equatorial Pacific by using planktic foraminifera and developed a scheme of twenty planktic species. The Late Miocene palaeoecology of this region were constructed by using planktonic and benthic foraminifera records from the Baratang Island in the Andaman Sea (Mazumdar and Sharma 1991; Sharma and

Mazumdar 1993). Saidova (2008) has distinguished eleven communities of benthic foraminifera based on the results of a qualitative analysis of the structure of the population at 94 stations in the Andaman Sea.

In the Andaman Sea, palaeoceanographic studies are mainly based on oxygen and carbon isotopes. The study by Duplessy (1982) demonstrates that during the LGM the ISM was weaker than present whereas winter monsoon was stronger. Naqvi et al. (1994) studied the deep-water circulation and glacial-interglacial CO_2 changes by using oxygen and carbon isotopes of benthic foraminifera. The study based on magnetic properties of sediments of the BoB and the Andaman Sea (Colin et al. 1998) suggests that rapid cold events of the North Atlantic during the last glacial stages are characterized by a weaker ISM rainfall over the Himalaya via atmospheric teleconnection. Glacial-interglacial changes in the surface water characteristics of the Andaman Sea using stable isotopic ratios of planktic foraminifera were studied for the late Quaternary sediments of the Andaman Sea (Naqvi et al. 1994; Ahmad et al. 2000; Marzin et al. 2013; Sijinkumar et al. 2016a; Kumar et al. 2018; Gebregiorgis et al. 2018). Geochemical and magnetic proxies from the Andaman sediments were widely used to understand the weathering history, provenance and ISM variability during late Quaternary (Colin et al. 1999, 2006; Alagarsamy et al. 2010; Achyuthan et al. 2014; Awasthi et al. 2014; Nagasundaram et al. 2014; Ali et al. 2015; Cao et al. 2015; Miriyala et al. 2017; Sebastian et al. 2019). Colin et al. (2006) have studied long-term evolution of weathering pattern in the Indo-Burman Ranges based on Sr and Nd isotopes of core collected from the Ayeyarwady River mouth. There are several studies based on planktonic foraminifera Mg/Ca and $\delta^{18}O$ for the reconstruction of ISM (Rashid et al. 2007; Marzin et al. 2013; Sijinkumar et al. 2016a; Gebregiorgis et al. 2016, 2018;). Their findings supported the hypothesis that rapid climate change during the last deglaciation and Holocene included substantial hydrologic changes in the ISM that was coherent with the larger Asian monsoon. However, none of the studies utilized a temporal variation of *MARs* of carbonate, $CaCO_3$ and organic carbon for understanding past ISM variation and associated productivity changes.

3 Geological and Oceanographic Settings of the Andaman Sea

The Andaman Sea is a semi-enclosed marginal sea, in the eastern part of the northeast Indian Ocean situated between the east of Andaman Nicobar Ridge and west of Malayan Peninsula, with a maximum water depth of 4400 m. It is 1200 km long, 650 km wide and has an area of 8,00,000 km^2 and its physiographic features are aligned to north-south and parallel to the rift valleys. The Andaman Sea exchanges water with the BoB in the northern Indian Ocean and marginal seas of western Pacific mainly via shallow channels (Naqvi et al. 1994). It is interconnected with the BoB by the Deep Prepares Channel, Ten Degree Channel, and the Great Channel. The oceanographic processes between the Andaman Sea and the BoB are however comparable only up to a depth of about 1000 m, as the deep-water exchange between these two regions is hampered by several sills which have a significant influence on intermediate to deep water circulation in the Andaman Sea (Naqvi et al. 1994). The Sunda Shelf provides a shallow route to the eastern seas and to the Pacific Ocean, which is the southeast extension of the continental shelf of Southeast Asia. The Strait of Malacca maintains the connection of the Pacific Ocean water flowing through the South China Sea to the BoB (Wyrtki 1971).

The BoB and the Andaman Sea show a sharp contrast to the Arabian Sea in terms of atmospheric flux and continental input with respect to the monsoon climate. The Arabian Sea is marked by positive annual net atmospheric divergence flux (E-P, 3–3.5 mm/day) related to strong evaporation (Trenberth and Guillemot 1998). Climatic differences lead to greater evaporation in the north-western parts of the Indian Ocean, while the north-eastern parts experience intense rainfall. Therefore, the Persian Gulf, the Red Sea, and the Arabian Sea are negative water bodies while the BoB and the Andaman Sea are positive (Trenberth and Guillemot 1998; Rashid et al. 2007). Annual outflow from the Ganges-Meghna-Brahmaputra-Ayeyarwady Rivers (GMBA) to the BoB and the Andaman Sea is ~0.05 Sv (Perry et al. 1996), which translates to approximately 2 mm/day (Delaygue et al. 2001). Therefore, annual GMBA River runoff is greater or comparable to the annual direct precipitation in the BoB/Andaman Sea region (Rashid et al. 2007). Similar to the Arabian Sea, the Andaman Sea experiences a seasonal reversal in surface circulation, driven by the ISM. In the Andaman Sea, ISM is active between mid-May and the end of September while the winter monsoon is active between December and February. In response to the monsoons, the oceanic flow of surface waters in the Andaman Sea changes direction twice a year with a cyclonic flow in spring and summer and anti-cyclonic for the rest of the year (Potemra et al. 1991). During the ISM, cyclonic circulation develops in the surface waters of the BoB and the Andaman Sea and water movements are from west to east. The wind direction is reversed during the northeast monsoon period; the gyre becomes anticyclonic and circulation turns from east to west.

Biological productivity in the offshore region is ~0.8 to 1.0 mgC/m^2/d compared to lesser values (<0.6 mgC/m^2/d) observed in the coastal areas (Janekarn and Hylleberg 1989). The Andaman Sea is characterized by a low surface salinity ranging between 31.8 and 33.4 practical salinity units (PSU), due to the influence of freshwater discharge from the Ayeyarwady-Salween river system (Varkey et al. 1996; Sarma and Narvekar 2001). The freshwater input induces estuarine characteristics with reduced surface salinities over most of the northern parts of the BoB and the Andaman Sea. It is estimated that Ayeyarwady River drains ~428 km^3 of fresh-water annually from the Ayeyarwady catchment region to the Andaman Sea (Milliman and Meade 1983) and transports between 226 and 334 million tons (MT) of suspended sediment (Robinson et al. 2007). More than 90% of the freshwater and suspended load is delivered between mid-June and mid-November (Ramaswamy et al. 2008). Ayeyarwady River is the third largest river in the world in terms of suspended sediment discharge (Robinson et al. 2007). Most of the outflow occurs during the summer to late fall, with a peak during October. The main source of freshwater and suspended sediment to the northern Andaman Sea is from the Ayeyarwady and the Salween Rivers which have a length of ~2000 and ~2800 km respectively. Comparing with Ayeyarwady, contributions from the Sittang and Tavoy rivers are minor. The annual freshwater flux of Salween River is approximately 211 km^3, which transports 188–206 MT of suspended load (Robinson et al. 2007). While the primary productivity in the BoB is much lower than that in the Arabian Sea (Gauns et al. 2005), the productivity in the Andaman Sea is higher than that of the BoB but lower than that of the Arabian Sea.

The average surface temperature of the Andaman Sea is 29 °C and is nearly homogenous up to a depth of 50 m leading to stratification, which hinders vertical mixing (Sarma and Narvekar 2001). The temperature decreases down to 13 °C at a depth of 200 m and 9 °C at a depth of 500–600 m and the total thermocline thickness is of 150 m (Saidova 2008). Deep Andaman Sea water is consistently warmer than that of the BoB with an approximate offset of 2 °C (Sarma and Narvekar 2001), which can be attributed to the enclosed nature of the Andaman Basin (Rao and Jayaraman 1968; Sengupta et al. 1981) or the sinking of intermediate BoB waters into the Andaman deep (Naqvi et al. 1994). The high temperature of the deeper water favours better carbonate preservation than the BoB. Presence of mid-depth intense oxygen minimum zone in the Andaman Sea is reported between 150 and 800 m water depths (Rodolfo 1969; Sengupta et al. 1981). In the BoB severe oxygen depletion (dissolved $O_2 < 0.2$ mL/L) is found between about 100–600 m in the north-western bay (Wyrtki 1971). The sediments in contact with the oxygen minimum zone exhibit higher organic carbon contents (>1%), particularly along the western slope of the Andaman Sea.

The Andaman basin is marked by prominent morphological features such as the Nicobar deep, Barren-Narcondam volcanic islands, Invisible bank, and Alcock and Sewell seamount complexes (Rodolfo 1969). In addition, new seamounts with a well-developed crater at the summit were discovered near to the centre of the Nicobar swarm (Kamesh Raju et al. 2012). The Andaman Sea has a complex bottom topography that is characteristic of back-arc basins, being separated by the Andaman-Nicobar Ridge (Fig. 1). The western Andaman Fault (WAF) is a major tectonic feature in the Andaman Sea which forms a lithospheric scale boundary and together with their tectonic elements modulate the occurrence of large earthquakes and their rupture pattern in the area (Kamesh Raju et al. 2007). WAF is connected to the back-arc spreading centre and the Sagaing Fault in the north, and Sumatran and Mentawai fault systems in the south (Kamesh Raju et al. 2007). The submarine delta has been built by the outflow of the Ayeyarwady and Salween rivers, which is connected to the eastern shallow shelves along the Malay Peninsula and the Malacca Strait, in the northern region. From the shelves, the sea bottom drops sharply into a large central and two smaller basins deeper than 2000 m extending along the North-South Island Arc.

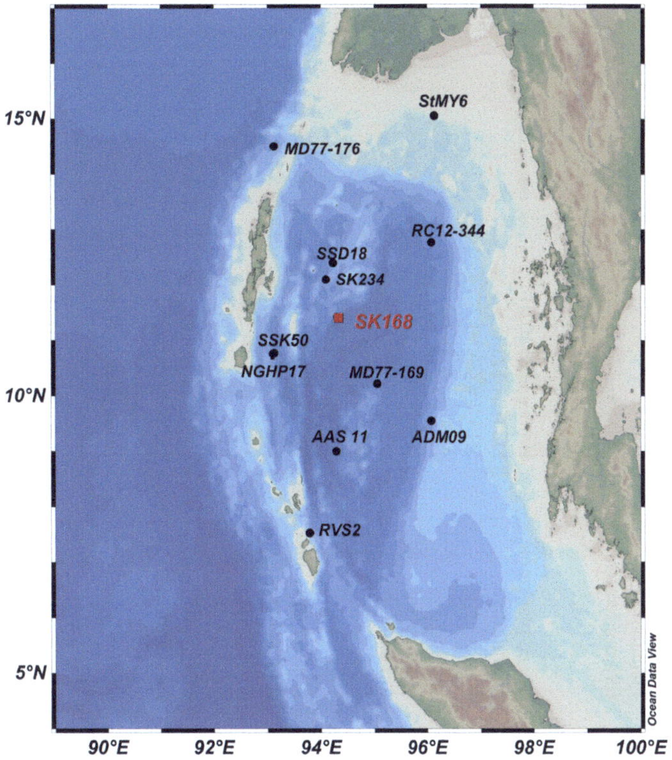

Fig. 1 Location of the sediment core SK 168 (filled red square) used in the present study along with available AMS ^{14}C dated cores (details with references are therein Table 1)

4 Materials and Methods

A sediment core SK 168/GC-1 was collected (Lat 11°42.463′ N; Long 94°29.606′ E, water depth: 2064 m, core length: 4.20 m) during the 168th cruise of ORV *Sagar Kanya* from the Alcock Seamount Complex in the Andaman Sea (Fig. 1). The core SK 168 was sub-sampled at a 2 cm interval down to 52 cm, 2–3 cm interval down to 200 cm and at every 5 cm interval down to 420 cm. The core comprises three distinct sediment layers, dark yellowish-brown coloured clays in the top 10 cm and olive-grey sediments between 30 and 420 cm (bottom of the core), while the 20 cm section of light olive grey sediments is sandwiched between these two layers. These sediments are extremely sticky and dominantly clayey in texture, containing tests of foraminifera. The details of other core samples used in the discussion are given in Table 1. The age model for the core was constructed mainly by using Accelerator Mass Spectrometer (AMS) ^{14}C dates of planktic foraminiferal tests (mixed

Table 1 Record of Late Quaternary sedimentation rate of the Andaman Sea (from N to S)

Core, water depth (m) and location (Lat and Long)	Core length and total record	Dating technique and No. of ^{14}C AMS age points	Material used	Sedimentation rate (cm/ka)	Reference
MD77-176 W.D: 1375 14°31′ N 93°08′ E	9.46 m 40 kyr	^{14}C AMS 31	Planktonic foraminifera-*G. ruber*	Holocene = 39.93 LGM = 17.69 MIS 2 = 15.87 Average = 23.75	Marzin et al. (2013)
RC 12-344 W.D: 2140 12°46′ N 96°04′ E	7.40 m 4.90 = 25 ka	^{14}C AMS 9	Mixed planktonic foraminifera	Holocene = 10.01 LGM = 17.80 MIS 2 = 23.00 Average = 19.60	Rashid et al. (2007)
SK 234-60 W.D: 2000 12°05′46″ 94°05′18″	4.0 m 77 ka	^{14}C AMS 5	Mixed planktonic foraminifera	Holocene = 5.6 LGM = 4.8 MIS 2 = 5.0 Average = 4.76	Awasthi et al. (2010)
SK 168 GC1 W.D: 2064 11°42.463′ N 94°29.606′ E,	4.2 m 54 ka	^{14}C AMS and OIS[a] 5	Mixed planktonic foraminifera	Holocene = 7.90 LGM = 7.84 MIS 2 = 8.10 Average = 7.79	Sijinkumar et al. (2010)
NGHP 17A W.D: 1356 10°46′ N 93°07′ E	6.20 = 56 ka	^{14}C AMS and OIS[a] 5	Mixed planktonic foraminifera	Holocene = 11.0 LGM = 16.0 MIS 2 = 13.1 Average = 11.0 (up to 400 cm)	Ali et al. (2015)
MD77-169 W.D: 2360 10°13′ N 95°03′ E	14.60 m 14.6 = 74 ka	^{14}C AMS and OIS[a] 13	Planktonic foraminifera-*G. ruber*	Holocene = 10.90 LGM = 10.90 MIS 2 = 10.90 Average = 10.90	Colin et al. (1998)

(continued)

Table 1 (continued)

Core, water depth (m) and location (Lat and Long)	Core length and total record	Dating technique and No. of ^{14}C AMS age points	Material used	Sedimentation rate (cm/ka)	Reference
ADM 09 W.D: 1242 09°33′ N 96°04′ E	2.70 m 2.70 = 26 ka	^{14}C AMS 10	Planktonic foraminifera-*N. dutertrei*	Holocene = 16 LGM = 5.4 MIS 2 = 8.0 Average = 13.0	Cao et al. (2015)
AAS 11GC 1 W.D: 2909 09. 00.0′ N 94°17′ E	4.28 m 48 ka	^{14}C AMS 12	Mixed planktonic foraminifera	Holocene = 8.36 LGM = 11.00 MIS 2 = 10.00 Average = 9.0	Sijinkumar et al. (2015)
RVS2 GC3 W.D: 2301 07°42.501′ N 93°58.001′ E	5.64 m 30 ka	^{14}C AMS 7	Mixed planktonic foraminifera	Holocene = 25.50 LGM = 16.15 MIS 2 = 16.24 Average = 23.23	Sijinkumar et al. (2011)

[a] *OIS*—Oxygen Isotope Stratigraphy

Globigerinoides ruber and *Globigerinoides sacculifer*). AMS ^{14}C ages were determined at five different intervals at NOSAMS facility at WHOI, USA (Sijinkumar et al. 2010).

4.1 Mass Accumulation Rate

Skeletal carbonates (no./cm^2/ka), $CaCO_3$ and organic carbon (g/cm^2/ka) were calculated to eliminate the dilution effect and to express actual changes of accumulation rates to the seafloor. *MARs* were calculated by using the total species abundance (Planktonic, benthic foraminifera, and pteropods) for skeletal carbonates and elemental concentration for carbonate and the organic carbon, linear sedimentation rate (LSR, cm/ka) and dry bulk density (DBD, g/cm^3) (e.g., Brummer and van Eijden 1992). LSR was calculated from the final age model defined by AMS radiocarbon dates and the marine isotopic stage boundaries (Sijinkumar et al. 2010). DBD was determined from a known volume of sediment, which was weighed wet, dried at 50 °C and reweighed to obtain the dry weight. DBD was taken as the ratio of dry weight/wet volume, including a correction for the salt content of 3.47%. The *MARs* eliminate the bias (due to dilution effect) associated with interpreting direct weight concentrations in sediments with varying physical properties and sedimentation rates. A total of 124 subsamples (>125 μm) were analyzed for total planktonic foraminifera, benthic foraminifera, and pteropods.

Mass accumulation rate (*MARs*) was computed as follows:

$$MARs = \text{species abundance or element concentration} \times (\text{LSR cm/ka}) \times \left(\text{DBD g/cm}^3\right).$$

5 Results and Discussion

5.1 Late Quaternary Chronostratigraphy of the Andaman Sea

In the Andaman Basin, fairly good sedimentation has been taking place since Miocene with a major contribution from the biogenic sources (Rodolfo 1969). The two deep drilling sites (NGHP-17 and U1448) in the Andaman Sea have recovered fairly long sedimentary record from this basin (Flores et al. 2014; Clemens et al. 2016). These studies show a significant hiatus in the Neogene records but there is no report on any break in the sedimentation history during the Quaternary period. One of the world's highest physical and chemical erosion rates are found in Himalayan Rivers and contribute enormous quantity of sediments to the BoB and the Andaman Sea. The Andaman sediments are the products of long-term weathering and erosional history of the Himalayan and Burman ranges (Colin et al. 1998; Miriyala et al. 2017) which could reflect the changes in the ISM. Data of this work and the available late Quaternary sedimentation records of the Andaman Sea are compiled (Table 1, Figs. 2, 3, and 4) which shows the spatial and temporal variability in sedimentation history of this marginal basin. The average sedimentation rate of the Andaman Basin is 13 cm/ka which is calculated by considering ten well-dated cores of greater than 1200 m water depths from north to south (Fig. 4). Rodolfo (1969) calculated the average accumulation rates of 15 cm/ka within the Andaman Basin and 10 cm/ka in the adjacent BoB. Rodolfo's estimate is based on different biostratigraphic events and single radiocarbon dates, whereas the present estimate is based on several AMS radiocarbon dates.

Contribution from the Ayeyarwady constitutes 86% of the total sediment in the central trough (Rodolfo 1969). The outer shelf and continental slope are characterized by comparatively low sedimentation rates and the marine contribution of the

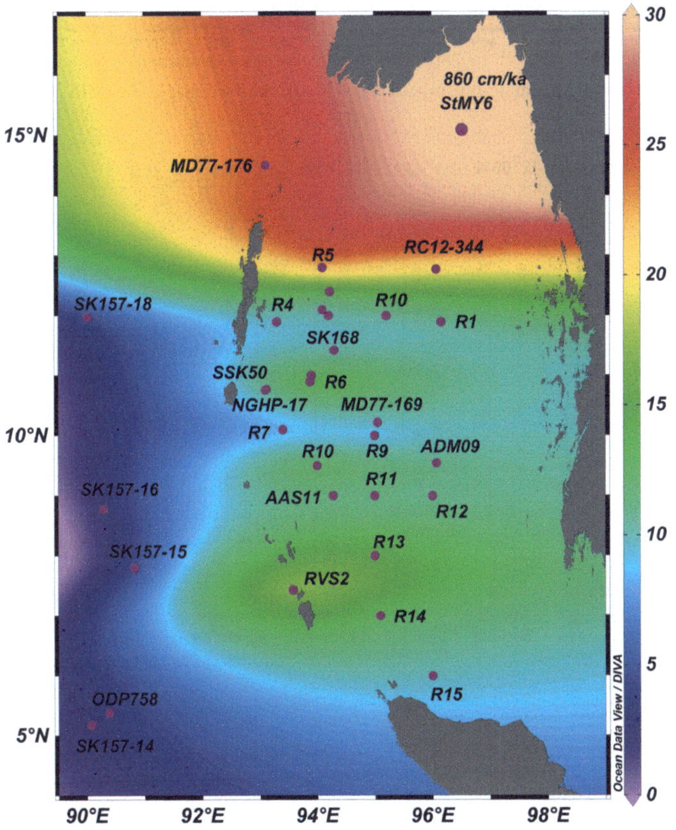

Fig. 2 Average sediment accumulation in the Andaman Sea (sample locations marked with R series are from Rodolfo 1969)

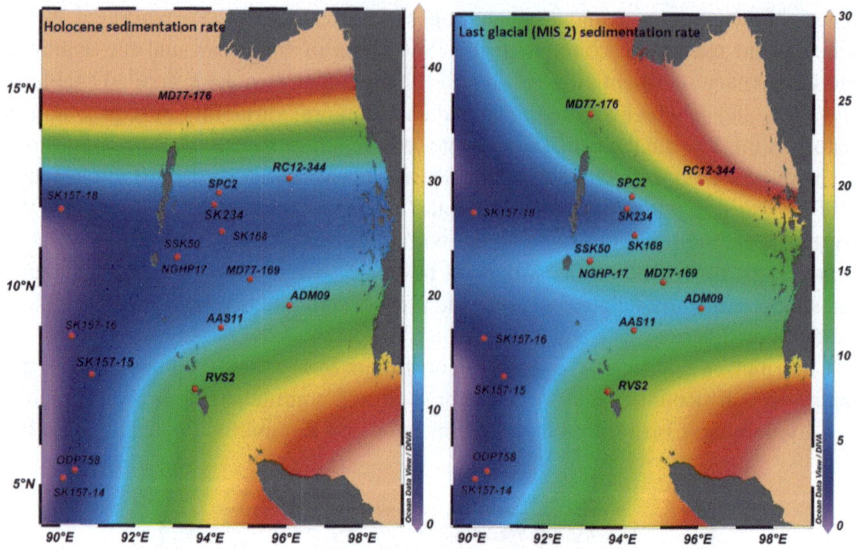

Fig. 3 Comparison of sedimentation rate between Holocene and last glacial (MIS 2) in the Andaman Sea

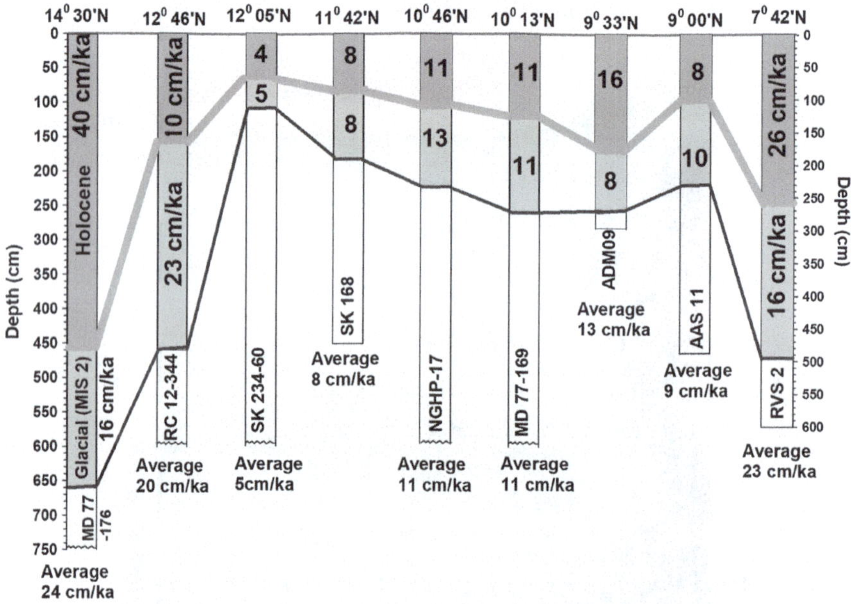

Fig. 4 Spatial and temporal variability in the sediment accumulation in the N-S Andaman Sea

organic matter becomes more dominant (Ramaswamy et al. 2004). The sediment accumulation and thickness varies from north to south, east to west and ridges/rises to basin. Sediment influx from the rivers are transported along the shelf by monsoon currents and carried to the deep sea-floor through submarine canyons (Ramaswamy et al. 2004; Rao et al. 2005). The controlling factors such as particle size, rate of sediment supply, currents and the geometry of the shelf influence sediment dispersal and transport pattern, with tidal and monsoon currents playing a major role in the Andaman Sea (Rao et al. 2005). The geology and tectonics of the Andaman Sea are extremely complex and the sediments of the Andaman Sea are largely derived from Ayeyarwady, Salween, Sittang and Tavoy rivers (Rodolfo 1969).

The high sedimentation rate is seen in the northern part of the basin is due to the proximity of river mouths. The variations in sedimentation is also seen between Holocene and last glacial, core MD77-176 in the northern part of the basin has shown a very high sedimentation rate during the Holocene over last glacial. Whereas sediment cores from the central basin have shown high sedimentation during the last glacial over Holocene (Figs. 3, and 4). The compiled available records show that the northernmost cores MD77-176 and RC 12-344 shows an average sedimentation rate of >20 cm/ka (Rashid et al. 2007; Marzin et al. 2013) and the southernmost core RVS 2 shows the high sedimentation rate of 23 cm/ka. The higher sedimentation rate of northern cores is possibly due to its proximity to the Ayeyarwady River mouth. Further close to the river mouth (~100 km from Ayeyarwady River mouth), very high sedimentation rate in the order of 860 cm/ka (sediment core StMY6) is reported (Ota et al. 2017). The eastern and inner Ayeyarwady delta-shelf is accumulating 90% of this sediment at a rate of 200 cm/100 years. The sedimentation rate of cores in the western and central part of the Andaman Sea ranges from 3.1 to 13 cm/ka as these locations are far from river mouths and positioned over ridges/rises. The rate of sedimentation and the total thickness in the Andaman Basin provides important clues to the age and tectonic history of the basin.

The sedimentation rate and the thickness of sediments in the Andaman Basin suggest that the basin is young (Rodolfo 1969). The longer record of sedimentation at site NGHP-17 has shown the varying rate of sedimentation since Miocene and the rates vary from 5 cm/ka during the Pleistocene to 13 cm/ka for the late Miocene (Flores et al. 2014). This estimation of 5 cm/ka during Pleistocene is corroborates with 5–6 cm/ka estimate for site U1448 (Clemens et al. 2016). Moderate to high sedimentation rate in this basin provides an opportunity to carry out high-resolution studies. The high sedimentation rate of RVS 2 (southern Andaman) shows that it acted as a local depositional centre close to the southern Andaman Nicobar ridges. Core RVS 2 also shows characteristic low sedimentation rate during last glacial which may be related to the sea-level low stand during LGM and related reduced runoff from the southern passages. Cores SK 168, RVS 2 and SSK 50 show high sedimentation rate during the Pleistocene to Holocene transition which is matching with increased Ayeyarwady River outflow during early Holocene (Rashid et al. 2007; Sijinkumar et al. 2016b; Sebastian et al. 2019) and enhanced ISM record (Fig. 5). Summer insolation is also high during Holocene last glacial transition marked by relatively high sedimentation rate (Fig. 4). The study also shows that there is high variability

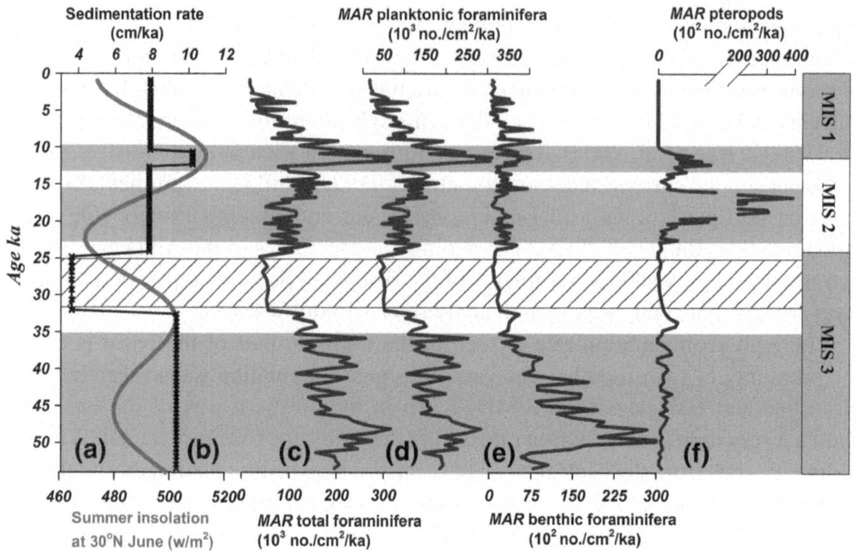

Fig. 5 Mass accumulation rate of skeletal carbonate (>125 μm) **a** summer insolation at 30° N (Berger 1978); **b** linear sedimentation rate; **c** mass accumulation rates of total foraminifera; **d** mass accumulation rates of planktonic foraminifera; **e** mass accumulation rates of benthic foraminifera; **f** mass accumulation rates of pteropods. Marine Isotope stages and periods of high and low *MARs* of carbonate were highlighted

in spatial as well as temporal sedimentation rate which is mainly related to the complex bottom topography with valleys and seamounts as well as ISM variability. The Andaman Sea has one of the highest sedimentation rates in the northern Indian Ocean and the higher sedimentation rate provides good and continuous palaeo-record of high-resolution palaeoclimatic studies.

5.2 *Record of* MAR *of Carbonate and Inferences on Productivity and ISM Variability*

The results from the Andaman Sea provide significant insights into the short- and long-term monsoon-controlled productivity history in the region. The present study has used a combination of geochemical (*MARs* of $CaCO_3$ and organic carbon) and *MARs* of carbonate skeletons (pteropods and foraminifera) to minimize uncertainties in palaeo-productivity reconstructions. *MARs* of organic carbon content (Müller and Suess 1979; Sarnthein et al. 1988), benthic foraminifer accumulation rate (Herguera 1992) and the species composition of foraminiferal assemblages (Mix 1989; Loubere 1991) were used in the past to quantify palaeoproductivity rates. Productivity proxies such as $CaCO_3$ and organic carbon, have shown a considerable variability with respect to last glacial to Holocene monsoon climate. The organic carbon content in the sediment is often closely related to the surface water productivity and thus the variations in oceanic sedimentary organic carbon can be used as a productivity proxy (Müller and Suess 1979; Pedersen and Calvert 1990).

The *MARs* of $CaCO_3$ increased drastically during the last glacial to Holocene transition when ISM was high. A sudden reduction in the *MARs* of carbonate, $CaCO_3$ and organic carbon was seen during the late glacial should have resulted from the weakening of ISM and resultant productivity in the Andaman Sea (Fig. 5). In the last 54 ka records, highest productivity is seen in last glacial-interglacial transition with an increase in ISM intensification whereas the winter monsoon must be responsible for medium to low productivity from LGM to late glacial. In contrast, this higher water column productivity during last glacial-interglacial transition is not reflected in the *MARs* of benthic foraminifera (Fig. 5) indicates only a small part of the organic matter produced in the upper ocean, reaches the abyssal floor (Heath et al. 1977). Calcium carbonate sedimentation is considerably more during the ISM and it has been documented, from the sediment trap studies that nearly 70% of carbonate deposited during the summer monsoon (Nair et al. 1989). In the Andaman Sea, no systematic changes are noticed in the productivity variations with respect to glacial and interglacial cycles. Geochemical proxies from the SE Arabian Sea exhibit higher productivity during interglacial than glacial (e.g., Pattan et al. 2003), whereas the NE Arabian Sea shows relatively high productivity during LGM (von Rad et al. 1999; Singh et al. 2011). According to Sarnthein et al. (1988), NE Arabian Sea productivity was higher during LGM than the early Holocene but comparable with that of late Holocene. Increased glacial productivity was also inferred from other cores of the SE Arabian Sea (Singh et al. 2011). The evidence from C_{37} alkenones and foraminiferal transfer function from the Western Arabian Sea also suggest that calcite productivity was higher during LGM than the early Holocene (Rostek et al. 1997). The low to moderate palaeo-productivity in the Andaman Sea during LGM may be the result of the strengthening of winter monsoonal winds which caused poor upwelling, good convective mixing and mixed layer deepening. In the eastern Arabian Sea, the high glacial productivity has driven by stronger glacial winter monsoon winds (Singh et al. 2011). Moderate to high productivity during glacial periods and low productivity during interglacial stages are evident from foraminiferal records of the non-upwelling area of the Gulf of Aden (Almogi-Labin et al. 2000).

The *MARs* of planktonic foraminifera were widely used for quantifying palaeo-productivity induced by monsoon variability. The temporal variations of *MARs* of planktonic foraminifera closely follow *MARs* of total foraminifera and range from 0.2 to 3.0×10^5 no./cm^2/ka (Fig. 5). In the Andaman Sea, a significant peak of high carbonate productivity is seen during the last glacial Holocene transition (10.5–12 cal ka BP) and marks the period of highest productivity. The ISM was maximum during the early Holocene in the Indian subcontinent too (Sukumar et al. 1993; Overpeck et al. 1996). Evidence from pollen studies also suggests higher monsoon precipitation during 11–9 ka BP and minimum during LGM in a record from southwest India (Van Campo 1986). ISM strengthening resulted in enhanced terrigenous inputs from the source region and increased sedimentation rates compared to periods prior to 11 ka.

This is consistent with the strengthened early Holocene monsoon induced massive sediment influx to the Ganges-Brahmaputra Delta (Karpytchev et al. 2018).

Pteropods also contribute a substantial amount of carbonate to the ocean floor but there are no sediment trap studies of pteropods from the BoB and the Andaman Sea. Singh and Conan (2008) have reported aragonite pteropod flux to the Somali basin by using sediment traps and estimated that about 22.5% of the total $CaCO_3$ in the Somali Basin was contributed by aragonitic pteropods (>125 μm). The *MARs* of pteropods in the Andaman Sea was completely absent during Holocene, driven by poor preservation due to increased ISM and shoaling of ACD during Holocene (Sijinkumar et al. 2010, 2015). A characteristic progressive reduction in the *MARs* of planktonic and benthic foraminifera was seen from 4 ka onwards, indicates a weakening of carbonate productivity which in turn is controlled by the deteriorating ISM. This decrease in skeletal carbonate *MARs* has corresponded well with lower *MARs* of $CaCO_3$, organic carbon and solar insolation (Fig. 5, and 6). The decrease in ISM during the late Holocene was also reported from the records from the Arabian Sea, the Andaman Sea and the BoB (Kudrass et al. 2001; Rashid et al. 2007; Achyuthan et al. 2014; Cao et al. 2015; Sijinkumar et al. 2016a).

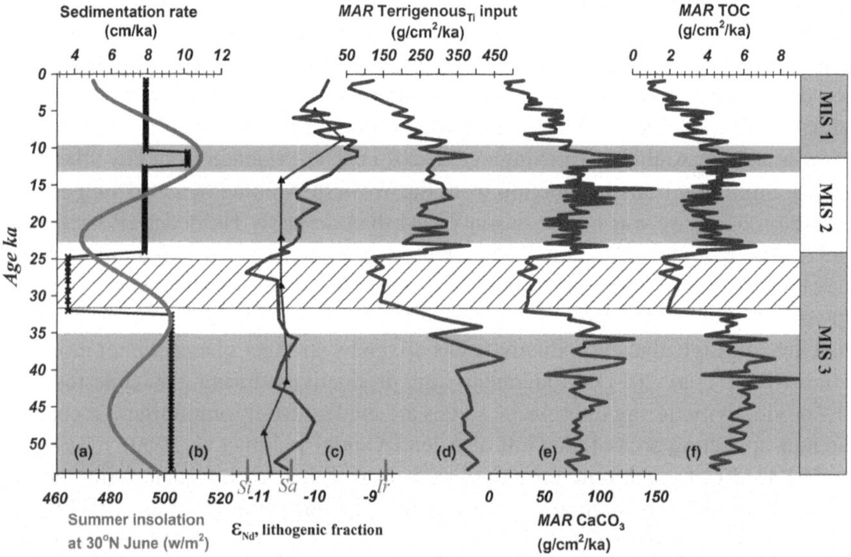

Fig. 6 Proxies for terrigenous input: **a** summer insolation at 30° N (Berger 1978); **b** linear sedimentation rate; **c** sediment provenance indicator ε_{Nd}, of the central Andaman Sea, MD77-169 (dark red, Colin et al. 2006) and SK168 (dark green, Miriyala et al. 2017); **d** mass accumulation rates of terrigenous input based on weight percentage titanium (Ti) in lithogenic fraction (see text for the lithogenic fraction); **e** and **f** Mass accumulation rates of calcium carbonate and total organic carbon content. The red markings on the x-axis of (**c**) indicates the ε_{Nd} isotopic values of surface sediments off river mouths in Myanmar, Ir-off Ayeyarwady, Sa-off Salween and Si-off Sittang

5.3 Late Quaternary Terrigenous Input to the Andaman Sea

The calculated *MARs* of terrigenous input varies in between 60 and 480 g/cm^2/ka and the temporal variation is very similar to the accumulation rates of total organic carbon and $CaCO_3$ suggesting the role of monsoon intensification in sediment removal processes (say erosion, weathering and runoff on land and deposition at sea) and the biological productivity in the oceans (Fig. 6). The major sediment sources to the Andaman Sea are fluvial associated with the Myanmar terrains, and can be related to physical and chemical erosion which can strongly depend on the ISM (Miriyala et al. 2017). While Colin et al. (1999) found a negligible contribution from the Andaman Sea volcanoes except for a minor input in the vicinity of Barren volcanoes, the contribution from volcanogenic sources to the Andaman Back Arc Basin was evident from sediment mineralogy (smectite crystallinity) and geochemistry (Kurian et al. 2008). The sediments of the northern Andaman Sea are sourced mainly from the discharge of the Ayeyarwady River, which is the longest river in Myanmar and ranks fifth in the world in terms of suspended load (Rashid et al. 2011; Awasthi et al. 2014). Ayeyarwady River discharge has been reported to be strongly influenced by ISM variations during the glacial-interglacial cycle (e.g., Colin et al. 1999; Miriyala et al. 2017; Sebastian et al. 2019).

The Ayeyarwady carries an annual suspended load of 364 MT and the Salween delivers nearly half of this (164 MT) to the Andaman Sea (Robinson et al. 2007). This is about 35% of the total suspended load delivered by the Ganges-Brahmaputra river system (Robinson et al. 2007). The high-resolution elemental and Sr-Nd isotopic study from a core retrieved from the Alcock Seamount Complex has shown that the Salween was the major source during glacial sea-level low stands, but with an increased influence of Ayeyarwady from deglacial to mid-Holocene as an effect of ISM intensification (Miriyala et al. 2017). The Neodymium isotopes (ε_{Nd}—provenance indicator) of shelf sediments recovered off Myanmar river mouths bracketing the entire variation of the central Andaman Sea sediment cores MD77-169 and SK168 at least since the last 280 ka (Fig. 6c), strongly suggests that the central Andaman Sea received terrigenous material mainly from Myanmar rivers like Ayeyarwady, Salween and Sittang (Miriyala 2017). These sediments, therefore, record the erosion and weathering history of the Indo-Burman Ranges, Central Myanmar basin through Ayeyarwady river watersheds and Sino-Burman ranges through Salween and Sittang river catchments (Colin et al. 1999; Miriyala et al. 2017; Sebastian et al. 2019). Rodolfo (1969) and Rao et al. (2005) proposed that most of the sediments discharged from Ayeyarwady are displaced towards Gulf of Martaban and not to Narcondam-Barren Basin as an effect of clock-wise movement of southwest monsoonal winds. Damodararao et al. (2016) have studied elemental and Sr-Nd isotopic studies on surface sediments of Martaban shelf which covers the northern and northeastern part of the basin and identified Mergui shelf has the lowest radiogenic (Nd) material of the basin and represents old crustal material deposited on the shelf.

The source area of the western Andaman Sea sediments is variously attributed. Using radiogenic isotopes, Ali et al. (2015), suggested a major contribution from the Indo-Burman Ranges and Central Myanmar basin to the Andaman Sea through Ayeyarwady River with a considerable contribution from the Andaman Islands. Awasthi et al. (2010) have found seven ash layers in a sediment core SK234/60 (sediment record of 71 ka) collected in the vicinity of the Barren Island, attributed mainly to the Barren volcanism. However, the absence of such ash layers in another sediment core SK168, in the vicinity, shows the limited role of volcanic material in terrigenous sediment budget of the basin. According to Awasthi et al. (2014), the provenance of western Andaman Sea sediments changed from an Ayeyarwady dominated to an Indo-Burman-Arakan to Andaman dominated regime since early MIS 2. An increase in silicate weathering with the initiation of interglacial warm climate at ~17.7 ka followed by a major depositional change at 15.5 ka was noticed by Miriyala et al. (2017). Based on multiproxy records, Sebastian et al. (2019), indicate a stronger ISM strength and increased chemical weathering during the deglacial, early Holocene and mid-Holocene periods.

6 Conclusions

The Andaman Sea has one of the highest sedimentation rates in the northern Indian Ocean and the higher sedimentation rate provides good and continuous palaeo-record of high-resolution palaeoclimatic studies. The study also shows that there is high variability in spatial as well as temporal sedimentation rate which is mainly related to the complex bottom topography with valleys and seamounts as well as ISM variability. The high sedimentation rate during the last glacial to Holocene transition marks the enhanced ISM intensity in the Andaman Sea. The average sedimentation rate of Andaman basin is calculated which works out to 13 cm/ka which is ideal for high-resolution Quaternary climate studies. The highest sedimentation rate of ~860 cm/ka is reported in the northern part of the Andaman Sea in the Ayeyarwady river mouth with a water depth of ~500 m, whereas the lower rate of sedimentation (~3.1 cm/ka) is found in the Western Andaman Sea. The sediments of this basin contain a substantial amount of carbonate which can widely be used for palaeoclimatic reconstructions. The *MARs* of carbonate were quantified and found large temporal variations in response to changes in ISM. Very high (~3.0×10^5 no./cm^2/ka) *MARs* of foraminifera was seen during last glacial Holocene transition, which is also coinciding with relatively high sedimentation rate and summer insolation in the 30°N. The *MARs* of skeletal carbonate were high during early to mid-MIS 3 and characteristically low during late MIS 3 and late Holocene owing to be driven by the ISM variability. The temporal variation in *MARs* of terrigenous input and the productivity indicators $CaCO_3$ and organic carbon were broadly synchronous suggesting the prevalence of unifying climatic conditions, viz., the ISM, responsible both for increased weathering on mainland Myanmar and the enhanced productivity in the Andaman Sea.

Acknowledgements Captain and crew of ORV Sagar Kanya are thanked for their help. Ministry of Earth Science Government of India has kindly provided the ship time. Directors of CSIR-NIO and CSIR-NGRI are acknowledged for permitting us to publish this work. AVS and PM acknowledge the support rendered by Head of the Department of the Marine Sciences, Goa University. We thank the two anonymous reviewers for their constructive comments, which helped us to improve the manuscript. AVS acknowledge SERB New Delhi for the ECR Grant (No. ECR/2017/000818).

References

Achyuthan H, Nagasundaram M, Gourlan AT, Eastoe C, Ahmad SM, Padmakumari VM (2014) Mid-Holocene Indian Summer Monsoon variability off the Andaman Islands, Bay of Bengal. Quat Int 349:232–244

Ahmad SM, Patil DJ, Rao PS, Nath BN, Rao BR, Rajagopalan G (2000) Glacial-interglacial changes in the surface water characteristics of the Andaman Sea: evidence from stable isotopic ratios of planktonic foraminifera. Proc Indian Acad Sci (Earth Planet Sci) 109:153–156

Alagarsamy R, You C-F, Nath BN, Sijinkumar AV (2010) Determination of rare earth, major and trace elements in authigenic fraction of Andaman Sea (North-eastern Indian Ocean) sediments by inductively coupled plasma-mass spectrometry. Microchem J 94:90–97

Ali S, Hathorne EC, Frank M, Gebregiorgis D, Stattegger K, Stumpf R, Kutterolf S, Johnson JE, Giosan L (2015) South Asian monsoon history over the past 60 kyr recorded by radiogenic isotopes and clay mineral assemblages in the Andaman Sea. Geochem Geophys Geosyst 16:505–521

Almogi-Labin A, Schmiedl G, Hemleben C, Siman-Tov R, Segl M, Meischner D (2000) The Influence of the NE winter monsoon on productivity changes in the Gulf of Aden, NW Arabian Sea, during the last 530 Ka as recorded by foraminifera. Mar Micropaleontol 40:295–319

Awasthi N, Ray JS, Laskar AH, Kumar A, Sudhakar M, Bhutani R, Sheth HC, Yadava MG (2010) Major ash eruptions of Barren Island volcano (Andaman Sea) during the past 72 kyr: clues from a sediment core record. Bull Volcanol 72:1131–1136

Awasthi N, Ray JS, Singh AK, Band ST, Rai VK (2014) Provenance of the Late Quaternary sediments in the Andaman Sea: implications for monsoon variability and ocean circulation. Geochem Geophys Geosyst 15:3890–3906

Berger WH (1978) Deep-Sea carbonate: pteropod distribution and the aragonite compensation depth. Deep-Sea Res 25:447–452

Berger WH (1989) Global maps of the ocean productivity, In: Berger WH, Smetacek VS, Wefer G (eds) Productivity of the ocean: present and past. Dahlem Workshop Reports, Wiley, Chichester, pp 429–455

Bhattacharjee D (2005) Pteropods preservation profiles in seabed sediments off Middle Andaman Island in Andaman Sea. Indian J Mar Sci 34:259–266

Bradley RS (1999) Paleoclimatology: reconstructing climates of the quaternary. Academic Press, San Diego

Brummer GJA, van Eijden AJM (1992) "Blue-ocean" paleoproductivity estimates from pelagic carbonate mass accumulation rates. Mar Micropaleontol 19:99–117

Cao P, Shi X, Li W, Liu S, Yao Z, Hu L, Khokiattiwong S, Kornkanitnan N (2015) Sedimentary responses to the Indian Summer Monsoon variations recorded in the southeastern Andaman Sea slope since 26ka. J Asian Earth Sci 114:512–525

Chernova TG, Rao PS, Pikovskii YuI, Alekseeva TA, Nath BN, Rao BR, Rao ChM (2001) The composition and source of hydrocarbons in sediments taken from the tectonically active Andaman Backarc Basin, Indian Ocean. Mar Chem 75:1–15

Clemens SC, Kuhnt W, LeVay LJ, Anand P, Ando T, Bartol M, Bolton CT, Ding X, Gariboldi K, Giosan L, Hathorne EC, Huang Y, Jaiswal P, Kim S, Kirkpatrick JB. Littler K, Marino G, Martinez P, Naik D, Peketi A, Phillips SC, Robinson MM, Romero OE, Sagar N, Taladay KB, Taylor

SN, Thirumalai K, Uramoto G, Usui Y, Wang J, Yamamoto M, Zhou L (2016) Expedition 353 summary. In: Clemens SC, Kuhnt W, LeVay LJ, The Expedition 353 Scientists (eds) Indian monsoon rainfall. Proceedings of the International Ocean Discovery Program 353, College Station, TX (IODP) http://dx.doi.org/10.14379/iodp.proc.353.101.2016

Colin C, Kissel C, Blamart D, Turpin L (1998) Magnetic properties of sediments in the Bay of Bengal and the Andaman Sea: impact of rapid North Atlantic Ocean climatic events on the strength of the Indian monsoon. Earth Planet Sci Lett 160:623–635

Colin C, Turoin L, Bertaux J, Desprairies A, Kissel C (1999) Erosional history of the Himalayan and Burman ranges during the last two glacial-interglacial cycles. Earth Planet Sci Lett 171:647–660

Colin C, Turpin L, Blamart D, Frank N, Kissel C, Duchamp S (2006) Evolution of weathering patterns in the Indo-Burman Ranges over the last 280 kyr: effects of sediment provenance on $^{87}Sr/^{86}Sr$ ratios tracer. Geochem Geophys Geosyst 7:Q03007. https://doi.org/10.1029/2005GC000962

Cullen JL, Prell WL (1984) Planktonic foraminifera of Northern Indian Ocean: distribution and preservation in surface sediments. Mar Micropaleontol 9:1–52

Damodararao K, Singh SK, Rai VK, Ramaswamy V, Rao PS (2016) Lithology, monsoon and sea-surface current control on provenance, dispersal and deposition of sediments over the Andaman continental shelf. Front Mar Sci 3(118). https://doi.org/10.3389/fmars.2016.00118

Delaygue G, Bard E, Rollion C, Jouzel J, Stievenard M, Duplessy JC, Ganssen G (2001) Oxygen isotope/salinity relationship in the northern Indian Ocean. J Geophys Res 106:4565–4574

Duplessy JC (1982) Glacial to interglacial contrasts in the northern Indian Ocean. Nature 295:494–498

Flores JA, Johnson JE, Mejía-Molina AE, Alvarez MC, Sierro FJ, Singh SD, Mahanti S, Giosan L (2014) Sedimentation rates from calcareous nannofossil and planktonic foraminifera biostratigraphy in the Andaman Sea, northern Bay of Bengal, and eastern Arabian Sea. Mar Petrol Geol 58:425–437

Frerichs WE (1968) Pleistocene-recent boundary and Wisconsin glacial biostratigraphy in the northern Indian Ocean. Science 1959:1456–1458

Frerichs WE (1970) Distribution and ecology of benthonic foraminifera in the sediments of the Andaman Sea. J Foramin Res 21:123–147

Frerichs WE (1971) Paleobathymetric trends of Neogene foraminiferal assemblages and sea floor tectonism in the Andaman Sea area. Mar Geol 11:159–173

Gauns M, Madhupratap M, Ramaiah N, Jyothibabu R, Fernandes V, Paul JT, Prasanna Kumar S (2005) Comparative accounts of biological productivity characteristics and estimates of carbon fluxes in the Arabian Sea and the Bay of Bengal. Deep Sea Res 52:2003–2017

Gebregiorgis D, Hathorne EC, Sijinkumar AV, Nath BN, Nürnberg D, Frank M (2016) South Asian summer monsoon variability during the last ~54 kyrs inferred from surface water salinity and river runoff proxies. Quat Sci Rev 138:6–15

Gebregiorgis D, Hathorne EC, Giosan L, Clemens S, Nürnberg D, Frank M (2018) Southern Hemisphere forcing of South Asian monsoon precipitation over the past ~1 million years. Nat Commun 9:4702. https://doi.org/10.1038/s41467-018-07076-2

Girishbai D, Varghese S, Dinesh AC, Bhadran A, Joshi RK (2017) Provenances of possible turbidites and volcanoclastic sediments in the northern part of Narcondam-Barren Basin. Andaman Sea. Arab J Geosci 10:39. https://doi.org/10.1007/s12517-016-2825-8

Heath GR, Moore TC, Dauphin JP (1977) Organic carbon in deep-sea sediments. In: Andersen NR, Malahoff A (eds) The fate of fossil fuel CO_2 in the oceans. Plenum Press, pp 605–625

Herguera JC (1992) Deep-sea benthic foraminifera and biogenic opal: glacial to postglacial productivity changes in the western equatorial Pacific. Mar Micropaleontol 19:79–98

Janekarn V, Hylleberg J (1989) Coastal and offshore primary production along the west coast of Thailand (Andaman Sea) with notes on physical chemical variables. Phuket Mar Biol Centre Res Bull 51:1–19

Kamesh Raju KA, Murty GP, Amarnath D, Kumar MM (2007) The west Andaman fault and its influence on the aftershock pattern of the recent megathrust earthquakes in the Andaman-Sumatra region. Geophys Res Lett 34(3). https://doi.org/10.1029/2006gl028730

Kamesh Raju KA, Ray D, Mudholkar A, Murty GPS, Gahalaut VK, Samudrala K, Paropkari AL, Ramachandran R, Prakash LS (2012) Tectonic and volcanic implications of a cratered seamount off Nicobar Island, Andaman Sea. J Asian Earth Sci 56:42–53

Karpytchev M, Ballu V, Krien Y, Becker M, Goodbred S, Spada G, Calmant S, Shum CK, Khan Z (2018) Contributions of a strengthened early Holocene monsoon and sediment loading to present-day subsidence of the Ganges-Brahmaputra delta. Geophys Res Lett 45:1433–1442. https://doi.org/10.1002/2017GL076388

Kudrass HR, Hofmann A, Doose H, Emeis K, Erlenkeuser H (2001) Modulation and amplification of climatic changes in the Northern Hemisphere by the Indian summer monsoon during the past 80 k.y. Geology 29:63–66

Kumar PK, Band ST, Ramesh R, Awasthi N (2018) Monsoon variability and upper ocean stratification during the last ~66 ka over the Andaman Sea: inferences from the δ18O records of planktonic foraminifera. Quatern Int 479:12–18

Kurian S, Nath BN, Ramaswamy V, Naman D, Rao TG, Kamesh Raju KA, Selvaraj K, Chen CTA (2008) Possible detrital, diagenetic and hydrothermal sources for Holocene sediments of the Andaman Backarc Basin. Mar Geol 247:178–193

Loubere P (1991) Deep-sea benthic foraminiferal assemblage response to a surface ocean productivity gradient: a test. Paleoceanography 6:193–204

Lyle M (1988) Climatically forced organic carbon burial in equatorial Atlantic and Pacific Oceans. Nature 335:529–532

Marzin C, Kallel N, Kageyama M, Duplessy J-C, Braconnot P (2013) Glacial fluctuations of the Indian monsoon and their relationship with North Atlantic climate: new data and modelling experiments. Clim Past 9:2135–2151

Mazumdar D, Sharma V (1991) Late Miocene (Neillian) planktonic foraminifera from Baratang Island, Andaman Sea. J Geol Soc India 37:482–491

Milliman JS, Meade RH (1983) World-wide delivery of river sediment to the oceans. J Geol 91:1–21

Miriyala P (2017) Late Quaternary geochemical history of lithogenic sedimentation in the Andaman Sea. Goa University

Miriyala P, Sukumaran NP, Nath BN, Ramamurty PB, Sijinkumar AV, Vijayagopal B, Ramaswamy V, Sebastian T (2017) Increased chemical weathering during the deglacial to mid-Holocene summer monsoon intensification. Sci Rep 7(44310):600. https://doi.org/10.1038/srep44310

Mix AC (1989) Influence of productivity variations on long-term atmospheric CO_2. Nature 337:541–544

Müller PJ, Suess E (1979) Productivity, sedimentation rate and sedimentary organic matter in the oceans. Deep Sea Res 26:1347–1362

Nagasundaram M, Achyuthan H, Ahmad SM (2014) Monsoonal changes inferred from the Middle to Late Holocene sediments off Landfall Island, North Andaman. Arab J Geosci 7:3513–3523

Nair RR, Ittekkot V, Manganini SJ, Ramaswamy V, Haake B, Degens ET, Desai BN, Honjo S (1989) Increased particulate flux to the deep ocean related to monsoon. Nature 338:749–751

Naqvi WA, Charles CD, Fairbanks RG (1994) Carbon and oxygen isotopic records of benthic foraminifera from the Northeast Indian Ocean: implications on glacial-interglacial atmospheric CO_2 changes. Earth Planet Sci Lett 121:99–110

Ota Y, Kawahata H, Murayama M, Inqoue M, Yokoyama Y, Miyari Y, Aung T, Hossain Z, Suzuki A, Kitamura A, Moe KT (2017) Effects of intensification of the Indian Summer Monsoon on northern Andaman Sea sediments during the past 700 years. J Quat Sci 32:528–539

Overpeck J, Anderson D, Trumbore S, Prell W (1996) The southwest Indian Monsoon over the last 18 000 years. Clim Dyn 12:213–225

Panchang R, Nigam R, Riedel F, Janssen AW, Hla UKY (2007) A review of the studies on pteropods from the northern Indian Ocean region with a report on the pteropods of Ayeyarwady continental shelf off Myanmar (Burma). Indian J Mar Sci 36:384–398

Paropkari AL, Prakashbabu C, Mascarenhas A (1993) New evidence for enhanced preservation of organic carbon in contact with oxygen minimum zone on the western continental slope of India. Mar Geol 111:7–13

Pattan JN, Masuzawa T, Naidu PD, Parthiban G, Yamamoto M (2003) Productivity fluctuations in the southeast Arabian Sea during the last 140 Ka. Palaeogeogr Palaeoclimatol Palaeoecol 193:575–590

Pedersen TF, Calvert SE (1990) Anoxia vs. Productivity: what controls the formation of organic carbon rich sediments and sedimentary rocks. Am Asso Petrol Geol Bull 74:454–466

Perry GD, Duffy PB, Miller NL (1996) An extended data set of river discharges for validation of general circulation models. J Geophys Res 101:21339–21349

Potemra JT, Luther ME, O'Brien JJ (1991) The seasonal circulation of the upper ocean in the Bay of Bengal. J Geophys Res 96:12667–12683

Prahl FG, de Lange GJ, Lyle M, Sparrow MA (1989) Post depositional stability of long-chain alkenones under contrasting redox conditions. Nature 341:434–437

Ramaswamy V, Rao PS, Rao KH, Thwin S, Srinivasa Rao N, Raiker V (2004) Tidal influence on suspended sediment distribution and dispersal in the northern Andaman Sea and Gulf of Martaban. Mar Geol 208:33–42

Ramaswamy V, Gaye B, Shirodkar PV, Rao PS, Chivas AR, Wheeler D, Thwin S (2008) Distribution and sources of organic carbon, nitrogen and their isotopic signatures in sediments from the Ayeyarwady (Ayeyarwady) continental shelf, northern Andaman Sea. Mar Chem 111:137–150

Rao VP (1983) Clay minerals in the sediments around the Andaman Islands. Indian J Mar Sci 12:17–20

Rao LVG, Jayaraman R (1968) Hydrographical features of the southern and central Bay of Bengal during the transition period between winter and summer. Bull National Sci India 38:184–205

Rao PS, Kamesh Raju KA, Ramprasad T, Nath BN, Rao BR, Rao ChM, Nair RR (1996) Evidence for hydrothermal activity in the Andaman Backarc Basin. Curr Sci 70:379–385

Rao PS, Ramaswamy V, Thwin S (2005) Sediment texture, distribution and transport on the Ayeyarwady continental shelf, Andaman Sea. Mar Geol 216:239–247

Rashid H, Flower BP, Poore RZ, Quinn TM (2007) A ~25 ka Indian Ocean monsoon variability record from the Andaman Sea. Quatern Sci Rev 26:2586–2597

Rashid H, England E, Thompson L, Polyak L (2011) Late glacial to Holocene Indian summer monsoon variability based upon sediment records taken from the Bay of Bengal. Terr Atmos Oceanic Sci 22:215–228

Robinson RAJ, Bird MI, Oo NW, Hoey TB, Aye MM, Higgitt DL, Swe A, Tun T, Win SL (2007) The Ayeyarwady River sediment flux to the Indian Ocean: the original nineteenth-century data revisited. J Geol 115:629–640

Rodolfo K (1969) Sediments of the Andaman Basin, north eastern Indian Ocean. Mar Geol 7:371–402

Roonwal GS, Glasby GP, Roelandts I (1997) Mineralogy and geochemistry of surface sediments from the Andaman Sea and upper Nicobar Fan, northeast Indian Ocean. J Indian Assoc Sedimentol 16:89–101

Rostek F, Bard E, Beaufort L, Sonzogni C, Ganssen G (1997) Sea surface temperature and productivity records for the past 240 ka in the Arabian Sea. Deep Sea Res Part II 44:1461–1480

Saidova KhM (2008) Benthic foraminifera communities of the Andaman Sea (Indian Ocean). Oceanology 48:517–523

Sarma VVSS, Narvekar PV (2001) A study on inorganic carbon components in the Andaman Sea during the post monsoon season. Oceanol Acta 24:125–134

Sarnthein M, Winn K, Duplessy JC, Fontugne MR (1988) Global variations of surface ocean productivity in low and mid latitudes: influence on CO_2 reservoirs of the deep ocean and atmosphere during the last 21,000 years. Paleoceanography 3:361–399

Sengupta R, Moraes C, George M, Kureishy T, Noronha R, Fondekar S (1981) Chemistry and hydrography of the Andaman Sea. Indian J Mar Sci 10:228–233

Shimmield GB, Derrick S, Mackensen A, Grobe H, Pudsey C (1994) The history of barium, biogenic silica and organic carbon accumulation in the Weddell Sea and Antarctic Ocean over the last 150,000 years. In: Zahn R (ed) Carbon cycling in the Glacial Ocean: constraints on the ocean's role in global change. Springer, New York, pp 555–574

Sebastian T, Nath BN, Venkateshwarlu M, Miriyala P, Prakash A, Linsy P, Kocherla M, Kazip A, Sijinkumar AV (2019) Impact of the Indian Summer Monsoon variability on the source area weathering in the Indo-Burman ranges during the last 21 kyr—a sediment record from the Andaman Sea. Palaeogeogr Palaeoclimatol Palaeoecol 516:22–34

Sharma V, Mazumdar D (1993) Palaeoecology of Late Miocene smaller benthic foraminifera from Baratang Island, Andaman Sea. J Palaeontol Soc India 138:7–15

Sijinkumar AV (2011) Paleoceanographic investigations of sediments from the Andaman Sea. Goa University

Sijinkumar AV, Nath BN, Gupta MVS (2010) Late Quaternary record of pteropod preservation from the Andaman Sea. Mar Geol 275:221–229

Sijinkumar AV, Nath BN, Possnert G, Aldahan A (2011) Pulleniatina minimum events in the Andaman Sea (NE Indian Ocean): implications for winter monsoon and thermocline changes. Mar Micropaleontol 81:88–94

Sijinkumar AV, Nath BN, Guptha MVS, Ahmad SM, Rao BR (2015) Timing and preservation mechanism of deglacial pteropods spike from the Andaman Sea, Northeastern Indian Ocean. Boreas 44:432–444

Sijinkumar AV, Clemens S, Nath BN, Prell W, Benshila R, Lengaigne M (2016a) $\delta^{18}O$ and salinity variability from the Last Glacial Maximum to recent in the Bay of Bengal and Andaman Sea. Quat Sci Rev 135:79–91

Sijinkumar AV, Nath BN, Clemens S (2016b) Rapid climate changes in the North Atlantic is reflected in the foraminiferal abundance in the Andaman Sea. Palaeogeogr Palaeoclimatol Palaeoecol 446:11–18

Singh AD, Conan SMH (2008) Aragonite pteropod flux to the Somali Basin, NW Arabian Sea. Deep Sea Res Part I 55:661–669

Singh AD, Jung SJA, Darling K, Ganeshram R, Ivanochko T, Kroon D (2011) Productivity collapses in the Arabian Sea during glacial cold phases. Paleoceanography Paleoclimatology 26:PA3210. https://doi.org/10.1029/2009pa001923

Srinivasan MS, Azmi RJ (1979) Correlation of Late Cenozoic marine sections in Andaman-Nicobar, northern Indian Ocean and the equatorial Pacific. J Paleontol 53:1401–1415

Sukumar R, Ramesh R, Pant RK, Rajagopalan G (1993) A ^{13}C record of Late Quaternary climate change from tropical peats in southern India. Nature 364:703–706

Trenberth KE, Guillemot CJ (1998) Evaluation of the atmospheric moisture and hydrological cycle in the NCEP/NCAR re-analyses. Clim Dyn 14:213–231

Tripathi SK (2014) Biogenic sediment distribution around south of Central Andaman Trough, Andaman Sea: signatures from micropaleontological studies. Indian J Geosci 68:337–346

Van Campo E (1986) Monsoon fluctuations in two 20,000-yr B.P. oxygen-isotope/pollen records off southwest India. Quatern Res 26:376–388

Varkey MJ, Murthy VSN, Suryanarayana A (1996) Physical oceanography of the Bay of Bengal and Andaman Sea. Oceanogr Mar Biol Annu Rev 34:1–70

Venkatesan MI, Ruth E, Rao PS, Nath BN, Rao BR (2003) Hydrothermal petroleum in the sediments of the Andaman Backarc Basin, Indian Ocean. Appl Geochem 18:845–861

von Rad U, Schulz H, Riech V, den Dulk M, Berner U, Sirocko F (1999) Multiple monsoon-controlled breakdown of oxygen-minimum conditions during the past 30,000 years documented in laminated sediments off Pakistan. Palaeogeogr Palaeoclimatol Palaeoecol 152:129–161

Wyrtki K (1971) Oceanographic Atlas of the International Indian Ocean Expedition, U.S. Government Printing Office, Washington, D.C

Mid to Late Holocene Reconstruction of the Southwest Monsoonal Shifts Based on a Marine Sediment Core, off the Landfall Island, Bay of Bengal

M. Nagasundaram, Hema Achyuthan and Jyotsna Rai

Abstract Undisturbed marine sediment cores raised from the oceans hold continuous records of sedimentation and palaeoenvironmental changes, hence it is suitable material for reconstructing past climate fluctuations. In the present study a marine sediment core about 124 cm raised from a depth of 250 m, near the Landfall Island, North Andaman, Bay of Bengal (BOB) was used to reconstruct the south west monsoonal shifts over the Indian sub-continent and BOB during the mid to late Holocene by using various proxies such as clay mineralogy, oxygen ($\delta^{18}O$) and carbon ($\delta^{13}C$) stable isotopes, and nanoplankton. Five organic carbon sediment samples were radiocarbon dated. The calibrated radiocarbon date ranges in age from (6078 to 1658 years BP) that is from the mid to late Holocene period. The texture analysis indicates that the sediments are predominantly clayey silt in nature. Smectite, illite, kaolinite and chlorite were the clay minerals present in the sediment core in which smectite and illite are dominant. The high smectite content reflects the weathering product of mafic rock contributed from the nearby island whereas illite is predominantly coming from the rivers such as Irrawaddy, Salween and Sittang rivers from Myanmar. Occurrence of nanoplankton such as *Gephyrocapsa oceanica, Emiliania huxleyi, and Ascidian* spicules in the calcareous nanoplankton assemblage suggest a mixed source also from the sedimentary rocks of the Mio-Pliocene age. The sediment core reveals layers of coarser sand flux since ~6500–6000 years BP and ~3300 years BP that reflects a strengthened South West Monsoon (SWM) in an overall weakening of the SWM from the middle Holocene (6000 yrs BP) to the late Holocene period (2000 years BP). This is also inferred from the high smectite, high C/I ratio and low K/C ratio. Within this period, an intense weakening of the SWM is noticed at ~4400–4200 years BP (Sub—Boreal Optimum), which is also supported by the $\delta^{18}O$ data of *G. ruber* becoming more positive from −3.39 to −2.33‰. A major wet phase was found reaching its maximum around 3400–3200 years BP and amelioration in

M. Nagasundaram
Geological Survey of India, Kolkata, India

H. Achyuthan (✉)
Department of Geology, Anna University, Chennai 600025, India
e-mail: hachyuthan0@gmail.com

J. Rai
Birbal Sahni Institute of Palaeosciences, Lucknow 226007, India

J. S. Ray and M. Radhakrishna (eds.), *The Andaman Islands and Adjoining Offshore: Geology, Tectonics and Palaeoclimate*, Society of Earth Scientists Series,
https://doi.org/10.1007/978-3-030-39843-9_15

climate ~2000 years BP with a warm dry phase ~2200–1800 years BP (Roman Warm Period) followed by the Medieval Warm Period (1000–800 years BP).

Keyword Landfall island · Marine sediment core · SWM · Mid-late Holocene

1 Introduction

Sediment deposition in the Bay of Bengal (BOB) is governed by the dominantly detrital input even though there are other two sources such as biogenic (e.g. calcareous ooze, carbonate ooze, petropods, Coccolithophores, microfossils) and diagenetic (deformation of sedimentary layers, solution structures, occurrence of pyrite crystals etc.). Its input reveals the nature of the physical and chemical erosion of the terrain and it is strongly dependent on the monsoon intensity, especially the southwest or summer monsoon (Bhushan et al. 2007; Sarin et al. 1979). Nath et al. (2005) used sedimentological and geochemical proxies in order to show the evidence of Himalayan erosional events in a sediment core from the equatorial Indian Ocean. Rare earth elements (REE) and other geochemical data (Kurian et al. 2008) suggest a strong influence of the Irrawaddy River as a major continental source of the Andaman Backarc Basin (ABB) sediments. Kessarkar et al. (2005) studied sediment characteristics and radiogenic isotope of the detritus in sediment cores from the distal Bengal Fan and reported changing sedimentary environment at about 12,000 years BP and Pleistocene hemiturbidites, older than 13,000 years BP, with major source contributions from the Himalayan drainage systems. The hemiturbidites were overlain by calcareous pelagic sediments in the past 12,000 years BP. The change in litho facies suggests that the sediment deposition by turbidity current activity ceased in the distal Bengal Fan at ~12,000 years BP, and this was perhaps due to rapid sea level rise during the melt water pulse of the Early Holocene period.

A detailed review on monsoonal shifts (Nigam and Hashimi 1995; Gupta and Thamban 2008; Staubwasser and Weiss 2006; Kuppusamy and Ghosh 2012; Ponton et al. 2012; Patnaik et al. 2012; Achyuthan et al. 2017) indicate that limited studies have been carried out in the BOB as compared to the Arabian Sea. Rashid et al. (2011) reconstructed the sea-surface temperature (SST) and salinity from paired $\delta^{18}O$ and Mg/Ca measurements in planktonic foraminifera *Globigerinoides ruber* from the western BOB core VM29-19. This indicated that the monsoon was stronger in the Bølling/Allerød period and weaker during the Younger Dryas. They observed; post ~5000 years BP the Indian summer monsoon significantly weakened, indicating less dilution of the sea surface waters by the Ganga-Brahmaputra-Meghna outflow and/or less direct rainfall (i.e. more saline sea surface waters). This was due to the prevailing late Holocene dry climate such as observed in the subsequent abandonment of the settlements of the great Indus Valley Civilizations. These studies highlight the utility of sediment cores from the BOB to study palaeomonsoon and its variability over different time scales and resolution (Ponton et al. 2012; Giosan et al. 2018). It is however appropriate to mention here that marine sediment cores can provide monsoon precipitation related information if the oceanic salinity changes can be calculated after delineating independently determined temperature from the oxygen isotope values of the planktic foraminifera. Likewise, surface run-off proxies such as the terrestrial sedimentary fractions can also provide precipitation related information. These are proxies to monsoon winds and as well as actual rainfall on the land and it depends on the proxies used to explain the different processes. This aspect has often been missed in interpreting higher upwelling to higher rainfall. To add on; an interesting study of the ancient texts such as Arthashastra, Nirukta, Ramayana etc., along with modern scientific data, it was observed that the periods of water stress as inferred from the texts seem to correlate with the periods of low monsoon winds as inferred from the oceanic records (Pandey et al. 2003).

The present work aims to study the influence of various global climate phenomena, such as Roman Warm Period (RWP, 2200–1800 years BP) (Desprat et al. 2003; Vollweiler et al. 2006; Wang et al. 2013), the Medieval Warm Period (MWP, 1100–700 years BP) (Hughes and Diaz 1994; Mann 2002a; Sinha et al. 2007), and the Little Ice Age (LIA, 600–200 years BP) (Mann 2002b) on the SWM during the mid to the late Holocene using marine sediment core collected off the Landfall Island, Andaman Islands.

During the monsoon season, depressions and cyclones, mainly originate in the BOB and are the chief rainfall producing synoptic weather systems over India (Dash et al. 2004). The Andaman Islands are located in the prime monsoon zone, sensitive to depression and cyclone formation and hence marine sediment cores retrieved from here could be very good palaeoclimate archives. The BOB receives large quantities of fresh water from the adjacent coastal mountain rivers ($1.6 \times 1012\ m^3yr^{-1}$, Subramanian 1993) and over 1350 million tonnes of suspended sediment every year (Milliman and Meade 1983; Subramanian 1985). The lithogenic sediments in the BOB derived from erosion of the Himalayas and Indo-Burma ranges, and brought by the Ganga-Brahmaputra and Irrawaddy rivers into the basin holds the signature in understanding monsoon variations, weathering patterns, and sediment provenance (Colin et al. 1999). Hence it has attracted a fair attention with regard to its sedimentation geology notably the origin and history of the Bengal sediments (Stewart et al. 1965), however the geochemical investigation of this vast expanse is limited and confined to the deep (Ramesh and Ramasamy 1997). Moreover, till date, no, high-resolution evidence on sediment weathering, sediment source, and their palaeoclimatic implications from the Andaman Forearc Basin, off the Landfall Island, BOB has been presented. In the BOB, Landfall Island is an ideal and suitable site to retrieve marine sediment core to study and reconstruct the relationships between continental weathering and palaeomonsoonal shifts. The main aim and objectives of the present study are: (a) To reconstruct a high resolution palaeoclimate record from Middle to late Holocene period. (b) To delineate the rate of sedimentation varying with southwest monsoonal variation.

2 The Study Area

The Landfall Island is located north of the Andaman Island forearc basin (Fig. 1). The geology of the area consists of conglomerate, sandstone and shale of the Mithakari Group (Pal et al. 2003) are underlain by volcanic rocks (the ophiolite suite). The ophiolitic suite is represented by andesite and dacite where plagioclase, orthopyroxene, clinopyroxene, olivine, titano-magnetite, hornblende minerals are the most common minerals with sometime some appearance of quartz grains (Ray et al. 2011). The area near the Landfall Island receives dominantly SWM and less intense rains during NE monsoon period. Meteorological data from Port Blair shows that the present-day climate is humid with a mean annual precipitation of 3168 mm (average of 30 years:

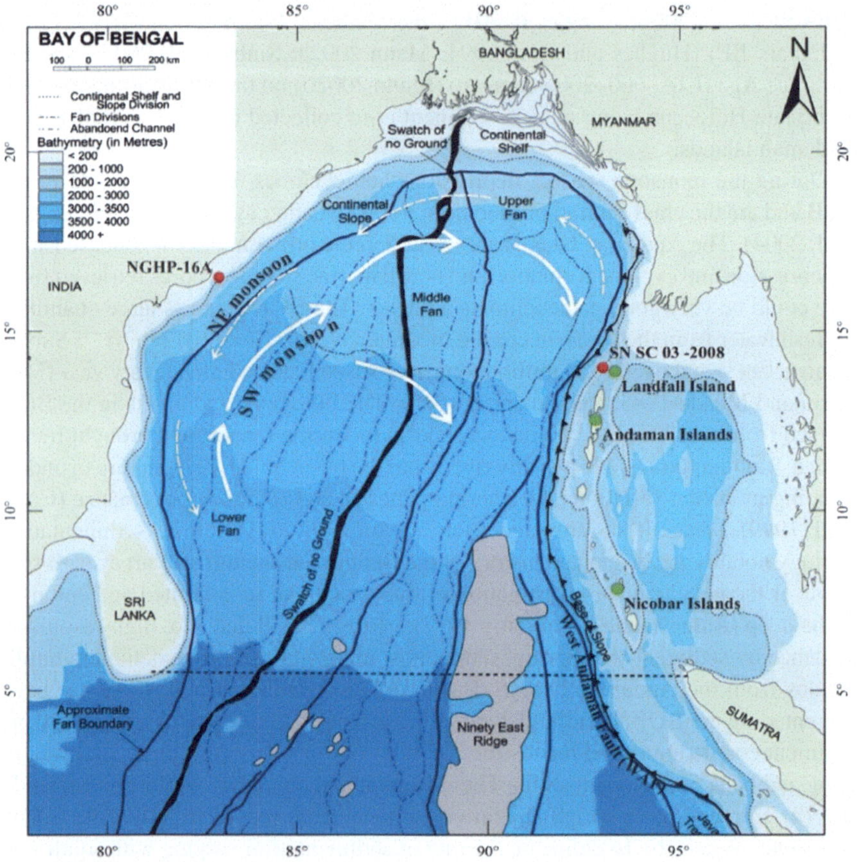

Fig. 1 Map showing the sediment core location (SN SC 03 2008) near the Landfall Island with the regional hydrography (prepared after Shetye et al. 1991, 1993; Varkey et al. 1996; Chauhan and Vogelsang 2006 and the site of the marine sediment core (NGHP-16A) collected in the Godhavari delta studied by Ponton et al. (2012)

1951–1980 CE). The rain mainly occurs between May and October due to SWM (Laskar et al. 2011). The long-term mean annual surface temperature and humidity are 26.6 °C and 82%, respectively. The BOB is an ideal site to decipher the palaeo-climate change in the Indian subcontinent (Chauhan and Suneethi 2001), since the BOB receives the largest freshwater influx and sediments from the Himalayas and the Indian peninsular rivers during the NE and the SWMs (Chauhan and Vogelsang 2006). This is because the clays and the sediment detritus are the weathering products of parent rocks and their formation and the discharge is dependent upon the geology, drainage and the prevalent climate (Weaver 1989) of a given region.

The mineral characteristics of the sediment influx from the Himalayan Rivers are markedly different in their composition and abundance than those dumped by the peninsular rivers and hence the mineralogy of the sedimentary inputs varies with climatic conditions and monsoonal intensity across the BOB. The circulation of SW and NE monsoon over the BOB is presented in Fig. 1.

3 Materials and Methods

A marine sediment gravity core nearly 124 cm long was retrieved from the BOB using a gravity corer, from the North of Andaman (13° 47′ 15.05″ N, 92° 53′ 47.96″ E), near the Landfall Island from a depth of 250 m (Fig. 1). The sediment core was collected during the Sagar Nidhi cruise organized by the National Institute of Ocean Technology (NIOT), Chennai, and the Institute of Ocean Management (IOM), Anna University, Chennai in June 2008. The sediment core was opened, described for its colour using the Munsell colour chart and also the grain size distribution and sediment texture (Fig. 2). The sediment core was sub- sampled at an interval of 2 cm,

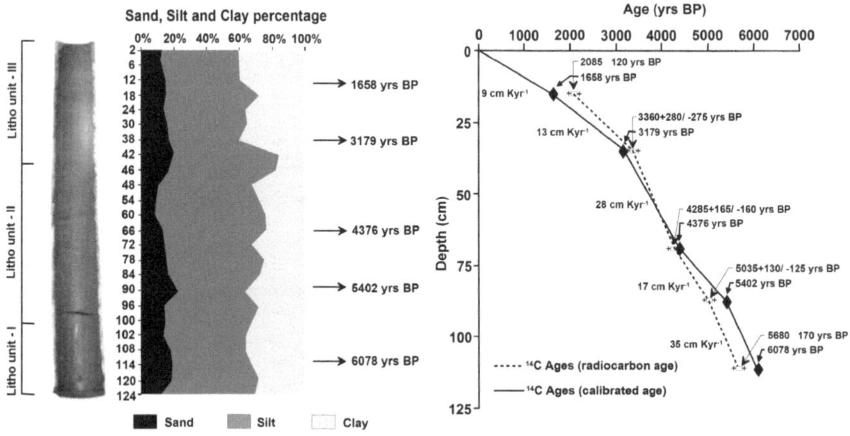

Fig. 2 Lithounits of the marine sediment core with variations in sand, silt and clay content with calibrated age. Note that the rate of sedimentation is non-linear

then dried at 50 °C, scooped in a zip lock plastic cover and preserved for further analyses.

The sediment samples (n = 21) were analysed for textural composition and grain size determination after treating the sediment samples with mild acetic acid (0.5 M) to remove the carbonate coatings and further treated with H_2O_2 to remove the organic matter. The washed, cleaned and dried samples were then analysed for grain size composition following pipette method of Krumbein and Pettijohn (1938). The grain size variations were then calculated for sand, silt and clay content to its weight percentage (Fig. 3). The same twenty-one samples that were selected for textural studies representing the entire sediment core was further analysed for clay mineral composition. For this purpose, clay particles were separated using Stoke's settling velocity principle. The separated clay fractions were made free of carbonate and organic matter by treating the samples with 5 ml acetic acid and 10 ml hydrogen peroxide, respectively. The slides were prepared by pipetting 1 ml of the concentrated clay suspensions onto glass slides. Extreme care was taken for the uniform distribution and to avoid size sorting (Rao and Rao 1995). These slides were then air dried and then kept in a desiccator for one hour under Ethylene glycol vapours at 100 °C. X-ray diffraction spectra were then carried out on these slides from 3° to 30° at 20/min on a Rigaku X-ray diffractometer using nickel-filtered Cu Ka radiation. Clay minerals were then identified and quantified following the semi-quantitative method of Biscaye (1965) (Fig. 4). Detailed clay mineral analyses were carried out in the sedimentology laboratory of the National Institute of Oceanography (NIO), Goa. Organic matter (OM) content was analysed following the protocol developed by Loring and Rantala (1992) and Calcium carbonate ($CaCO_3$) content was determined by calculating the mass difference after treating the sediment samples 10% HCl to the known weight of the sediment sample following Carver (1971). *Globigerinoides ruber* (*G. ruber*) was handpicked from all the sediment samples of the sediment core. For this purpose, the sediment samples were soaked in water overnight, then washed and passed through an ASTM 150 μm sieve mesh. The retained fractions on this sieve were oven dried at 60 °C and set aside for microscopic examination. About 10–12 tests of planktonic foraminifera (*G. ruber*) 250–315 μm size was selected for stable oxygen (O) and

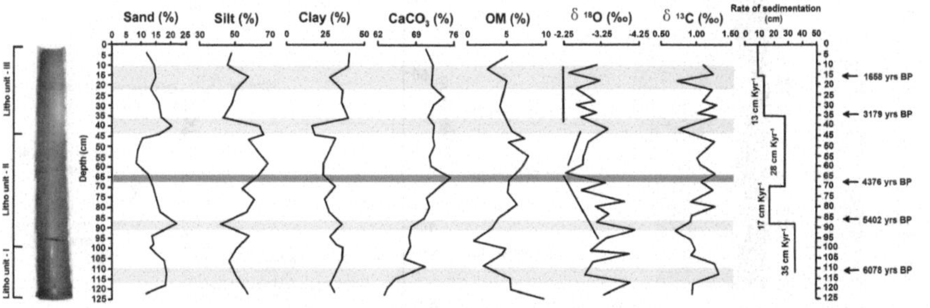

Fig. 3 Down core variation of sand, silt, clay, $CaCO_3$, organic matter (OM) (in %), $\delta^{18}O$ and $\delta^{13}C$ (‰) with calibrated ages in yrs BP

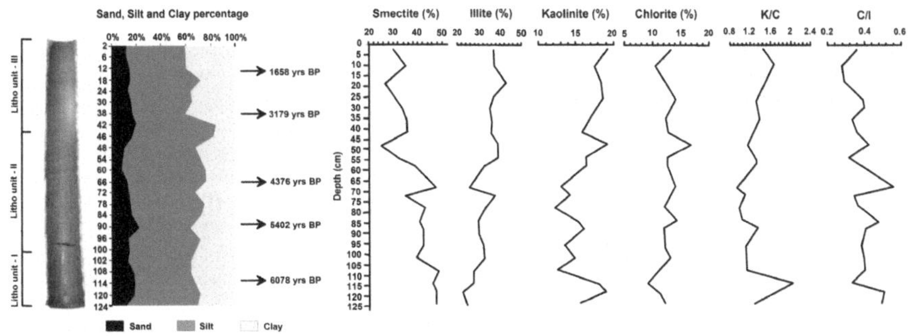

Fig. 4 Down sediment core variations in clay minerals, K/C and C/I

carbon (C) isotopic measurements. Microscopic examination revealed that the tests were well preserved with no apparent signs of dissolution and taphonomy. All the selected foraminiferal tests were cleaned ultrasonically in methanol to do away with clay coatings. Oxygen ($\delta^{18}O$) and carbon ($\delta^{13}C$) isotopic measurements were carried out at the National Geophysical Res Institute, Hyderabad, India, using a Delta plus advantage Isotope Ratio Mass Spectrometer (IRMS) coupled with a Kiel-IV automatic carbonate device following Ahmad et al (2008). The results of the Isotopic compositions are reported in δ notations as per mil deviation from VPDB standard. Analytical precision was better than 0.10‰ for $\delta^{18}O$ and 0.05‰ for $\delta^{13}C$ (Fig. 3). Calibration to the VPDB standard was achieved by repeated measurements of international reference standards NBS-19 and NBS-18. Twelve sediment samples were selected at regular intervals covering the entire sediment core for phytolith study. The extraction of phytoliths was undertaken in the laboratory by using standard methods with removal of carbonates and nitrates followed by heavy density separation (Piperno 1988). Slight modifications were required at times depending on the nature of the sediment type. Simultaneous extraction of Phytoliths, Diatoms and Sponge Spicules was undertaken. Up to 200 Phytoliths, inclusive of other microfossils from each sample were observed and counted so as to generate a quantified data (Figs. 5, 6). Observations were also made on the physical characters of phytoliths so as to get an idea about its general preservation condition. The classification used for phytolith analysis is a combination of—phytolith shape, anatomical origin and classification based on grass families.

The phytoliths were observed under 'Olympus' Res microscope and photomicrographs were taken under 45× magnifications. The basic phytolith classification adapted here is derived from grass families as suggested by Twiss (2001) with further modifications by including Phytoliths of anatomical origin and other non-diagnostic silicified cells, such as woody elements and perforated cells ensued from shrubs and trees (Eksambekar 2002). For the nanoplankton assemblages, a pinch of the powdered sample material was taken and kept in a covered crucible in which a few drops of distilled water were poured to make an even suspension by stirring it. With the help of a clean dropper, a smear of sediment suspension was spread on pre-heated slide. Two slides, one containing the fines, and the other containing comparatively coarse

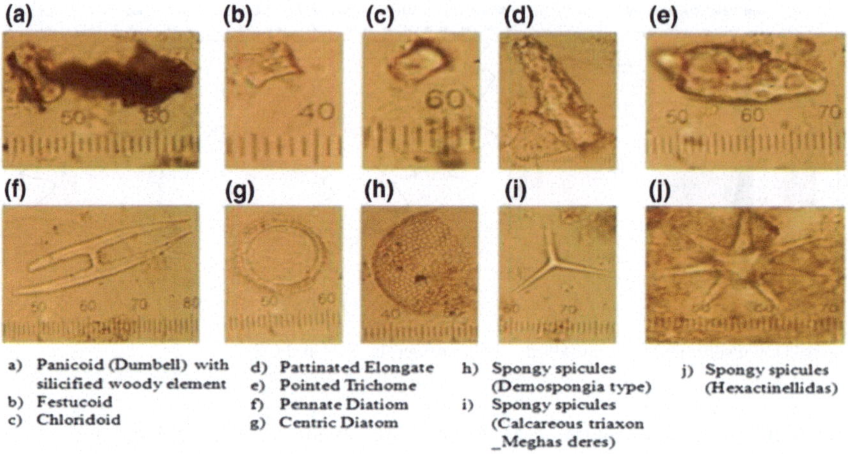

Fig. 5 Photomicrographs of various types of phytoliths, spongy spicules and diatoms of the present study (light Microscopic Photographs, scale = 10 μm)

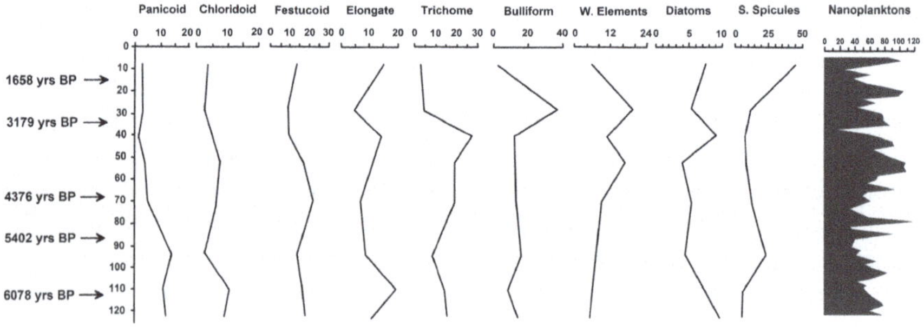

Fig. 6 Down sediment core variation of phytolith and nanoplankton assemblage

fraction of the suspension were prepared and allowed to dry on a hot plate. Drops of mounting medium (Canada Balsam) were poured with the help of a glass rod and a cover slip of 22 × 40 mm dimension was used to cover the slide containing dried suspension film. The cooked slide with the cover slip was picked with the help of a pincer and allowed to cool on a flat surface. The air bubbles were allowed to leave the surface. The cover slip was even pressed after taking out the slide from the hot plate so that extra Canada balsam may leave the slide while the slide is cooling and made permanent. These slides were examined for the presence of nanofossils under Leitz polarizing microscope (LM) with ×10 or ×12.5 oculars and an ×100 oil immersion objective. Photomicrographs were taken by using gypsum plate which helped in a precise identification of crystal elements of critical species. Nannofossils were identified and a checklist was prepared. The recovered nannofossils were compared with global marker charts and microphotographs from available publications for verification and confirmation (Figs. 7 and 8a–f). In the present study, organic carbon rich five

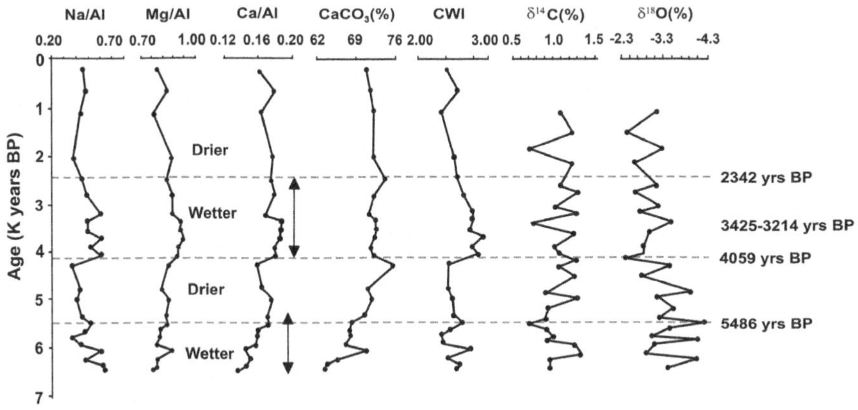

Fig. 7 Downcore variations of Elemental ratios with $CaCO_3$%, $\delta^{13}C$, $\delta^{18}O$ reveals variations wetter and drier conditions owing to the intensity of SWM

samples were radiocarbon dated following the acid-base-acid pre-treatment by liquid scintillation counting at the Department of Geosciences, University of Arizona, Tucson, USA. Calibrated and reservoir corrections (400 years) and calibrated using Calib 6.0 (Stuiver et al. 2010) following Stuiver and Reimer (1993) was applied and presented in Table 1 and Figs. 2 and 3.

4 Results

The sediment core samples were radiocarbon dated to have a continuous record of the Holocene past and these studies were conducted to reconstruct the mid—late Holocene palaeoenvironment variations. and the ages range in age from middle Holocene to late Holocene (5680 ± 170; −22.8‰ $\delta^{13}C$ to 2085 ± 120 years BP; −23.3‰ $\delta^{-13}C$) (Tables 1, 2, 3 and 4; Fig. 2). An age of 1658 years BP (−23.3‰ $\delta^{13}C$) was obtained from the core at the depth 15 cm. Based on the reservoir corrected ages the average sedimentation rates were calculated and the sedimentation rate varies 0.35 mm y^{-1} from 111 to 87 cm (6078 years BP to 5402 years BP), 0.17 mm y^{-1} from 87 to 69 cm (5402 years BP to 4376 years BP), 0.28 mm y^{-1} from 69 to 35 cm (4376 years BP to 3179 years BP), 0.13 mm y^{-1} from 35 to 15 (3179 years BP to 1658 years BP) and 0.09 mm y^{-1} from 15 cm to the sediment core surface (1658 years BP to present) indicating a non-linear phases of deposition (Figs. 2 and 3). Interpolations of ages as a function of depth were calculated by considering the sedimentation rates between the obtained radiocarbon ages (Table 1). Overall the sediment core exhibited sediment layers in shades of olive black to olive grey. Based on sediment texture and colour, the sediment core was divided into three distinct litho-units. Unit I (124–100 cm) is olive black (10Y 3/2) coloured, sandy to fine silt, Unit II (100–45 cm) is olive grey (10Y 4/2) coloured dominantly fine

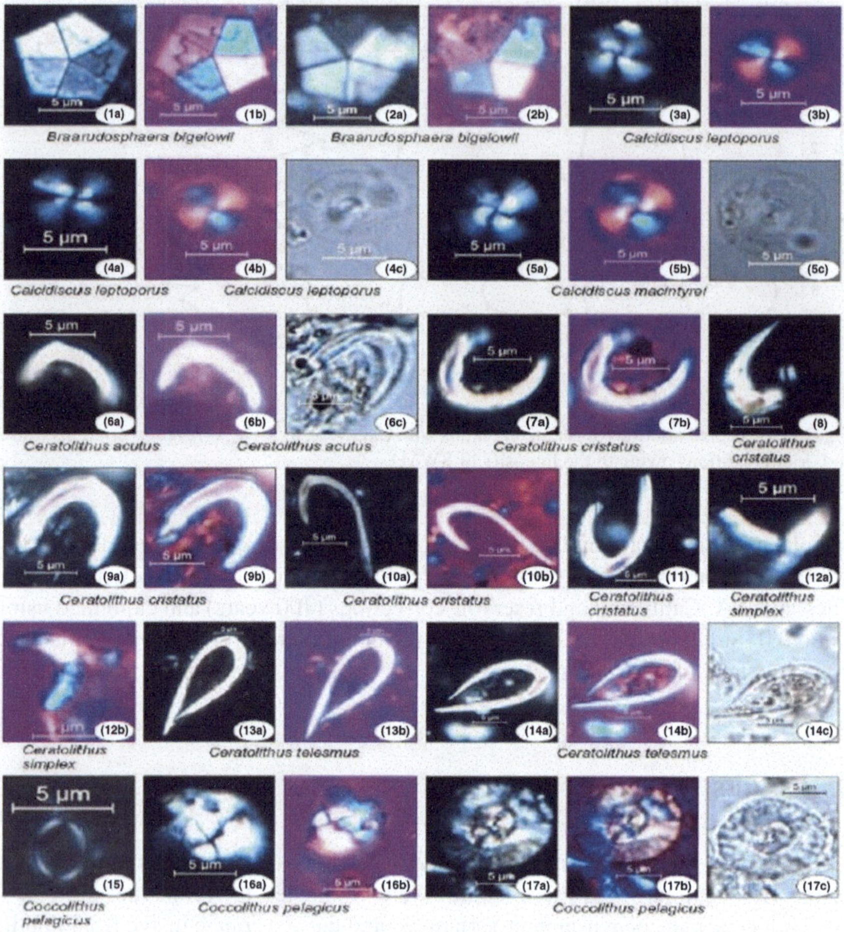

Fig. 8 **a** Photomicrographs of nanoplanktons under plane and crossed nicols. **b** Photomicrographs of nanoplanktons under plane and crossed nicols. **c** Photomicrographs of nanoplanktons under plane and crossed nicols. **d** Photomicrographs of nanoplanktons under plane and crossed nicols. **e** Photomicrographs of nanoplanktons under plane and crossed nicols. **f** Photomicrographs of nanoplanktons under plane and crossed nicols

to medium fine grained fine sandy silt. Fine silty layers were observed within this unit. Unit III (45–0 cm) is olive grey coloured (10Y 5/2), with the clayey fine silt of fine laminae and minor ripples were observed in the uppermost layers of this unit (Figs. 2 and 3). Moreover, there is no significant variation in $\delta^{13}C$ values as it varies marginally from −23.3 to −24.3‰, showing a difference of nearly −1.0‰ only, since 4376–1658 years BP (Figs. 2, 3).

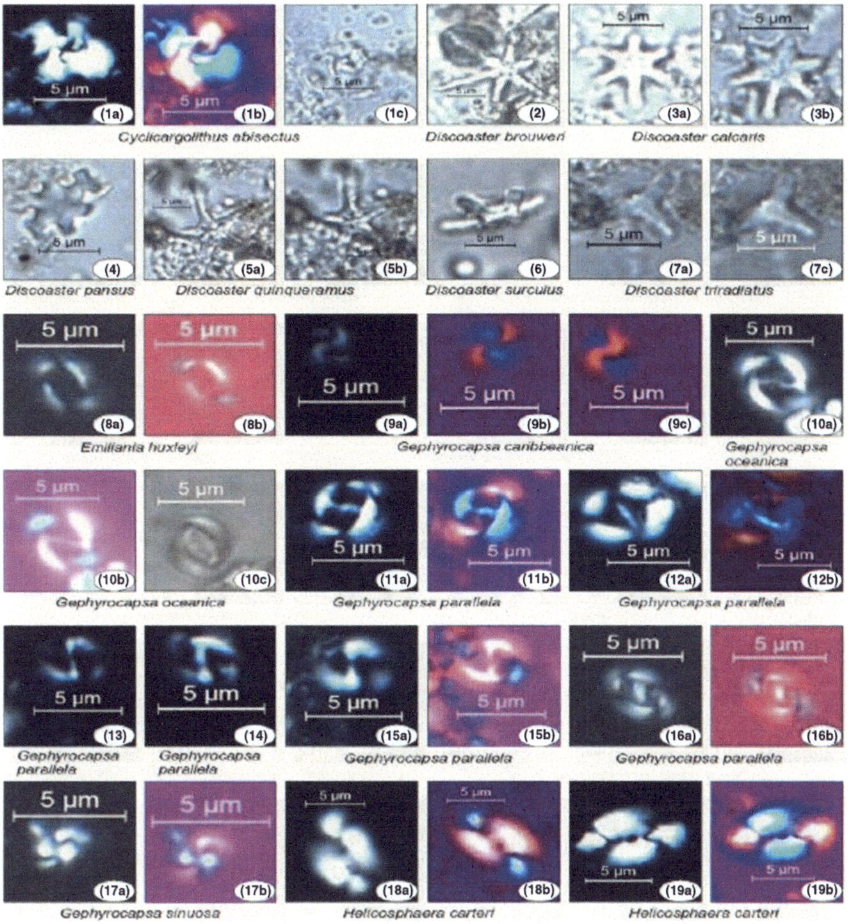

Fig. 8 (continued)

In this sediment core grain size analysis reveals a predominance of silt and clay over the sand content (Fig. 3; Table 2), indicating relatively deeper and calm environment of deposition. The sediment core is primarily composed of silt (41.88–68.05%) followed by clay (15.90–40.32%) and sand (8.28–22.04%) (Table 2). Based on Shepard (1954) classification using the ternary diagram that is divided into ten classes of sediment texture, the sediment samples of the present study are clayey silt in nature (Fig. 3). On the contrary, to the fine silt content, the sand content is higher at the depths 120–106 cm (18–17%; 6360–5937 years BP), 90–88 cm (22%, 5486 years BP) and 42–40 cm (20%, 3425 years BP) and least between 60 and 52 cm (8%, 4059 to 3848 years BP). The clay percentage is higher at the depth 36 cm (39%, 3285 years BP), and 40% at depth 6–4 cm (663 years BP) and exhibits a low at 42–40 cm depth (15.9%, 3425 years BP) (Table 2). The variation in clay composition, content, K/C

Fig. 8 (continued)

and C/I ratios for all the samples studied is presented in Table 3. Detailed clay mineral analyses indicate the dominant occurrence of smectite (range ~25–49%) and illite (range ~21–43%) followed by kaolinite (range ~12–19%) and chlorite (range ~9–17%). The down sediment core clay composition variation reveals smectite (S) (49%) and illite (I) (42%) peaks at the depth 108 cm and 18 cm respectively, while kaolinite (K) (19%) and chlorite (C) (16%) occur at the depths 118 cm and 48 cm respectively. Smectite shows an inverse trend with illite and chlorite. K/C ratio (0.93) shows a minimum value at the depth 66-68 cm while the C/I ratio (0.56) is maximum at the same depth (Table 3, Fig. 3). At the depth 112–114 cm K/C ratio is the maximum (2.03) followed by a minimum of C/I ratio (0.28) at the depth 8–10 cm (Fig. 4). The $\delta^{13}C$ and $\delta^{18}O$ isotopic records of planktonic foraminifera (*G. ruber*)

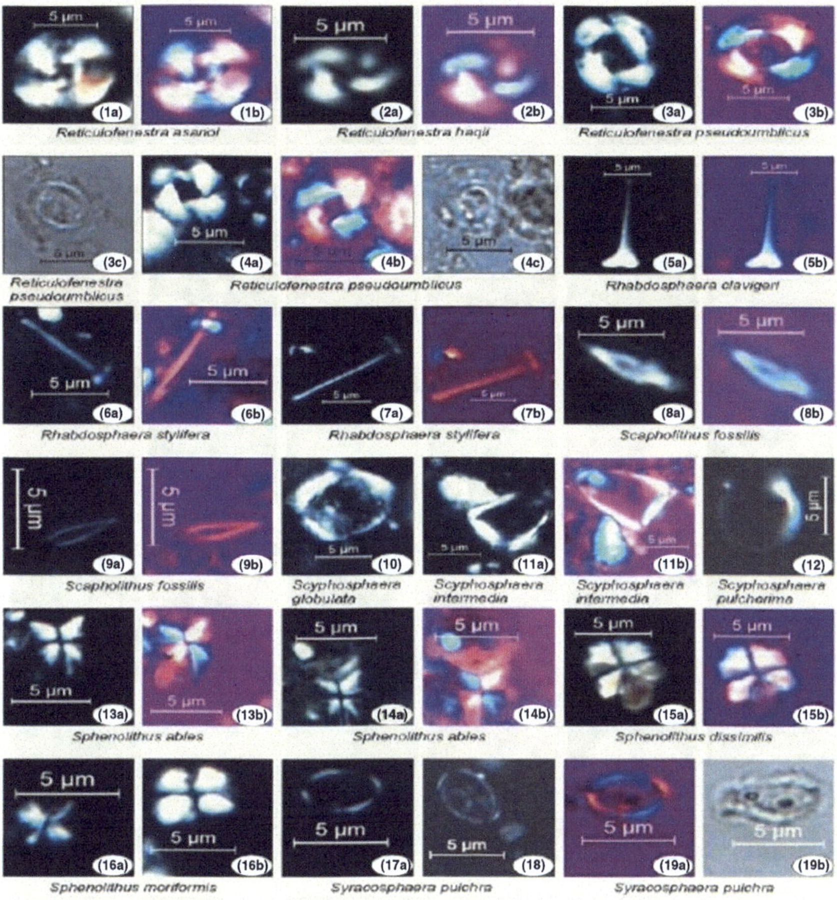

Fig. 8 (continued)

shows that the $\delta^{13}C$ in *G. ruber* vary significantly from 0.70 to 1.33‰ with an average of 1.06‰ (Table 4). Planktonic $\delta^{18}O$ values vary between −4.18 and −2.33‰. An average $\delta^{18}O$ value for the mid to late Holocene is −3.13‰ (Table 4; Fig. 3).

Different phytolith morphotypes were noted that included subtypes within each group. The phytolith morphotypes were grouped as follows: (1) Panicoid (e.g. dumbbells) (2) Chloridoid (e.g. saddles) (3) Festucoid (e.g. spherical, square) (4) Elongate (e.g. Rods and tracheids) (5) Trichome (e.g. epidermal hairs) (6) Bulliform (e.g. bulky phytoliths like fan shape) (7) Silicified cells (e.g. woody elements) (Fig. 5). Phytoliths observed in each sample were compared with the 'in house' phytolith database of phytolitharium collection housed at the "Phytolith Research Institute", Pune and other published references.

The phytolith morphotypes from the marine sediment core based on their anatomical origin and structural character suggest the presence of following vegetation in the

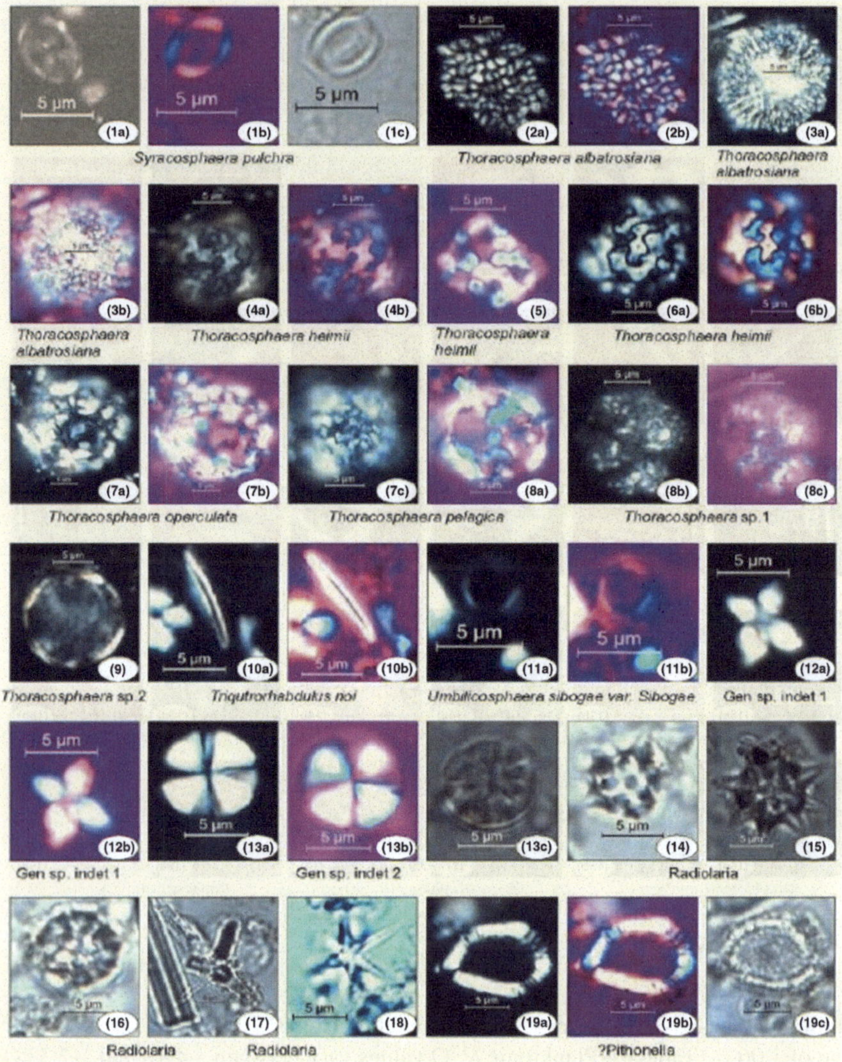

Fig. 8 (continued)

observed samples. High percentage of Panicoid (14%) with low Chloridoid (2.5%) was noted at 96–94 cm depth around 5599 years BP, a low Panicoid (2%) and Festucoid (9%) was found at 42–40 cm depth approximately around 3355 years BP with high content of Trichome (27.5%) (Fig. 6; Table 5). While high Festucoid was found at 70–68 cm depth around 4433 years BP. It was also observed that at 28–30 cm depth in the sediment core around 2646 years BP the Chloridoid (2.5%), Elongate (5%) and Festucoid (9%) content is low, with a higher percentage of Bulliform (37.5%)

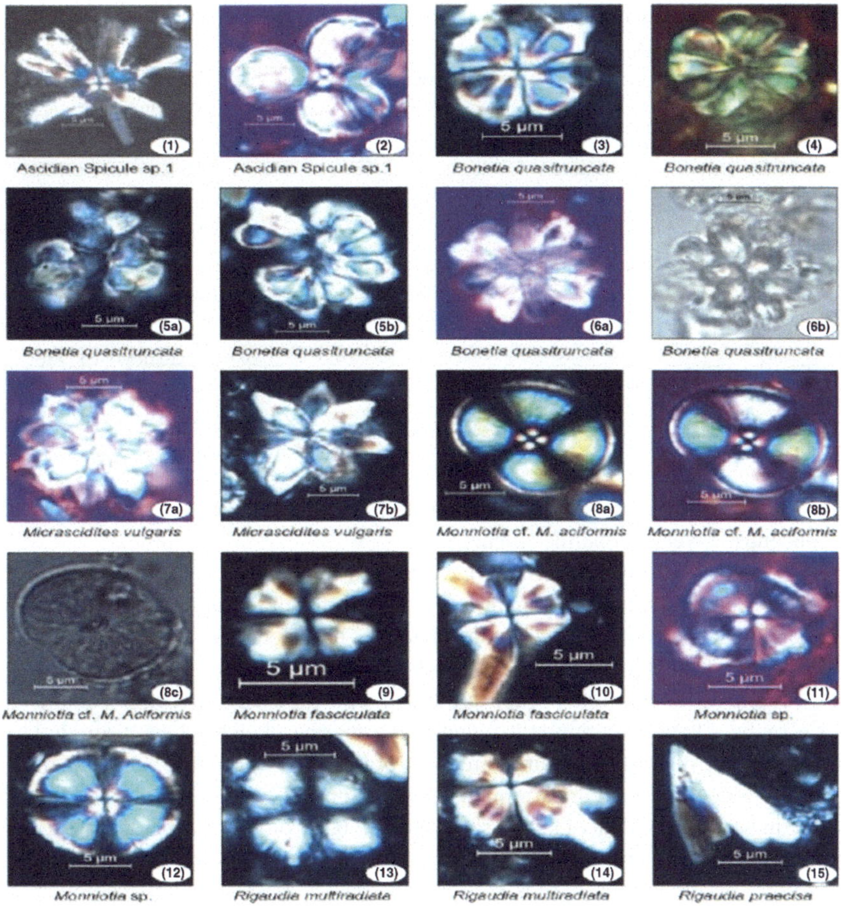

Fig. 8 (continued)

and Woody elements (19.5%). The low percentages of Bulliform (2.5%) and Trichome (3%) with high Spongy spicules (45.5%) occur at the depth 10–8 cm around 884 years BP (Table 5; Fig. 6).

Diatoms were observed to be high at 124–122 cm around 6416 years BP with low percentage of woody elements (5%) and Spongy spicules (5%). The low occurrence of diatoms was observed at the depth 54–52 cm ~3778 years BP. In the present study most, sponges and spicules were composed of calcium carbonate or silica (Figs. 5, 6).

In this study siliceous sponge spicules have been separated into two major categories based on their size as megascleres and microscleres. The siliceous spicules studied here are mainly of two types: megascleres (larger skeletal elements, >100 μ in length, existing either free or as part of a framework) and microscleres (smaller,

Table 1 Conventional and Calibrated ^{14}C ages of at selected depth intervals

S. No.	Lab ID	Depth (cm)	$\delta^{13}C$‰	Conventional radiocarbon ages (years BP)	Calibrated age with 2σ range (years BP) and median years
1	A15142	14–16	−23.3	2085 ± 120	1293–1871 Median yrs BP 1658
2	A 15143	34–36	−23.5	3360 + 280/−270	2418–3817 Median yrs BP 3179
3	A 15144	68–70	−24.3	4285 +165/−160	3858–4783 Median yrs BP 4376
4	A 15145	86–88	−23.1	5035 +130/−125	4900–5583 Median yrs BP 5402
5	A 15189	110–112	−22.8	5680 ± 170	5605–6363 Median yrs BP 6078

10–110 μ, always occurring as free elements). These siliceous spicules are formed of concentric layers of opaline silica.

In the present study the diatoms have been differentiated using full forms that are centric, i.e. circular, and pennate (Fig. 5), i.e. having bilateral form. The term 'pennate' denotes feather-like structures. Diatoms have been divided into solitary and colonial forms. Diatoms have been further sub-divided according to whether they lack a raphe (a median line or slot in the cell wall), a pseudoraphe, or completely lack a raphe. Diatoms commonly found in the marine plankton have been divided into the centric diatoms including three sub-orders based primarily on the shape of the cells, the polarity and the arrangement of the processes.

The nanoplankton data of all the samples (n = 62) from the marine sediment core of the present study show a significant variation in the productivity, presented as a down core variation (Table 6; Figs. 6, 7). It contains a consistent record of complete or disaggregated spicules (star-shaped spicules), belonging to the Didemnidae family, at intermediate depths. These prosobranch ascidians are essentially composed of shallow benthic coastal colonies, though same species have been recorded from abyssal depths. The nanoplankton identified and their abundance from the present study is shown in Fig. 6 and Table 6.

The dominant species in the calcareous nannofossil assemblage were *Gephyrocapsa oceanica*, small *Gephyrocapsa, Emiliania huxleyi,* Ascidian spicules and the species of *Calcidiscus leptoporus*, *Helicospahaera carteri*, *Ceratolithus telesmus, C. cristatus, C. acutus* are also abundant. Subordinate number of *H. hyalina, H. inversa, H. pavimentum* are also seen. Rare *Syracosphaera pulchera, Neosphaera ccolithomorpha* are in the assemblage and very rare *Scyphospheres, Rhabdoliths* and *Discoasters Braarudosphaera bigelowii were* also present in some samples (Fig. 8a–f). No specimen of *Florisphaera profunda,* which is a lower photic zone species, was observed. When nutricline is shallow, *coccolithophore* production in the upper euphotic zone is enhanced and the abundance *of F. profunda* is minimal. The species richness of *C. leptoporus, H. carteri, Gephyrocapsa* spp., which are typical to the

Table 2 Weight percentages of sand, silt, clay and organic matter (OM) content in the sediment core

Depth (cm)	Age years BP	Sand (%)	Silt (%)	Clay (%)	OM (%)
0–2	221	–	–	–	6.21
4–6	663	11.90	47.78	40.32	5.13
10–12	1326	14.43	46.05	39.53	2.56
16–18	1886	14.82	57.51	27.67	4.99
22–24	2342	13.35	50.80	35.86	4.45
28–30	2799	15.88	48.82	35.29	4.22
36–38	3285	16.97	43.11	39.92	5.00
40–42	3425	20.08	64.02	15.90	5.79
44–46	3566	16.27	65.87	17.86	7.35
46–48	3637	10.82	57.11	32.06	5.16
52–54	3848	8.85	62.98	28.17	7.97
58–60	4059	8.28	68.05	23.67	7.04
64–66	4270	12.5	63.69	23.81	5.32
70–72	4547	14.29	53.97	31.75	5.16
76–78	4889	15.34	60.76	23.90	6.41
82–84	5231	16.57	56.14	27.29	4.32
88–90	5486	22.04	41.88	36.07	3.10
94–96	5655	13.75	58.37	27.89	0.81
98–100	5768	14.17	53.89	31.94	4.99
100–102	5824	13.15	51.00	35.86	4.59
106–108	5993	17.66	46.63	35.71	2.02
112–114	6191	18.61	49.70	31.68	5.53
118–120	6360	18.24	53.71	28.06	5.67
122–124	6473	11.87	57.00	31.13	9.85

upper photic zone with high nutrient availability are also present in the assemblage. The ecological preference of H. carteri indicates its preference for nutrient enriched waters.

5 Discussion

5.1 Provenance of the Mid-Holocene Marine Sediments

SWM controls precipitation in BOB (Ding 1994) and the strengthened winter monsoon causes deep chemical weathering of the bedrocks (Wang 1999) resulting in

Table 3 Clay mineral composition and their ratio of the sediment core

Depth (cm)	Age (years BP)	S (%)	I (%)	K (%)	C (%)	K/C	C/I
0–2	221	30.34	37.08	19.22	13.36	1.44	0.36
8–10	1105	34.88	37.21	17.44	10.47	1.67	0.28
16–18	1886	26.40	42.70	18.37	12.53	1.47	0.29
24–26	2494	31.07	36.16	18.72	14.04	1.33	0.39
28–30	2799	33.53	34.73	18.14	13.60	1.33	0.39
34–36	3214	35.48	35.84	16.60	12.07	1.38	0.34
40–42	3425	35.39	35.96	15.77	12.88	1.23	0.36
46–48	3637	25.09	38.95	19.20	16.75	1.15	0.43
52–54	3848	31.96	39.18	16.36	12.51	1.31	0.32
58–60	4059	38.72	32.30	16.45	12.53	1.31	0.39
66–68	4341	47.57	25.24	13.09	14.10	0.93	0.56
70–72	4547	34.95	37.86	14.22	12.96	1.10	0.34
76–78	4889	43.54	32.39	12.11	11.96	1.01	0.37
82–84	5231	40.61	30.03	15.06	14.29	1.05	0.48
86–88	5430	42.41	29.57	16.14	11.87	1.36	0.40
94–96	5655	42.15	32.29	13.47	12.09	1.11	0.37
100–102	5824	39.47	32.57	14.91	13.05	1.14	0.40
106–108	5993	48.82	27.59	12.55	11.04	1.14	0.40
112–114	6191	46.07	27.14	17.93	8.85	2.03	0.33
116–118	6304	47.86	22.00	19.08	11.06	1.73	0.50
122–124	6473	47.69	24.62	15.58	12.12	1.29	0.49

Smectite (S), Illite (I), Kaolinite (K), Chlorite (C)

the intense weathering of the nearby continental margins of the Andaman Islands. Thus, the terrestrial detritus in the sediments in northern Andaman Islands exhibit stronger chemical weathering signals (Wei et al. 2003). The chemical, textural and isotopic signatures (C and O) observed at the Andaman Forearc Basin (AFB) sediment core near the Landfall Island reflects a mixture of sediments distinctly from a nearby source, i.e. from the Andaman terrain of ophiolite basalt suites and sediments from Irrawaddy, Salween and Sittang rivers that transport weathering products from the land masses to the AFB. This inference also supports the argument suggested by Moore et al. (1982) that the sediments are transported from the Irrawaddy delta to the south of Sumatra along the Andaman Fore arc basin. However, the sediment fluxes vary during the dry, warm/wet phases. High content of silt occurs at the depth 60–55 cm (4000–3900 years BP), while laminar clays and fine silt are deposited at the depth 35 cm (3200 years BP). Sediment texture and clay mineral composition reveal low clay content (16%), with higher coarser components at the 40 cm depth and this indicates a high energy event, unaffected chemical weathering of the Landfall Island bed rocks (conglomerates, sandstone, shale and the ophiolite suite

Table 4 $\delta^{18}O$ and $\delta^{13}C$ values of *Globigerina ruber*

Depth (cm)	Age (years BP)	No of specimen picked	$\delta^{13}C$ (‰)	$\delta^{18}O$ (‰)
8–10	1105	25	1.09	−3.11
12–14	1547	35	1.23	−2.39
16–18	1886	25	0.71	−3.21
20–22	2190	35	1.24	−2.58
24–26	2646	60	1.09	−3.09
28–30	2799	25	1.29	−2.57
32–34	3103	30	1.02	−3.15
34–36	3214	45	1.30	−2.68
40–42	3425	35	0.74	−3.43
46–48	3637	35	1.26	−2.90
54–56	3918	25	1.02	−2.77
58–60	4059	35	1.06	−2.76
62–64	4200	25	1.28	−2.33
66–68	4341	30	1.07	−3.39
70–72	4547	50	1.25	−2.73
76–78	4889	40	0.91	−3.89
78–80	5003	22	1.29	−3.10
82–84	5231	35	0.94	−3.47
86–88	5430	30	0.90	−3.13
90–92	5542	10	0.70	−4.18
94–96	5655	35	0.92	−3.39
100–102	5824	45	0.98	−2.96
102–104	5881	30	0.93	−4.02
106–108	5993	50	1.26	−3.04
112–114	6191	40	1.33	−2.82
116–118	6304	50	0.95	−4.03
122–124	6473	25	0.95	−3.34

of rocks). This clearly shows that the rate of sedimentations has been non-linear. It is also noted that the decrease in the sedimentation rate (Fig. 2) post 4300 years BP that points to probably a thicker vegetation canopy in the hinterland and with low carbonate content indicating strong chemical weathering of the Island bed rocks. The low values of calcium carbonate and organic matter also points to low productivity due to reduced SWM. Probably a very weak monsoon could have also resulted in reduced sediment input (Fig. 4, 5 and 9).

Studies carried out on the clay minerals (Rao 1983) near the study area reveal that the clays and the very fine sediments are rich in kaolinite, chlorite and illite whereas montmorillonite (smectite group) is less. Rao (1983) found two distinct clay mineral

Table 5 Types of phytoliths in percentage with climate index (Ic), phytolith index (Iph) and cold/warm ratio

Depth (cm)	Panicoid (%)	Chloridoid (%)	Elongate (%)	Festucoid (%)	Trichome (%)	Bulliform (%)
8–10	3.0	3.5	15.0	14.0	*3.0*	*2.5*
28–30	3.0	*2.5*	*5.0*	*9.0*	5.5	**37.5**
40–42	*2.0*	5.5	14.0	*9.0*	**27.5**	13.0
52–54	4.0	7.5	11.0	16.5	19.0	12.5
70–76	5.0	6.5	7.0	**22.0**	19.0	13.5
94–96	**14.0**	*2.5*	9.0	14.0	9.0	16.5
110–112	11.0	**10.5**	**19.0**	16.5	14.0	9.0
122–124	12.0	9.0	11.0	17.5	15.5	14.5
Depth (cm)	Woody elements (%)	Diatoms (%)	Spongy spicules (%)	Climate Index (Ic)	Phytolith Index (Iph)	Cold/Warm ratio
8–10	6.0	7.5	**45.5**	**68.29**	53.85	**2.31**
28–30	**19.5**	5.5	12.5	62.07	45.45	0.91
40–42	11.0	9.0	9.0	54.55	**73.33**	1.87
52–54	16.5	*4.0*	9.0	58.93	65.22	0.96
70–76	9.0	5.5	12.5	65.67	56.52	0.61
94–96	7.0	4.5	23.5	45.90	*15.15*	0.55
110–112	6.0	7.5	6.5	*43.42*	48.84	0.88
122–124	*5.5*	**9.5**	*5.5*	45.45	42.86	*0.52*

Bold numbers signify higher percentage of phytolith types at different depths while the italics numbers indicate the lower climate and phytolith indices

assemblages: (i) Fe-rich chlorite in the eastern side, (ii) Fe-poor chlorite in the western side of the Andaman Islands, however this study was not quantified.

In the present study smectite and illite are the two principal clay minerals present in the sediment core other than the kaolinite and chlorite (Fig. 4; Table 3). The high smectite content in the sediment core suggests that the sediments are derived mainly from nearby basic rock source (Griffin et al. 1998; Siby et al. 2008). Illite and chlorite are often derived mainly by physical erosion or degradation of metamorphic and granitic rocks or from the erosion of sedimentary rocks (Liu et al. 2005). The catchment area of the rivers Irrawaddy and Salween (which discharge from Myanmar and have catchment in the NE Himalayas) comprise metamorphic and intrusive rocks (Chauhan et al. 1993). Thus, it can be inferred that a considerable amount of fine grained illite and chlorite sediment deposition near the Landfall Island is controlled by the Irrawaddy and Salween Rivers.

Clay minerals such as chlorite and illite are often produced under the arid/cold climate while the smectite and kaolinite are formed under humid/warm conditions causing strong weathering coupled with intense SWM (Griffin 1968; Liu et al. 2004;

Table 6 List of the nanoplankton species identified in the marine sediment

Depth (cm)	Braarudosphaera *bigelowii*	Calcidiscus *leptoporus*	Ceratolithus *acutus*	Ceratolithus *cristatus*	Ceratolithus *simplex*	Ceratolithus *telesmus*
0–2			F			
2–4			F			F
4–6			F			
6–8						
8–10						
10–12						F
12–14						
14–16				R	R	F
16–18		R		R		R
18–20		R		R	R	F
20–22		R		R	R	R
22–24		R		R	R	
24–26					R	
26–28		R				
28–30						R
30–32		R				R
32–34		R	R		R	R
34–36		R	R	R		R
36–38		R		R		R
38–40	R	R	R	R		R
40–42		R	R			R

(continued)

Table 6 (continued)

Depth (cm)	Braarudosphaera *bigelowii*	Calcidiscus *leptoporus*	Ceratolithus *acutus*	Ceratolithus *cristatus*	Ceratolithus *simplex*	Ceratolithus *telesmus*
42–44			R	R		R
44–46	R		R	R		R
46–48			R			R
48–50		R	R	R		R
50–52		R	R	R	R	R
52–54		R	R	R		R
54–56		R				R
56–58			R	R		R
58–60	R	R		R		
60–62		R	R			R
62–64		F	R	R		R
64–66		F				R
66–68	R	F		R		R
68–70			R	R		
70–72						
72–74		R		R		R
74–76				R		R
76–78		R	R	R		R
78–80	R					
80–82		R			R	
82–84		R	R			

(continued)

Table 6 (continued)

Depth (cm)	Braarudosphaera *bigelowii*	Calcidiscus *leptoporus*	Ceratolithus *acutus*	Ceratolithus *cristatus*	Ceratolithus *simplex*	Ceratolithus *telesmus*
84–86	R		R			R
86–88				R		R
88–90	R	R		R		R
90–92		R		R		
92–94			R	R		
94–96			R	R		R
96–98		R	R			R
98–100	R	R		R		
100–102				R		R
102–104		R			R	R
104–106		R	R			R
106–108		R	R	R		
108–110	R			R		R
110–112						R
112–114		R	R	R		R
114–116	R	R	R	R		R
116–118		R		R		R
118–120	R	R		R		
120–122		R	R	R		R

(continued)

Table 6 (continued)

Depth (cm)	Braarudosphaera *bigelowii*	Calcidiscus *leptoporus*	Ceratolithus *acutus*	Ceratolithus *cristatus*	Ceratolithus *simplex*	Ceratolithus *telesmus*
122–124		R		R		R

(continued)

Table 6 (continued)

Depth (cm)	Coccolithus *pelagicus*	Cyclicargolithus *abisectus*	Discoaster *brouweri*	Discoaster *Calcaris*	Discoaster *pansus*	*Discoaster quinqueramus*
0–2	F					
2–4	F					
4–6						
6–8						
8–10	F					
10–12	F					
12–14						
14–16	R		R			R
16–18	R	R				
18–20	F	R				
20–22	R				R	
22–24						
24–26						
26–28						
28–30					R	
30–32						
32–34						

(continued)

Table 6 (continued)

Depth (cm)	Coccolithus *pelagicus*	Cyclicargolithus *abisectus*	Discoaster *brouweri*	Discoaster *Calcaris*	Discoaster *pansus*	*Discoaster quinqueramus*
34–36	R					
36–38						

(continued)

Table 6 (continued)

Depth (cm)	Coccolithus *pelagicus*	Cyclicargolithus *abisectus*	Discoaster *brouweri*	Discoaster *Calcaris*	Discoaster *pansus*	*Discoaster quinqueramus*
38–40						
40–42						
42–44	R			R		
44–46		R				
46–48	R	R		R		
48–50	R	R				
50–52	R	R				
52–54		R				
54–56						
56–58	R					
58–60		R				
60–62	R	R				
62–64		R				
64–66						
66–68	R					
68–70		R				
70–72	R					
72–74						
74–76	R					

(continued)

Table 6 (continued)

Depth (cm)	Coccolithus *pelagicus*	Cyclicargolithus *abisectus*	Discoaster *brouweri*	Discoaster *Calcaris*	Discoaster *pansus*	*Discoaster quinqueramus*
76–78		R				
78–80						

(continued)

Table 6 (continued)

Depth (cm)	Coccolithus *pelagicus*	Cyclicargolithus *abisectus*	Discoaster *brouweri*	Discoaster *Calcaris*	Discoaster *pansus*	*Discoaster quinqueramus*
80–82		R				
82–84						
84–86	R	R				
86–88		R				
88–90	R	R				
90–92						
92–94		R				
94–96						
96–98						
98–100						
100–102	R					
102–104						
104–106		R				
106–108						
108–110						

(continued)

Table 6 (continued)

Depth (cm)	Coccolithus *pelagicus*	Cyclicargolithus *abisectus*	Discoaster *brouweri*	Discoaster *Calcaris*	Discoaster *pansus*	*Discoaster quinqueramus*
110–112						
112–114	R	R				
114–116			R			
116–118		R				
118–120						
120–122	R					

(continued)

Table 6 (continued)

Depth (cm)	Coccolithus *pelagicus*	Cyclicargolithus *abisectus*	Discoaster *brouweri*	Discoaster *Calcaris*	Discoaster *pansus*	*Discoaster quinqueramus*
122–124						

Depth (cm)	Discoaster *surcutus*	Discoaster *triradiatus*	Emiliania *huxleyi*	Gephyrocapsa *caribbeanica*	Gephyrocapsa *oceanica*	Gephyrocapsa *parallela*
0–2			V		A	C
2–4			C		V	C
4–6			A		A	F
6–8			C		C	F
8–10			F		C	R
10–12			R		V	C

(continued)

Table 6 (continued)

Depth (cm)	Discoaster *surcutus*	Discoaster *triradiatus*	Emiliania *huxleyi*	Gephyrocapsa *caribbeanica*	Gephyrocapsa *oceanica*	Gephyrocapsa *parallela*
12–14			F		A	R
14–16		R	R	R	A	R
16–18			F	R	A	F
18–20			F	R	A	F
20–22	R		R	R	V	F
22–24			R		V	F
24–26			F		V	
26–28			F		A	
28–30			F		A	
30–32			F		C	
32–34			C	R	A	F
34–36			C	R	A	R

(continued)

Table 6 (continued)

Depth (cm)	Discoaster *surcutus*	Discoaster *triradiatus*	Emiliania *huxleyi*	Gephyrocapsa *caribbeanica*	Gephyrocapsa *oceanica*	Gephyrocapsa *parallela*
36–38			C	F	A	R
38–40			F	F	A	F
40–42			F	R	A	F
42–44			F	R	A	F
44–46			F	R	A	F
46–48			F	R	V	F
48–50			F	R	V	F
50–52			F	R	A	F
52–54			F	R	V	F
54–56			R			

(continued)

Table 6 (continued)

Depth (cm)	Discoaster *surcutus*	Discoaster *triradiatus*	Emiliania *huxleyi*	Gephyrocapsa *caribbeanica*	Gephyrocapsa *oceanica*	Gephyrocapsa *parallela*
56–58			R			
58–60			F	R	A	R
60–62			F	R	V	F
62–64			F	R	V	F
64–66			F	F	A	F
66–68			F	F	A	F
68–70			F	F	A	F
70–72			F	F	A	
72–74			F		A	
74–76			F	F	A	
76–78			C	F	A	F

(continued)

Table 6 (continued)

Depth (cm)	Discoaster *surcutus*	Discoaster *triradiatus*	Emiliania *huxleyi*	Gephyrocapsa *caribbeanica*	Gephyrocapsa *oceanica*	Gephyrocapsa *parallela*
78–80			C	F	A	F
80–82			F	F	A	C
82–84			C	F	A	
84–86			C		A	C
86–88			C		A	C
88–90			C	F	A	
90–92			C		A	F

(continued)

Table 6 (continued)

Depth (cm)	Discoaster *surcutus*	Discoaster *triradiatus*	Emiliania *huxleyi*	Gephyrocapsa *caribbeanica*	Gephyrocapsa *oceanica*	Gephyrocapsa *parallela*
92–94			C	F	V	F
94–96			C	F	A	F
96–98			F		C	
98–100			F		A	R
100–102			F		A	R
102–104			C	F	A	F
104–106			F	F	A	F
106–108			C	F	A	F
108–110			C	F	V	C
110–112			C	F	V	C
112–114			C	F	A	F
114–116			C	F	A	F
116–118			C	F	A	F

(continued)

Table 6 (continued)

Depth (cm)	Discoaster *surcutus*	Discoaster *triradiatus*	Emiliania *huxleyi*	Gephyrocapsa *caribbeanica*	Gephyrocapsa *oceanica*	Gephyrocapsa *parallela*
118–120			C	F	A	F

(continued)

Table 6 (continued)

Depth (cm)	Discoaster *surcutus*	Discoaster *triradiatus*	Emiliania *huxleyi*	Gephyrocapsa *caribbeanica*	Gephyrocapsa *oceanica*	Gephyrocapsa *parallela*
120–122			C	F	A	F
122–124			C	F	A	F
Depth (cm)	Gephyrocapsa *sinuosa*	Helicosphaera *carteri*	Helicosphaera *hyalina*	Helicosphaera *inversa*	Helicosphaera *pavimentum*	Helicosphaera *selli*
0–2	F	C	F			
2–4		C	F			
4–6		C	R			
6–8	F	C	F			
8–10		C	F			
10–12		C	F			
12–14		C	F			
14–16		C	F	R		
16–18		C		R		
18–20		A		R		R
20–22		C	F	R		
22–24		C	F	F		
24–26		C	F	C		
26–28		C	R			
28–30		C	R	R		
30–32		F	R	R		
32–34		F	R			

(continued)

Table 6 (continued)

Depth (cm)	Gephyrocapsa *sinuosa*	Helicosphaera *carteri*	Helicosphaera *hyalina*	Helicosphaera *inversa*	Helicosphaera *pavimentum*	Helicosphaera *selli*
34–36		F	R	R		
36–38	R	C	R	R		
38–40		C	R	R	R	
40–42	R	C		R	R	R
42–44		C	R	R	F	
44–46		C	R	R	F	
46–48	R	C	R	R	R	
48–50	R	C	C	R	R	
50–52	R	C	C	R	R	
52–54	R	C	C	R	R	
54–56						
56–58						

(continued)

Table 6 (continued)

Depth (cm)	Gephyrocapsa *sinuosa*	Helicosphaera *carteri*	Helicosphaera *hyalina*	Helicosphaera *inversa*	Helicosphaera *pavimentum*	Helicosphaera *selli*
58–60	R					
60–62	R	C	R	R	R	
62–64	R	A	R	R	R	
64–66	R	C	R	R		
66–68		C	R			
68–70	R	C			R	
70–72		C		R	R	
72–74		C	R	R	R	
74–76		C				

(continued)

Table 6 (continued)

Depth (cm)	Gephyrocapsa *sinuosa*	Helicosphaera *carteri*	Helicosphaera *hyalina*	Helicosphaera *inversa*	Helicosphaera *pavimentum*	Helicosphaera *selli*
76–78		F	R	R		
78–80	R	C	R			
80–82	R	C			R	
82–84	R	C	R	R	R	
84–86		C		R		
86–88		C	R			
88–90		F	R			R
90–92		F	R	R		
92–94	R	C	R	R		
94–96	R	C	R	R		
96–98	R	C	R	R	R	
98–100	R	C	R		R	
100–102	R	C	R		R	
102–104		C	R			

(continued)

Table 6 (continued)

Depth (cm)	Gephyrocapsa *sinuosa*	Helicosphaera *carteri*	Helicosphaera *hyalina*	Helicosphaera *inversa*	Helicosphaera *pavimentum*	Helicosphaera *selli*
104–106	R	C	R	R		
106–108	R	C		R	R	
108–110		C		R		
110–112	R	C	R	R	R	
112–114	R	C	R	R		R
114–116	R	C	R	R		
116–118		C	R	R		

(continued)

Table 6 (continued)

Depth (cm)	Gephyrocapsa *sinuosa*	Helicosphaera *carteri*	Helicosphaera *hyalina*	Helicosphaera *inversa*	Helicosphaera *pavimentum*	Helicosphaera *selli*
118–120	R	C	R	R		
120–122	R	C		R		
122–124		C				
Depth (cm)	Helicosphaera *wallichii*	Neosphaera *coccolithomorpha*	Pontosphaera *discopora*	Pontosphaera *jonesii*	Pontosphaera *multipora*	Pseudomiliania *lacunosa*
0–2						
2–4						
4–6						
6–8						
8–10						
10–12						
12–14						
14–16			R			F
16–18						R
18–20			R	R		R
20–22			R			R
22–24			R	R		

(continued)

Table 6 (continued)

Depth (cm)	Helicosphaera *wallichii*	Neosphaera *coccolithomorpha*	Pontosphaera *discopora*	Pontosphaera *jonesii*	Pontosphaera *multipora*	Pseudomiliania *lacunosa*
24–26						
26–28			R			
28–30			R			R
30–32						

(continued)

Table 6 (continued)

Depth (cm)	Helicosphaera *wallichii*	Neosphaera *coccolithomorpha*	Pontosphaera *discopora*	Pontosphaera *jonesii*	Pontosphaera *multipora*	Pseudomiliania *lacunosa*
32–34						
34–36			R		R	
36–38			R		R	R
38–40	R					
40–42						
42–44	R		R		R	
44–46	R					R
46–48	R				R	
48–50	R		R			R
50–52	R		R		R	
52–54					R	R
54–56						
56–58						R
58–60						
60–62	R		R			
62–64	R	F			R	
64–66	R					R
66–68		F	R			R
68–70	R					R

(continued)

Table 6 (continued)

Depth (cm)	Helicosphaera *wallichii*	Neosphaera *coccolithomorpha*	Pontosphaera *discopora*	Pontosphaera *jonesii*	Pontosphaera *multipora*	Pseudomiliania *lacunosa*
70–72						
72–74		R				

(continued)

Table 6 (continued)

Depth (cm)	Helicosphaera *wallichii*	Neosphaera *coccolithomorpha*	Pontosphaera *discopora*	Pontosphaera *jonesii*	Pontosphaera *multipora*	Pseudomiliania *lacunosa*
74–76		R				R
76–78		R				R
78–80	R		R			
80–82	R					
82–84						
84–86	R	R				
86–88						R
88–90		R				R
90–92						R
92–94						
94–96	R	R	R			
96–98	R	R				
98–100	R					
100–102						
102–104		R				
104–106			R			R
106–108						
108–110		R	R			
110–112	R	R	R			R
112–114						
114–116		R	R			R

(continued)

Table 6 (continued)

Depth (cm)	Helicosphaera *wallichii*	Neosphaera *coccolithomorpha*	Pontosphaera *discopora*	Pontosphaera *jonesii*	Pontosphaera *multipora*	Pseudomiliania *lacunosa*
116–118		R	R			R
118–120	R	R				
120–122						R
122–124						
Depth (cm)	Reticulofenestra *asnoi*	Reticulofenestra *haqii*	Reticulofenestra *pseudoumbilicus*	Rhabdosphaera clavigeri	Rhabdosphaera *stylifera*	Scapholithus *fossilss*
0–2		F		F		
2–4		F				
4–6	F					
6–8						
8–10	F	F				
10–12				F		
12–14	F					
14–16		F			R	R
16–18	R	F	F			
18–20						
20–22	R	R				
22–24						
24–26				F		
26–28					R	
28–30	R				R	

(continued)

Table 6 (continued)

Depth (cm)	Reticulofenestra *asnoi*	Reticulofenestra *haqii*	Reticulofenestra *pseudoumbilicus*	Rhabdosphaera clavigeri	Rhabdosphaera *stylifera*	Scapholithus *fossilss*
30–32			R			
32–34	R	R				
34–36						

(continued)

Table 6 (continued)

Depth (cm)	Reticulofenestra *asnoi*	Reticulofenestra *haqii*	Reticulofenestra *pseudoumbilicus*	Rhabdosphaera clavigeri	Rhabdosphaera *stylifera*	Scapholithus *fossilss*
36–38						
38–40	R					
40–42	R	R				
42–44	R			R		R
44–46	R	R	R			
46–48	R	R			R	
48–50		R		R		
50–52	R			R		
52–54	R	R				
54–56			R			
56–58						
58–60						
60–62	R		R			
62–64	R	R	R			
64–66					R	
66–68	R			R		
68–70						R
70–72	R		R			

(continued)

Table 6 (continued)

Depth (cm)	Reticulofenestra *asnoi*	Reticulofenestra *haqii*	Reticulofenestra *pseudoumbilicus*	Rhabdosphaera clavigeri	Rhabdosphaera *stylifera*	Scapholithus *fossilss*
72–74						
74–76		R				
76–78	R			R	R	
78–80						

(continued)

Table 6 (continued)

Depth (cm)	Reticulofenestra *asnoi*	Reticulofenestra *haqii*	Reticulofenestra *pseudoumbilicus*	Rhabdosphaera clavigeri	Rhabdosphaera *stylifera*	Scapholithus *fossilss*
80–82				R		
82–84						
84–86						
86–88			R			
88–90	R					
90–92		R		R		R
92–94						
94–96						
96–98			R			
98–100	R				R	
100–102				R		
102–104		R				
104–106						
106–108				R		
108–110						
110–112	R	R		R		
112–114			R			

(continued)

Table 6 (continued)

Depth (cm)	Reticulofenestra *asnoi*	Reticulofenestra *haqii*	Reticulofenestra *pseudoumbilicus*	Rhabdosphaera clavigeri	Rhabdosphaera *stylifera*	Scapholithus *fossilss*
114–116	R	R				
116–118	R				R	R
118–120	R		R	R		
120–122	R					
122–124						
Depth (cm)	Scyphosphaera *globulata*	Scyphosphaera intermedia	Scyphosphaera *pulcherima*	Sphenolithus *abies*	Sphenolithus *dissimilis*	Sphenolithus *moriformis*
0–2					F	
2–4					F	
4–6		R			F	
6–8					F	
8–10					F	
10–12					F	
12–14					F	
14–16			R		F	
16–18			R		F	
18–20		R			F	
20–22					R	
22–24					F	
24–26					R	
26–28						

(continued)

Table 6 (continued)

Depth (cm)	Scyphosphaera *globulata*	Scyphosphaera intermedia	Scyphosphaera *pulcherima*	Sphenolithus *abies*	Sphenolithus *dissimilis*	Sphenolithus *moriformis*
28–30					R	
30–32					R	
32–34						
34–36	R			R		
36–38				R		
38–40			R	R		
40–42				R		
42–44				R		
44–46				R		

(continued)

Table 6 (continued)

Depth (cm)	Scyphosphaera *globulata*	Scyphosphaera intermedia	Scyphosphaera *pulcherima*	Sphenolithus *abies*	Sphenolithus *dissimilis*	Sphenolithus *moriformis*
46–48						
48–50						R
50–52	R			R		
52–54				R	R	
54–56			R		R	
56–58					R	
58–60						
60–62				R	R	
62–64	R				R	R
64–66						
66–68						
68–70						

(continued)

Table 6 (continued)

Depth (cm)	Scyphosphaera *globulata*	Scyphosphaera intermedia	Scyphosphaera *pulcherima*	Sphenolithus *abies*	Sphenolithus *dissimilis*	Sphenolithus *moriformis*
70–72						
72–74				R		
74–76	R			R	R	
76–78			R			
78–80				R		
80–82						
82–84				R		
84–86						R
86–88					R	
88–90						
90–92				R		

(continued)

Table 6 (continued)

Depth (cm)	Scyphosphaera *globulata*	Scyphosphaera intermedia	Scyphosphaera *pulcherima*	Sphenolithus *abies*	Sphenolithus *dissimilis*	Sphenolithus *moriformis*
92–94	R					
94–96				R		
96–98				R		
98–100			R	R		R
100–102				R		
102–104					R	
104–106	R					
106–108						
108–110				R		
110–112						

(continued)

Table 6 (continued)

Depth (cm)	Scyphosphaera *globulata*	Scyphosphaera intermedia	Scyphosphaera *pulcherima*	Sphenolithus *abies*	Sphenolithus *dissimilis*	Sphenolithus *moriformis*
112–114	R			R	R	R
114–116				R		R
116–118						R
118–120	R		R	R	R	
120–122	R					R
122–124				R		R
Depth (cm)	**Syracosphaera *pulchra***	**Thoracosphaera *albatrosiana***	**Thoracosphaera *heimii***	**Thoracosphaera *operculata***	**Thoracosphaera *pelagica***	**Thoracosphaera *sp. 1***
0–2		F	C			C
2–4		F	C			C
4–6		C	F			C
6–8		F	C			F
8–10		F	F			F
10–12		F	F			F

(continued)

Table 6 (continued)

Depth (cm)	Syracosphaera *pulchra*	Thoracosphaera *albatrosiana*	Thoracosphaera *heimii*	Thoracosphaera *operculata*	Thoracosphaera *pelagica*	Thoracosphaera *sp. 1*
12–14		F	F			C
14–16		C	C	F	F	C
16–18		C	C	F	F	C
18–20		C	F	F	F	C
20–22		C	F	F		A
22–24		A	F	F	R	A
24–26		F	R			C

(continued)

Table 6 (continued)

Depth (cm)	Syracosphaera *pulchra*	Thoracosphaera *albatrosiana*	Thoracosphaera *heimii*	Thoracosphaera *operculata*	Thoracosphaera *pelagica*	Thoracosphaera *sp. 1*
26–28		R	R		R	F
28–30		F	F		R	R
30–32		F	F			R
32–34		R	R	R	R	R
34–36		R		R	R	
36–38		R	R	R	R	R
38–40		R	R	R	R	R
40–42	R	R	R	R	R	R
42–44		R	R	R	R	R
44–46		F	R	R	R	F
46–48	F	F	R	R	R	F
48–50	F	F	R	R	R	F
50–52	F	F			R	F
52–54		F	R	R	R	F
54–56						
56–58						

(continued)

Table 6 (continued)

Depth (cm)	Syracosphaera *pulchra*	Thoracosphaera *albatrosiana*	Thoracosphaera *heimii*	Thoracosphaera *operculata*	Thoracosphaera *pelagica*	Thoracosphaera *sp. 1*
58–60						
60–62		F	F	R	R	
62–64	R	F	R	R	R	
64–66						
66–68						

(continued)

Table 6 (continued)

Depth (cm)	Syracosphaera *pulchra*	Thoracosphaera *albatrosiana*	Thoracosphaera *heimii*	Thoracosphaera *operculata*	Thoracosphaera *pelagica*	Thoracosphaera *sp. 1*
68–70	R		R	R	R	
70–72		F				
72–74				R	R	
74–76	R	F			R	
76–78	R			R	R	
78–80			R	R	R	
80–82		F		R	R	
82–84	R	F				
84–86			R			
86–88	R	R		R	R	
88–90		R			R	
90–92	R		R	R	R	
92–94		F		R		
94–96						
96–98					R	
98–100	R	F			R	
100–102	R		R	R	R	
102–104						

(continued)

Table 6 (continued)

Depth (cm)	Syracosphaera *pulchra*	Thoracosphaera *albatrosiana*	Thoracosphaera *heimii*	Thoracosphaera *operculata*	Thoracosphaera *pelagica*	Thoracosphaera *sp. 1*
104–106		R		R	R	
106–108	R				R	
108–110				R		

(continued)

Table 6 (continued)

Depth (cm)	Syracosphaera *pulchra*	Thoracosphaera *albatrosiana*	Thoracosphaera *heimii*	Thoracosphaera *operculata*	Thoracosphaera *pelagica*	Thoracosphaera *sp. 1*
110–112	R	R				
112–114	R	R	R	R	R	
114–116	R	R		R	R	R
116–118		R	R	R	R	
118–120		R		R		
120–122	R	R	R	R	R	R
122–124	R		R		R	

Depth (cm)	Triqutrorhabdulus *rioi*	Umbilicosphaera *sibogae var. siboga*	? *Pithonella*	Ascidian *Spicules*	Foram *lining*
0–2	F			F	
2–4				F	
4–6				F	
6–8					
8–10					
10–12					
12–14				C	F
14–16	R	R		C	
16–18				C	
18–20				C	
20–22				R	
22–24				R	F

(continued)

Table 6 (continued)

Depth (cm)	Triqutrorhabdulus *rioi*	Umbilicosphaera *sibogae var. siboga*	? *Pithonella*	Ascidian *Spicules*	Foram *lining*
24–26				R	
26–28					
28–30				F	
30–32				R	R
32–34			R	C	
34–36	R			C	R
36–38		R		C	
38–40				C	
40–42		R	R	C	
42–44			R	C	
44–46		R		C	R
46–48		R			
48–50		R		A	
50–52			R	C	
52–54		R		A	F
54–56					F
56–58					
58–60					
60–62		R	R	A	F
62–64		R	R	A	
64–66				A	

(continued)

Table 6 (continued)

Depth (cm)	Triqutrorhabdulus *rioi*	Umbilicosphaera *sibogae var. siboga*	*? Pithonella*	Ascidian *Spicules*	Foram *lining*
66–68					
68–70		R		C	

(continued)

Table 6 (continued)

Depth (cm)	Triqutrorhabdulus *rioi*	Umbilicosphaera *sibogae var. siboga*	*? Pithonella*	Ascidian *Spicules*	Foram *lining*
70–72		R		C	R
72–74		R		C	
74–76				C	
76–78					
78–80			R	F	R
80–82		R			
82–84		R		F	
84–86					
86–88		R	R	F	R
88–90					
90–92				F	
92–94		R			
94–96		R		R	
96–98		R	R	R	
98–100				R	R
100–102					
102–104					
104–106			R	F	
106–108		R		F	F

(continued)

Table 6 (continued)

Depth (cm)	Triqutrorhabdulus *rioi*	Umbilicosphaera *sibogae var. siboga*	? *Pithonella*	Ascidian *Spicules*	Foram *lining*
108–110				F	
110–112				F	F
112–114		R		F	
114–116		R		F	

(continued)

Table 6 (continued)

Depth (cm)	Triqutrorhabdulus *rioi*	Umbilicosphaera *sibogae var. siboga*	? *Pithonella*	Ascidian *Spicules*	Foram *lining*
116–118		R	R	C	
118–120		R		F	
120–122		R		F	F
122–124			R	F	

A—Abundance, C—Common, F—Fair, R—Rare, V—Very rare and Blank-absence

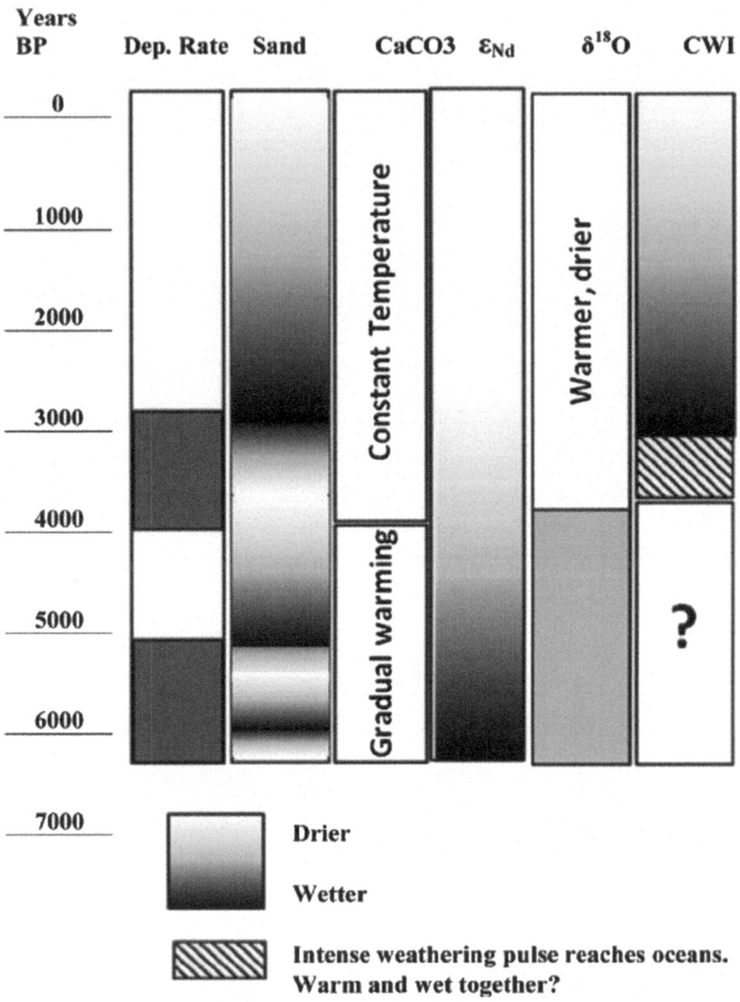

Fig. 9 SWM shifts inferred based on sediment rate, sand and $CaCO_3$ content, ∑Nd, $\delta^{18}O$ and chemical weathering Index. It is noted that each parameter has responded in different way to the SWM variability

Weaver 1989; Chauhan and Suneethi 2001; Chauhan et al. 2004). Occurrence of varying amounts of kaolinite, smectite, chlorite, illite, kaolinite/chlorite (K/C) and chlorite/illite (C/I) ratio indicates periods of intense wetter conditions (intense SWM) with incursions of shorter weaker monsoonal conditions from mid to late Holocene. High kaolinite, K/C ratio with reduced chlorite and low C/I ratio around 6500–6000 years BP (124–111 cm depth) points to intense SWM causing strong weathering of the bedrocks in the hinterland (Fig. 4). The reduced $CaCO_3$ content (72.32–70.63%) and occurrence of kaolinite, around mid-Holocene 3300 years BP and 6000 years

BP (depth 38–40 cm and 108–110 cm) respectively indicates strong weathering of bed rocks, intensified SWM that led to an upwelling resulting in higher surface biogenic productivity. This observation is also supported by the large coarser component influx at these depths. In the studied sediment core an overall fluctuating occurrence in the abundance of smectite, kaolinite, illite and chlorite supported by K/C and C/I ratio is observed from 6000 to 2000 years BP (111–20 cm depth) indicating an overall weakening of SWM. However, within this 4000-year period a short incursion of extreme weakening of SWM occurred from 4400 years BP to 4200 years BP i.e. from 70 to 65 cm depth as revealed by high smectite content, reduced kaolinite and illite clay minerals with high C/I ratio (Fig. 4). The characteristic clay mineral association of smectite and illite also implies dry to semi-drier conditions around 4400–4200 years BP. Subsequently an amelioration of SWM is observed with high percentage of kaolinite and chlorite with a higher C/I ratio from 3900 years BP to 3600 years BP (depth 55–48 cm). An occurrence of arid phase is noticed around ~2000–1700 years BP (20–15 cm) due to the presence of higher illite content and reduced C/I ratio (Fig. 4). Similar observations were made by Chauhan and Vogelsang (2006) using the clay mineral composition of the sediments from the western BOB. They reported four arid phases during the Holocene period (10,000–9600, 7300, 5800–4300 and 2200–1800 years BP) and our inferences gains support from the observations made by Chauhan and Vogelsang (2006) and Staubwasser and Weiss (2006).

5.2 *Sediment Texture, $CaCO_3$, OM, Stable Isotopes, Phytolith, Diatom and Nanoplankton Data*

In the present study, high sand content in the sediment core occurring at depths 114–112 cm (~6191 years), 90-88 cm (~5486 years) and 42–40 cm (~3425 years BP) (Table 2) with less $CaCO_3$ content indicates a terrigenous influx of sediments and wet conditions (intense monsoon) (Figs. 3, 7). The increasing OM content also point towards wetter conditions. Moreover, a virtual increase in sand component (from 8.28 to 20.02%) and the decrease in clay content (from 23.67 to 16.00%) between ~4059 and 3425 years BP suggests high terrigenous input from the surrounding islands and an unaffected chemical weathering of the Landfall Island bed rocks during the mid and late Holocene period. The sand layers with less silt component during this period have also pointed to the river influx such as Irrawaddy, Sittang and the Salween rivers. On the contrary, from the depths such as 96–94 cm (~5655 years), 60–58 cm (~4059 years) and 24–16 cm (~2038 years) the sediments are dominantly fine silt with a decreasing trend of sand content and elevated $CaCO_3$ percentage indicating warm and dry conditions (Figs. 3, 7).

The $\delta^{13}C$ of *G. ruber* varies from 0.70‰ (16–18 cm; ~1700 years BP) to 1.33‰ (112–114 cm; ~6100) years BP (Fig. 3; Table 4). These variations in the $\delta^{13}C$ are not coherent with the Holocene variations, however, we attribute the shifts in $\delta^{13}C$ to the variations in carbon isotope composition of surface waters due to the strengthening or weakening of SWM and wind induced variability. During enhanced high wind velocity period, subsurface water with low $\delta^{13}C$ might have been brought from to the surface, resulting in depletion of $\delta^{13}C$ values and vice versa (Ahmad et al. 2012). Our $\delta^{13}C$ values conform well with the values of planktonic foraminifera 0.74‰ to 1.25‰ from the sediment core SK157-14 (Ahmad et al. 2012) data.

Duplessy (1982) analysed the variation in the stable isotope *G. ruber* ($\delta^{18}O$) from the marine sediment cores collected from the BOB, Andaman Sea and the Arabian Sea. The data ranges between −3.70‰ and −1.58‰ for the Holocene times and for the glacial times it varied from +0.55 to −1.17‰. Chauhan (2003) also observed the heavy $\delta^{18}O$ values of *G. ruber* during 20,000–15, 000 years BP (−0.9‰) and lighter $\delta^{18}O$ value of *G. ruber* in the BOB around −3.38‰ in the Holocene period, which is much lighter compared to the Last Glacial Maxima (LGM). A comparison of stable isotope data reveals that the beginning of the Holocene is marked with much lighter $\delta^{18}O$ values which persisted for the entire Holocene, with two pauses at ~5000–4300 years BP and ~2000 years BP when compared to the last glacial maximum (LGM) (magnitude 2.44‰ for *G. ruber*).

The sediment influx data also corroborate well with the stable isotope data on the *G. ruber*. $\delta^{18}O$ isotope data on the *G. ruber* reveals fluctuations in the monsoonal regime from 6000 to 4300 years BP (−4.03 to −3.39‰). More positive $\delta^{18}O$ value occurs at ~5500 years BP and around 4200 years BP (Fig. 3) indicating weakened SWM during this period. However around 4400–4200 years BP; weakened monsoon with cooler conditions is discerned by the values of $\delta^{18}O$ moving from −3.39 to −2.33‰ (values becoming more positive). The lighter values of $\delta^{18}O$ in the species *G. ruber* could be due to local factors such as changes in sea surface salinity (SSS) and by higher fluvial influx from the Ganga-Brahmaputra system, and/or higher sea surface temperature (SST) in the BOB. Further, there is a slight amelioration in the monsoons as revealed by the $\delta^{18}O$ values ranging from −2.76 to −3.43‰ (~3425 years BP). Higher $\delta^{18}O$ values of *G. ruber* and an increase in its abundance (Fig. 3) can be strongly linked to an increase in SSS caused by a decrease in riverine runoffs strongly associated with reduced SWM.

An integration of the down core sediment texture, clay mineralogy and stable isotope data on *G. ruber* points that the decrease in sediment influx into the Bay of Bengal near the Landfall island reflects weakening SWM and higher salinity conditions of the sea water during (~6000–4400 years BP) (Figs. 7, 9). An increasing occurrence of kaolinite and illite indicate an intense chemical alteration of island bed rocks from ~4400 to 1700 years BP. The arid/dry episodes of 5000–4300 and ~2000 years BP largely coincide with similar events reported from the northern Indian Ocean (Von Rad et al. 1999; Chauhan et al. 2004; Chauhan and Suneethi 2001).

A high-resolution study of mineralogy and major element geochemistry combined with Sr, Nd and Oxygen isotopes ($\delta^{18}O$) was conducted by Colin et al. (1999, 2006) in two marine sediment cores collected off the Irrawaddy river mouth (MD77-180) in the Bay of Bengal and in the Andaman Sea (MD77-169). Pedogenic clays (smectite and kaolinite) to primary mineral (feldspar, quartz, illite and chlorite) ratios showed strong precessional cycles, suggesting a control due to the past changes in the summer monsoon intensity (Colin et al. 1999, 2006). They also demonstrated wet periods of summer monsoon reinforcement corresponding to an increase in the weathering of the Irrawaddy plain soils with a decrease in $^{87}Sr/^{86}Sr$ ratio.

Planktonic diatoms generally decrease with climate warming because of reduced nutrient redistribution and increasing sinking velocities. Climate warming exhibits a selection pressure on the diatom cell size, this favour small-sized diatoms that are able to out compete larger sized cells and expand under intensified stratification. This observation was made in this sediment core recovered from the Bay of Bengal High frequency of diatoms is found in lower zones as compared to the upper levels. This points also to a hydrodynamic condition that created a calmer environment, with strong sedimentation of clay particles with abundant plant remains such as branches, roots and leaves as indicated by phytolith morph types and silicified woody elements, forming an organic clay sedimentary deposit, colonized by small shrubs and vegetation. These environmental conditions probably contributed to the retention of organic muddy sediments that favoured the dominance of diatom species. The presence of freshwater sponge spicules in sediments obviously indicates an aquatic environment, but their gemmoscleres/ megasclare besides allowing for species identification, point to seasonal changes in temperature and sea water column levels.

Siliceous sponge spicules provide a basis for palaeoecological interpretation. Sponges require relatively clear, non-turbid water for living conditions because muddy waters clog their pores (Eksambekar 2002). Thus, it appears that only some spicules were from in situ sponges while the rest were probably re-deposited. The low spicule abundance in the sediment core resulted from local current conditions and the dilution effect through input of terrestrial sediment (Figs. 6, 7). Other possible explanations for the varying spicule abundance in the sediment core are due to the difference in the local fauna, such as coral reefs which usually have high diversity and abundance of sponges.

Picoplanktons (<2 μm) and nanoplankton (2–20 μm) can also contribute downward carbon flux (Gorsky et al. 1999), other than diatoms and the dinoflagellates (two chief component of phytoplankton) (Severdrup et al. 1942), Hence the calcareous nanoplankton is also used as one of the most important climate indicators among other proxy such as foraminifera, alkenones, dinoflagellate cysts, especially in the Mediterranean Sea (Triantaphyllou et al. 2009).

The nanoplankton assemblage in the studied marine sediment core (0–124 cm) reveals the abundance of *Emiliania huxleyi and Gephyrocapsa Oceanica* rare of Pleistocene to recent age (Bukry 1971). It is also noted that the presence of some important species of nanoplanktons of the Pliocene (*Discoaster brouweri*) and Miocene (*Coccolithus pelagicus*) suggests that the sediments are derived from the reworked sediment of Mio-Pliocene age (Fig. 8a–f; Table 6).

The nanoplankton data from the BOB sediment core (Table 6) revealed fluctuations in its abundance for the past 5680 years BP (Figs. 5, 6 and 7). This variation is associated with the fluctuations in the nutrient enriched/depleted water and changes in the thermocline/nutricline depth. A significant increase in the surface water productivity during the deglaciation period is evident from the phytolith proxy data.

The overall correlation between ascidian spicule abundance and the relative abundances of upper photic zone calcareous nannoplankton reflect a close connection between the enrichment in nutrients in the upper photic zone and an increment in bottom productivity. This supports the idea of benthic communities in this area strongly benefiting from surface water productivity (bentho-pelagic coupling). According to Xiao et al. (2006) $CaCO_3$ is a good indicator of temperature variation. Higher content $CaCO_3$ (75.56%) percentage indicates warmer and dry conditions in the hinterland (around 66-64 cm, 4270 years BP) (Fig. 3).

High content of sand with reduced silt, OM, and nanoplankton productivity ~6300–6000 years BP, 5600–5200 years BP suggests intense SWM. It is also observed that ~3600–3300 years BP the sand and silt content was higher with reduced nanoplankton and clay reveals high river influx due to the SWM. In contrast, two major phases of weakening SWM were found around 4200–3800 years BP and 2000–1800 years BP inferred from the reduced sand with increasing percentage of silt, OM and nanoplankton productivity (Fig. 7). It was in agreement with Jochem (1989), that the primary production of nanoplankton was lower in winter monsoon whereas it is higher during the summer season as a dominating size fraction which can, contribute 70–100% of primary production compared to the other plankton (Pico and microplankton).

The basis of palaeomonsoon reconstruction in the tropics is the amount effect (Yadava and Ramesh 2005). They observed significant amount effect in the plains of central India. However, exceptions were also observed in the tropical to subtropical SWM rainfalls in the hilly terrains of NE India. (Breitenbach et al. 2010; Laskar et al. 2013a) and some parts of Southern India (Yadava et al. 2007). Variations in the amount of past rainfall have been owing to factors, such as changes in the $\delta^{18}O$ of the ocean ("ice volume" effect), shifts in the source of moisture Bar-Matthews et al. (1999), Fleitmann et al. (2003, 2007), Fleitmann and Matter (2009), and changes in the seasonality of precipitation (e.g., change in the proportion of winter and summer precipitation).

5.3 *Reconstruction of Palaeomonsoon and a Comparison with Other Records from the Region*

It has been noted that in the last 4000 years BP, there is no significant change in the ocean $\delta^{18}O$, that can cause changes in rainwater and hence sedimentation. In the present study, a strong reduction in the monsoon during 4400–4000 and 2200–1800 years BP is observed (Figs. 7, 8). The period 2200–1800 years BP coincides with the RWP, previously thought to be restricted to higher latitude regions (Vollweiler et al. 2006; Desperat et al. 2003; Martinez-Cortizas et al. 1999; Laskar et al. 2013a, b). The BOB sediment core data indicate that the effect of RWP is evident in the tropics also. Weakening of the monsoon has also been observed around 1500 and 400–800-years BP, the latter period is the transition from the MWP to LIA. A significant reduction in the SWM during this period was also reported by several others (Fleitmann et al. 2004; Sinha et al. 2007).

Middle Holocene is an important period in Indian Prehistory as during this period river valley civilization, and establishment of rain fed agriculture got stabilized. Middle Holocene terrestrial and marine palaeoclimate proxy records overlap to a large extent indicating erosion from the hinterland and being deposited in the oceans (Patnaik et al. 2012). Till date marine record from Arabian Sea has the most consistent and continuous record of the monsoon (Sirocko et al. 1993; Kudrass et al. 2001; Thamban et al. 2002) whereas limited studies have been carried out in the BOB (Duplessy 1982; Chauhan et al. 1993, 2004; Rashid et al. 2011).

Based on the marine and terrestrial records, Patnaik et al. (2012) advocated that the Holocene period is characterized by a strong monsoonal phase known as "Holocene Climatic Optimum (HCO)" i.e. the middle Holocene period, that subsequently was followed by weak monsoonal phases around ~2500–1500, 1000, 650–450 years BP and the Little Ice Age (LIA) (1450–1850 AD). Chauhan et al. 2004 and Chauhan and Vogelsang (2006) found an intense weakening of SWM ~4300 and 2200–1800 years BP. Sarkar et al. (1990, 2000) and Yadava and Ramesh (2005) found an increase in the intensity of SWM around ~3200 years BP. Weakened monsoon during the Roman warm period is observed and this may be owing to cause and lag effect of the monsoons. However, it is generally observed that both in climate modelling and proxy data settings, monsoon increases during warm periods.

SWM significantly affected and fluctuated during the RWP. The reduction in the SWM during the RWP is evidently observed in the abundances of *Globigerina bulloides* (Gupta et al. 2003) and $\delta^{18}O$ of *G. ruber* and *G. Sacculifer* (Tiwari et al. 2006) in sediment cores from Arabian Sea. A significant decrease in monsoon strength is also observed during the period between 800 and 400 years BP, i.e. the transition period from the MWP to the LIA (Figs. 9, 10) in all the studies. A relatively dry episode at the beginning and a wet phase at the end of the LIA are also reported from the Himalayan region (Kotlia et al. 2012). The monsoon was significantly stronger as compared to the present during 1200–800 cal years BP, the MWP. Evidence for a stronger monsoon during this interval is also found in the records of varve thickness (Von Rad et al. 1999) and stalagmite $\delta^{18}O$ from the Indian subcontinent (Sinha et al. 2007). The SWM was similar to the present without any major change in the source, extent, and seasonality before 6000 years BP except two short-term arid phases around 4400–4000 years BP and 2000–1700 years BP (Sinha et al. 2007).

Xu et al. (2002) studied plant cellulose $\delta^{8}O$ variations in sediment cores retrieved from peat deposits at the northeastern edge of the Qinghai- Tibetan Plateau in China. Following the decline of the RWP, their data revealed the existence of three particularly cold intervals centered at approximately 500, 700 and 900 AD during the DA (cold period).

Yang et al. (2002) used nine separate proxy climate records, derived from peat, lake sediment, ice core, tree ring and other proxy sources, to compile a single weighted temperature history for China spanning the past two millennia. Their composite

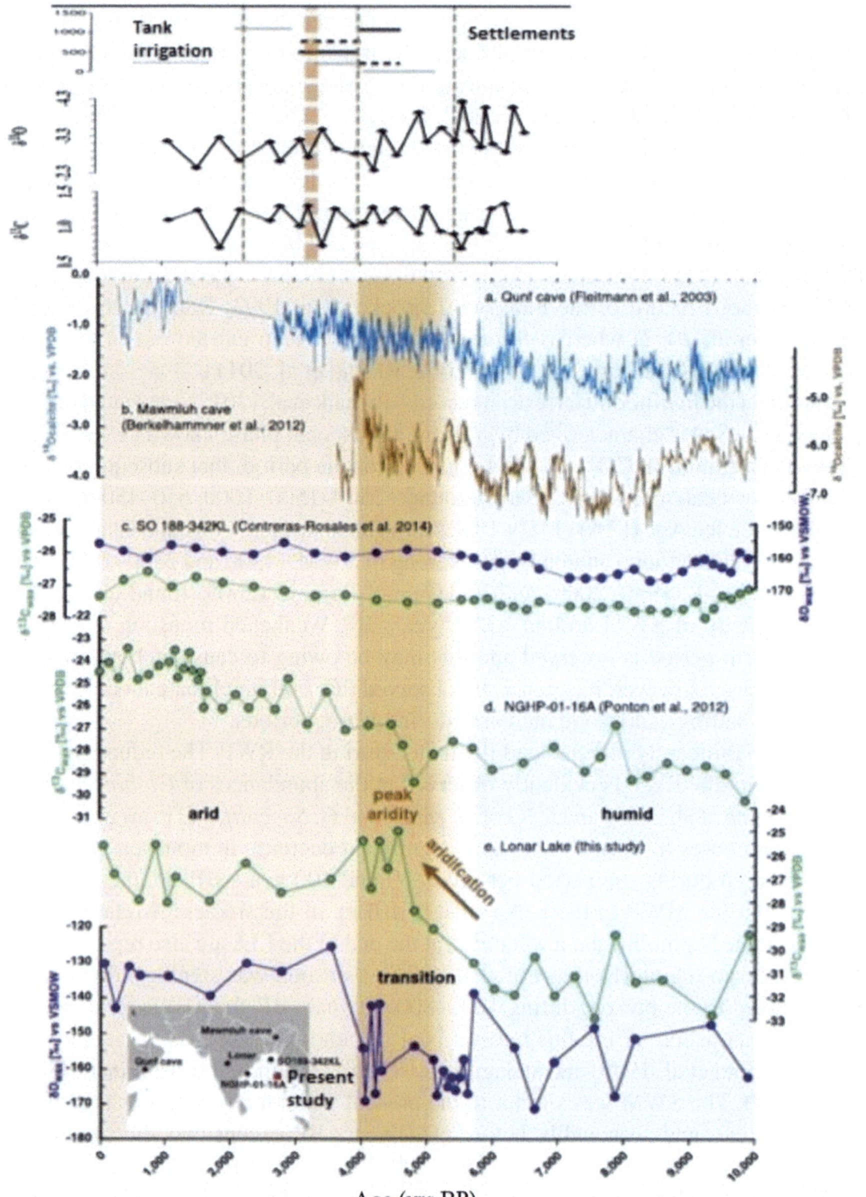

Fig. 10 A comparison of palaeomonoonal reconstruction based on several proxies prepared after Sarkar et al. (2015)

record revealed five distinct climate epochs: a warm stage from 0 to 240 AD (the tail-end of the RWP), a cold interval between 240 and 800 AD (the DA), a return to warm conditions from 800 to 1400 AD (which included the MWP between 800 and 1100 AD), a cool interval between 1400 and 1820 (LIA), and the current warm regime (MWP) that followed the increase in temperature that began in the early 1800s. Another important finding of this study was that warmer temperatures of the past two millennia were observed during the second and third centuries AD near the end of the RWP.

Ponton et al. (2012) reconstructed the Holocene palaeoclimate in the core monsoon zone (CMZ) of the Indian peninsula using a sediment core raised from the offshore; from the mouth of the Godavari River (Fig. 1). Carbon isotopes of sedimentary leaf waxes indicated a gradual increase in aridity-adapted vegetation from ~4000 until 1700 years ago and was subsequently trailed by the arid-adapted plants. A comparison with the oxygen isotopic composition of planktonic foraminifer *Globigerinoides ruber* revealed high salinity peaks in the Bay of Bengal over the last 3000 years, and especially after 1700 years ago, signifying that in the CMZ aridification intensified in the late Holocene through a series of sub-millennial dry episodes (Ponton et al. 2012).

Probable variances in seasonal monsoonal moisture sourcing during the 5000–4200 yrs BP transition are evident in other terrestrial (speleothem) records from the region. A gradual increase in $\delta^{18}O$ values over the Holocene from the Qunf record in Oman, which receives moisture exclusively from the Arabian sea reflect the decrease in monsoon strength due to changes in solar insolation (Fig. 10). A comparison of the Mawmluh cave record in NE India, which is closer to the BOB also receives the Arabian Sea sourced ISM moisture during the summer months with a mean $\delta^{18}O$ of −7.2‰ and also that the BO B moisture would affect NE India (Breitenbach et al. 2010). The transition from early Holocene wet to late Holocene dry conditions in the BOB was not gradual, but characterised by relatively rapid changes in hydroclimate conditions, with at times bringing about a complete failure of the summer monsoon especially between 5000 and 4200 yrs BP.

A mid-Holocene transition has been recognized throughout the Asian monsoon region, nonetheless its nature (i.e., abrupt vs. gradual) has been highly debated. Morrill et al. (2003) discussed an abrupt (century scale) change in ISM while Fleitmann (2007) argued for gradual weakening of the ISM during the Holocene. Recent leaf wax based stable isotopic records from the Indian Ocean basin (Bay of Bengal) revealed a similar overall pattern but with gradual vegetation and hydrological changes during mid-Holocene (Ponton et al. 2012; Contreras-Rosales et al. 2014), reflecting the decrease in monsoon strength due to changes in solar insolation. This strongly affected the river valley civilisations causing cultural changes and decline that occurred across the Indian subcontinent as the climate became more arid after ~4000 years. Sedentary agriculture took hold in the drying central and south India, while the urban Harappan civilization collapsed in the already arid Indus basin. The establishment of a more variable hydroclimate over the last ca. 1700 years may have led to the rapid proliferation of water-conservation technology in south India also (Fig. 10).

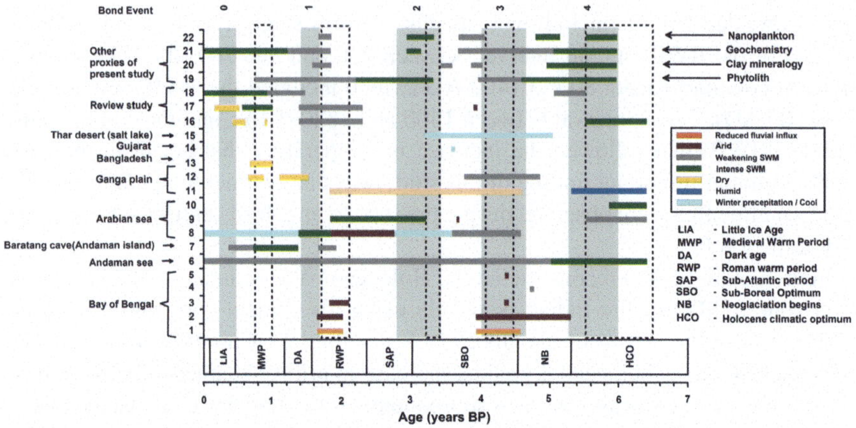

Fig. 11 A comparison of monsoonal shifts of the present study with global climate events, Bond events (shaded) and other palaeoclimate records, reviews from Arabian Sea, Bay of Bengal, Andaman Sea and Indian sub-continent. [References (1) Chauhan et al. (2004), (2) Chauhan and Vogelsang (2006), (3) Chauhan and Suneethi (2001), (4) Mathien and Bassinot (2008), (5) Chauhan et al. (2000), (6) Rashid et al. (2007), (7) Laskar et al. (2013a, b), (8) Lückge et al. (2001), (9) Sarkar et al. (2000), (10) Thamban et al. (2002), (11) Sharma et al. (2006), (12) Saxena et al. (2013), (13) Masud Alam et al. (2009), (14) Singh et al. (1990), (15), Enzel et al. (1999), (16) Patnaik et al. (2012), (17) Kuppusamy and Ghosh (2012), (18) Thamban et al. (2007) and 19, 20, 21 is our present data of phytolith, clay mineralogy and Geochemistry

To check whether our marine sediment core based chronology reflect local or broader regional rainfall variability, we compared our data with other marine and continental climate proxy records from some of the areas receiving SWM (Chauhan et al. 2004; Chauhan and Vogelsang 2006; Chauhan and Suneethi 2001; Mathien and Bassinot 2008; Rashid et al. 2007; Chauhan et al. 2006; Sharma et al. 2006; Saxena et al. 2013; Masud Alam et al. 2009; Enzel et al. 1999; Lückge et al. 2001; Sarkar et al. 2000; Thamban et al. 2002; Patnaik et al. 2012; Kuppusamy and Ghosh 2012; Thamban et al. 2007; Ponton et al. 2012; Laskar et al. 2013a, b) (Figs. 10, 11). An integration of all our proxy results shows that the variations in the sand/clay ratio influenced the ratio of the benthic/planktonic diatom species, sponge spicules and phytolith and this probably was caused by local hydrodynamic changes in SWM and also due to the fluctuations in the marine water column, temperature and salinity. Our data points to an overall weakening of SWM from 5000 years BP to 1200 AD (Figs. 9, 10). The Early middle Holocene period (~6500 years BP to 5000 years BP) experienced strong SWM that eventually tapered. The period from ~5000 to 3600 years BP witnessed a weakening of SWM rains, followed by fluctuations in its intensity from 3600 to 2400 years BP. Further the SWM conditions did not improve, but continued to weaken from 2400 years BP to 800 years BP. However, considerable variability exists in the continental records about the widespread climate of the Himalayas during the mid Holocene period. For example, in the Gangotri area, the Shivling Stage of glaciations has been reported around 5100 years BP (Sharma and

Owen 1996) and the pollen records from the Dokini Glacier-periglacial peat indicate prevalence of arid climate during 4000–3500 years BP (Phadtare 2000).

In our records this phase terminates at around 4300 years BP. Phytolith records from Deoria Tal (2770 m asl; Uttarkashi), Sharma et al. (1995) reported an occurrence of a humid period around 4000 years BP and a persistence of drier climate between 3200- and 1700-years BP. By and large, the variations observed in the marine sediment record are similar to the other proxy records, except with some mismatches and sharp changes that may be attributed to uncertainty in the chronologies and relative differences in sampling resolution.

The most common feature in all these records is the reduction in the SWM intensity during the RWP and MWP and this may be owing to a reduction in solar radiation reaching the Earth, which may have triggered a change in atmospheric circulation patterns (Wang et al. 2003) Further a reconstruction of ocean current patterns, based on the granularity of deep ocean sediment, Bianchi and McCave (1999) concluded that there was a Roman Warm Period, which peaked around AD 150. An analysis of oxygen isotopes in mollusk shells analysed from an Icelandic inlet indicated that Iceland experienced an exceptionally warm period from 230 BC to AD 40 (Patterson et al. 2010). Recent Res (Werner et al. 2019), based on a much larger dataset using palaeoclimate proxies has revealed that the RWP, along with other periods such as the MWP and the LIA were regional phenomena and not globally clear episodes (Fig. 10).

Our present study results based on the sediment core collected from the BOB near the Landfall Island clearly demonstrates that there exists an agreement of climate change in the region as observed in North Atlantic Ocean, distant Himalayas and East Asia and provides evidence for RWP, DA (Dark Ages), MWP (Fig. 11).

6 Conclusions

A marine sediment core about 124 cm retrieved at a depth of 250 m depth, near Landfall Island, North Andaman, Bay of Bengal, was studied using various proxies such as clay mineralogy, oxygen ($\delta^{18}O$) and carbon ($\delta^{13}C$) stable isotope, neodymium (Nd), geochemistry, phytolith and nanoplankton assemblage. Based on the data collected and integration of results following conclusions are drawn:

1. The sediment texture is predominantly represented by clayey silt. The sediment core reveals layers of coarser sand flux since ~6500–6000 and ~3300 years BP, that reflects a strengthened SWM in an overall weak SWM of the middle to late Holocene.
2. The study of grain size and isotopic data of these sediments show alternate dry/warm and wet (humid) conditions prevailing in the southern tropical areas since ~6500 years BP with weaker southwest monsoon during 4300–4000 years BP, strengthened SWM conditions ~3300 years BP and an amelioration in climate since ~2000 years BP. Due to the intense SWM, the short, wet conditions

resulted in intense chemical weathering of the Island rocks. The arid event with an upwelling record during 3500 years BP of the western Arabian Sea as reported by Naidu et al. (2012) is not prominent in the Bay of Bengal

3. The clay minerals and nannoplankton assemblages suggest that the sediments deposited in this region have been contributed by two major provenances. A nearby source, i.e. from Andaman sediments predominately derives from the Andaman terrain of ophiolite suites and sedimentary rocks, the nanoplanktons are reworked, followed by sediments from the Irrawaddy, Salween and Sittang rivers from Myanmar.
4. Smectite and illite are the two dominant clay minerals in this sediment core. High content of smectite reflects the weathering product of mafic rock contributed from the nearby island whereas illite is predominantly coming from the rivers such as Irrawaddy and Salween.
5. An overall weakening of SWM is observed from ~6000 to 2000 years BP, and this is inferred from the high smectite, high C/I ratio and low K/C ratio. Within this period; an incursion of a more intense weakening of SWM is noted from the elevated C/I ratio and lesser K/C ratio from 4400 to 4200 years BP, which is also supported by the $\delta^{18}O$ data of *G. ruber* (−3.39 to −2.33‰). The occurrence of high kaolinite percentage and K/C ratio points to strong weathering during 3900 to 3600 years BP and subsequently an arid phase is observed between 20 and 15 cm (~2000–1700 years BP). A major wet phase also found reaching its maximum at 3400–3200 years BP and amelioration in climate since ~2000 years BP to present. The intensity of SWM tends to repeat itself.
6. The phytolith analysis shows a significant variation in the SWM from ~6500 to present, with weakening of the SWM from 5000 years BP. Variations in the sand/clay ratio influenced the ratio of the benthic/planktonic diatom species, sponge spicules and phytolith and this probably was caused by local hydrodynamic changes in SWM and also due to the fluctuations in the marine water column, temperature and salinity.
7. Application of various proxies in present study reflects the significant signatures of intense and weakened SWM such as the wet phase during 6500 to 5600 years BP (HCO), 3400–3200 years BP (SBO) and warm, dry phase ~2200–1800 years BP (RWP) and 1000–800 years BP (MWP) with an extreme weakened monsoon ~4400–4000 years BP (Sub—Boreal Optimum).

Acknowledgements We thank Dr. Jyotiranjan Ray, PRL Ahmedabad for giving us an opportunity to submit this manuscript and the two anonymous reviewers who painstakingly read through our work and suggested constructive comments that helped in the presentation of our work.

References

Achyuthan H, Farooqui A, Gopal VEE, Phartiyal B, Lone AM (2017) Late Quaternary to Holocene southwest monsoon reconstruction: a review based on lake and wetland systems (studies carried out during 2011–2016). Proc Indian Nat Sci Acad 82(3):847–868

Ahmad T, Dragusanu V, Tanaka T (2008) Provenance of Proterozoic Basal Aravalli mafic volcanic rocks from Rajasthan, Northwestern India: Nd isotopes evidence for enriched mantle reservoirs. Precamb Resour 162:150–159

Ahmad SM, Zheng H, Raza W, Zhou B, Lone MA, Raza T, Suseela G (2012) Glacial to Holocene changes in the surface and deep waters of the northeast Indian Ocean. Mar Geol 329(331):16–23

Bar-Matthews M, Ayalon A, Kaufman A, Wasserburg GJ (1999) The eastern Mediterranean palaeoclimate as a reflection of regional events: Soreq Cave, Israel. Earth Planet Sci Lett 166:85–95

Bhushan R, Singh SK, Burr GS, Jull AJT (2007) Palaeoclimatic studies from sediments in the Bay of Bengal. In: Proceedings 12th ISMAS symposium cum workshop on mass spectrometry, IRP-3

Bianchi GG, McCave IN (1999) Holocene periodicity in North Atlantic climate and deep-ocean flow south of Iceland. Nature 397(6719):515–517

Biscaye PE (1965) Mineralogy and sedimentation of recent deep-sea clays in the Atlantic Ocean and adjacent seas and oceans. Bull Geol Soc Am 76:803–832

Breitenbach SFM, Adkins JF, Meyer H, Marwan N, Kumar KK, Haug GH (2010) Strong influence of water vapor source dynamics on stable isotopes in precipitation observed in Southern Meghalaya, NE India. Earth Planet Sci Lett 292:212–220

Bukry D (1971) Calcareouss nanofossils' Coccolith Stratigraphy leg 13, Deep sea drilling project, initial report. U. S. Geol Surv, La Jolla, California, pp 817–822

Carver RE (1971) Procedures in sediment petrol. Wiley and Sons. Inc., New York, p 653

Chauhan OS (2003) Past 20,000 years history of Himalayan aridity: evidence from the oxygen isotope record of Bay of Bengal. Curr Sci 84:90–93

Chauhan OS, Suneethi J (2001) 18 ka BP records of climatic changes, Bay of Bengal: isotopic and sedimentological evidences. Curr Sci 81:1231–1234

Chauhan OS, Vogelsang E (2006) Climate induced changes in the circulation and dispersal patterns of the fluvial sources during late quaternary in the middle Bengal Fan. J Earth Syst Sci 115(3):379–386

Chauhan OS, Borole DV, Gujar AR, Antonio M, Mislanker PG, Rao ChM (1993) Evidences of climatic variations during Late Pleistocene-Holocene in the eastern Bay of Bengal. Curr Sci 65(7):558–562

Chauhan OS, Jayakumar S, Menezes AAS, Rajawat AS, Nayak, SR (2006) Anomalous inland influx of the River Indus, Gulf of Kachchh, India. Mar Geol 229:91–100. https://doi.org/10.1016/j.margeo.2005.12.003

Chauhan OS, Patil SK, Suneethi J (2004) Fluvial influx and weathering history of the Himalayas since Last Glacial Maxima; isotopic, sedimentological and magnetic records from the Bay of Bengal. Curr Sci 87(4):509–515

Chauhan OS, Sukhija BS, Gujar AR, Nagabhushanam N, Paropkari AL (2000) Late quaternary variations in clay mineral along the SW continental margin of India: evidence of climatic variations. Geo Mar Lett 20:118–122. https://doi.org/10.1007/s003670000043

Colin C, Turpin L, Bertauz J, Desprairies A, Kissel C (1999) Erosional history of the Himalayas and Burman ranges during the last two glacial-interglacial cycles. Earth Planet Sci Lett 171:647–660

Colin C, Turpin L, Blamart D, Frank N, Kissel C, Duchamp S (2006) Evolution of weathering patterns in the Indo-Burman ranges over the last 280 kyr: effects of sediment provenance on 87Sr/86Sr ratios tracer. Geochem Geophys Geosys 7(3):Q03007. https://doi.org/10.1029/2005GC000962

Contreras-Rosales A, Jennerjahn T, Tharammal T, Lückge A, Paul A, Schefuß E (2014) Evolution of the Indian Summer Monsoon and terrestrial vegetation in the Bengal region during the past 18 ka. Quat Sci Rev 102:133–148

Dash SK, Kumar JR, Shekhar MS (2004) On the decreasing frequency of monsoon depressions over the Indian region. Curr Sci 86:1404–1411

Desprat S, Sánchez-Goñi MF, Loutre MF (2003) Revealing climatic variability of the last three millennia in northwestern Iberia using pollen influx data. Earth and Planet Sci Lett 213:63–78

Ding YH (1994) Monsoons over China. Kluwer Academic Publishers, pp 432

Duplessy JG (1982) Glacial to interglacial contrast in the northern Indian Ocean. Nature 295:494–498

Eksambekar SP (2002) Contribution of the study of phytoliths to Bioarchaeology. Unpublished Ph.D. thesis, Deccan College PGRI (Deemed University), Pune, India

Enzel Y, Ely LL, Mishra S, Ramesh R, Amit R, Lazar B, Rajaguru SN, Baker VR, Sandler A (1999) High-resolution Holocene environmental changes in the Thar Desert. Northwestern India. Science 284(5411):125–128

Fleitmann D (2007) Holocene ITCZ and Indian monsoon dynamics recorded in stalagmites from Oman and Yemen (Socotra). Quat Sci Rev 26:170–188

Fleitmann D, Matter A (2009) The Speleothem record of climate variability in Southern Arabia. Comptes Rendus Geosci 341:633–642

Fleitmann D, Burns SJ, Mudelsee M, Neff N, Kramers J, Mangini A, Matter A (2003) Holocene forcing of the Indian monsoon recorded in a stalagmite from Southern Oman. Science 300:1737–1739

Fleitmann D, Burns SJ, Neff U, Mudelsee M, Mangini A, Matter A (2004) Palaeoclimatic interpretation of high-resolution oxygen isotope profiles derived from annually laminated speleothems from Southern Oman. Quat Sci Rev 23:935–945

Giosan L, Orsi W, Collen MJL, Wuchter C, Dunlea AG, Thirumalai K, Munoz SE, Clift P, Fuller DQ (2018) Neoglacial climate Anomalies and the Harappan Metamorphosis. Climate Past, Under Rev. https://doi.org/10.5194/cp-2018-37

Gorsky G, Chertiennot-Dinet MJ, Blanchot J, Palazzoli I (1999) Picoplankton and nanoplankton aggregate on by appendicularian fecal pellet contents of Megalocercus huxleyi in the equatorial Pacific. J Geophys Oceans 104:3381–3390

Griffin RE (1968) Clay mineralogy. McGraw Hill, New York, p 565

Griffin WL, Zhang AD, O'Reilly SY, Ryan CG (1998) Phanerozoic evolution of the lithosphere beneath the Sino-Korean Craton. Mantle Dyn Plate Interact East Asia 27:107–126. In: Flower MFJ, Chung SL, Lo CH, Lee TY (eds) Amer Geophys Union, Geodynam Ser

Gupta AK, Thamban M (2008) Holocene Indian monsoon variability. Glimpses Geosci Res India, pp 28–31

Gupta AK, Anderson DM, Overpeck JT (2003) Abrupt changes in the Asian southwest monsoon during the Holocene and their links to the North Atlantic Ocean. Nature 421:354–357

Hughes MK, Diaz HF (1994) Was there a 'Medieval Warm Period. and if so, where and when. Clim Change 26:109–142

Jochem F (1989) Distribution and importance of autotrophic ultra-plankton in a boreal inshore area (Kiel Bight, Western Baltic). Marine Ecol Progr Ser 53:153–169

Kessarkar PM, Rao VP, Ahmad SM, Patil SK, Anil Kumar A, Anil Babu G, Chakraborty S, Soundar Rajan R (2005) Changing sedimentary environment during the Late Quaternary: sedimentological and isotopic evidence from the distal Bengal Fan. Deep Sea Res Part I 52:1591–1615

Kotlia BS, Ahmed SM, Zhao J, Raza W, Collerson KD, Joshi LM, Sanwal J (2012) Climatic fluctuations during the LIA and post-LIA in the Kumaun Lesser Himalaya, India: evidence from a 400 y old stalagmite record. Quat Intl 263:129–138

Krumbein WC, Pettijohn FJ (1938) Manual of sedimentary petrography. Appleton-Century-Crofts, Inc, New York

Kudrass HR, Hofman A, Doose H, Emeis K, Erlenkeuser H (2001) Modulation and amplification of climatic changes in the Northern Hemisphere by the Indian summer monsoon during the past 80 ky. Geology 29:63–66

Kuppusamy M, Ghosh P (2012) Cenozoic climatic record for monsoonal rainfall over the Indian region. Mod Climat. In: Wang SY (ed). In Tech. ISBN 978-953-51-0095-9. https://doi.org/10.5772/36206

Kurian S, Nath BN, Ramaswamy V, Naman D, Rao TG, Kamesh Raju KA, Selvaraj K, Chen CTA (2008) Possible detrital, diagenetic and hydrothermal sources for Holocene sediments of the Andaman backarc basin. Mar Geol 247:178–193

Laskar AH, Raghav S, Yadava MG, Jani RA, Narayana AC, Ramesh R (2011) Potential of stable carbon and oxygen isotope variations of speleothems from Andaman Islands, India, for paleomonsoon reconstruction. J Geolog Res 2011. https://doi.org/10.1155/2011/272971

Laskar AH, Yadava MG, Sharma N, Ramesh R (2013a) Late Holocene climate in the Lower Narmada valley, Gujarat, Western India, inferred using sedimentary carbon and oxygen isotope ratios. Holocene 23(8):1115–1122

Laskar AH, Yadava MG, Ramesh R, Polyak VJ, Asmerom Y (2013b) A 4 kyr stalagmite oxygen isotopic record of the past Indian Summer Monsoon in the Andaman Islands. Geochem Geophys Geosys 14(9):3555–3566

Liu ZT, Colin C, Trentesaux A, Blamart D, Bassinot F, Siani G, Sicre MA (2004) Erosional history of the eastern Tibetan Plateau over the past 190 kyr: clay mineralogical and geochemical investigations from the southwestern South China Sea. Mar Geol 209:1–18

Liu ZT, Colin C, Trentesaux A, Siani G, Frank N, Blamart D, Farid S (2005) Late quaternary climatic control on erosion and weathering in the eastern Tibetan Plateau and the Mekong Basin. Quat Resou 63(3):316–328

Loring DH, Rantala RTT (1992) Manual for the geochemical analyses of marine sediments and suspended particulate matter. Regional seas, reference methods for marine pollution studies, United Nations Environment Programme. Earth-Sci Rev 32(63):235–283

Lückge A, Doose-Rolinski H, Khan AA, Schulz H, von Rad U (2001) Monsoonal variability in the northeastern Arabian Sea during the past 5000 years: geochemical evidence from laminated sediments. Palaeogeogr Palaeoclimatol Palaeoecol 167:273–286

Mann ME (2002a) Medieval climatic optimum, in Encyclopedia of global environmental change. In: MacCracken MC, Perry JS (eds) Wiley, Chichester, U.K, pp 514–516

Mann ME (2002b) Little ice age. In: MacCracken MC, Perry JS (eds) Wiley, Chichester, U.K, pp 504–509

Martinez-Cortizas A, Pontedevedra-Pombal X, Garcia-Rodeja E, Novoa-Munoz JC, Shotyk W (1999) Mercury in a Spanish peat bog archive of climate change and atmospheric metal deposition. Science 284:939–942

Masud Alam AKM, Xie S, Wallis LA (2009) Reconstructing late Holocene palaeoenvironments in Bangladesh: phytolith analysis of archaeological soils from Somapura Mahavihara site in the Paharpur area, Badalgacchi Upazila, Naogaon District, Bangladesh. J Archaeol Sci 36:504–512

Mathien E, Bassinot F (2008) Abrupt hydrographic changes in the Bay of Bengal during the Holocene. Geophys Res Abst, EGU2008-A-09423

Milliman JD, Meade RH (1983) World-wide delivery of river sediment to the oceans. J Geol 9:1–19

Moore GF, Curray JR, Emmel FJ (1982) Sedimentation in the Sunda trench and forearc region. Geol Soc London, Spl Publ 10:245–258. In: Legget JK (ed) Trench-forearc geology: sedimentation and tectonics on modern and ancient active plate margins

Morrill C, Overpeck JT, Cole JE (2003) A synthesis of abrupt changes in the Asian summer monsoon since the last deglaciation. Holocene 13:465–476

Naidu PD, Patil JS, Narale DD, Anil AC (2012) A first look at the dinoflagellate cysts abundance in the Bay of Bengal: implications on Late Quaternary productivity and climate change. Curr Sci 102(3):495–499

Nath BN, Gupta SM, Mislankar P.G, Rao BR, Parthiban G, Roelands I, Patil SK (2005) Evidence of Himalayan erosional event at approx. 0.5 Ma from a sediment core from the equatorial Indian ocean in the vicinity of ODP Leg 116 sites. Deep Sea Res 52(2):2061–2077

Nigam R, Hashimi NH (1995) Marine sediments and palaeoclimatic variations since the Late Pleistocene: an overview of Arabian sea. Mem Geol Soc India 32:380–390

Pal T, Chakraborty PP, Duttagupta T, Singh DC (2003) Geodynamic evolution of the outer-arc-forearc belt in the Andaman Islands, the central part of the Burma-Java subduction complex. Geol Mag 140(3):289–307
Pandey DN, Gupta AK, Anderson DM (2003) Rainwater harvesting as an adaptation to climate change. Curr Sci 85:46–59
Patnaik R, Gupta AK, Naidu PD, Yadav RR, Bhattacharyya A, Kumar M (2012) Indian Monsoon variability at different time scales: marine and terrestrial proxy records. Procs Indian Nat Sci Acad 78(3):535–547
Patterson WP, Dietrich KA, Holmden C, Andrews JT (2010) Two millennia of North Atlantic seasonality and implications for Norse colonies. Procs Nat Acad Sci USA 107(12):5306–5310
Phadtare NR (2000) Sharp decrease in summer monsoon strength 4000-3500 cal yr BP in the central Higher Himalaya of India based on pollen evidence from alpine peat. Quat Res 53:122–129
Piperno DR (1988) Phytolith analysis: an archaeological and geological perspective. Academic Press, New York, p 280
Ponton C, Giosan L, Eglinton TI, Fuller DQ, Johnson JE, Kumar P, Collett TS (2012) Holocene aridification of India. Geophys Res Lett 39:L03704. https://doi.org/10.1029/2011GL050722
Ramesh S, Ramasamy S (1997) Rare earth element geochemistry of a sediment core from lower Bengal Fan. J Geol Soc India 50:339–406
Rao VP (1983) Clay minerals in the sediments around the Andaman Islands. Indian J Mar Sci 12:17–20
Rao VP, Rao BR (1995) Provenance and distribution of clay minerals in the continental shelf and slope sediments of the west coast of India. Continental Shelf Res 15:1757–1771
Rashid H, England E, Thompson L, Polyak L (2011) Late Glacial to Holocene Indian Summer Monsoon variability based upon sediment records taken from the Bay of Bengal. Terrestrial Atmospheric Ocean Sci 22(2):215–228
Rashid H, Flower BP, Poore RZ, Quinn TM (2007) A ~25 ka Indian Ocean monsoon variability record from the Andaman Sea. Quat Sci Rev 26:2586–2597
Ray D, Rajan S, Ravindra R, Jana A (2011) Microtextural and mineral chemical analyses of andesite-dacite from Barren and Narcondam islands: evidences for magma mixing and petrological implications. J Earth Syst Sci 120(1):145–155
Sarin MM, Borole DV, Krishnaswami S (1979) Geochemistry and geochronology of sediments from the Bay of Bengal and the equatorial Indian Ocean. Procs Indian Acad Sci 88:131–154
Sarkar S, Prasad S, Wilkes H, Niedel R, Stibich M, Basavaiah N, Sachse D (2015) Monsoon source shifts during the drying mid-Holocene: Biomarker isotope based evidence from the core 'monsoon zone' (CMZ) of India. Quaternary Sci Rev 123:144–157
Sarkar A, Ramesh R, Bhattacharya SK, Rajagopalan G (1990) Oxygen isotope evidence for a stronger winter monsoon current during the last glaciations. Nature 343:549–551
Sarkar A, Ramesh R, Somayajulu BLK, Agnihotri R, Jull AJT, Burr GS (2000) High resolution Holocene palaeomonsoon record from Eastern Arabian Sea. Earth Plan Sci Lett 177:209–218
Saxena A, Prasad V, Singh IB (2013) Holocene palaeoclimate reconstruction from the phytoliths of the lake-fill sequence of Ganga Plain. Curr Sci 104(8):1054–1062
Severdrup HU, Johnson MW, Fleming RH (1942) The oceans their physics, chemistry and general biology. Prentice Hall, New York, p 1087
Sharma S, Joachimski MM, Tobschall HJ, Singh IB, Sharma C, Chauhan MS (2006) Correlative evidence of monsoon variability, vegetation change and human habitation in Senai lake deposit, Ganga plain. Curr Sci 90:973–978
Sharma C, Chauhan MS, Gupta A, Rajagopalan G (1995) In: Symposium volume, "recent advances in geological studies of Northwest Himalaya and Foredeep". Geological Survey of India, Lucknow, pp 90
Sharma MC, Owen LA (1996) Quaternary glacial history of the NW Garhwal, central Himalayas, India. Quat Sci Rev 15:335–365
Shepard F (1954) Nomenclature based on sand-silt-clay ratios. J Sed Petrol 24:151–158

Shetye SR, Shenoi SSC, Gouveia AD, Michael GS, Sundar D, Nampoothri G (1991) Wind driven coastal upwelling along the western boundary of the Bay of Bengal during southwest monsoon. Continental Shelf Res 11:1397–1408

Shetye SR, Gouveia A, Shenoi SSC, Sundar D, Michael GS, Nampoothri G (1993) The western boundary current in the seasonal subtropical gyre in the Bay of Bengal. J Geophys Res 98:945–954

Siby K, Nath BN, Ramaswamy V, Naman D, Gnaneshwar Rao T, Kamesh Raju KA, Selvaraj K, Chen CTA (2008) Possible detrital, diagenetic and hydrothermal sources for Holocene sediments of the Andaman backarc basin. Mar Geol 247(3–4):178–193

Singh G, Wasson RJ, Agrawal DP (1990) Vegetational and seasonal climate changes since last full glacial in the Thar Desert. Rev Palaeobot Palynol 64:351–358

Sinha A, Cannariato KG, Stott LD, Cheng H, Edwards RL, Yadava MG, Ramesh R, Singh IB (2007) A 900-year (600 to 1500 AD) record of the Indian summer monsoon precipitation from the core monsoon zone of India. Geophys Res Lett 34:L16707. https://doi.org/10.1029/2007GL030431

Sirocko F, Sarnthein M, Erlenkeuser H, Lange H, Arnold M, Duplessy JC (1993) Century scale events in monsoon climate over the past 24,000 years. Nature 364:322–324

Staubwasser M, Weiss H (2006) Holocene climate and cultural evolution in late prehistoric-early historic West Asia. Quat Res. https://doi.org/10.1016/j.yqres.2006.09.001

Stewart RA, Pilkey OH, Nelson BW (1965) Sediments of the Northern Arabian Sea. Mar Geol 3:411–427

Stuiver M, Reimer PJ (1993) Extended ^{14}C data base and revised CALIB 3.0 ^{14}C age calibration program. Radiocarbon 35:215–230

Stuiver M, Reimer PJ, Reimer RW (2010) CALIB 6.0. Available online from http://calib.qub.ac.uk/calib

Subramanian V (1985) Geochemistry of river basins in the Indian Subcontinent, part I: water chemistry, chemical erosion and water-mineral equilibria. Transp Carbon Min Major World Rivers, Part 3 58: 495–512. In: Degens ET, Kempe S (eds) Mittilungen aus dem Geologisch-Palaeontologischen, Institut der Universitat Hamburg

Subramanian V (1993) Sediment load of Indian rivers. Curr Sci 64:928–930

Thamban M, Kawahata H, Rao VP (2007) Indian summer monsoon variability during the Holocene as recorded in sediments of the Arabian Sea: timing and implications. J Oceanogr 63:1009–1020

Thamban M, Rao VP, Schneider RR (2002) Reconstruction of late quaternary monsoon oscillations based on clay mineral proxies using sediment cores from the western margin of India. Mar Geol 186:527–539

Tiwari MR, Ramesh MG, Yadava BLK, Somayajulu AJ, Jull T, Burr S (2006) Is there a persistent control of monsoon winds by precipitation during the late Holocene. Geochem Geophys Geosys 7:Q03001. https://doi.org/10.1029/2005GC001095

Triantaphyllou MV, Antonarakou A, Dimiza M, Anagnostou C (2009) Calcareous nannofossil and planktonic foraminiferal distributional patterns during deposition of Sapropels S6, S5 and S1 in the Libyan Sea (Eastern Mediterranean). Geo-Mar Lett 30(1):1–13

Twiss PCA (2001) Curmudgeons view of Grass Phytolithology. Phytoliths: Appl Earth Sci Human History. In: Meunier JD, Colin F (eds) A. A. Balkema Publishers, Lisse, pp 7–25

Varkey MJ, Murty VSN, Suryanarayana A (1996) Physical oceanography of the Bay of Bengal and Andaman sea. Oceanogr Mar Biol 34:1–70

Vollweiler N, Scholz D, Muhlinghaus C, Mangini A, Spotl C (2006) A precisely dated climate record for the last 9 kyr from three high alpine stalagmites, Spannagel Cave, Austria. Geophys Res Lett 33:L20703. https://doi.org/10.1029/2006GL027662

Von Rad U, Schaaf M, Michels KH, Schulz HW, Berger H, Sirocko F (1999) A 5000-yr record of climate change in varved sediments from the oxygen minimum zone off Pakistan, Northeastern Arabian Sea. Quat Res 51:39–53

Wang PX (1999) Response of western Pacific marginal seas to glacial cycles: palaeoceanographic and sedimentological features. Mar Geol 16:5–39

Wang T, Surge D, Walker K Jo (2013) Seasonal climate change across the Roman Warm Period/Vandal Minimum transition using isotope sclerochronology in archaeological shells and otoliths, southwest Florida, USA. Quat Int 308–309:230–241. https://doi.org/10.1016/j.quaint.2012.11.013

Wang B, Wu R, Li T (2003) Atmosphere–warm ocean interaction and its impact on Asian-Australian monsoon variation. J Clim 16:1195–1211

Weaver CE (1989) Clays, muds and shales, vol 44. Developments in: Sedimentology. Elsevier, Amsterdam, p 819

Wei GJ, Liu Y, Li XH, Shao L, Liang X (2003) Climatic impact on Al, K, Sc and Ti in marine sediments: evidence from ODP site 1144, South China Sea. Geochem J 37:593–602

Werner JP, Wang J, Gómez-Navarro JJ, Steiger N, Neukom R (2019) No evidence for globally coherent warm and cold periods over the preindustrial Common Era. Nature 571:550–554. https://doi.org/10.1038/s41586-019-1401-2

Xiao JL, Wu JT, Si B, Liang WD, Nakamura T, Liu BL, Inouchi Y (2006) Holocene climate changes in the monsoon/arid transition reflected by carbon concentration in Daihai Lake of Inner Mongolia. Holocene 16:551–560

Xu H, Hong Y, Lin Q, Hong B, Jiang H, Zhu Y (2002) Temperature variations in the past 6000 years inferred from 618O of peat cellulose from Hongyuan, China. Chinese Sci Bull 7:1578–1584

Yadava MG, Ramesh R (2005) Monsoon reconstruction from radiocarbon dated tropical Indian Speleothem. Holocene 15:48–59

Yadava MG, Ramesh R, Pandarinath K (2007) A positive 'amount effect' in the Sahayadri (Western Ghats) rainfall. Curr Sci 93:560–564

Yang B, Braeuning A, Johnson KR, Yafeng S (2002) General characteristics of temperature variation in China during the last two millennia. Geophys Res Lett, vol 29. https://doi.org/10.1029/2001gl014485

The manufacturer's authorised representative in the EU is Springer Nature Customer Service Centre GmbH, Europaplatz 3, 69115 Heidelberg, Germany. If you have any concerns regarding our products, please contact ProductSafety@springernature.com

Printed and bound by CPI Group (UK) Ltd, Croydon, CR0 4YY
30/09/2026
02225602-0014